THERMODYNAMICS
FOR CHEMISTS

By

SAMUEL GLASSTONE

*Formerly Research Associate of Princeton University
and Professor of Chemistry in the University of Oklahoma*

NEW YORK

D. VAN NOSTRAND COMPANY, Inc.

250 Fourth Avenue

1947

COPYRIGHT, 1947
BY
D. VAN NOSTRAND COMPANY, INC.

All Rights Reserved

This book, or any parts thereof, may not be reproduced in any form without written permission from the author and the publisher.

PRINTED IN THE UNITED STATES OF AMERICA

To V.

PREFACE

The object of the present book is to provide an introduction to the principles and applications of thermodynamics for students of chemistry and chemical engineering. All too often it appears that such students tend to regard the subject as an academic and burdensome discipline, only to discover at a later date that it is a highly important tool of great practical value. The writer's purpose has been to explain the general structure of thermodynamics, and to give some indication of how it may be used to yield results having a direct bearing on the work of the chemist.

More than one hundred illustrative numerical problems are worked out in the text, and a total of about three hundred and sixty exercises of a variety of types have been included for solution by the reader. In the hope of imparting the whole subject with an aspect of reality, much of the material for this purpose has been taken from the chemical literature, to which references are given. In order to economize space, and at the same time to test the reader's grasp of thermodynamics, the derivations of a number of interesting results have been set as exercises. To this extent, at least, the exercises are to be considered as part of the text, although their solution should in any event be regarded as essential to any adequate course in chemical thermodynamics.

In treating the various topics in this book the particular method employed has been determined in each case by considerations of simplicity, usefulness and logical development. In some instances the classical, historical approach has been preferred, but in others the discussion follows more modern lines. Whenever feasible the generalized procedures, involving reduced temperatures and pressures, which have been evolved in recent years chiefly by chemical engineers, are introduced. As regards statistical methods, the author feels that the time has come for them to take their place as an essential part of chemical thermodynamics. Consequently, the applications of partition functions to the determination of heat capacities, entropies, free energies, equilibrium constants, etc., have been introduced into the text in the appropriate places where it is hoped their value will be appreciated.

The symbols and nomenclature are essentially those which have been widely adopted in the American chemical literature; however, for reasons given in the text, and in accordance with a modern trend, the Gibbs symbol μ and the shorter term "chemical potential" are employed for the partial molar free energy. Because atmospheric pressure is postulated for the conventional standard state of a liquid, some confusion has resulted from the use of the same symbol for the standard state as for the liquid at an arbitrary pressure. Hence, the former state is indicated in the text in the usual manner, by a zero (or circle), but the latter is distinguished by a small square as superscript.

The writer would like to take this opportunity to acknowledge his indebtedness to certain books, namely, F. H. Macdougall, "Thermodynamics and Chemistry"; L. E. Steiner, "Introduction to Chemical Thermodynamics"; B. F. Dodge, "Chemical Engineering Thermodynamics"; and, in particular, G. N. Lewis and M. Randall, "Thermodynamics and the Free Energy of Chemical Substances." He is also sincerely grateful to Dr. Allen E. Stearn, University of Missouri, and Dr. Roy F. Newton, Purdue University, for reading the manuscript of this book and for making numerous suggestions which have helped materially to clarify and improve the treatment. Finally, the author wishes to express his thanks to his wife for reading the proofs, and for her continued aid and encouragement.

SAMUEL GLASSTONE

BERKELEY, CALIF.
November 1946

CONTENTS

CHAPTER		PAGE
	PREFACE...	v
I.	HEAT, WORK AND ENERGY................................	1
	1. Introduction, 1; 2. Temperature, 2; 3. Work, Energy and Heat, 5.	
II.	PROPERTIES OF THERMODYNAMIC SYSTEMS.................	14
	4. Thermodynamic Systems, 14; 5. Equations of State, 18.	
III.	THE FIRST LAW OF THERMODYNAMICS......................	33
	6. The Conservation of Energy, 33; 7. Energy and Heat Changes, 36; 8. Reversible Thermodynamic Processes, 41.	
IV.	HEAT CHANGES AND HEAT CAPACITIES.....................	47
	9. The Heat Content, 47; 10. Adiabatic Processes, 55; 11. The Joule-Thomson Effect, 60.	
V.	THERMOCHEMISTRY..	68
	12. Heat Changes in Chemical Reactions, 68; 13. Flame and Explosion Temperatures, 84; 14. Calculation of Heat of Reaction, 89.	
VI.	CALCULATION OF ENERGY AND HEAT CAPACITY.............	95
	15. Classical Theory, 95; 16. Quantum Statistical Theory of Heat Capacity, 99; 17. Heat Capacity of Solids, 120.	
VII.	THE SECOND LAW OF THERMODYNAMICS....................	129
	18. Conversion of Heat into Work, 129; 19. Entropy, 141.	
VIII.	ENTROPY RELATIONSHIPS AND APPLICATIONS...............	154
	20. Temperature and Pressure Relationships, 154; 21. Energy and Heat Capacity Relationships, 163; 22. The Joule-Thomson Effect, 171.	
IX.	ENTROPY DETERMINATION AND SIGNIFICANCE...............	178
	23. The Third Law of Thermodynamics, 178; 24. Statistical Treatment of Entropy, 183.	
X.	FREE ENERGY...	201
	25. The Free Energy and Work Functions, 201; 26. Chemical Potential, 213.	
XI.	PHASE EQUILIBRIA..	222
	27. Systems of One Component, 222; 28. Systems of More Than One Component, 237.	

CHAPTER	PAGE
XII. FUGACITY AND ACTIVITY	250

29. Fugacity of a Single Gas, 250; 30. Mixtures of Gases, 261; 31. Liquid Mixtures, 268.

XIII. FREE ENERGY AND CHEMICAL REACTIONS ... 273

32. The Equilibrium Constant, 273; 33. Free Energy Change in Chemical Reactions, 282.

XIV. THE PROPERTIES OF SOLUTIONS ... 317

34. Ideal Solutions, 317; 35. Nonideal Solutions, 330; 36. Dilute Solutions, 337.

XV. ACTIVITIES AND ACTIVITY COEFFICIENTS ... 350

37. Standard States, 350; 38. Determination of Activities, 356.

XVI. SOLUTIONS OF ELECTROLYTES ... 378

39. Activities and Activity Coefficients, 378.

XVII. THE DEBYE-HÜCKEL THEORY ... 407

40. Ionic Interaction in Solution, 407; 41. Applications of the Debye-Hückel Equation, 412.

XVIII. PARTIAL MOLAR PROPERTIES ... 427

42. Determination of Partial Molar Properties, 427; 43. Partial Molar Volumes, 433; 44. Partial Molar Thermal Properties, 437.

XIX. E.M.F.'s AND THE THERMODYNAMICS OF IONS ... 462

45. E.M.F.'s and Electrode Potentials, 462; 46. Thermodynamics of Ions in Solution, 486.

APPENDIX:

Table 1.	Constants and Conversion Factors	501
Table 2.	Properties of Gases and Liquids	502
Table 3.	Heat Capacities of Gases	503
Table 4.	Heat Capacities of Solids	504
Table 5.	Standard Free Energies and Heats of Formation, and Entropies	505
Table 6.	Integral Heats of Solution of Salts	506

INDEX ... 509

CHAPTER I

HEAT, WORK AND ENERGY

1. INTRODUCTION

1a. Scope and Limitations of Thermodynamics.[1]—The subject matter of thermodynamics is based essentially on two fundamental postulates (or laws) which summarize actual experience with regard to the interconversion of different forms of energy. These are called the first and second laws of thermodynamics. There is also another postulate dealing with certain aspects of the second of these laws; it is often referred to as the third law of thermodynamics, but its use is more restricted than is that of the other laws. By the application of relatively simple and well established mathematical procedures to the two basic laws, it has been possible to derive results which have proved of fundamental importance to chemistry, physics and engineering. For example, equations have been developed giving the variation with temperature and pressure of certain physical properties of substances. Of more direct interest to the chemist, however, is the derivation of the exact conditions for spontaneous chemical reaction and for chemical equilibrium.

Although the great practical value of thermodynamics is undeniable, as will be shown in the subsequent pages, there are certain limitations that must be borne in mind. The methods of thermodynamics are independent of atomic and molecular structure, and also of reaction mechanism. Consequently, the results throw no direct light on problems related to these subjects. The conclusions of thermodynamics may be correlated with those of the kinetic theory of matter, for example, but the distinction between the two approaches to the study of physical problems must be clearly understood. Thus, the observable thermodynamic property of a body known as its "temperature" may be regarded as being determined by the average kinetic energy of the molecules. However, the concept of temperature as used in thermodynamics is independent of any theories concerning the existence of molecules. As will be seen shortly, temperature, like other thermodynamic variables of state, is based on experimental, macroscopic observation of the body as a whole.

One aspect of thermodynamics is the prediction of relationships between various quantities that are directly observable or which can be derived from observable properties, but thermodynamics alone cannot give any indication of the actual values of these quantities. In order to obtain the information it is possible to invoke certain procedures, such as the kinetic theory of matter, statistical mechanics and the Debye-Hückel theory, which really

---

[1] P. W. Bridgman, "The Nature of Thermodynamics," 1941.

lie outside the scope of thermodynamics. Nevertheless, because such methods and theories provide a means for the calculation of thermodynamic properties, they may be regarded as complementary to thermodynamics. They are, however, not essential, for the observable quantities under consideration can usually be obtained by experimental methods without recourse to theory.

As far as chemical reactions are concerned, thermodynamics can indicate whether a particular process is possible or not, under a given set of circumstances, e.g., temperature and concentrations of reactants and products. However, no information can be obtained from pure thermodynamics as to the rate at which the reaction will take place. For example, it can be shown by means of thermodynamics that hydrogen and oxygen gases should combine to form liquid water at ordinary temperatures and pressures, but it is not possible to state whether the reaction will be fast or slow. Actually, in the absence of a catalyst, the combination is so slow as to be undetectable in many years. In effect, thermodynamics deals quantitatively with equilibrium conditions, that is, conditions which do not change with time, and it does not take into account the rate of approach to the equilibrium state.[2]

2. Temperature

2a. The Concept of Temperature.—Since the laws of thermodynamics deal with the interconversion of energy, it is necessary to consider the significance of energy, and of the related quantities, heat and work. Before doing so, however, it is desirable to examine the concept of **temperature**. The ability to distinguish broadly between hot and cold is a familiar faculty of the human senses. *A substance which is hot is said to have a higher temperature than one which is cold.* If a hot body, such as a piece of metal, is placed in contact with a similar but colder body, then after a short time the senses show that one is neither hotter nor colder than the other. That is to say, the two bodies have attained a state of thermal (or temperature) equilibrium, and they are both said to be at the same temperature. The body which was originally hot will now feel colder to the touch, whereas the colder one will now feel hotter. The temperature of the hot body has consequently decreased while that of the cold body has been raised until at equilibrium the two temperatures are the same. It would appear, therefore, that something —actually energy—is transferred from the hotter to the colder body until the two bodies have equal temperatures; that which is apparently transferred in this manner is called "heat." Thus, **heat** may be defined as *that which passes from one body to another solely as the result of a difference in temperature*. The quantity of heat transferred in this manner depends, in the first place, on the change of temperature of each body. Another factor, namely the heat capacity, will be considered later.

Although the human senses can detect temperature differences to some extent, they are obviously not adequate for precise measurement; neither

[2] S. Glasstone, K. J. Laidler and H. Eyring, "The Theory of Rate Processes," 1941.

can they be used to associate a definite number with each temperature. For these purposes it is necessary to have an instrument or device known as a "thermometer." The principle of the thermometer is based on the fact that certain properties, such as volume or electrical resistance, vary with temperature. In the ordinary mercury thermometer, for example, use is made of the change in volume of mercury with temperature; in this case, however, the volume changes are observed by the alteration in length of the column of mercury in a narrow glass tube. Suppose such a thermometer is placed in contact with a body until thermal equilibrium is attained; the position of the mercury in the glass tube is then said to represent an arbitrary temperature of t degrees ($t°$). It has been found experimentally, in agreement with expectation, that *if two bodies are each in thermal equilibrium with a third, they are in thermal equilibrium with one another.* This fact renders possible the use of the thermometer as an indicator of temperature. It means that whenever the mercury reaches a certain point the temperature of any body is always $t°$, provided thermal equilibrium is attained between the body and the thermometer.

2b. Thermometric Scales: The Centigrade Scale.—In order to use a thermometer for the quantitative expression of temperature and of temperature differences, two things are necessary. First, a zero point of the temperature scale must be chosen, absolutely or arbitrarily, and second, the size of the unit, i.e., the degree, must be defined. On the centigrade temperature scale the zero is taken as the "ice point," that is, the freezing point of water in contact with air at standard atmospheric pressure (76.00 cm. of mercury).* The size of the degree is then defined by postulating that the "steam point," that is, the boiling point of water at standard atmospheric pressure, shall be taken as exactly 100°.

If X represents any physical property which varies with temperature, X_0 and X_{100} are the values at 0° and 100°, respectively, on the centigrade scale; the degree is then represented by the change $\frac{1}{100}(X_{100} - X_0)$ in the given property. If X is the value of the property at any temperature † the magnitude of that temperature is then $t°$, as given by

$$t = \frac{X - X_0}{\frac{1}{100}(X_{100} - X_0)}. \tag{2.1}$$

In the common mercury thermometer X is the length of a mercury column in a glass tube, and the distance between the positions representing X_0 and X_{100} is divided into one hundred equal parts, in order to facilitate evaluation of the temperature in accordance with equation (2.1).

* It should be noted that 0° C is taken as the freezing point of water in equilibrium, and hence saturated, with air at 1 atm. pressure, and not that of pure water; the freezing point of the latter is + 0.0023° C.

† By the expression "the value of the property at any temperature" is implied the value of the property of the thermometric substance when it is in thermal equilibrium with a body at the given temperature.

Since the property X of different thermometric substances, or the different thermometric properties of a given substance, do not vary in an identical manner with temperature, various thermometers, all of which have been standardized at 0° and 100° C, may indicate different temperatures when in thermal equilibrium with the same body at an intermediate point. A mercury thermometer and a toluene thermometer, for example, which agree at 0° and 100° C, would differ by several degrees in the vicinity of 50° C. Even mercury thermometers in tubes made of various types of glass indicate slightly different temperatures.[3]

2c. The Absolute Ideal Gas Scale.—Gases have frequently been used as thermometric substances; thus X may represent the volume of a given mass of gas at constant pressure, or the pressure at constant volume. However, here again, the variation with temperature of the volume (or pressure) of a gas depends somewhat on the nature of the gas. For gases, such as hydrogen and helium, which do not depart greatly from ideal behavior (§ 5c) under ordinary conditions, the temperatures recorded do not differ very appreciably, and the differences become less marked as the pressure of the gases is decreased. It appears, from experimental observation, that at sufficiently low pressures or, better, if the results are extrapolated to zero pressure, the temperature of a given body as recorded by a gas thermometer would always be the same irrespective of the nature of the gas.[4]

It follows, therefore, that when gases approximate to ideal behavior, i.e., at very low pressures, the differences in their thermometric properties disappear. This fact presents the possibility of devising a temperature scale which shall be independent of the thermometric substance, the latter being a hypothetical "ideal gas." Such a scale is the so-called "absolute ideal gas scale," in which *the (absolute) temperature is taken as directly proportional to the volume of a definite mass of an ideal gas at constant pressure*, or to the pressure at constant volume. For convenience, the magnitude of the degree on the absolute scale is usually taken to be the same as on the centigrade scale (§ 2b), so that the absolute temperature T on the ideal gas scale is given by

$$T = \frac{V}{\frac{1}{100}(V_{100} - V_0)}, \qquad (2.2)$$

where V is the volume of the ideal gas at this temperature, and V_{100} and V_0 are the corresponding volumes at the steam point and ice point, respectively, all the volumes being determined at the same pressure.

The value of the ice point T_0 on the absolute scale may be determined by setting V equal to V_0 in equation (2.2), so that

$$T_0 = \frac{V_0}{\frac{1}{100}(V_{100} - V_0)}. \qquad (2.3)$$

By making measurements on various gases, at constant pressures, and

---

[3] International Critical Tables, Vol. I, p. 55.
[4] Cf., Wensel, *J. Res. Nat. Bur. Stand.*, **22**, 375 (1939).

extrapolating the results to zero pressure, it has been found that the ice point T_0, as given by equation (2.3), is 273.16°. Temperatures on the absolute ideal gas scale are thus obtained by adding 273.16° to the temperature of the ideal gas thermometer on the ordinary centigrade scale, i.e., with the ice point taken as 0° C [equation (2.1)]. It will be noted from equation (2.2) that on the absolute scale the volume of an ideal gas should become zero at a temperature of zero degrees. This hypothetical point, which should be − 273.16° C, is known as the **absolute zero** of temperature; it presumably represents the lowest conceivable temperature. There are reasons for believing that the absolute zero is unattainable (§ 18j), although temperatures within less than 0.005° of it have been realized.

It will be seen in Chapter VII (§ 18k) that it is possible to develop an absolute temperature scale, also independent of the nature of the thermometric substance, based on the second law of thermodynamics. This is sometimes called the **Kelvin scale**, in honor of its originator, Lord Kelvin (William Thomson). Actually, the thermodynamic scale can be shown to be identical with the absolute ideal gas scale, as defined above; hence, temperatures on the latter, as well as the former, scale are represented by the symbol ° K. The ice point is consequently 273.16° K. It may be noted, incidentally, that the thermodynamic derivation of the absolute temperature scale provides a more definite interpretation of the absolute zero, i.e., the lowest limit of temperature, than is possible by means of the ideal gas thermometer.*

2d. Practical Temperature Scale.—For practical purposes gas thermometers are not satisfactory; consequently a number of fixed points have been chosen by international agreement which can be used for the determination of experimental temperatures on the centigrade scale. The addition of 273.16 then gives the corresponding absolute temperatures. The following temperatures are taken as primary standards: boiling point of oxygen, − 182.97° C; ice point, 0.00° C; steam point, 100.00° C; boiling point of sulfur, 444.60° C; melting point of silver, 960.5° C; and melting point of gold, 1063° C, all at standard atmospheric pressure. A number of subsidiary fixed points, some of which extend the scale to 3400° C, and others of which determine various intermediate temperatures, have also been proposed.[5]

3. Work, Energy and Heat

3a. Work and Energy.—Mechanical **work** is done whenever *the point of application of a force is displaced in the direction of the force.* If F is the magnitude of the force and l is the displacement of its point of application, in the direction in which the force acts, then the mechanical work done is equal to the product $F \times l$, expressed in appropriate units. In addition to

---

* Engineers frequently express temperatures on the Rankine (absolute) scale, using the Fahrenheit degree; the temperature is then given by $t°(F) + 459.69°$ R.

[5] Burgess, *J. Res. Nat. Bur. Stand.*, **1**, 635 (1928); Roeser and Wensel, *ibid.*, **14**, 247 (1935).

mechanical work, other forms of work are possible, e.g., electrical work, but in each case the work done is equal to the product of a generalized force, sometimes referred to as the **intensity factor**, and a generalized displacement, the **capacity factor**. In electrical work, for example, the generalized force is the so-called electromotive force, i.e., the E.M.F., and the generalized displacement is the quantity of electricity. The work done is thus equal to the product of the applied E.M.F. and the quantity of electricity which is passed.

Energy is defined as *any property which can be produced from or converted into work*, including, of course, work itself. There are thus various manifestations of energy, as represented by the different forms of work, but each can be expressed as the product of an intensity factor and a capacity factor. Further, because of the connection between work and energy, all forms of the latter can be expressed in work units. The derivation of the concept of energy through those of force and work, as given above, follows historical tradition. However, because heat, which is a form of energy, cannot be completely converted into work, some writers prefer to describe *energy as that which can be transformed into heat*, as defined in § 2a, including heat itself. Although this approach to the subject of energy has some advantages, the one used here is somewhat simpler when the problems of dimensions and units of energy come up for consideration.

3b. Dimensions and Units of Energy.—By Newton's laws of motion, mechanical force is equal to the product of mass and acceleration; hence force has the dimensions of mass $\times$ length/(time)^2, i.e., mlt^{-2}. Since work is equal to force multiplied by length, as stated above, the dimensions of work are ml^2t^{-2}, and all forms of energy must consequently have these dimensions. The units of mass, length and time usually employed in scientific work are the gram (g.), centimeter (cm.) and second (sec.), respectively, constituting what is known as the c.g.s. system. The meter, or 100.000 cm., was originally defined as 10^{-7} times the length of the earth's quadrant, from north pole to equator, passing through Paris. The standard meter is, however, the distance between two marks on a bar at 0° C, kept at Sèvres, near Paris. The standard kilogram, i.e., 1000.00 grams, is taken as the mass of a lump of platinum, also at Sèvres.* It was intended to be equal to the weight of exactly 1000 cc. of pure water at its temperature of maximum density, i.e., 4° C, but there is actually a small difference. The liter is the volume occupied by 1 kilogram of water at 4° C and 1 atm. pressure; it was meant to be 1000 cc., but is actually 1000.028 cc. Because of this discrepancy it is becoming the practice to express volumes of liquids and gases in terms of the milliliter (ml.), which is exactly one-thousandth part of a liter. For most purposes, however, the difference between the ml. and the cc. is not significant. The second is defined as 1/86,400 part of a mean solar day, the latter being the average interval between successive transits of the sun across the meridian at any given place.

The unit of energy in the c.g.s. system is the **erg**; it is the work done when a force of 1 dyne acts through a distance of 1 cm., the dyne being the force which acting for 1 sec. on a mass of 1 g. produces in it a velocity of motion of 1 cm. per

---

* Exact duplicates of the standard meter and kilogram are kept at the National Bureau of Standards, Washington, D. C.

sec. Since the erg is so small, a subsidiary unit, the **joule**, is defined as 10^7 ergs. Unfortunately, confusion has arisen because of the necessity for distinguishing between the **absolute joule** as thus defined, and the **international joule**, based on electrical measurements. The latter is the work done when 1 (international) coulomb, i.e., 1 ampere-second, of electricity flows under the influence of a potential, i.e., E.M.F., of 1 (international) volt. The (international) coulomb is the quantity of electricity which will cause the deposition of 1.11800 milligrams of silver in the electrolysis of a solution of silver nitrate under certain specified conditions. The (international) volt is defined in terms of the E.M.F. of the Weston standard cell, which is taken to be exactly 1.0183 (international) volts at 20° C. On this basis, the international joule is apparently equal to 1.0002 absolute joules.[6]

Problem: A quantity of 26.45 coulombs of electricity flows through a conductor under the influence of a fall of potential of 2.432 volts. Calculate the energy expenditure in (a) international joules, (b) ergs.

The electrical work done, i.e., the energy expended, is equal to the product of the quantity of electricity and the applied potential difference. Since the former is given in int. coulombs and the latter in int. volts, the result will be in int. joules; thus,

$$\text{Energy expenditure} = 26.45 \times 2.432 = 64.33 \text{ int. joules.}$$

The int. joule is equal to 1.0002 abs. joules, and hence to 1.0002×10^7 ergs, so that

$$\text{Energy expenditure} = 64.33 \times 1.0002 \times 10^7 = 64.34 \times 10^7 \text{ ergs,}$$

to four significant figures.

3c. Heat and Energy.—In view of the definitions of energy given above, heat must be regarded as a form of energy, since most forms of work can be readily converted into heat, and heat can be, at least partially, converted into work. Heat is produced from mechanical work, for instance, by means of friction, and electrical work is transformed into heat by the passage of electricity through a resistance. On the other hand, by means of a suitable machine, viz., a heat engine, a certain amount of heat can be converted into work. From the standpoint of thermodynamics, *heat is energy in transit;* it is the form in which energy is transferred from one body to another, either by direct contact or by means of radiation, as the result of a difference of temperature. In evaluating the quantity of heat energy that has passed to a given body, the intensity factor is the change of temperature, and the capacity factor is the heat capacity (§ 3d); the product of these two quantities, which can be stated in ergs if required (§ 3e), is a measure of the heat energy transferred.

3d. Heat Capacity.—As a general rule, heat is expressed in terms of a unit known as the (15°) **calorie**, which is defined as the quantity of heat required to raise the temperature of 1 gram of water by 1° in the vicinity of 15° C. The actual temperature is specified because the quantity of heat that must be supplied to raise the temperature of a given amount of water, or of any other substance for that matter, by 1° depends to some extent on the temperature itself.

[6] Birge, *Rev. Mod. Phys.*, **13**, 233 (1941); Curtis, *J. Res. Nat. Bur. Stand.*, **33**, 235 (1944).

The **heat capacity** of a body is the property which multiplied by the temperature change gives the quantity of energy which has entered or left the body as heat when it is brought into contact with another body having a different temperature. Thus, if the temperature of the body is raised from T_1 to T_2 by the passage to it of an amount of heat Q, the heat capacity C of the body is given by

$$Q = C(T_2 - T_1). \tag{3.1}$$

According to this equation,

$$C = \frac{Q}{T_2 - T_1}, \tag{3.2}$$

so that the heat capacity is sometimes defined as the quantity of heat required to raise the temperature of a body by 1°. From the definition of the calorie given above, it is evident that the heat capacity of 1 gram of water, in the vicinity of 15° C, is equal to 1 (15°) calorie.

The heat capacity is a property which is proportional to the quantity of matter present, and this must consequently be stated. Two particular quantities are commonly employed. The heat capacity is frequently referred to 1 gram of material; it is then called the **specific heat**. Thus, the specific heat of water is exactly 1 calorie per degree per gram in the region of 15° C. From the standpoint of the chemist, a more useful form of the heat capacity is that referred to 1 mole, i.e., the molecular (or formula) weight in grams.* The quantity of heat is then known as the **molar heat capacity**.

As already indicated, the heat capacity of a body or substance usually varies with the temperature. The heat capacity as defined by equation (3.2) is thus the mean value for the temperature range from T_1 to T_2. In order to define the heat capacity at a given temperature it is necessary to make the temperature interval as small as possible; thus, if the temperature increase $T_2 - T_1$ is represented by ΔT, the true heat capacity is given by the expression

$$C = \lim_{T_2 \to T_1} \frac{Q}{\Delta T}. \tag{3.3}$$

Since the quantity of heat Q is usually stated in calories and ΔT in degrees, it is evident that the dimensions of heat capacity are calories/degrees for the given amount of material, e.g., cal. deg.$^{-1}$ g.$^{-1}$ or cal. deg.$^{-1}$ mole^{-1}, that is, calories per degree per gram (or per mole).

It will be seen later (§ 9b) that the heat capacity of a substance can have a definite value only under certain specified conditions. The two most important are constant pressure and constant volume. It is necessary to indicate, therefore, the particular conditions under which heat is transferred to or from the given substance.

* Unless otherwise stated, the term "mole" will refer to the *simple* formula weight in grams; thus, the mole of acetic acid is 60.05 g., in spite of the fact that a large proportion of $(C_2H_4O_2)_2$ molecules may be present.

3e. The Measurement of Heat: The Defined Calorie.—Heat quantities are usually measured by means of a **calorimeter**. In its simplest form this consists of a vessel, which is insulated as well as possible in order to prevent loss or gain of heat, and which contains a calorimetric liquid, usually water. The heat capacity of the calorimeter and its contents is determined by placing an electrical conductor of known resistance in the liquid, and passing a definite current for a specified time. The accompanying increase of temperature of the calorimeter is then determined by means of a thermometer placed in the liquid. From the resistance, current and time the amount of electrical energy expended can be calculated in terms of int. joules. If it is assumed that this energy is completely transferred from the conductor to the calorimetric liquid in the form of heat, the heat capacity of the calorimeter can be determined in int. joules per degree.

Problem: A current of 0.565 amp. is passed for 3 min. 5 sec. through a coil of resistance 15.43 ohms contained in a calorimeter. The rise of temperature is observed to be 0.544° C. Assuming that no heat is lost to the surroundings, calculate the heat capacity of the calorimeter.

By Ohm's law, the fall of potential (or E.M.F.) across the resistance coil is equal to $I \times R$, where I is the current and R the resistance. The electrical work is equal to the product of the E.M.F. and the quantity of electricity, and since the latter is given by $I \times t$, where t is the time for which the current is passed, it follows that

$$\text{Electrical Work} = I^2 \times R \times t.$$

If I is in int. amps., R in int. ohms * and t in sec., the work will be given in int. joules. In the present case, therefore, taking the heat generated as equal to the electrical work, it is seen that

$$\text{Heat generated} = (0.565)^2 \times 15.43 \times 185 \text{ int. joules.}$$

Since the rise of temperature of the calorimeter is 0.544° C, it follows that

$$\text{Heat capacity} = (0.565)^2 \times 15.43 \times 185/0.544$$
$$= 1675 \text{ int. joules deg.}^{-1}.$$

Once the heat capacity of the calorimeter is known, the value of any quantity of heat transferred to it, as the result of a chemical reaction, for example, can be readily determined by equation (3.1). Similarly, the heat capacity of any substance can be found from the amount of heat it transfers to a calorimeter; the change in temperature of the substance must also be known. In actual practice various devices are used to improve the accuracy of the results, but the foregoing description indicates the fundamental principle involved in modern calorimetric work.

In order to convert the results obtained by the electrical heating method into calories, it is necessary to know the relationship between joules and calories. That there is such an exact connection is really an aspect of the first law of thermodynamics (§ 6a), which will be tacitly assumed for the present. Because of a slight uncertainty, of about two parts in 10,000, concerning the relationship between the standard (15°) calorie, as defined in § 3d, and the int. joule (§ 3b), a **defined calorie,**

---

* An int. ampere is the current which flowing for 1 sec. gives 1 int. coulomb of electricity, and an int. ohm is the resistance of a conductor through which passes 1 int. ampere when 1 int. volt is applied.

equivalent to 4.1833 int. joules has been proposed.[7] This differs to a very small extent from the standard calorie, and is equal to the specific heat of water at about 17° C.

Problem: A chemical reaction was allowed to take place in a calorimeter and the rise of temperature was observed. In order to obtain the same increase of temperature it was necessary to pass a current of 0.565 amp. for 185 sec. through a heating coil of resistance 15.43 ohms. Calculate the heat in defined calories evolved in the reaction.

The heat produced in the reaction is here exactly equal to that obtained from the electric current. Hence, from the preceding problem it follows that

Heat evolved in the reaction = 1675 int. joules
= 1675/4.183 = 400.4 calories.

3f. Heat and Work.—It will be seen in Chapter VII that heat differs from all other forms of energy in one highly significant respect. Whereas all other forms of energy are completely convertible into work, at least in principle, heat cannot be completely transformed into work without leaving some change in the system or its surroundings. In the continuous conversion of heat into work, by means of a heat engine, for example, part of the heat taken up at a particular temperature is converted into work, and the remainder is given out at a lower temperature (see Chapter VII). For this reason, it is convenient in the thermodynamic treatment to distinguish between the energy that enters or leaves a body as heat, i.e., due to a temperature difference, and energy that is transferred in other ways. The latter is usually considered under the general description of "work," of which various types, e.g., mechanical, electrical, surface, etc., are possible. *The gain or loss in energy of a body may thus be defined in terms of the heat transferred to or from it, and the work done upon or by it.*

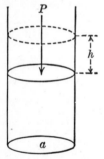

FIG. 1. Work of expansion

3g. Work of Expansion.—In most processes of interest to the chemist, the only work involved is that due to a change of volume against an external pressure, e.g., that of the atmosphere. This is frequently referred to as **work of expansion**. Consider any substance, which may be gaseous, liquid or solid, contained in a cylinder of cross-sectional area a (Fig. 1), fitted with a piston* upon which is exerted a *constant pressure P*; the total force acting on the piston is then the product of the pressure and the area, i.e., $P \times a$. Suppose the substance in the cylinder expands, and as a result the piston is raised through a distance h. The work done against the external pressure, that is, the work of expansion, is equal to the force $P \times a$ multiplied by the

---

[7] Rossini, *J. Res. Nat. Bur. Stand.*, **22**, 407 (1939); Mueller and Rossini, *Am. J. Phys.*, **12**, 1 (1944).

* It is assumed that the piston is weightless and moves without friction, so that no work is lost in the actual motion of the piston itself.

displacement h of its point of application in the direction of the force (§ 3a), i.e., $P \times a \times h$. The quantity $a \times h$ represents the increase in volume in the expansion process, and this may be replaced by $V_2 - V_1$, where V_1 is the initial volume and V_2 is the final volume. It follows, therefore, that

$$\text{Work of expansion} = P(V_2 - V_1) \qquad (3.4)$$

$$= P\Delta V, \qquad (3.5)$$

where ΔV, equal to $V_2 - V_1$, is the increase of volume against the external pressure P, which is supposed to remain constant throughout the expansion.

The units employed for expressing pressure and volume determine the units in which the work of expansion is obtained from equation (3.4) or (3.5). If P is given in dynes per sq. cm. and V is in cc., the work will be obtained in ergs, and this can be readily converted into abs. joules, i.e., 10^7 ergs, and calories, if required. For some purposes, it is convenient to express the pressure P in atmospheres, and the volume change ΔV in liters; the work is then in liter-atm.

Problem: A substance expands by 1 liter against a constant pressure of 1 atm. Calculate the work done in (*a*) liter-atm. units, (*b*) ergs, (*c*) int. joules, (*d*) defined calories.

(*a*) In this case P is 1 atm. and ΔV is 1 liter, so that the work of expansion is 1 liter-atm.

(*b*) To obtain the value in ergs, the pressure must be expressed in dynes cm.$^{-2}$ and the volume in cm.^{3}, so that the product is in dynes cm., i.e., ergs. A pressure of 1 atm. is equivalent to exactly 76 cm. of mercury, the density of which is 13.595 at 0° C; hence,

$$1 \text{ atm.} = 76.000 \times 13.595 \times 980.66 \text{ dynes cm.}^{-2}$$
$$= 1.0132_5 \times 10^6 \text{ dynes cm.}^{-2},$$

where the factor 980.66 cm. sec.$^{-2}$ is the acceleration due to gravity, at sea level and a latitude of 45°.

Since 1 liter is equal to 1000.028 cc. (§ 3b), the work of expansion is given by equation (3.5) as $1.0132_5 \times 10^6 \times 1.000028 \times 10^3$, i.e., 1.0133×10^9 ergs, to five significant figures.

(*c*) The int. joule is equal to 1.0002 abs. joules (§ 3b), and hence to 1.0002×10^7 ergs. The 1.0133×10^9 ergs is thus equivalent to 1.0131×10^2 int. joules.

(*d*) The defined calorie is 4.1833 int. joules, and so the work of expansion is $1.0131 \times 10^2/4.1833 = 24.218$ defined cal.

The results may be summarized: 1 liter-atm. $= 1.0133 \times 10^9$ ergs $= 1.0131 \times 10^2$ int. joules $= 24.218$ def. cal.

3h. Conversion Factors.—Because quantities having the same dimensions, e.g., energy, are frequently expressed in various units, confusion is liable to ensue unless all quantities are labelled correctly and completely by the appropriate units. One way of diminishing the risk of error is to express all quantities in c.g.s. units (§ 3b) before undertaking any calculations, but this procedure is sometimes unnecessarily tedious. The treatment can often be considerably simplified by employing conversion factors, such as those derived in the problem in § 3g, and by utilizing the fact that various units can be multiplied and divided, and that similar units cancel one another if they appear in the numerator and denominator of a fraction. A number of conversion factors which will be required from time to time are collected in Table 1 of the Appendix at the end of the book.

In applying these factors, a useful device which avoids the possibility of error in the less simple cases, is to convert them into a dimensionless form. For example, as seen above, 1 liter-atm. is equivalent to 24.218 cal.; hence, 24.218 cal./liter-atm., i.e., 24.218 cal. liter^{-1} atm.$^{-1}$, is equal to unity, without dimensions. It is then permissible to multiply one side of an expression by unity and the other side by 24.218 cal. liter^{-1} atm.$^{-1}$. The subsequent cancellation of identical units with opposite exponents makes conversion from one set of units to another a relatively simple matter. The application of the foregoing ideas will be illustrated in subsequent portions of the book.[8]

EXERCISES

1. A mercury-in-glass thermometer is standardized at 0° and 100° C. The change of volume of mercury between these temperatures is represented by

$$v_t = v_0(1 + 1.8146 \times 10^{-4}t + 9.205 \times 10^{-9}t^2),$$

where v_t is the volume at the correct centigrade temperature $t°$. Neglecting the expansion of the glass, what temperature will the thermometer read at exactly 50° C?

2. A current of 1.243 amp. is passed for 15 min. 45 sec. through a resistance of 20.18 ohms. Calculate the heat generated in (i) joules, (ii) ergs.

3. The heat generated in the preceding exercise is liberated in a calorimeter, and the temperature rises by 0.287° C. Evaluate the heat capacity of the calorimeter in (defined) cal. deg.$^{-1}$.

4. When an electric current was passed through a heating coil of resistance 16.49 ohms, the fall of potential across the coil was 3.954 volts. Neglecting loss of heat by radiation, etc., determine the rate of increase of temperature, in deg. sec.$^{-1}$, of a calorimeter with a heat capacity of 125.4 (defined) cal. deg.$^{-1}$.

5. The heat capacity of a calorimeter was found to be 218.4 cal. deg.$^{-1}$ at 23° C. When a piece of metal weighing 19.46 g., previously heated to 100.00° C, was dropped into the calorimeter, the temperature of the latter was found to rise from 22.45° to 23.50° C. What is the mean heat capacity of the metal in cal. deg.$^{-1}$ g.$^{-1}$?

6. A quantity of 2.500 g. of a metal of mean heat capacity 0.0591 (defined) cal. deg.$^{-1}$ g.$^{-1}$ is cooled from the steam point to the ice point. Determine the total heat evolved in (i) ergs, (ii) abs. joules.

7. The volume of a gas, corrected to ideal behavior, is 442.6 ml. at the ice point. At a certain temperature the volume has increased by 22.42 ml., at constant pressure. What is the temperature on the centigrade scale?

8. Justify the statement in the text that if the degree is taken as that on the Fahrenheit scale, absolute temperatures on the Rankine scale are equal to $t°$ F + 459.69°.

[8] R. F. Newton, private communication.

9. The specific volume of liquid acetic acid exceeds that of the solid at the melting point by 0.1595 cc. g.$^{-1}$ at 1 atm. pressure. Determine the work of expansion, in defined cal., accompanying the fusion of 1 mole of acetic acid at atmospheric pressure.

10. Calculate the work done against a constant pressure of 740.0 mm. when 1 mole of water evaporates completely at 100° C; the specific volume of water is approximately 1 cc. g.$^{-1}$ and that of the vapor is 1720 cc. g.$^{-1}$. Express the results in (i) ergs, (ii) liter-atm., (iii) abs. joules, (iv) int. joules, (v) defined cal.

11. The density of mercury at 0° C and 1 atm. pressure is 13.595 g. cc.$^{-1}$. Using the expression in Exercise 1, determine the work of expansion in ergs when the temperature of 100 g. of mercury is raised from 0° to 100° C at 1 atm. pressure. Given that the mean specific heat in this range is 0.0330 cal. deg.$^{-1}$ g.$^{-1}$, what proportion of the total amount of heat supplied to the mercury is the work of expansion?

12. The faraday is usually given as 96,500 coulombs per g. equiv.; show that it is also equal to 23,070 cal. volt^{-1} g. equiv.$^{-1}$, and that the quantity 23,070 cal. volt^{-1} g. equiv.$^{-1}$ faraday^{-1} is dimensionless and equal to unity.

CHAPTER II

PROPERTIES OF THERMODYNAMIC SYSTEMS

4. Thermodynamic Systems

4a. Types of System.—In order to develop the consequences of the laws of thermodynamics, which will be considered shortly, it is necessary to define the terms of reference. The portion of the universe which is chosen for thermodynamic consideration is called a **system**; it usually consists of a definite amount (or amounts) of a specific substance (or substances). A system may be **homogeneous**, that is, completely uniform throughout, such as a gas or a mixture of gases, or a pure liquid or solid, or a liquid or solid solution. When a system is not uniform throughout it is said to be **heterogeneous**; it then consists of two or more phases which are separated from one another by definite bounding surfaces. A system consisting of a liquid and its vapor, or of two immiscible (or partially miscible) liquids, or of two or more solids, which are not a homogeneous solid solution, are examples of heterogeneous systems. There are, of course, numerous other kinds of heterogeneous systems, as is well known to students of chemical equilibrium and the phase rule.[1]

A system may be separated from its surroundings, which consist, in effect, of the remainder of the universe,* by a real or imaginary boundary through which energy may pass, either as heat or as some form of work. *The combination of a system and its surroundings* is sometimes referred to as an **isolated system**.

4b. State of a System.—The thermodynamic or macroscopic state or, in brief, the **state**, of a system can be defined completely by four observable properties or "variables of state"; these are the composition, pressure, volume and temperature.† If the system is homogeneous and consists of a single substance, the composition is, of course, fixed, and hence the state of the system depends on the pressure, volume and temperature only. If these properties are specified, all other physical properties, such as mass, density, viscosity, refractive index, dielectric constant, etc., are thereby definitely fixed. *The thermodynamic properties thus serve to define a system completely.*

In actual practice it is not necessary to state the pressure, the volume and the temperature, for experiment has shown that these three properties of a simple homogeneous system of definite mass are related to one another.

---

[1] S. Glasstone, "Textbook of Physical Chemistry," 2nd ed., 1946, Chapters V, X and XI.

* For thermodynamic purposes the "surroundings" are usually restricted to a limited portion of the universe, e.g., a thermostat.

† Electrical, magnetic, surface, gravitational and similar effects are neglected.

The value of any one of these properties thus depends on the values of the other two. The relationship between them is called an **equation of state**, but its precise form lies, strictly speaking, outside the province of pure thermodynamics; an equation of state must be derived from molecular (kinetic) theory or from direct experiments on the system under consideration. For example, the equation of state for an ideal gas, namely $PV = RT$, where P is the gas pressure, V is the volume of 1 mole, T is the absolute temperature, and R is a constant, is based partly on the kinetic theory of gases and partly on experiment. The van der Waals equation represents a modification of the ideal gas equation, derived by means of molecular theory. Other equations of state, particularly those involving several empirical constants, are determined from experimental data, although their general form may have a theoretical basis. The derivation of such equations is not possible by means of thermodynamics, but the results of thermodynamics may be applied to them with interesting consequences, as will be evident in the subsequent discussion. In any event, it may be accepted that, in general, the pressure, volume and temperature of a system are not independent variables, and consequently *the thermodynamic state of a simple, homogeneous system may be completely defined by specifying two of these properties*.

4c. Thermodynamic Equilibrium.—The results stated above, namely that only two of the three properties of a system, viz., pressure, volume and temperature, are independently variable, and that a homogeneous system of definite mass and composition is completely defined by these two properties, are based on the tacit assumption that the *observable properties* of the system are not undergoing any change with time. Such a system is said to be in **thermodynamic equilibrium**. Actually this term implies three different types of equilibrium which must exist simultaneously. First, there must be **thermal equilibrium**, so that the temperature is the same throughout the whole system. Second, if the system consists of more than one substance there must also be **chemical equilibrium**, so that the composition does not vary with time. Finally, the system must be in a state of **mechanical equilibrium**; in other words, there must be no macroscopic movements within the system itself, or of the system with respect to its surroundings. Disregarding the effect of gravity, mechanical equilibrium implies a uniformity of temperature and pressure throughout the system; if this were not the case, it would, of course, be impossible to describe the system in terms of the pressure, volume and temperature.

Systems in which diffusion or chemical reaction is taking place at an appreciable rate are not in thermodynamic equilibrium, and consequently their state cannot be completely specified in a simple manner. Certain systems which are not in true equilibrium may nevertheless be treated by thermodynamic methods, provided the approach to equilibrium is so slow as to be undetectable over a considerable period of time. An instance of this type is represented by a mixture of hydrogen and oxygen gases under normal conditions of temperature and pressure. As mentioned earlier, reaction should take place with the formation of liquid water, so that the system is not really in chemical equilibrium. However, the reaction is so slow that in the absence of a catalyst it behaves as if it were in thermodynamic equilibrium, provided, of course, that thermal and mechanical equilibria are established. The conditions of chemical equilibrium will, naturally, not apply to such a system.

4d. Properties of a System.—The physical properties of a system may be divided into two main types. There are first, the **extensive properties** which depend on the quantity of matter specified in the system. Mass and volume are two simple examples of extensive properties. The total value of an extensive property is equal to the sum of the values for the separate parts into which the system may, for convenience, be divided. It will be seen later that several properties of thermodynamic interest, such as the energy of a system, are extensive in nature.

The other group of properties are the **intensive properties**; these are characteristic of the substance (or substances) present, and are independent of its (or their) amount. Temperature and pressure are intensive properties, and so also are refractive index, viscosity, density, surface tension, etc. It is because pressure and temperature are intensive properties, independent of the quantity of matter in the system, that they are frequently used as variables to describe the thermodynamic state of the system. It is of interest to note that an extensive property may become an intensive property by specifying unit amount of the substance concerned. Thus, mass and volume are extensive, but density and specific volume, that is, the mass per unit volume and volume per unit mass, respectively, are intensive properties of the substance or system. Similarly, heat capacity is an extensive property, but specific heat is intensive.

The properties of a system in thermodynamic equilibrium depend only on the state, in the sense defined in § 4c, and not on the previous history of the system. If this were not the case, the properties would have no significance, for they would be determined not only by the actual temperature and pressure, but also by the temperature and pressure the system may have had in the past. This is clearly not the case. It follows, therefore, as a consequence that *the change in any property due to a change in the thermodynamic (equilibrium) state depends only on the initial and final states of the system*, and not on the path followed in the course of the change.

4e. Thermodynamic Properties and Complete Differentials.—If any quantity G, such as a thermodynamic property, is a single-valued function of certain variables $x, y, z, \ldots$, which completely determine the value of G, that is,

$$G = f(x, y, z, \ldots), \tag{4.1}$$

then the change in G resulting from a change in the variables from $x_A, y_A, z_A, \ldots$, in the initial state, to $x_B, y_B, z_B, \ldots$, in the final state, is given by

$$\Delta G = G_B - G_A = f(x_B, y_B, z_B, \ldots) - f(x_A, y_A, z_A, \ldots). \tag{4.2}$$

As a mathematical consequence, it is possible to write for a small increase dG in the property G,

$$dG = \left(\frac{\partial G}{\partial x}\right)_{y,z,\ldots} dx + \left(\frac{\partial G}{\partial y}\right)_{x,z,\ldots} dy + \left(\frac{\partial G}{\partial z}\right)_{x,y,\ldots} dz + \cdots, \quad (4.3)$$

where the partial differential symbol, e.g., $(\partial G/\partial x)_{y,z,\ldots}$ represents the rate of change of G with the variable x, while all the other variables, y, z, ..., remain constant. Any differential dG, defined by equation (4.3), of a function G, represented by (4.1), is called a **complete differential**, or an **exact differential**, of that function.

As seen above, a thermodynamic property of a homogeneous system of constant composition is completely determined by the three thermodynamic variables, pressure, volume and temperature. Since only two of these three variables are independent, it is possible to write

$$G = f(P, T),$$

where G may be the energy, volume, or other property to be considered more fully later. This expression is equivalent to equation (4.1), and hence it follows from (4.3) that

$$dG = \left(\frac{\partial G}{\partial P}\right)_T dP + \left(\frac{\partial G}{\partial T}\right)_P dT. \quad (4.4)$$

The physical significance of this result can be understood from the following considerations. When the pressure and temperature of the system are P and T, respectively, the value of the thermodynamic property under consideration is G, but when the variables are changed to $P + dP$ and $T + dT$, it becomes $G + dG$. Since the value of the property is completely determined by the pressure and temperature, the change dG will be independent of the path between the initial and final states. Hence, any convenient method for carrying out the change from P and T to $P + dP$ and $T + dT$ may be chosen for the purpose of calculating dG. Suppose the change is carried out in two stages: (i) in which the temperature remains constant at T while the pressure is changed from P to $P + dP$, and (ii) in which the pressure is held constant at $P + dP$ and the temperature is changed from T to $T + dT$. In stage (i) the rate of change of G with pressure, at the constant temperature T, is $(\partial G/\partial P)_T$, and since the actual pressure change is dP, the change in G for this stage is equal to $(\partial G/\partial P)_T \times dP$; this is seen to be identical with the first term on the right-hand side of equation (4.4). In stage (ii), the rate of change of G with temperature at the constant pressure $P + dP$, which is very close to P, may be written as $(\partial G/\partial T)_P$. Since the actual temperature change is dT, the change in G for this stage is $(\partial G/\partial T)_P \times dT$, which corresponds to the second term on the right-hand side of equation (4.4). The sum of the two terms just derived gives the total change dG for the given process, in accordance with (4.4).

In the foregoing discussion, G has been treated as a single-valued function of the pressure and temperature. It is equally permissible to choose as variables any two of the three thermodynamic properties, pressure, volume and temperature. In some cases it is convenient to take pressure and volume as the independent variables; the complete differential of a thermodynamic property G may then be written as

$$dG = \left(\frac{\partial G}{\partial P}\right)_V dP + \left(\frac{\partial G}{\partial V}\right)_P dV. \tag{4.5}$$

Alternatively, if the volume and temperature are chosen as the variables,

$$dG = \left(\frac{\partial G}{\partial V}\right)_T dV + \left(\frac{\partial G}{\partial T}\right)_V dT. \tag{4.6}$$

Various forms of equations (4.4), (4.5) and (4.6) will be utilized in the treatment of thermodynamic properties.

Another important result, which will be required later, may be derived from the equations given above. Since the volume of a homogeneous system of constant composition is a single-valued function of the pressure and temperature, it is possible to write

$$dV = \left(\frac{\partial V}{\partial P}\right)_T dP + \left(\frac{\partial V}{\partial T}\right)_P dT. \tag{4.7}$$

If this value of dV is substituted in equation (4.6), the result is

$$\begin{aligned}dG &= \left(\frac{\partial G}{\partial V}\right)_T \left[\left(\frac{\partial V}{\partial P}\right)_T dP + \left(\frac{\partial V}{\partial T}\right)_P dT\right] + \left(\frac{\partial G}{\partial T}\right)_V dT \\ &= \left(\frac{\partial G}{\partial V}\right)_T \left(\frac{\partial V}{\partial P}\right)_T dP + \left[\left(\frac{\partial G}{\partial V}\right)_T \left(\frac{\partial V}{\partial T}\right)_P + \left(\frac{\partial G}{\partial T}\right)_V\right] dT.\end{aligned} \tag{4.8}$$

For a given infinitesimal change of thermodynamic state, dG must have a definite value, no matter how it is calculated; the coefficients of dP and dT, respectively, in equations (4.4) and (4.8) must therefore be identical. Hence,

$$\left(\frac{\partial G}{\partial P}\right)_T = \left(\frac{\partial G}{\partial V}\right)_T \left(\frac{\partial V}{\partial P}\right)_T \tag{4.9}$$

and

$$\left(\frac{\partial G}{\partial T}\right)_P = \left(\frac{\partial G}{\partial V}\right)_T \left(\frac{\partial V}{\partial T}\right)_P + \left(\frac{\partial G}{\partial T}\right)_V. \tag{4.10}*$$

5. Equations of State

5a. The Ideal Gas Equation.—The practical value of the results of thermodynamics is frequently greatly enhanced when an equation of state, relating the pressure, volume and temperature of the system, is available.

---

* Some writers derive equation (4.10) directly from (4.6) by "dividing through by dT," and then imposing the constant pressure condition; this procedure is, however, open to criticism on mathematical grounds.

No satisfactory relationship of this type is known for liquids and solids, but for gaseous systems certain moderately simple equations of state have been proposed. All gases actually differ in their behavior, and so the problem is approached by postulating the properties of an ideal gas, and then considering deviations from ideal behavior.

An ideal gas is one which satisfies the equation

$$PV = RT \tag{5.1}$$

for 1 mole at all temperatures and pressures; P and T are the pressure and the absolute temperature, respectively, V is the molar volume, and R is the molar (ideal) gas constant. It will be observed that at constant pressure, the volume of an ideal gas is directly proportional to its absolute temperature, in agreement with the postulate in § 2c. Attention may also be called to the fact that equation (5.1) implies that Boyle's law and Gay-Lussac's (or Charles's) law are both applicable to an ideal gas.

At a given temperature and pressure, the volume of any gas, ideal or not, will be proportional to its mass, or to the number of moles, contained in the system. Since equation (5.1) applies to 1 mole of an ideal gas, it follows that for n moles,

$$PV = nRT, \tag{5.2}$$

where V is now the *total volume* occupied by the gas; R is, however, still the molar gas constant.

The gas constant R is frequently encountered in thermodynamics, and so its value will be determined. Use is made of the fact, derived by extrapolating experimental data for a number of gases to zero pressure, that 1 mole of an ideal gas occupies 22.414 liters at 1 atm. pressure and a temperature of 273.16° K. It follows, therefore, that in equation (5.1), P is 1 atm., V is 22.414 liters mole^{-1} and T is 273.16° K; hence,

$$R = \frac{PV}{T} = \frac{1 \times 22.414}{273.16}$$
$$= 0.082054 \text{ liter-atm. deg.}^{-1} \text{ mole}^{-1}.$$

It will be observed that since the product PV has the dimensions of energy, R must be expressed in energy per degree; since the value of R is invariably given for 1 mole of ideal gas, it is stated in terms of energy deg.$^{-1}$ mole^{-1}. In the present case the energy is in liter-atm.

Since 1 liter contains 1000.028 cc., it follows that

$$R = 82.057 \text{ cc.-atm. deg.}^{-1} \text{ mole}^{-1}.$$

By utilizing the conversion factors in Table 1 (Appendix), the value of R can also be expressed in ergs deg.$^{-1}$ mole^{-1} and in cal. deg.$^{-1}$ mole^{-1}. Thus, since 1 liter-atm. is equivalent to 1.0133×10^9 ergs, it is seen that

$$R = 0.082054 \times 1.0133 \times 10^9$$
$$= 8.3144 \times 10^7 \text{ ergs deg.}^{-1} \text{ mole}^{-1}.$$

Finally, 1 liter-atm. is equivalent to 24.218 (defined) cal., so that

$$R = 0.082054 \times 24.218$$
$$= 1.9872 \text{ cal. deg.}^{-1} \text{ mole}^{-1}.$$

5b. Mixture of Ideal Gases.—In a mixture of ideal gases, it is to be expected that each gas will behave independently of the others. If the partial pressure p_i of any ideal gas in a mixture is defined as the pressure this particular gas would exert if it alone occupied the whole available volume V, then by equation (5.2)

$$p_i V = n_i R T, \tag{5.3}$$

where n_i is the number of moles of the gas present in the system. Since each gas in the mixture will exert its pressure independently of the others, it follows that the total pressure P is equal to the sum of the partial pressures of the constituent gases; thus,

$$P = p_1 + p_2 + \cdots + p_i + \cdots. \tag{5.4}$$

Upon introducing the values of $p_1, p_2, \ldots$, etc., as given by equation (5.3), it can be readily derived from (5.4) that

$$\begin{aligned} PV &= (n_1 + n_2 + \cdots + n_i + \cdots)RT \\ &= nRT, \end{aligned} \tag{5.5}$$

where n is the total number of moles of all the ideal gases present in the mixture. By combining equations (5.3) and (5.5), it is seen that

$$p_i = \frac{n_i}{n} P. \tag{5.6}$$

The fraction n_i/n, that is, the ratio of the number of moles (or molecules) of any constituent of a homogeneous mixture—gaseous, liquid or solid—to the total number of moles (or molecules) is called the **mole fraction** of that constituent; it is represented by the symbol N_i, so that

$$\text{Mole Fraction } \mathrm{N}_i = \frac{n_i}{n_1 + n_2 + \cdots + n_i + \cdots} = \frac{n_i}{n}. \tag{5.7}$$

Utilizing the definition of the mole fraction in conjunction with equation (5.6), it follows that the partial pressure p_i of any ideal gas in a mixture is related to the total pressure by

$$p_i = \mathrm{N}_i P. \tag{5.8}$$

Another expression for the partial pressure can be derived from equation (5.3), based on the fact that n_i/V represents the molar concentration, i.e., moles per unit volume, of the particular gas in the mixture. If this concentration is represented by c_i, equation (5.3) gives

$$p_i = c_i RT \tag{5.9}$$

for an ideal gas.

5c. Real Gases.—According to equation (5.1) or (5.2) the product PV for a given mass of an ideal gas, at constant temperature, should be constant at all pressures. Actual gases, however, exhibit considerable deviations from ideal behavior. At low temperatures the value of PV, instead of remaining constant, at first decreases as the pressure is increased; it then passes through a minimum and then increases. As the temperature is increased, the minimum becomes less marked, and at sufficiently high temperatures the value of PV increases continuously with increasing pressure.

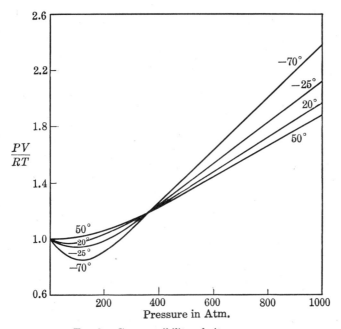

Fig. 2. Compressibility of nitrogen gas

The general nature of the experimental results can be seen from the curves in Fig. 2, which show the variation with pressure of PV/RT for nitrogen at a number of temperatures. For an ideal gas, all the curves would fall on the same horizontal line. The actual temperature at which the minimum in the curve disappears varies with the nature of the gas. For the gases which are difficult to liquefy, e.g., hydrogen and helium, there is no sign of the minimum at ordinary temperatures, but for nitrogen it is observed up to about 50° C. For a readily liquefiable gas, such as carbon dioxide, the minimum in the PV curve persists up to temperatures in the region of 400° C. The equation of state for a real gas must evidently account for the variations of PV with pressure and temperature, as described above, and also for the different behavior of different gases.[2]

[2] S. Glasstone, ref. 1, Chapter III; see also Beattie and Stockmayer, *Rep. Prog. Phys.* (Phys. Soc. London), **7**, 195 (1940).

5d. The van der Waals Equation.

—One of the earliest successful attempts to modify the ideal gas equation so as to make it applicable to real gases is that of J. D. van der Waals (1873), who proposed the equation

$$\left(P + \frac{a}{V^2}\right)(V - b) = RT \tag{5.10}$$

for 1 mole, where a and b are constants for a given gas. The constant a is determined by the attractive forces between the molecules, while b is dependent on their effective volume, which represents a balance between attractive and repulsive forces. By choosing appropriate values for a and b, the van der Waals equation (5.10) is found to represent moderately well the actual behavior of real gases. There is, however, an important weakness: if the equation is assumed to be exact, a and b are found to vary with the temperature. In spite of its approximate nature, the van der Waals equation is frequently employed, for it lends itself readily to mathematical treatment, and even if the results obtained from it are not exact, they are at least qualitatively correct.

By multiplying out, it can be readily seen that the van der Waals equation is a cubic in V, so that there are, under suitable conditions, three values of V for each pressure, at a given temperature (Fig. 3, curves I and II). The region in which this occurs corresponds to that in which liquefaction of the gas is possible. At higher temperatures, e.g., curve IV, two of the roots are always imaginary, only one being real. At a certain intermediate temperature (curve III), which should correspond to the **critical temperature**, the three values of V should become identical, at the point X. At this point the P-V curve will exhibit a horizontal inflection, so that both the first and second derivatives of the pressure with respect to the volume, at constant temperature, will be equal to zero. Thus, writing the van der Waals equation (5.10), for 1 mole of gas, in the form

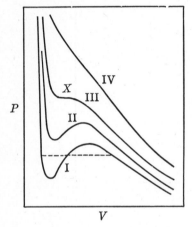

Fig. 3. Van der Waals isothermal curves

$$P = \frac{RT}{V - b} - \frac{a}{V^2}, \tag{5.11}$$

it is found that

$$\left(\frac{\partial P}{\partial V}\right)_T = -\frac{RT}{(V - b)^2} + \frac{2a}{V^3} \tag{5.12}$$

and

$$\left(\frac{\partial^2 P}{\partial V^2}\right)_T = \frac{2RT}{(V - b)^3} - \frac{6a}{V^4}. \tag{5.13}$$

At the critical point X both these expressions are equal to zero; hence, writing V_c and T_c to represent the molar critical volume and temperature, respectively, it follows that

$$-\frac{RT_c}{(V_c - b)^2} + \frac{2a}{V_c^3} = 0 \quad \text{and} \quad \frac{2RT_c}{(V_c - b)^3} - \frac{6a}{V_c^4} = 0.$$

From these two equations, and (5.11), it is readily found that

$$\text{(i)} \quad V_c = 3b; \quad \text{(ii)} \quad T_c = \frac{8a}{27Rb}; \quad \text{(iii)} \quad P_c = \frac{a}{27b^2}. \quad (5.14)$$

If the values of a and b which make the van der Waals equation represent the P-V relationship of a gas at a particular temperature are inserted in equation (5.14), the critical pressure, volume and temperature obtained are only in moderate agreement with the experimental results. This is not unexpected for, as already pointed out, a and b are not strictly constant, if the van der Waals equation is assumed to hold. If the P-V data from which a and b are derived are obtained at a temperature that is some distance from the critical, they will clearly not prove satisfactory for the evaluation of critical constants. In actual practice the procedure adopted is to calculate a and b from the observed critical data * by means of the equations (5.14). A number of the results obtained in this manner are given in Table I; if the pressure P of the gas is in atm. and the molar volume V in liter mole^{-1}, a will be in liter^2 atm. mole^{-2} since a/V^2 must have the same dimensions as P, and b will be in liter mole^{-1}, since this has the dimensions of V.

TABLE I. VAN DER WAALS CONSTANTS

Gas	a liter^2 atm. mole^{-2}	b liter mole^{-1}	Gas	a liter^2 atm. mole^{-2}	b liter mole^{-1}
Acetylene	4.40	5.15 × 10^{-2}	Hydrogen chloride	3.68	4.09 × 10^{-2}
Ammonia	4.17	3.72	Hydrogen cyanide	10.8	8.25
Argon	1.35	3.23	Hydrogen iodide	7.72	5.31
Benzene	18.0	11.5	Hydrogen sulfide	4.43	4.30
n-Butane	14.5	12.3	Methane	2.26	4.30
Carbon dioxide	3.60	4.28	Methanol	9.54	6.71
Carbon monoxide	1.49	4.00	Methyl chloride	7.47	6.51
Chlorine	6.50	5.64	Nitric oxide	1.34	2.80
Cyanogen	7.68	6.93	Nitrogen	1.39	3.92
Ethane	5.46	6.47	Nitrous oxide	3.79	4.43
Ethyl ether	17.4	13.5	Oxygen	1.36	3.19
Ethylene	4.47	5.73	Propane	8.66	8.47
Helium	0.034	2.38	Propylene	8.38	8.30
Hydrogen	0.245	2.67	Sulfur dioxide	6.72	5.65
Hydrogen bromide	4.46	4.44	Water	5.46	3.30

The values of a and b recorded in Table I may be used to calculate the pressure or volume of a gas, at a specified volume or pressure, respectively,

* For critical temperatures and pressures of a number of substances, see Table 2 at the end of the book.

to a moderate degree of accuracy. The results are by no means exact, but they serve as an approximation when no other information is available, especially at high pressures.

Problem: Compare the pressures given by the ideal gas and van der Waals equations for 1 mole of carbon dioxide occupying a volume of 0.381 liter at 40° C.

The ideal gas equation gives, for 1 mole, $P = RT/V$; taking V as 0.381 liter, R as 0.0820 liter-atm. deg.$^{-1}$ mole^{-1}, and T as $273 + 40 = 313°$ K, it is found that

$$P = \frac{0.0820 \times 313}{0.381} = 67.4 \text{ atm.}$$

Utilizing the van der Waals equation in the form of (5.11), with $a = 3.60$ and $b = 4.28 \times 10^{-2}$ (Table I),

$$P = \frac{0.0820 \times 313}{(0.381 - 0.043)} - \frac{3.60}{(0.381)^2} = 51.1 \text{ atm.}$$

The experimental value is 50 atm.

Problem: Compare the volumes given by the ideal gas and van der Waals equations for 1 mole of nitrogen at 400 atm. pressure and 0° C.

According to the ideal gas equation $V = RT/P$ for 1 mole; R is 0.0820 liter-atm. deg.$^{-1}$ mole^{-1}, P is 400 atm., T is 273° K, and hence

$$V = \frac{RT}{P} = \frac{0.0820 \times 273}{400} = 0.0560 \text{ liter mole}^{-1}.$$

In order to determine the volume from the van der Waals equation it is necessary to solve a cubic equation, and this is most simply done by the method of trial. Neglecting a/V^2 in comparison with P, equation (5.10) reduces to $P(V - b) = RT$, so that

$$V = \frac{RT}{P} + b.$$

From Table I, b for nitrogen is 0.0392, and for 400 atm. V is 0.0952 liter mole^{-1}. If this is inserted in equation (5.11), with a equal to 1.39, P is found to be 247 atm. Similarly, if V is taken as 0.0560 liter mole^{-1}, P is found to be 891 atm. Since the molar volume of 0.0560 liter corresponds to a van der Waals pressure of 891 atm., and 0.0952 liter corresponds to 247 atm., it is evident that the correct volume, for 400 atm., is about 0.075 liter. By a series of approximations, the volume is found to be 0.0732 liter. (The experimental value is 0.0703 liter. Although the van der Waals result is not correct, it is very much better than the ideal gas value.)

5e. Reduced Equation of State.—If the pressure, molar volume and temperature of a gas are expressed in terms of the critical pressure, volume and temperature, respectively, i.e.,

$$P = \pi P_c, \quad V = \phi V_c \quad \text{and} \quad T = \theta T_c, \tag{5.15}$$

and these expressions are introduced into the van der Waals equation (5.10), the result for 1 mole of gas is

$$\left(\pi P_c + \frac{a}{\phi^2 V_c^2}\right)(\phi V_c - b) = R\theta T_c.$$

Upon insertion of the values of P_c, V_c and T_c given by the equations (5.14), it is found that

$$\left(\pi + \frac{3}{\phi^2}\right)(3\phi - 1) = 8\theta. \tag{5.16}$$

The quantities π, ϕ and θ, which are equal to P/P_c, V/V_c and T/T_c, respectively, are called the **reduced pressure, volume** and **temperature**, and (5.16) is a **reduced equation of state**. The interesting fact about this equation is that it is completely general, for it does not involve a and b, and hence contains no reference to any specific substance. Consequently, if equimolar amounts of any two gases, whose P-V-T behavior may be represented by an expression of the form of the van der Waals equation, are at the same reduced pressure π, and have the same reduced volume ϕ, then they must be at the same reduced temperature θ. The two gases are then said to be in **corresponding states**, and equation (5.16) is taken as an expression of the **law of corresponding states**.

Experimental studies have shown that the principle of corresponding states derived above is, at least approximately, a valid one, although equation (5.16) does not give the correct quantitative relationship between π, ϕ and θ. It should be pointed out, however, that any equation of state containing two arbitrary constants, such as a and b, in addition to R, can be converted into a relationship involving the reduced quantities π, ϕ and θ, in agreement with the law of corresponding states. Because the law is not exact, however, it would appear that more than three empirical constants are necessary to obtain an exact equation of state.

5f. The Berthelot Equation.—Although it is not employed in connection with P-V-T relationships, the equation of state proposed by D. Berthelot has found a number of applications in thermodynamics. The van der Waals equation is first modified by changing the a/V^2 term to a/TV^2; thus,

$$\left(P + \frac{a}{TV^2}\right)(V - b) = RT.$$

Upon multiplying out, this becomes

$$PV = RT + Pb - \frac{a}{TV} + \frac{ab}{TV^2}. \tag{5.17}$$

Neglecting the last term, since it contains the product of two small quantities, and replacing V in a/TV, as a first approximation, by the ideal gas value RT/P, equation (5.17) becomes

$$PV = RT + Pb - \frac{aP}{RT^2}$$

$$= RT\left(1 + \frac{Pb}{RT} - \frac{aP}{R^2 T^3}\right). \tag{5.18}$$

The factors a, b and R are now replaced by expressions involving the critical constants, based upon experimental results; these are

$$a = \tfrac{16}{3} P_c V_c^2 T_c, \qquad b = \tfrac{1}{4} V_c \qquad \text{and} \qquad R = \tfrac{32}{9} \cdot \frac{P_c V_c}{T_c}.$$

Consequently, equation (5.18) becomes

$$PV = RT \left[1 + \frac{9}{128} \cdot \frac{PT_c}{P_c T} \left(1 - 6 \frac{T_c^2}{T^2} \right) \right], \qquad (5.19)$$

which is the **Berthelot equation**.

5g. The Beattie-Bridgeman Equation.—Another modification of the van der Waals equation, having a partial theoretical basis, is that proposed by J. A. Beattie and O. C. Bridgeman (1927). It takes the form

$$P = \frac{RT(1 - C)}{V^2} (V + B) - \frac{A}{V^2}, \qquad (5.20)$$

where

$$A = A_0 \left(1 + \frac{a}{V} \right), \qquad B = B_0 \left(1 - \frac{b}{V} \right) \qquad \text{and} \qquad C = \frac{c}{VT^3},$$

A_0, B_0, a, b and c being arbitrary constants. The chief application of the Beattie-Bridgeman equation lies in its use to represent the experimental P-V-T relationships of gases, the five constants being derived from actual observations. The values of these constants have been obtained for a number of gases and are recorded in the literature.[3] The results are useful for interpolation of P-V-T data, within the limits of applicability of the constants, and also for a number of thermodynamic purposes.

5h. General Equation of State.—Although numerous equations of state have been proposed from time to time, few of these have been used in thermodynamic studies. Mention may, however, be made of a purely empirical equation which takes the form of a power series in the pressure; thus,

$$PV = RT + aP + bP^2 + cP^3 + dP^4. \qquad (5.21)$$

The factors a, b, c and d are dependent upon the temperature, and the variations have been expressed by the relationships

$$a = a_1 + a_2 T^{-1} + a_3 T^{-3},$$
$$b = b_1 T^{-1} + b_2 T^{-4} + b_3 T^{-6},$$
$$c = c_1 T^{-1} + c_2 T^{-4} + c_3 T^{-6},$$
$$d = d_1 T^{-1} + d_2 T^{-4} + d_3 T^{-6}.$$

The equation thus involves twelve empirical constants, but these can be derived from the experimental P-V-T data without difficulty. Once they are known various uses can be made of equation (5.21), which is capable of relatively easy mathematical manipulation.[4]

[3] Beattie and Bridgeman, *J. Am. Chem. Soc.*, **49**, 1665 (1927); **50**, 3133 (1928); *Proc. Am. Acad. Arts Sci.*, **63**, 229 (1928); *Z. Physik*, **62**, 95 (1930); Beattie, et al., *J. Am. Chem. Soc.*, **52**, 6 (1930); **59**, 1587, 1589 (1937); **61**, 26 (1939); **64**, 548 (1942); *J. Chem. Phys.*, **3**, 93 (1935); see also, Deming and Shupe, *J. Am. Chem. Soc.*, **52**, 1382 (1930); **53**, 843 (1931); Maron and Turnbull, *Ind. Eng. Chem.*, **33**, 408 (1941).

[4] See, for example, Maron and Turnbull, *Ind. Eng. Chem.*, **34**, 544 (1942).

5i. Compressibility Factors.

Although such analytical expressions as the Beattie-Bridgeman equation and, to a less precise extent, the van der Waals equation are useful for solving problems relating to the pressure, volume and temperature of gases, much labor could be saved if complete compressibility diagrams, such as Fig. 2, were available for all gases. The necessity for a separate diagram for each gas would be a complicating factor, and so it is fortunate that the law of corresponding states has permitted the development of a single generalized compressibility diagram which is (approximately) applicable to all gases. Except for readily liquefiable gases, results can be obtained from this diagram which agree with the observed data within a few per cent.

The **compressibility factor** κ of a gas is defined by

$$\kappa = \frac{PV}{RT}, \qquad (5.22)$$

where P is the pressure of the gas, V is its molar volume and T is the absolute temperature. For an ideal gas κ would obviously be unity under all circumstances, but for a real gas it may be less or greater than unity, as may be seen from Fig. 2, which is really a plot of the compressibility factor against the pressure of nitrogen at several temperatures. It follows, therefore, from equation (5.22) that for 1 mole of gas

$$PV = \kappa RT, \qquad (5.23)$$

where κ is the compressibility factor in the given state of the gas. For n moles of gas, the volume would, of course, be increased n-fold.

If the pressure, volume and temperature in equation (5.22) are replaced by their respective reduced properties, i.e., P is replaced by πP_c, V by ϕV_c and T by θT_c, it follows that

$$\kappa = \frac{P_c V_c}{RT_c} \cdot \frac{\pi \phi}{\theta}. \qquad (5.24)$$

It will be seen from the three relationships of equation (5.14) that $P_c V_c/RT_c$ should be constant, equal to $\frac{3}{8}$, for all van der Waals gases. Experimental observation has shown that for many substances $P_c V_c/RT_c$ has, in fact, almost the same constant value, although it is more nearly $\frac{3}{11}$ than $\frac{3}{8}$. Since this quantity is constant, however, it follows from equation (5.24) that for all gases, as a first approximation,

$$\kappa = c \frac{\pi \phi}{\theta}, \qquad (5.25)$$

where c is a universal constant. According to the law of corresponding states (§ 5e), if the reduced pressure π and the reduced temperature θ have the same value for different gases, their reduced molar volumes ϕ must be equal. It follows, therefore, from equation (5.25), that their compressibility factors κ must then also be the same, irrespective of the nature of the gas. In other words, if the compressibility factor is plotted against the

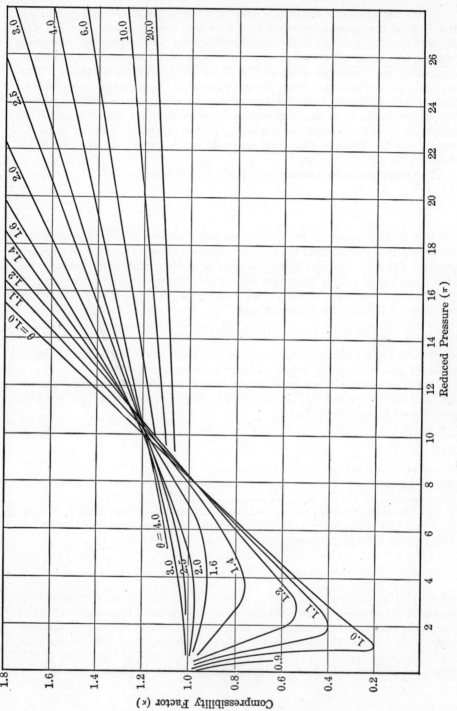

Fig. 4. Generalized compressibility curves

reduced pressure, for a given reduced temperature, the results for all gases will fall on the same curve.*

The conclusion just reached forms the basis of the generalized, or reduced, compressibility curves (Fig. 4).[5] From actual experiments on a number of gases, the mean observed compressibility factors at various temperatures and pressures have been derived, and the values of κ are plotted against the corresponding reduced pressures, with the reduced temperature as parameter. From these curves it is possible to derive, with a fair degree of accuracy, the value of either the pressure, volume or temperature of any gas, if the other two variables are given. The determination of the volume can be achieved directly from Fig. 4, but the evaluation of either pressure or temperature is not quite as simple.†

Problem: Utilize Fig. 4 to determine the volume of 1 mole of nitrogen at 400 atm. pressure and 0° C (cf. second problem in § 5d).

The critical pressure of nitrogen is 33.5 atm. and the critical temperature is 126° K; hence the reduced pressure and temperature are given by

$$\pi = \frac{400}{33.5} = 11.9; \quad \theta = \frac{273}{126} = 2.17.$$

From Fig. 4 it is found directly that for these values of π and θ, the compressibility factor κ is about 1.27. Utilizing equation (5.23), with $P = 400$ atm., $R = 0.0820$ liter-atm. deg.$^{-1}$ mole^{-1} and $T = 273°$, it is found that

$$V = \kappa \frac{RT}{P} = \frac{1.27 \times 0.0820 \times 273}{400} = 0.071 \text{ liter mole}^{-1}.$$

(The experimental value is 0.0703 liter mole^{-1}.)

Problem: Utilize Fig. 4 to determine the pressure of carbon dioxide gas when 1 mole occupies 0.381 liter at 40° C (cf. first problem in § 5d).

Substituting the value of $V = 0.381$ liter mole^{-1}, $T = 273 + 40 = 313°$ K, and R into equation (5.23), it follows that

$$P = \kappa \frac{RT}{V} = \kappa \frac{0.0820 \times 313}{0.381} = 67.4\kappa.$$

The critical pressure of carbon dioxide is 72.9 atm., so that

$$\pi = \frac{P}{P_c} = \frac{67.4}{72.9} \kappa = 0.925\kappa.$$

* For hydrogen and helium better results are obtained by adding 8 to the critical pressure and temperature when calculating the reduced values; thus, for these gases, $\pi = P/(P_c + 8)$ and $\theta = T/(T_c + 8)$.

[5] Cope, Lewis and Weber, *Ind. Eng. Chem.*, **23**, 887 (1931); Brown, Souders and Smith, *ibid.*, **24**, 513 (1932); Dodge, *ibid.*, **24**, 1353 (1932); Lewis and Luke, *ibid.*, **25**, 725 (1933); Lewis, *ibid.*, **28**, 257 (1936); Kay, *ibid.*, **28**, 1014 (1936); Maron and Turnbull, *ibid.*, **34**, 544 (1942); for a convenient nomograph applicable below the critical point, see Thompson, *ibid.*, **35**, 895 (1943).

† Generalized (reduced) compressibility diagrams are often referred to in the literature as "μ-charts," the symbol μ being used for the compressibility factor. Since μ is employed later for another purpose, it has been replaced by κ here.

The expression $\pi = 0.925\kappa$ is the equation for a straight line passing through the origin of Fig. 4. The point representing the system under consideration is that at which this line intersects the curve in Fig. 4 for the prescribed reduced temperature. The critical temperature of carbon dioxide is 304.1° K, and so the reduced temperature θ corresponding to 40° C, i.e., 313° K, is 313/304 = 1.03. The value of π at which the line $\pi = 0.925\kappa$ intersects the curve for $\theta = 1.03$ is found to be 0.7 from Fig. 4. Since P_c is 72.9, it follows that P is $0.7 \times 72.9 = 51$ atm. (The experimental value is 50 atm.)

5j. Mixtures of Real Gases: Additive Pressure Law.

The rule that the total pressure of a mixture of gases is equal to the sum of the pressures exerted by each gas if it alone occupied the whole of the available volume (§ 5b) does not apply to real gases. The total pressure is thus not equal to the sum of the partial pressures defined in the usual manner. However, for some purposes it is convenient to define the partial pressure of a gas in a mixture by means of equation (5.8), i.e., $p_i = N_i P$, where p_i is the partial pressure and N_i is the mole fraction of any constituent of the mixture of gases of total pressure P.

A simple modification of the law of partial pressures as applied to ideal gases has been proposed for mixtures of real gases (E. P. Bartlett, 1928).[6] If P'_i is the pressure which would be exerted by a constituent of a gas mixture when its molar volume is the same as that of the mixture, then it is suggested that the total pressure P is given by

$$P = N_1 P'_1 + N_2 P'_2 + \cdots + N_i P'_i + \cdots. \quad (5.26)$$

This rule has been found to give results in fair agreement with experiment. If compressibility data for the individual gases are not available, the values of P'_1, P'_2, etc., can be obtained with the aid of Fig. 4, provided the molar composition of the gas, its volume and temperature are known.

Problem: A mixture of $\frac{1}{4}$ mole nitrogen and $\frac{3}{4}$ mole hydrogen occupies 0.0832 liter at 50° C. Calculate the total pressure.

The molar volume of the mixture is 0.0832 liter; by using Fig. 4 in conjunction with the known critical temperature and pressure of nitrogen and hydrogen (see footnote, p. 29), the pressure of nitrogen gas at 50° C for this molar volume is found to be 404 atm. and that for hydrogen is 390 atm. The calculation is identical with that in the second problem of § 5i. By equation (5.26), the total pressure is given by

$$P = (\tfrac{1}{4} \times 404) + (\tfrac{3}{4} \times 390) = 394 \text{ atm.}$$

(The experimental value is 400 atm. The agreement is partly fortuitous, because Fig. 4 cannot be read to this degree of accuracy.)

5k. Additive Volume Law.

The additive pressure law, as given by equation (5.26), is useful for the calculation of the approximate pressure exerted by each gas, and the total pressure, in a mixture of real gases, when the volume is known. If the total pressure is given, however, the evaluation of the volume is somewhat more complicated, involving a series of trial solutions. An alternative approxi-

---

[6] Bartlett, Cupples and Tremearne, *J. Am. Chem. Soc.*, **50**, 1275 (1928).

mate method is available which makes use of the law of additive volumes (E. H. Amagat, 1893; A. Leduc, 1898).[7] It can be readily shown that for a mixture of ideal gases, the total volume should be equal to the sum of the volumes which the constitutent gases would occupy at the total pressure of the mixture, at the same temperature. This rule has been found to hold, with a fair degree of accuracy, for mixtures of real gases. If complete experimental P-V-T data for the individual gases in the mixture are available, it is a simple matter to make use of the additive volume law. For approximate purposes the generalized compressibility diagram (Fig. 4) may be utilized.

Problem: Calculate the volume of a mixture of $\frac{1}{4}$ mole nitrogen and $\frac{3}{4}$ mole hydrogen at 400 atm. pressure and 50° C. (This is the reverse of the problem in § 5j.)

It is found from Fig. 4 that at 400 atm. and 50° C, 1 mole of nitrogen would occupy 0.083 liter; under the same conditions, the volume of 1 mole of hydrogen would be 0.081 liter. The method of calculation is, in each case, identical with that used in the first problem in § 5h. By the Amagat law of additive volumes, the volume of the given mixture would be

$$(\tfrac{1}{4} \times 0.083) + (\tfrac{3}{4} \times 0.081) = 0.082 \text{ liter.}$$

(The experimental value is 0.0832 liter.)

EXERCISES

1. A liquid mixture consists of equal parts by weight of water and sulfuric acid, what is the mole fraction of each constituent?

2. The composition of dry air by weight is as follows: nitrogen 75.58%, oxygen 23.08%, argon 1.28%, carbon dioxide 0.06%, with negligible traces of other gases. Calculate the partial pressure of each of these four gases in air at exactly 1 atm. pressure, assuming ideal behavior.

3. Suppose 10.0 liters of gas A, measured at 0.50 atm., and 5.0 liters of gas B, at 1.0 atm., are passed into a vessel whose capacity is 15.0 liters. What is the resulting total pressure, if the gases behave ideally and the temperature remains constant?

4. A gas collected over water at 25° C becomes saturated with water vapor, its partial pressure being 23.8 mm. The measured volume of the moist gas is 5.44 liters, at a total pressure of 752.0 mm. Calculate the volume the dry gas would occupy at a pressure of 760 mm., assuming ideal behavior of the gas and of the water vapor.

5. Prove that for a mixture of ideal gases the total volume is equal to the sum of the volumes which the constituent gases would occupy at the total pressure of the mixture at the same temperature (Amagat's rule).

6. At 0° C and 400 atm. pressure a mixture containing 0.75 mole fraction of nitrogen and 0.25 mole fraction of hydrogen was found to occupy a volume of 71.5 ml. mole^{-1}. Utilizing the generalized compressibility diagram, determine whether Amagat's volume rule or Bartlett's partial pressure rule is in better agreement with experiment.

[7] Cf., Leduc, *Ann. chim. phys.*, **15**, 5 (1898); see also, Masson, *et al.*, *Proc. Roy. Soc.*, **A103**, 524 (1923); **A122**, 283 (1929); **A126**, 248 (1930).

7. Show that at moderate and low pressures the van der Waals equation may be written in the form $PV = RT(1 - BP)$, where $B = \dfrac{1}{RT}\left(\dfrac{a}{RT} - b\right)$. (Multiply out the van der Waals equation, neglect the term ab/V^2, which is small if the pressure is not large, and replace V in a/V by RT/P.)

Express B in terms of T_c and P_c, and compare the result with that given by the Berthelot equation (5.19) for hydrogen and oxygen at 25° C. (The experimental values are -5.9×10^{-4} and 6.8×10^{-4} atm.$^{-1}$.)

8. A cylinder, of capacity 100.0 liters, contains methane gas originally at 200 atm. and 25° C. Determine by means of the compressibility chart the weight of gas used up when the pressure in the cylinder has fallen to 50.0 atm. Compare the result with that for (i) an ideal gas, (ii) a van der Waals gas.

9. The critical temperature and pressure of ethane are 305.2° K and 48.8 atm., respectively. Taking R as 0.08205 liter-atm. deg.$^{-1}$ mole^{-1}, determine the van der Waals constants. What would be the approximate molar volume at the critical point?

10. Compare the pressures given by (i) the ideal gas equation, (ii) the van der Waals equation, (iii) the compressibility diagram, for 1 mole of ethylene occupying a volume of 0.279 liter at 40.0° C.

11. Compare the volumes given by (i) the ideal gas equation, (ii) the van der Waals equation, (iii) the compressibility diagram, for 1 mole of hydrogen at 500 atm. and 0° C.

12. Assuming Bartlett's rule and using the compressibility diagram, determine the total pressure of a mixture consisting of 23.0 g. of oxygen and 77.0 g. of nitrogen occupying 155 ml. at 0° C. What would be the pressure if the gases behaved ideally?

13. By means of Amagat's rule and the compressibility chart, calculate the volume occupied by the mixture in Exercise 12 at a total pressure of 500 atm. at 0° C. Compare the result with that which would be found if the gases were ideal.

14. It has been suggested [Kay, *Ind. Eng. Chem.*, **28**, 1014 (1936)] that the compressibility factor of a mixture of gases may be determined by treating the mixture as a single gas with a critical temperature equal to $N_1 T_{c1} + N_2 T_{c2} + \cdots$, and a critical pressure of $N_1 P_{c1} + N_2 P_{c2} + \cdots$, where N_1, N_2, $\cdots$ are the mole fractions of the various gases in the mixture, and T_{c1}, T_{c2}, $\cdots$ are the critical temperatures and P_{c1}, P_{c2}, $\cdots$ the critical pressures. Apply this rule to Exercises 12 and 13.

15. If G is a function of P, V and T, prove that

$$\left(\frac{\partial G}{\partial P}\right)_T = \left(\frac{\partial G}{\partial V}\right)_P \left(\frac{\partial V}{\partial P}\right)_T + \left(\frac{\partial G}{\partial P}\right)_V,$$

and derive an analogous expression for $(\partial G/\partial T)_V$.

16. Verify the units of the van der Waals a and b given in Table I.

CHAPTER III

THE FIRST LAW OF THERMODYNAMICS

6. The Conservation of Energy

6a. The Equivalence of Work and Heat.—The relationship between mechanical work and heat was first clearly seen by Count Rumford (Benjamin Thompson) in 1798. As a result of his observations made on the heat developed during the boring of a cannon, he concluded that the heat produced was related to the mechanical work expended in the boring process. Some experiments carried out by H. Davy (1799) appeared to indicate the connection between work and heat, but the most important results were those obtained by J. P. Joule in an extended series of observations, commenced about 1840 and lasting for nearly forty years. In a number of carefully planned and executed experiments, Joule converted known amounts of work into heat, and measured the amount of heat thus produced by determining the rise in temperature of a calorimeter of known heat capacity. Among the methods used for converting heat into work, mention may be made of the following: stirring water or mercury by a paddle wheel, compression or expansion of air, forcing water through capillary tubes, passage of an electric current through wires of known resistance, and passage of induced current through a coil of wire rotated in a magnetic field.

As a result of his studies, Joule came to the highly significant conclusion that *the expenditure of a given amount of work, no matter what its origin, always produced the same quantity of heat.* This fact is the basis of the concept of a definite **mechanical equivalent of heat**, that is, of a definite and constant ratio between the number of ergs of mechanical work done and the number of calories produced by the conversion of this work into heat. According to the most recent experiments, one standard (15°) calorie is equivalent to 4.1858×10^7 ergs, i.e., 4.1858 abs. joules, of work, irrespective of the source or nature of this work. Assuming that 1 int. joule is equivalent to 1.0002 abs. joules (§ 3b), it follows that a standard calorie is equivalent to 4.1850 int. joules. However, because heat quantities are usually determined at the present time by comparison with the heat produced by electrical work, the value of which is known in int. joules, it has become the practice by chemists to adopt the relationship between work and heat given in § 3e. Thus, a defined calorie is equivalent to 4.1833 int. joules. The difference between this value and the one given above, i.e., 4.1850 int. joules, is, at least partly, due to the difference between the standard and defined calories.

6b. The Mechanical Equivalent from Heat Capacities.—It is of interest to mention that while Joule's experiments were in progress, J. R. Mayer (1842) calculated, from the specific heats of air at constant pressure and constant volume (§ 9b), the heat change accompanying the expansion of air, and compared the result with the work done against the external pressure. The mechanical equivalent of heat derived in this manner was close to that obtained by Joule in an entirely different manner. If modern specific heat data are used, the results of Mayer's calculation are in excellent agreement with the accepted value of the relationship between ergs and calories.

Problem: The difference between the heat capacities of 1 g. of air at constant pressure (1 atm.) and constant volume is 0.0687 cal. deg.$^{-1}$ at 0° C. The volume of 1 g. of air at 0° C and 1 atm. is 773.4 cc. Assuming the difference in the heat capacities to be entirely due to the work done in the expansion of the air at constant pressure, which is not strictly true, calculate the relationship between the erg and the calorie.

By Gay-Lussac's law of the expansion of a gas at constant pressure [cf. equation (2.2)], the volume of a gas expands by 1/273.16 of its volume at 0° C for every 1° rise of temperature. In the present case, therefore, the increase in volume of 1 g. of air is 773.4/273.16 cc. for 1° increase of temperature. According to equation (3.5), the work done in the expansion against the constant pressure of the atmosphere is equal to the product of the pressure, i.e., 1 atm. or 1.013×10^6 dynes cm.$^{-2}$ (§ 3f), and the increase of volume. If the pressure is in dynes cm.$^{-2}$ and the volume in cc., the work will be expressed in ergs; thus,

$$\text{Work of expansion} = 1.013 \times 10^6 \times \frac{773.4}{273.16} = 2.87 \times 10^6 \text{ ergs.}$$

This work should be equivalent to the difference between the heat required to raise the temperature of 1 g. of air by 1° at constant pressure (1 atm.) and constant volume, in the vicinity of 0° C. This difference is given as 0.0687 cal.; hence,

$$0.0687 \text{ cal.} = 2.87 \times 10^6 \text{ ergs,}$$
$$1 \text{ cal.} = 4.18 \times 10^7 \text{ ergs.}$$

6c. The Conservation of Energy: The First Law of Thermodynamics.—The belief that "perpetual motion of the first kind," * that is, the production of energy of a particular type without the disappearance of an equivalent amount of energy of another form, was not possible has long been accepted by scientists. No success had attended the many attempts to construct a machine which would produce mechanical work continuously without drawing upon energy from an outside source, and without itself undergoing any change. The fundamental significance of this accepted view was not widely realized until 1847 when H. von Helmholtz showed that the failure to achieve perpetual motion and the equivalence of work and heat, described above, were aspects of a wide generalization which has become known as the **law of conservation of energy**. This law has been stated in various forms, but its fundamental implication is that *although energy may be converted from one form to another, it cannot be created or destroyed*; in other words, *whenever a quantity of one kind of energy is produced, an exactly equivalent amount of another kind (or kinds) must be used up.*

* For an explanation of "perpetual motion of the second kind," see § 18d.

It should be clearly understood that the law of conservation of energy is purely the result of experience, for example, the failure to construct a perpetual motion machine and the constancy of the mechanical equivalent of heat. It is upon the assumption that such experience is universal that the **first law of thermodynamics** is based. This law is, in fact, identical with the principle of conservation of energy, and can consequently be stated in either of the forms given above; other ways of stating the law will be mentioned later. It may be pointed out that the first law of thermodynamics can be regarded as valid as long as perpetual motion of the first kind is found not to be possible, as is apparently the case in the universe of which the earth is a part. If, in another universe or under other circumstances, the creation of energy were proved to be feasible, the laws of thermodynamics would not be applicable. However, this contingency appears to be so remote, as far as human observation is concerned, that it may be completely disregarded.

Following historical precedent, the validity of the first law of thermodynamics has here been associated with the impossibility of perpetual motion of the first kind and the constancy of the mechanical equivalent. The law has nevertheless a much firmer basis, for it leads to a wide variety of conclusions, as will be seen in this and later chapters, which have been found to be in complete agreement with actual experience.

6d. Isolated Systems and the First Law.—It was stated in § 4a that a combination of a system and its surroundings may be referred to as an isolated system. The first law of thermodynamics requires that *the total energy of an isolated system must remain constant, although there may be changes from one form of energy to another*. This means that any loss or gain of energy by a system must be exactly equivalent to the gain or loss, respectively, of energy by the surroundings. The forms of energy are not necessarily the same, but if there is to be a net conservation of energy the amounts must be equivalent. The conclusion derived from the first law, that the energy change in a system must be exactly compensated by that of its surroundings, is of great importance, and it will be applied in later developments.

6e. Energy and Mass.—Brief reference may be made here to circumstances under which the principle of the conservation of energy appears to fail; the failure is apparent, however, rather than real. According to the theory of relativity *there is an equivalence of mass and energy;* the loss or gain of energy by a body must be accompanied by a corresponding change of mass. The change of energy ΔE is related to the change of mass Δm by the relationship

$$\Delta E = c^2 \Delta m, \tag{6.1}$$

where c is the velocity of light. In certain reactions between atomic nuclei there is a liberation of large amounts of energy without the apparent disappearance of an equivalent quantity of energy of another kind. However, a study of the masses of the nuclei concerned shows that there is a loss of mass which corresponds exactly to the energy set free, as required by equation (6.1). The energy liberated ultimately becomes associated with some form of matter, and there is a gain of mass of the latter identical with the loss suffered by the nuclei. The equivalence of mass and energy means that the laws of conservation of mass and energy can both be true, but each is a mere corollary of the other. Since nuclear reactions are not considered in chemical thermodynamics, however, the problems associated with the relationship between mass and energy will not arise.

7. Energy and Heat Changes

7a. The Energy Content.—The energy possessed by a system, that is to say, its energy content, may be regarded as falling into two main categories. There is first, the energy which is a characteristic property of the system itself; this may include the translational energy of the moving molecules, the energy of vibration and rotation of the molecules, as well as the energy of the electrons and nuclei. On the other hand, there is the energy which is determined by the position of the system in a force field, e.g., magnetic, electric or gravitational, and by the motion of the system as a whole, e.g., through space. In thermodynamics the energy in this second category is usually ignored, and it is only that in the first which is taken into consideration. The latter is sometimes called the internal energy, to distinguish it from the energy due to external factors, but for present purposes it may be referred to simply as the **energy** or the **energy content** of the system. The term internal energy can then be used when it is desired to describe the energy associated with the motions, e.g., vibration, rotation, etc., within the molecule, as distinct from the translational energy resulting from the motion of the molecule as a whole.*

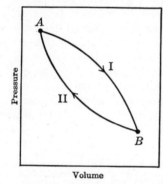

Fig. 5. Energy changes in direct and reverse paths

The energy content of a system must depend on its thermodynamic state. An increase of temperature at constant volume, for example, brought about by the transfer of heat to the system from the surroundings, must result in an increase of its energy. It will now be shown, by utilizing the first law of thermodynamics, that the energy content is a property of the system of the kind that is determined only by the state of the system, and not by the manner in which it reached that state. In other words, it will be proved that *the energy is a single-valued function of the thermodynamic variables of the system.* It is consequently a property to which the results obtained in § 4e may be applied.

Consider any system represented by the state A in Fig. 5, in which the coordinates are the observable properties, e.g., pressure and volume, that determine the energy content. Suppose the conditions are now altered, along the path I, so that the state of the system is represented by the

---

* This motion of the molecule is, of course, not to be confused with the motion of the *system* as a whole, which, as stated above, is not included in the thermodynamic energy content.

point B; the system is then returned to its exact original state by a series of changes indicated by the path II.* It is then a direct consequence of the first law of thermodynamics that the total energy change of the system in path I must be identical in magnitude, but opposite in sign, to that in path II, provided the surroundings remain unchanged; if this were not so a perpetual motion machine would be possible. Imagine, for example, that the increase of energy in path I were greater than the decrease in returning by II; then by carrying out the change $A \to B$ by path I and the reverse change $B \to A$ by path II, the system would be brought back to its original state, but there would be a residuum of energy if the surroundings were unchanged. In other words, energy would have been created, without the disappearance of an equivalent amount of another kind. Since this is contrary to the first law of thermodynamics, it must be concluded that the energy change in the process $A \to B$, by path I, must be numerically equal to that involved in the reverse process $B \to A$, by path II.

It is possible to proceed from A to B by various paths, I, I', I'', etc. (Fig. 6), returning in each case by path II. From the preceding argument, it is evident that the energy changes in I, I', I'', etc., must each be numerically equal to that in II. It follows, therefore, that the energy changes in the different paths I, I', I'', etc., between the two given states A and B, must all be equal to each other. The conclusion to be drawn, therefore, is that *the change in energy of a system, associated with the passage from one thermodynamic state to another, depends only on the initial and final states, and is independent of the path followed.* This statement may be regarded as a form of the first law of thermodynamics.

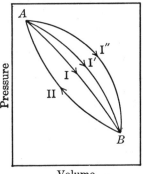

Fig. 6. Energy change independent of path

It is clear from these considerations that the energy of a system is a "property" of the system in the sense described in § 4e. It is, therefore, possible to ascribe a definite value E to the energy or energy content of a particular system in a given state; this value will depend only on the state, and not on the previous history, of the system. If E_A represents the energy in the thermodynamic state A, and E_B that in the state B, then the increase of energy ΔE† in passing from A to B is given by

$$\Delta E = E_B - E_A, \tag{7.1}$$

and is independent of the path taken in the change of state $A \to B$.

---

* In thermodynamics the word "change" is frequently used to imply "change in thermodynamic state"; the terms "path" or "process" then refer to the means whereby the given change in state is accomplished. As will be seen below, a particular change may be achieved by following various different paths, i.e., by different processes.

† The symbol Δ as applied to a change in a thermodynamic property accompanying a given process represents the *algebraic increase* in the property. It is always equal to the

Since the energy content of a system obviously depends on the quantity of material contained in the system, it is apparent that the energy E is an extensive property, as defined in § 4d. If the mass of the system is altered, the energy content will be affected in the same proportion. Similarly, the value of ΔE for any process depends upon the amount of material contained in the system which undergoes the change.

7b. Work, Heat and Energy Changes.—The energy change accompanying the change in thermodynamic state of a system may be due to the performance of work on or by the system, and also to the transfer of energy to or from the system in the form of heat. The term "work" is used here in the general sense referred to in § 3a, and may include work of expansion or any other form of mechanical work, as well as electrical work, etc. If W is the magnitude of the work done, then by convention W is *positive* if work is done *by* the system. In other words, by the convention W is taken as positive when the system loses energy as the result of doing some form of work. On the other hand, W is *negative* if work is done *on* the system, so that it gains energy.

The amount of heat transferred is represented by Q, and this is regarded as *positive*, by convention, if energy is transferred to, that is, *taken up* by, the system in the form of heat; thus, Q is positive when the system gains energy as heat. Similarly, if heat is transferred from, that is, is *given up* by, the system, so that there is a corresponding decrease of energy, Q is *negative*. It should be noted that the conventions for W and Q given above are those which are widely adopted in the study of chemical thermodynamics. There is no particular reason why these, or any other, conventions should be used, but once the conventions have been decided upon *they must be adhered to throughout* and never changed, if confusion is to be avoided.

Since the system loses energy W, because of work done, and gains energy Q, by the transfer of heat, the net gain of energy is $Q - W$. By the first law of thermodynamics, this must be identical with the increase ΔE in the energy content of the system; thus,

$$\Delta E = Q - W. \tag{7.2}$$

This equation may be regarded as a form of the first law of thermodynamics or, alternatively, it may be used as a means of defining the energy content E. Thus, *the difference between the heat absorbed by a system and the total work done by the system may be defined as equal to the increase in a property of the system called its energy content.*

7c. Energy Change in Cyclic Process.—If, as a result of a series of processes, a system returns to its exact original state, its energy content will be unchanged so that ΔE must be zero. In this event, it follows from equa-

value in the final state minus that in the initial state of the process. If ΔG defined in this manner is positive, the property G has a larger value in the final than in the initial state; on the other hand, if ΔG is found to be negative, G is larger in the initial state. The significance of $-\Delta G$ is, of course, the reverse of that of ΔG.

tion (7.2), that the work done by the system is equal to the heat absorbed in the process, i.e.,

$$Q = W. \tag{7.3}$$

This equation is an expression of the impossibility of "perpetual motion of the first kind," as stated in § 6c. Provided the system does not undergo any net change, i.e., ΔE is zero, it is impossible to produce work without drawing upon energy, namely heat, from an outside source. *A process, or series of processes, as a result of which the system returns exactly to its original thermodynamic state*, is referred to as a **cyclic process** or **cycle**. In any cycle ΔE is zero, and the heat absorbed Q is equal to W the work done by the system, in accordance with equation (7.3).

7d. Dependence of Heat and Work on the Path.—The heat and work terms involved in any change of thermodynamic state differ from the total change of the energy content in one highly important respect. As seen above, the quantity ΔE for a given change has a definite value depending only on the initial and final states of the system, and independent of the path taken in the process. This is, however, not the case for Q and W; *the values of the heat and work terms vary with the path followed in proceeding from the initial to the final state*. That this is the case may be shown in a number of ways.

It was seen in § 3g that when a system expands the work done is equal to the product of the external pressure and the increase of volume of the system. Obviously, the work done for a given volume change will thus depend on the external pressure, the magnitude of which may vary according to circumstances. The value of W will consequently not be determined solely by the initial and final states of the system; it will also depend on the manner in which the change is carried out. According to the first law of thermodynamics, the increase ΔE in the energy content depends only on the initial and final states, but since ΔE is equal to $Q - W$ by equation (7.2), and W varies with the path, Q, the heat absorbed in the process, must vary similarly. Incidentally, it should be borne in mind that although Q and W depend on the path, the difference $Q - W$, which is equal to ΔE, does not vary with the path taken, for it is determined only by the initial and final states of the system. It is seen, therefore, that even though Q and W are both variable, they are not independent, for their difference must have a definite value for a given change of thermodynamic state.

The variability of heat and work terms will be evident from other considerations. For example, a certain mass of hydrocarbon may be completely burnt in air, at constant volume, as in a combustion bomb (§ 12g). All the energy lost by the system appears in the form of heat, no work being done. On the other hand, in an internal combustion engine a large proportion of the energy of the hydrocarbon-air system is converted into mechanical work, and the remainder into heat. Another instance of a similar type is provided by a system consisting of zinc and dilute sulfuric acid. These substances may be made to react with one another in a calorimeter, so that the energy decrease resulting from their conversion into zinc sulfate and hydrogen is accounted for almost entirely by the heat evolved. However, by combining the zinc with an inert metal, such as copper or platinum, a galvanic cell can be formed with the sulfuric acid as electrolyte. Electrical energy is then produced, and this may be used to drive a machine, thus doing mechanical work. The inert metal is unaffected in the process, and so the energy, appearing partly as work and partly as heat, is that associated with the reaction between zinc and sulfuric acid to form zinc sulfate and hydrogen.

It should be understood that although the heat and work changes are variable quantities, for a given change in thermodynamic state, the values of Q and W for any *specified path* are quite definite. It will be seen later that an adequate specification of the path can often be given in a simple manner, e.g., constant pressure or constant volume. The results obtained are then useful for thermodynamic purposes.

7e. Heat and Work Not Properties of a System.—The conclusion to be drawn from the preceding discussion is that unlike the energy content, *heat and work are not thermodynamic properties of a system*. They merely represent different ways in which energy can enter or leave the system, the respective amounts varying with the method whereby the change from the initial to the final state is carried out. In other words, the performance of work and the flow of heat represent ways in which the energy of a system can be changed, but the energy content cannot be regarded as consisting of definite "work" and "heat" portions.[1]

A process involving an appreciable change in the thermodynamic state of a system is often conveniently considered as made up of a series of infinitesimal stages. For each of these stages, equation (7.2) is sometimes written in the form

$$dE = dQ - dW, \qquad (7.4)$$

but this equation is liable to be misleading. The energy E is a thermodynamic property of the system, in the sense of § 4d, and dE is a complete differential, which may be treated in the manner described in § 4e. The integral of dE between the limits of the initial (A) and final (B) states of a process has the definite value $E_B - E_A$ or ΔE [cf. equation (7.1)]. This is, however, not true for the quantities dQ and dW in equation (7.4); these quantities are not complete differentials, for Q and W are not properties of the thermodynamic state of the system. In order to avoid the misunderstandings which may arise from the use of such symbols as dQ and dW, they will not be employed in this book. Instead, equation (7.4) will be written as

$$dE = q - w, \qquad (7.5)$$

where dE is the increase in the energy content of the system accompanying an infinitesimal stage of an appreciable thermodynamic process; q is the quantity of heat absorbed and w is the work done by the system at the same

[1] For discussions of this and related topics, see Tunell, *J. Phys. Chem.*, **36**, 1744 (1932); Brønsted, *ibid.*, **44**, 699 (1940); MacDougall, *ibid.*, **44**, 713 (1940).

time. In other words, q may be taken as the energy entering the system as heat, and w is the amount leaving it as work done during an infinitesimal stage of a process. The algebraic sums of the q and w terms for all these infinitesimal stages will give the total heat and work changes, Q and W, respectively, to which equation (7.2) is applicable.

8. Reversible Thermodynamic Processes

8a. Thermodynamic Reversibility.—A particular type of path between two thermodynamic states is of special interest. This is the kind of path for which it is postulated that *all changes occurring in any part of the process are exactly reversed when it is carried out in the opposite direction.* Further, when the given process has been performed and then reversed, both *the system and its surroundings must be restored exactly to their original state.* A process of this kind is said to be **thermodynamically reversible.** In general, in order to follow a reversible path, it is necessary that *the system should always be in a state of virtual equilibrium,* and this requires that *the process be carried out infinitesimally slowly.*

A simple illustration of a reversible process is provided by isothermal evaporation carried out in the following manner. Imagine a liquid in equilibrium with its vapor in a cylinder closed by a frictionless piston, and placed in a constant temperature bath, i.e., a large thermostat. If the external pressure on the piston is increased by an infinitesimally small amount, the vapor will condense, but the condensation will occur so slowly that the heat evolved, i.e., the latent heat, will be taken up by the thermostat. The temperature of the system will not rise, and the pressure above the liquid will remain constant. Although condensation of the vapor is taking place, the system at every instant is in a state of virtual thermodynamic equilibrium. Similarly, if the external pressure is made just smaller than the vapor pressure, the liquid will vaporize extremely slowly, and again the temperature and pressure will remain constant. The system is changing, since vaporization is taking place, but the process may be regarded as a series of thermodynamic equilibrium states. Rapid evaporation or condensation, by the sudden decrease or increase of the external pressure, will lead to temperature and pressure gradients within the system, and thermodynamic equilibrium will be disturbed. Processes of this kind are not thermodynamically reversible.

The isothermal expansion of a gas can be carried out reversibly by placing the cylinder of gas in a thermostat, as described above, and adjusting the external pressure so as to be less than the pressure of the gas by an infinitesimally small amount. As the gas expands, however, its own pressure decreases, since the temperature is maintained constant. Hence, if the process is to be thermodynamically reversible, it must be supposed that the external pressure is continuously adjusted so as to be always infinitesimally less than the pressure of the gas. The expansion will then take place extremely slowly, so that the system is always in virtual thermodynamic equilibrium. The heat required by the gas, to balance the energy expended in the form of work against the external pressure, is taken up from the thermostat, but since the process is carried out extremely slowly the absorption of energy as heat keeps pace with the loss as work and the temperature of the system remains constant. If at any instant during the expansion the external pressure is adjusted so that it is maintained just infinitesimally greater than the gas pressure, the process will be reversed, and the gas will be compressed. At every stage in the compression the system and surroundings will be, apart from infinitesimal differences, in exactly the same thermodynamic state as they were at the corresponding point in the expansion.

If the expansion were carried out rapidly, e.g., by suddenly releasing the pressure on the piston, or the compression were rapid, e.g., by a sudden and large increase of the external pressure, the processes would not be reversible. The changes would not involve a continuous succession of equilibrium states of the system, and hence they could not be reversible. There would be both temperature and pressure gradients which would be different in the expansion and compression; the conditions for a thermodynamically reversible process would thus not be applicable.

The discussion given above has referred in particular to isothermal changes, but reversible processes are not necessarily restricted to those taking place at constant temperature. A reversible path may involve a change of temperature, as well as of pressure and volume. It is necessary, however, that the process should take place in such a manner that the system is always in virtual thermodynamic equilibrium. *If the system is homogeneous and has a constant composition,** two thermodynamic variables, e.g., pressure and volume, will completely describe its state at any point in a reversible process.

8b. Reversible Work of Expansion.—A general expression for the work of expansion accompanying a reversible process may be readily derived, and its complete solution is possible in certain cases. If P is the pressure within a system undergoing a reversible process, then from what has been stated in § 8a it follows that the external pressure must be $P - dP$, where dP is a very small quantity. The work w done by the system when it increases its volume by an infinitesimal amount dV is equal to the product of the external pressure and the volume change (§ 3f); thus,

$$w = (P - dP)dV.$$

Neglecting the very small, second order, product $dPdV$, it follows that

$$w = PdV. \qquad (8.1)$$

The total work W done in the process will then be equal to the sum of a continuous series of PdV terms, as the volume changes from its value in the initial state (V_1) to that in the final state (V_2). Since P and V are definite properties of the system, which is always in a virtual state of thermodynamic equilibrium, dV is a complete differential. The sum of the PdV terms may

---

* Such a system will sometimes be referred to as a "simple system" (cf. § 4b).

thus be replaced by the definite integral between the limits of V_1 and V_2, so that

$$W = \int_{V_1}^{V_2} P\,dV. \tag{8.2}$$

As may be expected from the discussion in § 8a, the value of the integral in equation (8.2) depends on the path between V_1 and V_2. This may be seen more explicitly in another manner. Since the state of a simple system at any point of a reversible process can be described completely by the pressure and volume, the path of the process can be represented by a curve on a P-V diagram, such as Fig. 5 or Fig. 6. According to equation (8.2) the work done in such a path is equal to the area enclosed by the curve and the two ordinates at the initial and final volumes. Since various paths, such as I, I', I'', etc., in Fig. 6, between the initial and final state are possible, the work of expansion will clearly be a variable quantity.

By specifying the reversible path it is possible to determine the actual value of the work of expansion from equation (8.2). Two simple cases may be considered. In the *isothermal vaporization* of a liquid, the pressure, which is equal to the vapor pressure at the specified constant temperature, remains constant throughout. The equation (8.2) for the work of expansion, then becomes

$$W = P \int_{V_1}^{V_2} dV = P(V_2 - V_1)$$
$$= P\Delta V, \tag{8.3}$$

where $V_2 - V_1$ is equal to ΔV, the increase in volume of the system, and P is the vapor pressure.

Another case of interest is that in which the system consists of an *ideal gas*. For n moles, the equation of state is $PV = nRT$, so that $P = nRT/V$. For an *isothermal expansion* involving n moles of an ideal gas, this value of P may be substituted in equation (8.2); since T and R are constant,

$$W = nRT \int_{V_1}^{V_2} \frac{dV}{V} = nRT \ln \frac{V_2}{V_1}. \tag{8.4}$$

An alternative form of this result is often convenient. For an ideal gas P_1V_1 is equal to P_2V_2, at constant temperature, and so V_2/V_1 is equal to P_1/P_2; hence equation (8.4) can also be written as

$$W = nRT \ln \frac{P_1}{P_2}, \tag{8.5}$$

for the isothermal, reversible expansion of n moles of an ideal gas.

The results just derived were based on the supposition that the change was one involving an increase of volume, i.e., expansion; nevertheless, the equations (8.4) and (8.5) are applicable to both expansion and compression. All that is necessary is to take V_1 and P_1 as representing the system in the initial state of the process, while V_2 and P_2 represent the final state. If there is an expansion, V_2 is greater than V_1, and hence W, the work done by the system, is positive. On the other hand, if a compression occurs, V_2 is less than V_1, and W is seen to be negative by equation (8.4); this is to be expected, for work is now done on the system. Although the volume change in a process may result in expansion or compression, the work done, given by equation (8.4) or (8.5), is always referred to as "work of expansion." It is seen to be positive when there is an increase of volume, and negative when the volume decreases.

Problem: Calculate the work of expansion in ergs when the pressure of 1 mole of an ideal gas is changed reversibly from 1.00 atm. to 5.00 atm. at a constant temperature of 25.0° C.

The energy units of W in equations (8.4) and (8.5) are determined by those of R; since the work of expansion is required in ergs, the value of R to be used is 8.314×10^7 ergs deg.$^{-1}$ mole^{-1}. The absolute temperature T is $273.2 + 25 = 298.2°$ K, the initial pressure P_1 is 1.00 and the final pressure P_2 is 5.00 atm.; the pressure units are immaterial, since the work depends on the ratio of the pressures. Substituting these data in equation (8.5), the work of expansion for 1 mole of gas is given by

$$W = RT \ln \frac{P_1}{P_2} = 8.314 \times 10^7 \times 298.2 \times 2.303 \log \tfrac{1}{5}$$
$$= -3.99 \times 10^{10} \text{ ergs mole}^{-1}.$$

Since W is negative, the work is done on the gas; this is to be expected, for the pressure of the gas increases in the process, and hence the volume is decreased.

In the two special cases of *isothermal, reversible expansion* considered above, the work done, as given by equations (8.3) and (8.4) or (8.5), is evidently a definite quantity depending only on the initial and final states, e.g., pressure or volume, at a constant temperature. Since there is always an exact relationship between P and V, it follows from equation (8.2) that the work done in any isothermal, reversible expansion must have a definite value, irrespective of the nature of the system. For an isothermal, reversible process in which the work performed is exclusively work of expansion, it is apparent, therefore, from equation (7.2), that both W and Q will be determined by the initial and final states of the system only, and hence they will represent definite quantities. Actually, this conclusion is applicable to *any isothermal, reversible change* (cf. § 25a), even if work other than that of expansion is involved.

8c. Maximum Work in Isothermal Reversible Processes.—A notable fact concerning an isothermal, reversible expansion is that the work done is the maximum possible for the given increase of volume. If the external pressure were made appreciably less, instead of infinitesimally less, than the gas pressure, the work done by the expanding gas would clearly be less than the value given by equation (8.2). The only possible way of increasing the work would be to make the external pressure greater than in the reversible, isothermal expansion. If this were done, however, the external pressure would be either equal to or greater than the gas pressure and hence expansion would be impossible. It follows, therefore, that the reversible work of isothermal expansion is the *maximum work of this type* possible for the given change in thermodynamic state. The work of isothermal expansion done by a system in an irreversible process is thus always less than for the same increase of volume when the change is carried out reversibly. In the case of an isothermal compression, the reversible work of expansion is (numerically) equal to the minimum amount of work which must be done by the surroundings *on* the system in order to bring about the specified change in volume (or pressure).

In general, the work that can be obtained in an isothermal change is a maximum when the process is performed in a reversible manner. This is true, for example, in the production of electrical work by means of a voltaic cell. Cells of this type can be made to operate isothermally and reversibly by withdrawing current extremely slowly (§ 33l); the E.M.F. of a given cell then has virtually its maximum value. On the other hand, if large currents are taken from the cell, so that it functions in an irreversible manner, the E.M.F. is less. Since the electrical work done by the cell is equal to the product of the E.M.F. and the quantity of electricity passing, it is clear that the same extent of chemical reaction in the cell will yield more work in the reversible than in the irreversible operation.

As reversible processes must take place infinitesimally slowly, it follows that infinite time would be required for their completion. Such processes are, therefore, not practicable, and must be regarded as ideal. Nevertheless, in spite of their impracticability, the study of reversible processes by the methods of thermodynamics is of great value, even in engineering, because the results indicate the maximum efficiency obtainable in any given change. In this way, the ideal to be aimed at is known. In chemistry, too, the state of equilibrium is important, for it shows to what extent a particular reaction can proceed; hence, thermodynamics provides information of considerable chemical significance, as will appear in later sections. It will be seen that the results derived from a study of reversible processes can be applied to reactions as a whole, even if they are actually carried out in an irreversible manner.

EXERCISES

1. In the combustion of 1 mole of sucrose ($C_{12}H_{22}O_{11}$), approximately 1.35×10^6 cal. are liberated. Should the accompanying loss of mass of the system be detectable?

2. Evaluate the energy, in calories, set free in the formation of 1 g. atom of helium nuclei, of mass 4.0028, from two protons, each of mass 1.0076, and two neutrons, each of mass 1.0090.

3. Determine the maximum work that can be done in the expansion of 5 moles of an ideal gas against an external pressure of 1 atm. when its temperature is increased by $t°$ C. Express the result in (i) ergs, (ii) defined cal.

4. Assuming ideal behavior, what is the minimum amount of work in ergs required to compress 1 kg. of air, consisting of 21% by volume of oxygen and 79% of nitrogen, from 1 atm. to 200 atm. at 0° C?

5. A vessel containing 1 liter of hydrogen at 1 atm. pressure is to be evacuated at 25° C to a pressure of (i) 0.01 atm., (ii) 0.001 atm. Compare the minimum amount of work required in each case, assuming ideal behavior.

6. An ideal gas is compressed isothermally and reversibly from 10 liters to 1.5 liters at 25° C. The work done by the surroundings is 2,250 cal. How many moles of gas are present in the system?

7. The vapor pressure of water at 25° C is 23.76 mm. Calculate the reversible work of expansion in (i) liter-atm., (ii) defined cal., when the pressure of 100 g. of water vapor is decreased isothermally to 0.001 atm., assuming ideal behavior.

8. Show that for a van der Waals gas, the isothermal, reversible work of expansion for 1 mole is given by

$$W = RT \ln \frac{V_2 - b}{V_1 - b} - a \left(\frac{1}{V_1} - \frac{1}{V_2} \right),$$

where V_1 is the initial and V_2 the final volume.

9. Determine the reversible work, in liter-atm., required to compress 1 mole of carbon dioxide isothermally from an initial pressure of 200 atm. to a final pressure of 1 atm. at 50° C, assuming van der Waals behavior. (Instead of solving the cubic equation to obtain the initial and final volumes, they may be obtained more simply by means of the generalized compressibility diagram.) Compare the result with that to be expected for an ideal gas.

10. Explain how the compressibility diagram alone could be used to determine the isothermal, reversible work of expansion of any gas. Apply the method to Exercise 9.

CHAPTER IV

HEAT CHANGES AND HEAT CAPACITIES

9. The Heat Content

9a. Heat Changes at Constant Volume and Constant Pressure.—Although the heat change is, in general, an indefinite quantity, there are certain simple processes, apart from isothermal, reversible changes, for which the paths are precisely defined. For such processes the heat changes will have definite values, dependent only on the initial and final states of the system.

Writing the first law equation (7.2) in the form

$$Q = \Delta E + W, \tag{9.1}$$

it will be supposed, as is the case in most thermodynamic processes which do not involve the performance of electrical work, that the work W is only *mechanical work due to a change of volume, i.e., work of expansion*. For a process occurring at *constant volume*, there is no expansion or contraction, and hence W will be zero; it follows then from equation (9.1), using the subscript V to indicate a constant volume process, that

$$Q_V = \Delta E_V, \tag{9.2}$$

so that the heat absorbed at constant volume, i.e., Q_V, is equal to the energy increase ΔE_V accompanying the process. Since the latter quantity depends only on the initial and final states of the system, the same must be true for the heat change at constant volume. The work term will also be definite, for it is equal to zero.

At *constant pressure* P, the work of expansion W may be replaced by $P\Delta V$, where ΔV is the increase of volume; representing constant pressure conditions by the subscript P, equation (9.1) takes the form

$$Q_P = \Delta E_P + P\Delta V. \tag{9.3}$$

The increase ΔE_P in the energy content is equal to $E_2 - E_1$, where E_1 and E_2 are the values for the initial and final states, respectively; similarly, the accompanying increase of volume ΔV may be represented by $V_2 - V_1$, so that equation (9.3) becomes

$$\begin{aligned} Q_P &= (E_2 - E_1) + P(V_2 - V_1) \\ &= (E_2 + PV_2) - (E_1 + PV_1). \end{aligned} \tag{9.4}$$

Since P and V are properties of the state of the system, it follows that the quantity $E + PV$, like the energy E, is dependent only on the thermodynamic state, and not on its previous history. The extensive, thermodynamic property $E + PV$ is called the **heat content**,* and is represented by the symbol H, i.e., by definition,

$$H = E + PV. \tag{9.5}$$

Consequently, equation (9.4) can be put in the form

$$Q_P = H_2 - H_1 = \Delta H_P, \tag{9.6}$$

where H_1 and H_2 are the values of the heat content in the initial and final states, respectively, and ΔH_P is the increase in the heat content of the system at constant pressure. The value of ΔH_P is seen by equation (9.6) to be equal to Q_P, the heat absorbed under these conditions; since the former quantity depends only on the initial and final states of the system, and not on the path taken, the same will be true for the heat term Q_P. For a process at constant pressure, therefore, the heat absorbed has a definite value.

The work W done in the constant pressure process, which is equal to $P\Delta V$, is also definite, since the pressure P is constant, and the volume change ΔV, equal to $V_2 - V_1$, is independent of the path. Aside from this argument, it is obvious, since ΔE and Q_P are here determined by the initial and final states, that W must now also have a definite value.

9b. Heat Capacities at Constant Volume and Constant Pressure.—As seen in § 3d, the heat capacity C may be defined as the ratio of the heat Q absorbed by a system to the resulting increase of temperature $T_2 - T_1$, i.e., ΔT. Since the heat capacity usually varies with the actual temperature, it is preferable to define it in terms of the limiting value as T_2 approaches T_1, that is, as ΔT is made very small; thus [cf. equation (3.3)],

$$C = \lim_{T_2 \to T_1} \frac{Q}{\Delta T} = \frac{q}{dT}, \tag{9.7}$$

where q is the quantity of heat absorbed for a small increase dT in the temperature of the system from T to $T + dT$. Since the heat absorbed is not, in general, a definite quantity depending on the initial and final states of the system, the unusual notation q/dT is used to define the heat capacity at the temperature T. In view of the dependence of q on the path taken, it is evident that the heat capacity will also be uncertain, unless conditions are specified, such as constant volume or constant pressure, which define the path.

At *constant volume*, for example, equation (9.7) may be written as

$$C_V = \frac{q_V}{dT}, \tag{9.8}$$

where q_V, and hence the heat capacity C_V, has a definite value. According to equation (9.2), the heat Q_V absorbed in an appreciable process at constant

---

* Engineers and physicists usually refer to it as the **enthalpy**. The use of the term "heat content," however, must not be interpreted as implying that the system possesses a definite amount of heat energy.

volume is equal to the increase ΔE_V of the energy content in that process. For an infinitesimal stage in a constant volume process q_V is equal to dE_V, and so the heat capacity at constant volume, as defined by equation (9.8), becomes

$$C_V = \frac{dE_V}{dT}$$
$$= \left(\frac{\partial E}{\partial T}\right)_V. \qquad (9.9)$$

The partial differential notation is used because E is a function of the volume (or pressure) as well as the temperature; the subscript V as applied to the partial derivative in equation (9.9) indicates that in this case the volume is maintained constant. *The heat capacity of a system at constant volume is therefore equal to the rate of increase of the energy content with temperature at constant volume.*

At *constant pressure*, equation (9.7) takes the form

$$C_P = \frac{q_P}{dT}, \qquad (9.10)$$

and since q_P is definite, the heat capacity C_P at constant pressure is also definite. Making use of the fact that according to equation (9.6), Q_P, the heat absorbed in an appreciable process at constant pressure, is equal to ΔH_P, the increase of heat content in the process, it can be shown in a manner exactly similar to that used in deriving (9.9), that

$$C_P = \left(\frac{\partial H}{\partial T}\right)_P. \qquad (9.11)$$

The heat capacity of a system at constant pressure is consequently equal to the rate of the increase of heat content with temperature at constant pressure.

It should be mentioned that the important equations (9.9) and (9.11), defining heat capacities at constant volume and constant pressure, respectively, are applicable to any homogeneous system of constant composition. The (simple) system may be gaseous, liquid or solid, and it may consist of a single substance or of a solution whose composition does not vary. As already seen, it is for such systems that the energy is dependent upon only two thermodynamic variables of state, e.g., pressure and temperature or volume and temperature.

9c. Heat Capacity Relationships.—From the results already given, it is possible to derive general equations connecting the heat capacity of a system at constant volume with that at constant pressure. Since the energy E of a homogeneous system of definite composition is a single-valued function of the volume and temperature, dE is a complete differential which can be represented by

$$dE = \left(\frac{\partial E}{\partial V}\right)_T dV + \left(\frac{\partial E}{\partial T}\right)_V dT \qquad (9.12)$$

[cf. equation (4.6)], and hence, by equation (4.10),

$$\left(\frac{\partial E}{\partial T}\right)_P = \left(\frac{\partial E}{\partial V}\right)_T \left(\frac{\partial V}{\partial T}\right)_P + \left(\frac{\partial E}{\partial T}\right)_V. \qquad (9.13)$$

According to equation (9.9), $(\partial E/\partial T)_V$ is equal to C_V, the heat capacity of the system at constant volume, so that

$$\left(\frac{\partial E}{\partial T}\right)_P = \left(\frac{\partial E}{\partial V}\right)_T \left(\frac{\partial V}{\partial T}\right)_P + C_V. \qquad (9.14)$$

By definition [equation (9.5)], $H = E + PV$, and since C_P is equal to $(\partial H/\partial T)_P$, it follows that

$$C_P = \left(\frac{\partial H}{\partial T}\right)_P = \left[\frac{\partial(E + PV)}{\partial T}\right]_P$$

$$= \left(\frac{\partial E}{\partial T}\right)_P + P\left(\frac{\partial V}{\partial T}\right)_P. \qquad (9.15)$$

Combination of this result with equation (9.14) yields immediately

$$C_P - C_V = \left[P + \left(\frac{\partial E}{\partial V}\right)_T\right]\left(\frac{\partial V}{\partial T}\right)_P. \qquad (9.16)$$

By starting with the relationship

$$dH = \left(\frac{\partial H}{\partial P}\right)_T dP + \left(\frac{\partial H}{\partial T}\right)_P dT, \qquad (9.17)$$

which is permissible since the heat content of a simple homogeneous system, like the energy content, is dependent only on two thermodynamic variables, e.g., the pressure and the temperature, it is possible to derive the equation

$$C_P - C_V = \left[V - \left(\frac{\partial H}{\partial P}\right)_T\right]\left(\frac{\partial P}{\partial T}\right)_V, \qquad (9.18)$$

by a procedure analogous to that used in obtaining equation (9.16).

As already indicated, the foregoing results will apply to any homogeneous system of constant composition. In certain cases, however, some simplification is possible, and this is especially true when an equation of state, relating pressure, volume and temperature, is available, as will be seen below. A particularly simple system is that involving an ideal gas, and this type of system will now be considered.

9d. Energy Content of Ideal Gas.—In the experiments made by J. L. Gay-Lussac (1807) and J. P. Joule (1844) it was found that when a gas was allowed to expand into a vacuum, there was no gain or loss of heat. Two similar copper globes, one containing air under pressure and the other evacuated, were connected by a wide tube with a stopcock. When the latter was opened the temperature of the globe which originally contained the air fell, but that of the other globe rose by an equal amount. It appeared, therefore, that there was no net heat change, i.e., the value of Q was zero, when the volume of the gas was increased in the manner described. Subsequent experiments, carried out along somewhat different lines by J. P. Joule and

W. Thomson (Lord Kelvin), between 1852 and 1862, showed that a temperature difference, indicating a value other than zero for Q, should have been observed. However, it appears that the more closely the gas approximates to ideal behavior the smaller is the heat effect, and so it is probable that for an *ideal gas* the value of Q in the free expansion described above would be actually zero.

Since the gas expands into a vacuum, that is, against no external pressure, the work of expansion W is also zero. It follows, therefore, from equation (7.2), that ΔE must be zero in the process; in other words, *there is no change in the energy content of an ideal gas as a result of free expansion*, i.e., a volume increase in which no external work is done. The energy of the gas depends on two variables, e.g., volume and temperature; hence, the conclusion just reached may be represented mathematically in the form

$$\left(\frac{\partial E}{\partial V}\right)_T = 0, \qquad (9.19)$$

the energy content being independent of the volume, at any constant temperature. The result expressed by equation (9.19) is taken as one of the criteria of an ideal gas. It will be shown later (§ 20d), by utilizing the second law of thermodynamics, that (9.19) is a necessity for a gas obeying the $PV = RT$ relationship at all temperatures and pressures.

There is an important consequence of the fact that the energy content of an ideal gas, at constant temperature, is independent of the volume. When an ideal gas expands against an appreciable external pressure, W has a finite value, but ΔE is zero; it follows, therefore, from equation (7.2) that for the isothermal expansion of an ideal gas,

$$Q = W, \qquad (9.20)$$

the heat absorbed by the system being equal to the work done by it. If the expansion is carried out reversibly, the work done is given by equation (8.4) or (8.5), and hence the heat Q absorbed in the isothermal, reversible expansion of n moles of an ideal gas is determined by the same expression, viz.,

$$Q = nRT \ln \frac{V_2}{V_1} = nRT \ln \frac{P_1}{P_2}. \qquad (9.21)$$

9e. Effect of Pressure and Temperature on Heat Capacity of Ideal Gas.
—Since the energy content of an ideal gas, at constant temperature, is independent of its volume, it must also be independent of the pressure. The energy of a given quantity of the gas thus varies only with the temperature; it is consequently possible to write equation (9.9) in the form

$$C_V = \frac{dE}{dT}, \qquad (9.22)$$

so that *the heat capacity at constant volume of an ideal gas is independent of the volume, or pressure, of the gas.* The same can be shown to be true for the heat capacity at constant pressure. Thus, if equation (9.5), i.e., $H = E + PV$, is differentiated with respect to volume, at constant temperature, the result is

$$\left(\frac{\partial H}{\partial V}\right)_T = \left(\frac{\partial E}{\partial V}\right)_T + \left[\frac{\partial (PV)}{\partial V}\right]_T.$$

For an ideal gas, $(\partial E/\partial V)_T$ is zero, as seen above, and $[\partial(PV)/\partial V]_T$ is also zero, since PV is equal to RT and hence is constant. It follows, therefore, that *the heat content of an ideal gas is dependent only on the temperature, and is independent of its volume or pressure.* The heat capacity at constant pressure can then be represented by equation (9.11) in the form

$$C_P = \frac{dH}{dT}, \qquad (9.23)$$

indicating that C_P *for an ideal gas does not vary with the volume or pressure.* For a real gas, the energy and heat contents are known to vary with the pressure, and hence equations (9.22) and (9.23) cannot be employed. Some expressions for the variation of C_P and C_V with pressure or volume will be derived in later sections (§§ 11e, 21d, 21e).

It must be realized that the conclusion that the heat capacities of an ideal gas are independent of pressure is not based on thermodynamics alone. It has been necessary to introduce the result that $(\partial E/\partial V)_T$ is zero, which may be regarded either as based on experiment or on the postulated equation of state $PV = RT$ for an ideal gas. Similar considerations are applicable to the problem of the variation of heat capacity of an ideal gas with temperature; thermodynamics alone cannot supply the answer, and it is necessary to introduce other information. According to the kinetic theory of gases, the heat capacity of an ideal gas, whose molecules possess translational energy only, should be independent of temperature (§ 15b). Gases which might be expected to approximate to ideal behavior in this respect are those containing only one atom in the molecule. Experimental observations have shown that the heat capacities of some monatomic gases, e.g., helium and argon, are almost constant over a very considerable range of temperature. In general, however, all gases do not behave in the same manner in respect to the effect of temperature on heat capacity as ideal behavior is approached, i.e., at low pressures, and so it is inadvisable to make a general postulate in this connection.

9f. Heat Capacity-Temperature Relationships.—For reasons which will be made clear in Chapter VI, the heat capacities of all gases containing two or more atoms in the molecule must vary with temperature. The form of the expression which represents this variation cannot be predicted by means of thermodynamics, and so purely empirical formulae are used. One of these, which has been widely employed, is a power series of the form

$$C_P = \alpha + \beta T + \gamma T^2 + \delta T^3 \cdots, \qquad (9.24)$$

where α, β, γ, δ, etc., are constants, which must be derived from the experimentally determined heat capacities over a range of temperatures, and T is the absolute temperature. The values of these constants for a number of common gases are recorded in Table II [1]; when inserted in equation (9.24) they give the heat capacities of the respective gases, usually within 1 per cent or less, in the temperature range from 273° to about 1500°K. The constants in Table II cannot be safely used for extrapolation outside this range.

TABLE II.* HEAT CAPACITIES OF GASES (C_P) AT 1 ATM. PRESSURE IN CAL. DEG.$^{-1}$ MOLE^{-1}

Gas	α	$\beta \times 10^3$	$\gamma \times 10^6$	$\delta \times 10^9$
H_2	6.947	-0.200	0.4808	—
D_2	6.830	0.210	0.468	—
O_2	6.095	3.253	-1.017	—
N_2	6.449	1.413	-0.0807	—
Cl_2	7.576	2.424	-0.965	—
Br_2	8.423	0.974	-0.3555	—
CO	6.342	1.836	-0.2801	—
HCl	6.732	0.4325	0.3697	—
HBr	6.578	0.955	0.1581	—
H_2O	7.219	2.374	0.267	—
CO_2	6.396	10.100	-3.405	—
H_2S	6.385	5.704	-1.210	—
HCN	5.974	10.208	-4.317	—
N_2O	6.529	10.515	-3.571	—
SO_2	6.147	13.84	-9.103	2.057
SO_3	6.077	23.537	-9.687	—
NH_3	6.189	7.787	-0.728	—

* For more accurate values and data for other gases, see Table 3 at end of book.

Because of the variation of heat capacity with temperature, it is sometimes convenient to use the mean value of the heat capacity over a range of temperatures; this value is then taken as constant for purposes of calculation. The mean heat capacity $\bar{C}_P$ in the temperature range from T_1 to T_2 is given by

$$\bar{C}_P = \frac{\int_{T_1}^{T_2} C_P dT}{T_2 - T_1},$$

and if the value of C_P from equation (9.24) is introduced, it is found upon integration that

$$\bar{C}_P = \frac{1}{T_2 - T_1} [\alpha(T_2 - T_1) + \tfrac{1}{2}\beta(T_2^2 - T_1^2) + \tfrac{1}{3}\gamma(T_2^3 - T_1^3) + \cdots]$$
$$= \alpha + \tfrac{1}{2}\beta(T_1 + T_2) + \tfrac{1}{3}\gamma(T_1^2 + T_1 T_2 + T_2^2) + \cdots. \quad (9.25)$$

It is then possible to calculate the mean heat capacity of any gas, using the constants in Table II.

[1] Spencer, et al., J. Am. Chem. Soc., 56, 2311 (1934); 64, 250 (1942); 67, 1859 (1945); see also, Bryant, Ind. Eng. Chem., 25, 820, 1022 (1933).

It appears that in certain cases the variation of heat capacity with temperature is better represented by a function containing a term in T^{-2} in place of T^2 in equation (9.24); thus,[2]

$$C_P = \alpha + \beta T + \gamma T^{-2}, \qquad (9.26)$$

the constants α, β and γ being quite different from those in Table II. Since neither equation (9.24) nor (9.26) can be regarded as representing exactly the dependence of heat capacity on temperature, the simple power series equation (9.24) will be used whenever possible, because of its more straightforward nature. In some instances, however, it will be found necessary to employ a function of the type of equation (9.26).

The heat required to raise the temperature of a system from T_1 to T_2, at constant pressure, i.e., the increase of heat content, is obtained by the integration of equation (9.11), using (9.24) to express C_P as a function of T; the result is

$$H_2 - H_1 = \int_{T_1}^{T_2} C_P dT = \int_{T_1}^{T_2} (\alpha + \beta T + \gamma T^2 + \cdots) dT$$
$$= \alpha(T_2 - T_1) + \tfrac{1}{2}\beta(T_2^2 - T_1^2) + \tfrac{1}{3}\gamma(T_2^3 - T_1^3) + \cdots, \qquad (9.27)$$

constant pressure being understood. The same result can, of course, be obtained upon multiplying the mean heat capacity $\bar{C}_P$, as given by equation (9.25), by the increase of temperature.

Problem: How much heat is required to raise the temperature of 1 mole of oxygen gas from 27° C to 127° C at 1 atm. pressure?

From Table II,

$$C_P = 6.095 + 3.253 \times 10^{-3} T - 1.017 \times 10^{-6} T^2 \text{ cal. deg.}^{-1} \text{ mole}^{-1},$$

and since T_1 is $273 + 27 = 300°$ K, and T_2 is $273 + 127 = 400°$ K, it follows that

$$H_{400} - H_{300} = \int_{300}^{400} (6.095 + 3.253 \times 10^{-3} T - 1.017 \times 10^{-6} T^2) dT$$
$$= 6.095(400 - 300) + \tfrac{1}{2} \times 3.253 \times 10^{-3}[(400)^2 - (300)^2]$$
$$\qquad - \tfrac{1}{3} \times 1.017 \times 10^{-6}[(400)^3 - (300)^3]$$
$$= 710 \text{ cal mole}^{-1}.$$

9g. Difference of Molar Heat Capacities.—Since $(\partial E/\partial V)_T$ is zero for an ideal gas, equation (9.16) takes the simple form

$$C_P - C_V = P\left(\frac{\partial V}{\partial T}\right)_P. \qquad (9.28)$$

The quantity $(\partial V/\partial T)_P$ represents the rate of increase of volume with temperature, at constant pressure, and hence the right-hand side of equation (9.28) may be taken as equal to the work of expansion when the temperature of the ideal gas is raised by 1° at constant pressure. The difference in the

[2] Maier and Kelley, *J. Am. Chem. Soc.*, **54**, 3243 (1932); Kelley, *U. S. Bur. Mines Bull.*, 371 (1934); see also, Chipman and Fontana, *J. Am. Chem. Soc.*, **57**, 48 (1935).

heat capacities of an ideal gas at constant pressure and constant volume may thus be attributed to the work of expansion, as a result of the increase of volume at constant pressure; at constant volume there is, of course, no work of expansion. It will be understood that this condition can hold only if $(\partial E/\partial V)_T$ is zero, i.e., for an ideal gas. For real gases the difference between C_P and C_V must include an allowance for the change in energy content due to the increase of volume, in addition to that due to the work of expansion. Actually this is quite small for most gases under ordinary conditions as is apparent from the value of the mechanical equivalent of heat derived in the problem in § 6b; this calculation is based on the tacit assumption that $(\partial E/\partial V)_T$ is zero for air at 0° C.

Since for 1 mole of an ideal gas $PV = RT$, it is seen that $(\partial V/\partial T)_P$ is equal to R/P, and hence

$$P\left(\frac{\partial V}{\partial T}\right)_P = R,$$

so that from equation (9.27)

$$C_P - C_V = R, \tag{9.29}$$

where C_P and C_V are now the *molar* heat capacities of the ideal gas. The values for constant pressure and constant volume, respectively, thus differ by a constant amount, equal to the molar gas constant, i.e., 1.987 cal. deg.$^{-1}$ mole^{-1}. For real gases $C_P - C_V$ will not be quite equal to R, although the discrepancy should not be large, except perhaps at high pressures and low temperatures when departure from ideal behavior is considerable. Further theoretical consideration of this subject will be given in Chapter VIII.

10. Adiabatic Processes

10a. Reversible Adiabatic Expansion and Compression.—An **adiabatic process** is defined as *one in which no heat enters or leaves the system*, at any stage. For every infinitesimal stage of the process q is zero, and hence, by equation (7.5), dE is equal to $-w$. If the work is restricted to work of expansion, w is given by PdV, so that

$$dE = -PdV.$$

In this equation, P represents, strictly speaking, the external pressure. However, if the adiabatic process is carried out *reversibly*, the actual pressure of the system is virtually identical with the external pressure (§ 8a), so that under these conditions P is the pressure of the system. For an *ideal gas*, dE may be replaced by $C_V dT$, as shown by equation (9.22), since the energy content is independent of the volume, so that

$$C_V dT = -PdV. \tag{10.1}$$

It will be observed from this result that in a reversible, adiabatic process the signs of dV and dT are opposite; thus, if the volume of the gas is increased, in such an adiabatic expansion, the temperature must fall, whereas if the volume is decreased, in a compression, the temperature will rise. This is to be expected from general considerations: a gas does work when it expands, and as no heat enters or leaves the system, the energy content of the gas must decrease. Since the energy is independent of the volume at constant temperature, this decrease of energy is accompanied by a fall of temperature.

10b. Temperature Changes in Reversible Adiabatic Processes.—The relationship between the pressure (or volume) and the temperature of an *ideal gas* in a reversible, adiabatic expansion or compression may be derived in the following manner. If 1 mole of the gas is considered, PV is equal to RT, so that substitution for P in equation (10.1) gives

$$C_V dT = -RT \frac{dV}{V},$$

$$C_V \frac{dT}{T} = -R \frac{dV}{V},$$

that is

$$C_V d \ln T = -R d \ln V, \tag{10.2}$$

where C_V is the *molar* heat capacity of the ideal gas at constant volume. If C_V is *assumed to be independent of temperature*, it is possible to integrate equation (10.2), the limits being T_1 and T_2, the initial and final temperatures, and V_1 and V_2, the corresponding volumes; the result is

$$C_V \ln \frac{T_2}{T_1} = -R \ln \frac{V_2}{V_1} = R \ln \frac{V_1}{V_2}. \tag{10.3}$$

For 1 mole of an ideal gas, R is equal to $C_P - C_V$, by (9.29), so that, after making this substitution and converting the logarithms, equation (10.3) becomes

$$\log \frac{T_2}{T_1} = \left(\frac{C_P}{C_V} - 1\right) \log \frac{V_1}{V_2}$$

$$= (\gamma - 1) \log \frac{V_1}{V_2}, \tag{10.4}$$

where the symbol γ is used to represent C_P/C_V, the ratio of the heat capacities at constant pressure and constant volume. The equation (10.4), or its equivalent,

$$\frac{T_2}{T_1} = \left(\frac{V_1}{V_2}\right)^{\gamma-1} = \left(\frac{V_1}{V_2}\right)^{R/C_V} \tag{10.5}$$

or

$$TV^{\gamma-1} = TV^{R/C_V} = \text{constant},$$

can be utilized to determine the temperature change in a reversible, adiabatic process with an ideal gas.

Problem: A quantity of air at 25° C is compressed adiabatically and reversibly from a volume of 10 liters to 1 liter. Assuming ideal behavior, and taking C_V for air as 5 cal. deg.$^{-1}$ mole^{-1}, calculate the (approximate) final temperature of the air.

Utilizing equation (10.5), T_1 is $273 + 25 = 298°$ K; V_1 is 10 liters, V_2 is 1 liter, C_V is 5.0 cal. deg.$^{-1}$ mole^{-1}, and R may be taken as 2.0 cal. deg.$^{-1}$ mole^{-1} with sufficient accuracy (note that C_V and R must be expressed in the same units); hence,

$$\frac{T_2}{298} = \left(\frac{10}{1}\right)^{2/5},$$

$$T_2 = 298 \times 10^{2/5} = 749°\ \text{K}.$$

The final temperature is thus $749 - 273 = 476°$ C.

For many purposes it is more useful to develop an expression relating the temperature to the pressure in a reversible, adiabatic change. Since an ideal gas is under consideration, it follows that $P_1V_1 = RT_1$ and $P_2V_2 = RT_2$; if these equations are combined with (10.5) so as to eliminate V_1 and V_2, it is found that

$$\frac{T_2}{T_1} = \left(\frac{P_2}{P_1}\right)^{(\gamma-1)/\gamma} = \left(\frac{P_2}{P_1}\right)^{R/C_P} \tag{10.6}$$

or

$$\frac{T}{P^{(\gamma-1)/\gamma}} = \frac{T}{P^{R/C_P}} = \text{constant}.$$

The results of equations (10.5) and (10.6) may be summarized in the convenient form

$$\frac{1}{R}\log\frac{T_2}{T_1} = \frac{1}{C_P}\log\frac{P_2}{P_1} = \frac{1}{C_V}\log\frac{V_1}{V_2}.$$

Problem: A quantity of air at 25° C is allowed to expand adiabatically and reversibly from 200 atm. to 20 atm. Assuming ideal behavior, calculate the (approximate) final temperature.

In this case T_1 is $273 + 25 = 298°$ K, P_1 is 200 atm., and P_2 is 20 atm. Since C_V was given above as 5.0 cal. deg.$^{-1}$ mole^{-1}, and assuming $C_P - C_V = R$, as for an ideal gas, C_P is 7.0 cal. deg.$^{-1}$ mole^{-1}. Hence, by equation (10.6),

$$\frac{T_2}{298} = \left(\frac{20}{200}\right)^{2/7}$$

$$T_2 = 298 \times (\tfrac{1}{10})^{2/7} = 154°\ \text{K}.$$

The final temperature is thus $154 - 273 = -119°$ C.

The marked fall of temperature accompanying adiabatic, or approximately adiabatic, expansion is used to some extent for cooling purposes in connection with the liquefaction of gases.

10c. Pressure-Volume Relationship in Reversible Adiabatic Process.—
By combining equations (10.5) and (10.6) so as to eliminate T_1 and T_2, it is found that

$$P_1V_1^\gamma = P_2V_2^\gamma \quad \text{or} \quad P_1V_1^{C_P/C_V} = P_2V_2^{C_P/C_V}, \tag{10.7}$$

or

$$PV^\gamma = PV^{C_P/C_V} = \text{constant}.$$

The relationship between the pressure and volume at any instant in a reversible, adiabatic process with a given mass of an ideal gas * is thus represented by $PV^\gamma = $ constant, which may be compared with the constancy of the simple product PV for an isothermal process. Since C_P is always greater than C_V, the ratio C_P/C_V, i.e., γ, is greater than unity; hence, the increase of volume for a given decrease of pressure will be less in an adiabatic than in an isothermal expansion. That is to say, the plot of the pressure against the volume, i.e., the pressure-volume curve (Fig. 7), will be steeper for an adiabatic than for an isothermal process, starting at the same point. The reason for the smaller volume increase in an adiabatic expansion is to be attributed to the accompanying fall of temperature, as explained above, which will tend to diminish the volume. In the isothermal process, the temperature is, of course, constant.

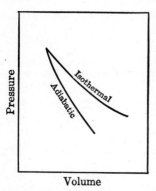

Fig. 7. Adiabatic and isothermal processes

10d. Work of Expansion in Reversible Adiabatic Process.—The work of expansion w in an infinitesimal stage of a reversible, adiabatic process, i.e., PdV, is given by equation (10.1) as equal to $-C_V dT$; hence, for an appreciable process the total work of expansion W, derived from equation (8.2), is

$$W = \int_{V_1}^{V_2} PdV = - \int_{V_1}^{V_2} C_V dT.$$

If C_V may be taken as constant, it follows that

$$W = - C_V(T_2 - T_1) = C_V(T_1 - T_2). \tag{10.8}$$

The negative sign means that work is done *on* the gas when $T_2 > T_1$, that is, in an adiabatic compression. Equation (10.8) may be combined with equations (10.4) or (10.6), so as to eliminate T_1 and T_2 and obtain expressions for the work of expansion in terms of the volumes or pressures, respectively.

Problem: Calculate the work of expansion in ergs when the pressure of 1 mole of an ideal gas at 25° C is changed adiabatically and reversibly from 1.0 atm. to 5.0 atm. The molar heat capacities may be taken as equal to those of air. (Compare the problem in § 8b, which is for an isothermal expansion between the same pressure limits.)

This problem may be solved by calculating T_2 as in the second problem in § 10b, and then substituting the result in (10.8), T_1 being known. Alternatively, by combining equation (10.6) with (10.8) it is readily found that

$$W = C_V T_1 \left[1 - \left(\frac{P_2}{P_1} \right)^{R/C_P} \right], \tag{10.9}$$

* The assumption made in the integration of equation (10.2), namely, that the heat capacity is independent of the temperature, should be borne in mind.

from which W can be derived directly by means of the available data. The energy unit of W will be that of the factor C_V; this is 5.0 cal. deg.$^{-1}$ mole^{-1}, or $5.0 \times 4.18 \times 10^7$ ergs deg.$^{-1}$ mole^{-1}. In the exponent the units of R and C_V are immaterial as long as they are consistent; hence, R may be taken as 2.0 and C_P as 7.0 cal. deg.$^{-1}$ mole^{-1}. Substituting these values in equation (10.9), with $T_1 = 273 + 25 = 298°$ K, $P_1 = 1.0$ and $P_2 = 5.0$ atm., the result is

$$W = 5.0 \times 4.18 \times 10^7 \times 298 \left[1 - \left(\frac{5.0}{1.0} \right)^{2/7} \right]$$
$$= -3.64 \times 10^{10} \text{ ergs mole}^{-1}.$$

The work required for the adiabatic process is seen to be somewhat less *numerically* than for the isothermal expansion, because of the smaller volume change.

10e. Ratio of Heat Capacities.—The results obtained in the preceding sections have been applied in determining the value of γ, the ratio of the heat capacities of a gas. In the method of F. Clément and C. B. Desormes (1812), the gas at a pressure greater than atmospheric is placed in a large vessel provided with a stopcock and a manometer for indicating the pressure. The initial pressure P_1 is observed, and the stopcock is opened to allow the pressure to fall to that of the atmosphere P_2; the stopcock is then immediately closed. During the expansion, which is virtually adiabatic, the gas is cooled from the initial temperature T_1 to the lower value T_2, and as it warms up to its original temperature, the pressure rises to P_1'. If V_1 is the volume of 1 mole of an ideal gas at pressure P_1, and V_2 is the volume after the adiabatic expansion when the pressure is P_2, then by equation (10.7), $P_1 V_1^\gamma = P_2 V_2^\gamma$. When the original temperature T_1 is restored, after the adiabatic expansion, the pressure of the gas is P_1', and since 1 mole still occupies the volume V_2, because the vessel was closed after the expansion, it follows that

$$P_1 V_1 = P_1' V_2, \tag{10.10}$$

where the left-hand side refers to the initial state and the right-hand side to the final state, at the same temperature. Eliminating V_1 and V_2 from equations (10.7) and (10.10), the result is

$$\gamma = \frac{\log P_1 - \log P_2}{\log P_1 - \log P_1'}. \tag{10.11}$$

From the three pressure measurements, therefore, the ratio of the heat capacities can be calculated.

In a modification of the foregoing procedure (O. Lummer and E. Pringsheim, 1891), the stopcock is allowed to remain open after the adiabatic expansion, and the temperatures before (T_1) and immediately after (T_2) expansion are measured by a sensitive thermometer. Since the corresponding pressures P_1 and P_2 are known, γ can be obtained from equation (10.6) in the form

$$\gamma = \frac{\log P_1 - \log P_2}{\log (P_1/P_2) - \log (T_1/T_2)}. \tag{10.12}$$

It will be recalled that equations (10.11) and (10.12) are both based on the assumption that the gas behaves ideally; for actual gases, however, the results must be corrected to ideal behavior by the use of a suitable equation of state.

For an ideal gas, the value of γ should be constant, but for real gases, other than monatomic, it varies appreciably with pressure and temperature, as shown by the results for air, given in Table III.[3] In general, the ratio

TABLE III. VARIATION OF RATIO C_P/C_V FOR AIR WITH TEMPERATURE AND PRESSURE

Temperature	Pressure			
	1	50	100	200 atm.
$-79°$ C	1.41	1.77	2.20	2.33
$0°$	1.403	1.53	1.65	1.83
$100°$	1.401	1.45	1.50	1.60

of the heat capacities decreases with increasing temperature. As the pressure is increased γ increases, passing through a maximum and then decreasing (cf. § 21d); for air at $-79°$ C the maximum was observed at 150 atm., but at $0°$ and $100°$ C it occurs at higher pressures than are given in the table.

When actual data are not available, a useful approximate rule for *ordinary temperatures and pressures*, is to take γ as 1.67 for monatomic gases, 1.40 for diatomic gases, 1.30 for simple polyatomic gases, such as water, carbon dioxide, ammonia and methane. It may be noted that the heat capacity ratio for hydrogen gas increases at low temperatures toward the value for a monatomic gas. This matter will be explained in Chapter VI.

11. THE JOULE-THOMSON EFFECT

11a. Expansion Through a Throttle.—In the experiments of Joule and Thomson, referred to in § 9d, a gas was allowed to stream from a higher to a lower pressure through a tube containing a "throttle," consisting of a porous plug of silk or cotton (Fig. 8). By the use of the throttle the expansion took place slowly, and *the pressure on each side of the plug was maintained virtually constant.*

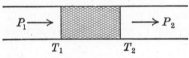

FIG. 8. Throttled expansion of a gas

The tube was made of a material having a low heat conductance, e.g., boxwood, and the conditions were made as nearly adiabatic as possible.

Suppose P_1 is the constant pressure of the gas before passing through the throttle, and P_2 is the constant pressure after the passage; the corresponding temperatures are T_1 and T_2. The volume of 1 mole of the gas at P_1 and T_1 is V_1, whereas at P_2 and T_2 the molar volume is V_2. The work done W by 1 mole of the gas as it streams through the plug is equal to $P_2V_2 - P_1V_1$, for the volume is increased by V_2 at the pressure P_2 while decreasing by the volume V_1 at the pressure P_1. Since the whole process is assumed to be adiabatic, so that Q is zero, it follows that the loss of energy from the system in the form of work, i.e., W, must be equal to the decrease $-\Delta E$ in the energy content, by equation (7.2). If E_1 and E_2 are the energy con-

[3] Data mainly from International Critical Tables, Vol. V, pp. 81–82.

tents of 1 mole of the gas in the initial and final states, i.e., before and after passage through the throttle, then

$$-(E_2 - E_1) = P_2V_2 - P_1V_1, \tag{11.1}$$

the left-hand side representing the decrease of the energy content, and the right-hand side the work done. By rearrangement of equation (11.1), it is seen that

$$E_1 + P_1V_1 = E_2 + P_2V_2, \tag{11.2}$$

and since, by definition, the heat content H is equal to $E + PV$, it follows from (11.2) that $H_1 = H_2$, so that *in the throttled expansion process the heat content of the system remains constant*.

11b. The Joule-Thomson Coefficient.—As seen in § 9a, the heat content of a system is determined precisely by the thermodynamic state of the system, so that dH is a complete differential; thus, taking pressure and temperature as the variables,

$$dH = \left(\frac{\partial H}{\partial P}\right)_T dP + \left(\frac{\partial H}{\partial T}\right)_P dT. \tag{11.3}$$

For a process in which the heat content is constant, equation (11.3) may be equated to zero, so that, inserting the subscript H to indicate constant heat content,

$$\left(\frac{\partial H}{\partial P}\right)_T dP_H + \left(\frac{\partial H}{\partial T}\right)_P dT_H = 0$$

or

$$\left(\frac{\partial T}{\partial P}\right)_H = -\left(\frac{\partial H}{\partial P}\right)_T \bigg/ \left(\frac{\partial H}{\partial T}\right)_P. \tag{11.4}$$

The quantity $(\partial T/\partial P)_H$ is called the **Joule-Thomson coefficient**, and is represented by the symbol $\mu_{\text{J.T.}}$; it is equal to the rate of change of temperature with the pressure in a streaming process through a plug or throttle. According to equation (9.11), $(\partial H/\partial T)_P$ is the heat capacity of the gas at constant pressure, i.e., C_P, so that (11.4) is equivalent to

$$\mu_{\text{J.T.}} = -\frac{1}{C_P}\left(\frac{\partial H}{\partial P}\right)_T. \tag{11.5}$$

Utilizing the relationship $H = E + PV$, this expression takes the form

$$\mu_{\text{J.T.}} = -\frac{1}{C_P}\left[\left(\frac{\partial E}{\partial P}\right)_T + \left(\frac{\partial(PV)}{\partial P}\right)_T\right],$$

and since E is a function of P, V and T, which are not independent, it follows from equation (4.9) that

$$\mu_{\text{J.T.}} = -\frac{1}{C_P}\left[\left(\frac{\partial E}{\partial V}\right)_T\left(\frac{\partial V}{\partial P}\right)_T + \left(\frac{\partial(PV)}{\partial P}\right)_T\right]. \tag{11.6}$$

Equation (11.6) is quite general and should apply to any gas,* for its derivation is based entirely on the first law of thermodynamics without assuming any specific properties of the system. However, for *an ideal gas*, $(\partial E/\partial V)_T$ is zero, as seen earlier, and since $PV = RT$, it follows that $[\partial(PV)/\partial P]_T$ is also zero; hence, since C_P is finite, it is seen from equation (11.6) that for an ideal gas $\mu_{\text{J.T.}}$ must be zero.† The Joule-Thomson coefficient of an ideal gas should thus be zero, so that there should be no change of temperature when such a gas expands through a throttle.‡

11c. Sign and Magnitude of the Joule-Thomson Effect.—The quantity $(\partial E/\partial V)_T$ has the dimensions of energy/volume, which is equivalent to force/area and hence to pressure; it is, therefore, frequently referred to as the **internal pressure**, especially in connection with liquids. Physically, it may be regarded as a measure of the pressure in the interior of a fluid, i.e., gas or liquid, resulting from the attractions and repulsions of the molecules; it has, in fact, been identified approximately with the a/V^2 term of the van der Waals equation (cf. § 20d). It is because the energy content of a real gas, unlike that of an ideal gas, is not independent of the volume, that it is necessary to introduce the a/V^2, or analogous, term in the equation of state for the real gas.

There are several methods for obtaining an indication of the magnitude of the internal pressure, and these all show that $(\partial E/\partial V)_T$ is usually positive for real gases. Further, since it is approximately equal to a/V^2, it increases with increasing pressure. The factor $(\partial V/\partial P)_T$ in equation (11.6), on the other hand, is always negative, since increasing pressure, at constant temperature, is invariably accompanied by a decrease of volume. It can be readily seen from an examination of the pressure-volume isotherm of a gas that the *numerical value* of the slope $(\partial V/\partial P)_T$ is large at low pressures, and diminishes as the pressure is increased. Hence, it follows that the first term in the bracket on the right-hand side of equation (11.6) is usually negative, its numerical value being approximately independent of the pressure.

Turning to the term $[\partial(PV)/\partial P]_T$, it is seen that this is the slope of the plot of PV against P, as in Fig. 2. At ordinary temperatures, this is negative for all gases, except hydrogen and helium, at low pressures, but it becomes positive at high pressures. At low and moderate pressures, therefore, both terms in the bracket of equation (11.6) are negative, and since the heat capacity C_P is always positive, it follows that the Joule-Thomson coefficient $\mu_{\text{J.T.}}$ will have a positive value. In other words, since $\mu_{\text{J.T.}}$ is equal to $(\partial T/\partial P)_H$, most gases experience a decrease of temperature as the result of a Joule-Thomson (throttled) expansion, at moderate and low pressures and

* The foregoing equations are really applicable to any *fluid*, i.e., liquid or gas, but they are usually employed for gases.

† The same conclusion may be reached directly from equation (11.4) or (11.5), since $(\partial H/\partial P)_T$ is zero for an ideal gas (cf., § 9e).

‡ It may be noted that for real gases $\mu_{\text{J.T.}}$ does not necessarily approach zero at very low pressures, in spite of the fact that $(\partial E/\partial V)_T$ approaches zero and PV approaches RT (cf. § 22a).

ordinary temperatures. As the gas pressure is increased, the numerical value of the first term in the bracket remains approximately constant; at the same time the second term decreases numerically, and eventually becomes positive. This means that the Joule-Thomson coefficient should decrease with increasing pressure, passing through a zero value, and finally changing sign. At sufficiently high pressures, therefore, a reversal of the Joule-Thomson effect should be observed, the coefficient $\mu_{J.T.}$ becoming negative; under these conditions a throttled expansion will be accompanied by an increase of temperature.

At lower temperatures not only is the numerical value of $[\partial(PV)/\partial P]_T$ larger at low pressures, but higher pressures must be attained before it changes sign. It follows, therefore, that at low pressures the Joule-Thomson coefficient $\mu_{J.T.}$ should increase as the temperature is diminished, and higher pressures will be required before a reversal occurs. The temperature at which the Joule-Thomson effect changes sign, at a given pressure, is called the **inversion temperature**; at this temperature $\mu_{J.T.}$ is, of course, zero. From what has just been stated, the inversion temperature for any gas should increase with the pressure.

The foregoing conclusions, derived from the purely thermodynamic equation (11.6) in conjunction with experimental data on the compressibility of gases, are in complete agreement with direct observations of the Joule-Thomson effect, as may be seen from the data for nitrogen recorded in Table IV.[4] At 200 atm. pressure, there is evidently an inversion tem-

TABLE IV. JOULE-THOMSON COEFFICIENT FOR NITROGEN IN DEG. ATM.$^{-1}$

Temperature	Pressure				
	1	20	60	100	200 atm.
200° C	0.0540	0.0460	0.0365	0.0260	0.0075
100°	0.125	0.114	0.0955	0.0760	0.0415
50°	0.179	0.166	0.141	0.115	0.066
25°	0.214	0.200	0.169	0.138	0.078
0°	0.257	0.242	0.204	0.166	0.090
− 50°	0.384	0.362	0.299	0.231	0.094
− 100°	0.628	0.578	0.443	0.281	0.062
− 150°	1.225	1.097	0.062	0.0215	− 0.0255

perature between − 100° and − 150° C, actually about − 126° C, but it decreases with decreasing pressure. It will be observed, especially at the higher pressures, that the value of the Joule-Thomson coefficient passes through a maximum as the temperature is changed. A high-temperature inversion point, e.g., just above 200° C at 200 atm., is thus to be expected; such inversion points have actually been observed. This subject will be considered more fully from another standpoint in § 22b.

[4] Roebuck and Osterberg, *Phys. Rev.*, **48**, 450 (1935); for other determinations of the Joule-Thomson effect see Roebuck, *et al.*, *Proc. Am. Acad. Arts Sci.*, **60**, 537 (1925); **64**, 287 (1930); *Phys. Rev.*, **43**, 60 (1933); **45**, 322 (1934); **46**, 785 (1934); **55**, 240 (1939); *J. Chem. Phys.*, **8**, 627 (1940); *J. Am. Chem. Soc.*, **64**, 400 (1942); Sage, Lacey, *et al.*, *Ind. Eng. Chem.*, **28**, 601, 718 (1936); **31**, 369 (1939).

For hydrogen and helium the value of PV increases with pressure at all pressures, at ordinary temperatures; thus, $[\partial(PV)/\partial P]_T$ is positive at all pressures, and consequently the Joule-Thomson coefficient is negative under these conditions. This means that the (upper) inversion temperature for hydrogen and helium is below the ordinary atmospheric temperature at all pressures. The difference in behavior exhibited by these gases, as compared with that of other gases at normal temperatures, can thus be accounted for. At sufficiently low temperatures, hydrogen and helium behave like other gases, having a positive value for the Joule-Thomson coefficient at moderate and low pressures; it decreases and eventually changes sign as the pressure is increased.

11d. Cooling by the Joule-Thomson Effect.—The fall of temperature that occurs when a gas undergoes throttled expansion, at suitable temperatures and pressures, has been utilized in the industrial liquefaction of gases. As seen above, the Joule-Thomson coefficient has the largest positive values, and hence the cooling accompanying the expansion will be most marked, at low temperatures, and preferably at low pressures. In order to obtain the maximum efficiency, therefore, the temperature of the gas to be liquefied is first reduced either by the performance of work in a gas engine, where it undergoes something approaching adiabatic expansion (§ 10d), or by utilizing the cooling effect of another portion of the gas which has been passed through a throttle.

Because of the variation of the Joule-Thomson coefficient with both temperature and pressure it is not easy to calculate the change of temperature resulting from a given throttled expansion, even when such data as in Table IV are available. This can be done, however, by a series of approximations. By estimating a rough average for the Joule-Thomson coefficient, some indication of the fall of temperature can be obtained.

Problem: Estimate the final temperature accompanying the throttled expansion of nitrogen, initially at 25° C, from 200 atm. to 1 atm. pressure.

An examination of Table IV shows that at 25° C, the mean value of $\mu_{J.T.}$ in the range from 200 atm. to 1 atm. is about 0.14° atm.$^{-1}$, so that the decrease of pressure by 199 (approx. 200 atm.) means a fall of temperature of 28°. As a first approximation, therefore, the final temperature is seen to be $-3.0°$ C. It is now possible to estimate a more accurate value of $\mu_{J.T.}$ in the pressure range of 200 atm. to 1 atm. and temperatures of 25° to $-3.0°$ C. This is seen to be about 0.155° atm.$^{-1}$, so that the fall of temperature is 31°, leading to a final temperature of $-6°$ C.

11e. Effect of Pressure on Heat Capacity.—Since C_P is equal to $(\partial H/\partial T)_P$, it follows that the variation of the heat capacity with pressure, at constant temperature, is given by

$$\left(\frac{\partial C_P}{\partial P}\right)_T = \frac{\partial^2 H}{\partial T \partial P}. \tag{11.7}$$

The heat content of a simple system is a single-valued function of the thermodynamic state, e.g., of the temperature and pressure, and hence the order of differentiation on the right-hand side of equation (11.7) is immaterial. Utilizing equation (11.5), therefore, it follows that

$$\left(\frac{\partial C_P}{\partial P}\right)_T = \left[\frac{\partial}{\partial T}\left(\frac{\partial H}{\partial P}\right)_T\right]_P = -\left[\frac{\partial(\mu_{\text{J.T.}} C_P)}{\partial T}\right]_P$$

$$= -\mu_{\text{J.T.}}\left(\frac{\partial C_P}{\partial T}\right)_P - C_P\left(\frac{\partial \mu_{\text{J.T.}}}{\partial T}\right)_P, \quad (11.8)$$

giving a relationship for the variation of heat capacity with pressure which is applicable to all fluids. For an ideal gas $\mu_{\text{J.T.}}$ and $(\partial \mu_{\text{J.T.}}/\partial T)_P$ are both zero, so that the heat capacity does not vary with pressure; this conclusion is in agreement with that reached in § 9e.

From data on the variation of the Joule-Thomson coefficient and heat capacity with temperature it is thus possible to determine the variation of the heat capacity of a fluid with pressure. This procedure has been utilized in certain instances.

Problem: From the data in Tables II and IV estimate the variation of the molar heat capacity of nitrogen with pressure, at ordinary temperatures and pressures.

From Table II, for nitrogen

$$C_P = 6.45 + 1.41 \times 10^{-3} T - 0.81 \times 10^{-7} T^2 \text{ cal. deg.}^{-1} \text{ mole}^{-1},$$

$$\left(\frac{\partial C_P}{\partial T}\right)_P = 1.41 \times 10^{-3} - 2 \times 0.81 \times 10^{-7} T \text{ cal. deg.}^{-2} \text{ mole}^{-1}.$$

At about 300° K, therefore,

$$C_P = 6.87 \text{ cal. deg.}^{-1} \text{ mole}^{-1},$$

$$\left(\frac{\partial C_P}{\partial T}\right)_P = 1.36 \times 10^{-3} \text{ cal. deg.}^{-2} \text{ mole}^{-1}.$$

From Table IV, $\mu_{\text{J.T.}}$ at ordinary temperatures and pressures is seen to be about 0.21° atm.$^{-1}$ and $(\partial \mu_{\text{J.T.}}/\partial T)_P$ is approximately -1.5×10^{-3} atm.$^{-1}$; if these results are inserted into equation (11.8), it is seen that

$$\left(\frac{\partial C_P}{\partial P}\right)_T = -0.21 \times 1.36 \times 10^{-3} + 6.87 \times 1.5 \times 10^{-3}$$

$$= 1.0 \times 10^{-2} \text{ cal. deg.}^{-1} \text{ mole}^{-1} \text{ atm.}^{-1}.$$

The heat capacity of nitrogen should thus increase by approximately 1.0×10^{-2} cal. deg.$^{-1}$ mole^{-1} for an increase of 1 atm. in the pressure, at ordinary temperatures and pressures.

The subject of the influence of pressure on heat capacity will be considered from another thermodynamic point of view in § 21d.

EXERCISES

1. Give the complete derivation of equation (9.18).
2. By combining equations (10.5) and (10.8) derive an expression for the work of reversible, adiabatic expansion of an ideal gas in terms of the initial and final volumes. Determine the work done in liter-atm. when 1 mole of a diatomic gas at 0° C expands from 10 ml. to 1 liter.

3. Determine the values of ΔH and ΔE in calories when 1 mole of liquid water is heated from 25° to 100° C at 1 atm. pressure. The mean heat capacity of water may be taken as 1.001 cal. deg.$^{-1}$ g.$^{-1}$ and the mean coefficient of cubical expansion as 4.0×10^{-4} deg.$^{-1}$; the density of water at 25° C is 0.9971 g. cc.$^{-1}$.

4. Calculate the molar heat capacity at 1 atm. pressure of nitrogen at 0°, 500° and 1000° C. Compare the results with the mean heat capacity in the range from 0° to 1000° C.

5. Evaluate the amount of heat in ergs which must be supplied to an ideal gas in the isothermal, reversible expansion from a pressure of 1.00 atm. to 25.0 atm. at 25° C.

6. How much heat, in calories, must be supplied to 100 g. of carbon dioxide in order to raise the temperature from 27° C to 727° C at 1 atm. pressure?

7. In the treatment of adiabatic processes in the text, the heat capacity has been assumed to be independent of temperature. How could allowance be made for the variation of C_P with temperature in equation (10.6)? (The gas may be assumed to be ideal in other respects.) Use this method to estimate the work done, in calories, and the final temperature in the reversible, adiabatic compression of 1 mole of oxygen from 10 liters to 1 liter, the initial temperature being 25° C. What would be the result if C_P were taken as having the constant (mean) value of 7 cal. deg.$^{-1}$ mole^{-1}?

8. A diatomic gas is to be compressed adiabatically from 1 atm. to 25 atm., the initial temperature being 25° C. The compression may be carried out in one stage or it may be performed in two stages, from 1 atm. to 5 atm., and then from 5 atm. to 25 atm., the gas being allowed to regain its original temperature between the two stages. Which process can be carried out with the smaller expenditure of energy?

9. When a gas is compressed adiabatically to half its initial volume, its temperature is observed to increase from 25° to 200° C. Estimate the (approximate) mean molar heat capacity of the gas at constant pressure.

10. Compare the values of ΔH and ΔE for the vaporization of 1 mole of water at 100° C and 1 atm. pressure. The heat of vaporization of water is 539 cal. g.$^{-1}$ at the normal boiling point; the specific volume of the vapor is then 1675 cc. g.$^{-1}$, and that of the liquid is approximately unity. Suggest why the value of ΔE in this case is sometimes called the "internal heat of vaporization."

11. Assuming $(\partial E/\partial V)_T$ for a van der Waals gas to be equal to a/V^2, show that the increase of the energy content in the isothermal expansion of 1 mole of the gas, under such conditions that no work is done, is given by

$$\Delta E = a\left(\frac{1}{V_1} - \frac{1}{V_2}\right).$$

Correlate this result with the analogous term in the expression for W in Exercise 8, Chapter III.

12. Calculate the (approximate) increase of internal energy in calories when 1 mole of carbon dioxide is expanded isothermally from 1 liter to 10 liters, assuming van der Waals behavior.

13. With the aid of equation (9.16) and the result derived in Exercise 11, show qualitatively that the value of $C_P - C_V$ for a gas such as nitrogen should pass through a maximum as the pressure is increased (cf. § 21a). (Note from Fig. 2 that $(\partial V/\partial T)_P$ is larger at low pressures and smaller at higher pressures than for an ideal gas.)

EXERCISES

14. If a is independent of temperature, show that C_V for a van der Waals gas should be independent of the pressure (or volume) at constant temperature, and that both C_P and C_P/C_V should pass through a maximum with increasing pressure, as in Table III (cf. § 21e). Use the results of Exercises 11 and 13.

15. Combine equations (9.18) and (11.5) to derive an expression for $C_P - C_V$ at the Joule-Thomson inversion point. What is the value for a van der Waals gas?

16. By utilizing a relationship of the form of equation (4.9), together with the result of Exercise 11, and the van der Waals equation in the low-pressure form obtained in Exercise 7, Chapter II, show that

$$\left(\frac{\partial E}{\partial P}\right)_T = -\frac{a}{RT} \quad \text{and} \quad \left[\frac{\partial(PV)}{\partial P}\right]_T = b - \frac{a}{RT}.$$

By introducing $H = E + PV$, show that

$$\left(\frac{\partial H}{\partial P}\right)_T = b - \frac{2a}{RT},$$

and hence derive the result

$$\mu_{\text{J.T.}} = \frac{1}{C_P}\left(\frac{2a}{RT} - b\right)$$

for the Joule-Thomson coefficient of a van der Waals gas at moderate pressures (cf. § 22a). Show that for a real gas, the Joule-Thomson coefficient is not necessarily zero even at zero pressure.

17. The variation of the Joule-Thomson coefficient of air with the absolute temperature T at 1 atm. pressure is given by

$$\mu_{\text{J.T.}} = -0.1975 + \frac{138.3}{T} - \frac{319.0}{T^2} \text{ deg. atm.}^{-1}$$

[Hoxton, *Phys. Rev.*, **13**, 438 (1919)]. The dependence of C_P on temperature at 1 atm. is given approximately by

$$C_P = 6.50 + 0.0010T \text{ cal. deg.}^{-1} \text{ mole}^{-1}.$$

Determine the rate of change of C_P for air with pressure in cal. deg.$^{-1}$ atm.$^{-1}$ mole^{-1} in the region of 27° C and 1 atm. pressure.

18. At 25° C, the variation of C_P for carbon dioxide with pressure is represented by

$$C_P = 8.90 + 0.0343P + 0.0123P^2,$$

with P in atm., and the variation with temperature, at 1 atm. pressure, is given (approximately) by

$$C_P = 6.80 + 0.0072T.$$

In the vicinity of 1 atm. pressure, the Joule-Thomson coefficient of the gas is 1.08° atm.$^{-1}$. Determine the rate of change of $\mu_{\text{J.T.}}$ with temperature at ordinary pressures. (The experimental value is about -0.010 atm.$^{-1}$).

19. Show that for a van der Waals gas $(\partial E/\partial V)_T$ is proportional to P^2 (cf. Exercise 11), and $(\partial V/\partial P)_T$ is approximately proportional to $1/P^2$, at not too high pressures; hence, justify the statement in § 11c that the first term in the bracket on the right-hand side of equation (11.6) is approximately independent of pressure.

CHAPTER V

THERMOCHEMISTRY

12. Heat Changes in Chemical Reactions

12a. Heat of Reaction.—The science of thermochemistry is concerned with the heat changes associated with chemical reactions; in other words, it deals essentially with the conversion of chemical energy into heat energy, and vice versa. Thermochemistry is, therefore, to be regarded as a branch of thermodynamics, especially since, as will be seen shortly, the subject is based largely on the first law. Further, the data obtained in thermochemical studies are utilized in the evaluation of properties of thermodynamic interest.

The heat change associated with a chemical reaction, like that for any other process, is an indefinite quantity depending on the path taken. However, as seen in § 9a, if the process is carried out at constant pressure or constant volume, the heat change has a definite value, determined only by the initial and final states of the system. It is for this reason that heat changes of chemical reactions are measured under constant pressure or constant volume conditions; processes involving liquids and gases are usually studied at constant (atmospheric) pressure, whereas combustion reactions are carried out at constant volume, e.g., in an explosion bomb. For the purpose of recording and tabulating the results, the data are quoted directly as, or are converted into, those for constant pressure, generally 1 atm. Although the reactants and products of a chemical reaction might well be at different temperatures, the situation is considerably simplified by determining heat changes for the condition in which all the substances concerned are at the same temperature. This practice is invariably adopted in making thermochemical measurements.

Since Q_P, the heat absorbed in any process at constant pressure, is equal to ΔH_P, the increase of heat content at constant pressure (§ 9a), the heat change accompanying a chemical reaction under these conditions is equal to the difference between the total heat content of the products and that of the reactants, at constant temperature and pressure. This is the quantity usually referred to as the **heat of reaction**; thus,

Heat of reaction = Heat content of products − Heat content of reactants,

or

$$Q_P = \Delta H_P = \sum H(\text{products}) - \sum H(\text{reactants}), \qquad (12.1)$$

where the H, i.e., heat content, values all refer to a specified temperature, e.g., 25° C, and pressure, e.g., 1 atm. The summation signs imply that the total heat contents, allowing for the different numbers of molecules that may be involved, of all the products or of all the reactants, respectively, are to be included. It may be remarked that in evaluating the heat of reaction it is assumed that the reactants are completely converted into the products. In other words, it is postulated that the reaction, as represented by the appropriate chemical equation, proceeds to completion. If this condition is not realized experimentally, the observed heat change is adjusted so as to give the result that would have been obtained for the complete reaction.

12b. Symbols and Units.—The heat change accompanying a reaction, for example, that between solid carbon (graphite) and hydrogen gas to form liquid benzene, is represented in the form of a thermochemical equation, viz.,

$$6C(s) + 3H_2(g) = C_6H_6(l), \quad \Delta H_{298.16} = 11.7 \text{ kcal.,}$$

where the symbols g, l and s indicate gaseous, liquid and solid state, respectively. It is the common practice in thermochemical work to express heat changes in **kilocalories**, abbreviated to kcal., where 1 kcal. = 1000 cal., because the statement of the result in calories would suggest a greater degree of accuracy than has actually been attained. The equation given above implies that the formation of 1 mole of liquid benzene from 6 gram-atoms of solid carbon (graphite) and 3 moles of hydrogen gas results in the absorption of 11.7 kcal., i.e., about 11,700 cal., at 25° C or 298.16° K, and 1 atm. pressure. The heat content of 1 mole of liquid benzene is thus 11.7 kcal. greater than that of the carbon (solid) and hydrogen (gas) of which it may be regarded as composed; that is,

$$\Delta H = H(C_6H_6, l) - [H(6C, s) + H(3H_2, g)] = 11.7 \text{ kcal. at } 25° \text{ C.}$$

It may be noted, in general, that if the heat content of the products exceeds that of the reactants, i.e., ΔH is positive, the reaction is accompanied by an absorption of heat. On the other hand, if the reverse is the case, so that ΔH is negative, heat is evolved when the reaction takes place.

If all gases behaved ideally, the value of ΔH would be independent of the pressure (§ 9e), but as this is not the case it is necessary, strictly speaking, to specify the pressure in connection with any reaction involving gases. Unless otherwise stated, the pressure is usually 1 atm., although for certain purposes it is desirable to express the data for the condition in which the gas behaves ideally, i.e., at very low pressures. In the great majority of cases, however, the difference between the observed value of ΔH at 1 atm. pressure and that corrected for departure of the gases from ideal behavior is considerably less than the experimental error, and so may be ignored (cf. § 20e).

For a reaction taking place in solution, it is necessary to specify the concentrations of the various reactants and products. If the solutions are so dilute that the addition of further solvent, e.g., water, results in no appreciable change in the heat of reaction, i.e., the heat of dilution is zero, the symbol aq is employed; thus,

$$HCl(aq) + NaOH(aq) = NaCl(aq) + H_2O, \quad \Delta H_{298} = -13.50 \text{ kcal.,}$$

for the reaction between hydrochloric acid and sodium hydroxide in dilute aqueous solution. The subjects of heat of solution and heat of dilution will be treated below (see also Chapter XVIII).

12c. Heat Changes at Constant Pressure and Constant Volume.—The heat change Q_V at constant volume is equal to ΔE_V the increase of the energy content at constant volume (§ 9a). For a reaction involving ideal, or approximately ideal, gases, ΔE would be independent of the volume, at constant temperature; for such a reaction, therefore, it is possible to identify Q_V with ΔE at the given temperature, without specifying constant volume conditions for the latter. Although this is strictly true only if the gases involved are ideal, it can be taken as being approximately true for all reactions. The change in the energy content due exclusively to a volume change, apart from the work of expansion, is invariably so small, particularly in comparison with the values of ΔE involved in chemical reactions, that it may be ignored. It will be assumed, therefore, that the value of ΔE, which is equal to Q_V, depends only on the temperature, and not on the actual volume or pressure. Bearing in mind these arguments, equation (9.3) can be written as

$$Q_P \approx \Delta E + P\Delta V$$
$$\approx Q_V + P\Delta V, \qquad (12.2)$$

which gives the relationship between the heat changes of a reaction at constant pressure and constant volume. The difference between these quantities is thus equal to $P\Delta V$, which is the work of expansion when the process is carried out at constant pressure.

For a reaction in which gases take part, the volume change ΔV may be appreciable, and its value can be determined with sufficient accuracy by assuming ideal behavior of the gases concerned. If n_1 is the number of moles of gaseous reactants, and n_2 is the number of moles of gaseous products of the reaction, the process is accompanied by an increase of $n_2 - n_1 = \Delta n$ moles of gas. If V is the volume of 1 mole of any (ideal) gas at the experimental temperature and pressure, then the increase of volume ΔV in the reaction will be equal to $V\Delta n$. For ideal gases, PV is equal to RT, so that

$$P\Delta V = PV\Delta n = RT\Delta n,$$

and substitution in equation (12.2) gives

$$Q_P \approx Q_V + RT\Delta n. \qquad (12.3)$$

From this expression the value of the heat of reaction at constant pressure can be calculated if that at constant volume is known, or vice versa. An important use of equation (12.3) is in the determination of the ΔH values for combustion reactions, since the actual thermochemical measurements are made in an "explosion bomb" at constant volume.

If the reaction involves solids and liquids only, and no gases, the volume change ΔV is usually so small that the $P\Delta V$ term in (12.2) may be neglected. In cases of this kind the heats of reaction at constant pressure and constant volume may be taken as identical, within the limits of experimental error. Since the volume changes due to solids and liquids are negligible, the value of ΔV for a reaction, in which gases as well as solids or liquids take part, applies to the gases only. The factor Δn in equation (12.3) then refers to *the gaseous molecules only*. For any reaction in which the same total number of gaseous molecules occurs on both sides of the chemical equation Δn is zero, and hence Q_P and Q_V are equal.

Problem: When 1 mole of solid naphthalene was burnt in oxygen gas to yield carbon dioxide gas at constant volume, at 25° C, the heat absorbed (Q_V) was found to be $-1{,}227.0$ kcal. Calculate ΔH for this reaction, assuming it to be independent of pressure.

The reaction is

$$C_{10}H_8(s) + 12O_2(g) = 10CO_2(g) + 4H_2O(l),$$

so that the number of molecules of gaseous reactants (n_1) is 12, and the number of molecules of gaseous products (n_2) is 10; hence, $\Delta n = n_2 - n_1 = -2$. The temperature is 25° C, i.e., 298° K, and R may be taken as 2 cal. or 2×10^{-3} kcal. deg.$^{-1}$ mole^{-1}; hence, by equation (12.3),

$$Q_P = -1{,}227.0 + (2 \times 10^{-3} \times 298 \times -2)$$
$$= -1{,}227.0 - 1.2 = -1{,}228.2 \text{ kcal.}$$

The value of ΔH for the reaction may thus be taken as $-1{,}228.2$ kcal. at 25° C.

12d. Thermochemical Laws.—Two important laws of thermochemistry are based on the principle of conservation of energy. According to A. L. Lavoisier and P. S. Laplace (1780), *the quantity of heat which must be supplied to decompose a compound into its elements is equal to the heat evolved when the compound is formed from its elements*. This experimental result is, of course, in direct agreement with the first law of thermodynamics, for otherwise it would be possible to create heat energy by making a compound from its elements, and then decomposing it, or vice versa. The law of Lavoisier and Laplace may be extended into the general form that *the heat change accompanying a chemical reaction in one direction is exactly equal in magnitude, but opposite in sign, to that associated with the same reaction in the reverse direction*. This conclusion follows directly from the fact, derived from the first law of thermodynamics, that the heat content of a substance, or system, has a definite value at a given temperature and pressure (§ 9a). The total heat content of the reacting substances must, therefore, differ from that of the products by a precise amount; there is thus a definite increase ΔH in one direction, and a numerically equal decrease $-\Delta H$ in the opposite direction of the chemical reaction [cf. equation (12.1)]. As a consequence of the foregoing conclusion, thermochemical equations can be reversed, provided the sign of ΔH is changed; thus,

$$CH_4(g) + 2O_2(g) = CO_2(g) + 2H_2O(l), \quad \Delta H_{298} = -212.80 \text{ kcal.}$$
$$CO_2(g) + 2H_2O(l) = CH_4(g) + 2O_2(g), \quad \Delta H_{298} = 212.80 \text{ kcal.}$$

Another important law of thermochemistry was discovered empirically by G. H. Hess (1840); it is known as the **law of constant heat summation**. This law states that *the resultant heat change, at constant pressure or constant volume, in a given chemical reaction is the same whether it takes place in one or in several stages.* This means that the net heat of reaction depends only on the initial and final states, and not on the intermediate states through which the system may have passed. The law of Hess thus also follows from the first law of thermodynamics for, as already seen, it leads to the conclusion that ΔH (or ΔE) depends only on the initial and final states of the system and is independent of the path connecting them. One result of Hess's law is that *thermochemical equations can be added and subtracted*, just like algebraic equations. The physical significance of these operations is that the heat contents of the various elements and compounds concerned are definite (extensive) quantities, and the addition or removal of a substance (or substances) from a system means that the heat content is increased or decreased, respectively, by a specific amount (or amounts). A useful practical application of this result is in the calculation of heat changes for reactions which cannot be studied directly.

Problem: Given the following heats of reaction at 25° C:

(i) $C_2H_4(g) + 3O_2(g) = 2CO_2(g) + 2H_2O(l)$, $\Delta H = -337.3$ kcal.
(ii) $H_2(g) + \frac{1}{2}O_2(g) = H_2O(l)$, $\Delta H = -68.3$
(iii) $C_2H_6(g) + 3\frac{1}{2}O_2(g) = 2CO_2(g) + 3H_2O(l)$, $\Delta H = -372.8$

determine the heat change of the reaction $C_2H_4(g) + H_2(g) = C_2H_6(g)$ at 25° C.

The required result can be obtained very simply by adding (i) and (ii) and subtracting (iii), so that

$$C_2H_4(g) + H_2(g) = C_2H_6(g), \quad \Delta H_{298} = -32.8 \text{ kcal.}$$

12e. Heat of Formation.—The **heat of formation** of a compound is *the increase of heat content ΔH when 1 mole of the substance is formed from its elements at a given temperature and pressure.* The value of ΔH depends on the physical state and condition of the substances involved (cf. § 12i), and so it is generally postulated that the elements are in their so-called **standard states**. For liquids and solids, the standard states are usually taken as the stable forms at the atmospheric temperature and a pressure of 1 atm. For gases, the standard state is chosen as 1 atm., although in certain cases the ideal gas is postulated; as indicated earlier, the difference between these two states does not have any considerable effect on the ΔH value. Where accurate thermochemical data are available, however, the distinction between the two states is significant (§ 20e). When all the substances concerned in a reaction are in their respective standard states, the change of heat content is indicated by the symbol ΔH^0.

As an example of the standard heat of formation, reference may be made to the reaction

$$C(s) + O_2(g) = CO_2(g), \quad \Delta H^0_{298} = -94.05 \text{ kcal.}$$

Here the carbon is in the form of graphite, for this is the stable form, i.e., the standard state, at 25° C and 1 atm.; the gases, oxygen and carbon dioxide, are both at 1 atm. pressure. The standard heat of formation of carbon dioxide at 25° C is then said to be − 94.05 kcal. mole^{-1}. The heat content of 1 mole of carbon dioxide is thus less than that of 1 g. atom of graphite and 1 mole of oxygen by 94.05 kcal. at a temperature of 25° C and a pressure of 1 atm.

The standard heats of formation of a number of compounds at 25° C are recorded in Table V.[1]

TABLE V.* STANDARD HEATS OF FORMATION AT 25° C IN KCAL. PER MOLE

Inorganic Compounds

Substance	ΔH	Substance	ΔH	Substance	ΔH
$HF(g)$	− 64.5	$N_2O(g)$	19.7	$HgO(s)$	− 21.6
$HCl(g)$	− 22.06	$NO(g)$	21.6	$FeO(s)$	− 64.3
$HBr(g)$	− 8.6	$NO_2(g)$	8.0	$Fe_2O_3(s)$	− 195
$HI(g)$	5.9	$H_2SO_4(l)$	− 194	$Al_2O_3(s)$	− 380
$H_2O(l)$	− 68.32	$HNO_3(l)$	− 42	$NaCl(s)$	− 98.3
$H_2S(g)$	− 5.3	$Na_2O(s)$	− 99	$KCl(s)$	− 104.3
$NH_3(g)$	− 11.03	$K_2O(s)$	− 86	$AgCl(s)$	− 30.3
$SO_2(g)$	− 70.9	$NaOH(s)$	− 102	$Hg_2Cl_2(s)$	− 62
$SO_3(g)$	− 94.4	$KOH(s)$	− 102	$Na_2SO_4(s)$	− 330
$CO(g)$	− 26.42	$CuO(s)$	− 38.5	$K_2SO_4(s)$	− 343
$CO_2(g)$	− 94.05	$Ag_2O(s)$	− 7.0	$PbSO_4(s)$	− 219

Organic Compounds

Substance	ΔH	Substance	ΔH
Methane (g), CH_4	− 17.89	Methanol (g), CH_3OH	− 48.1
Ethane (g), C_2H_6	− 20.24	Methanol (l), CH_3OH	− 57.0
Propane (g), C_3H_8	− 24.82	Ethanol (g), C_2H_5OH	− 56.3
Ethylene (g), C_2H_4	12.56	Ethanol (l) C_2H_5OH	− 66.4
Propylene (g), C_3H_6	4.96	Phenol (s), C_6H_5OH	− 38.4
Acetylene (g), C_2H_2	54.23	Aniline (l), $C_6H_5NH_2$	7.3
Benzene (l), C_6H_6	11.7	Urea (s), $CO(NH_2)_2$	− 79.4
Cyclohexane (l), C_6H_{12}	− 14.8	Benzoic acid (s), C_6H_5COOH	− 93.2
Naphthalene (s), $C_{10}H_8$	14.4	Sucrose (s), $C_{12}H_{22}O_{11}$	− 533.4

* For further data, see Table 5 at end of book.

12f. Heat Content and Heat of Formation.—There is a useful connection between the heat of a reaction, in general, and the heats of formation of the compounds involved. It will be evident that in thermochemical studies no

---

[1] Data mainly adapted from F. R. Bichowsky and F. D. Rossini, "The Thermochemistry of the Chemical Substances," 1936; Landolt-Börnstein, Physikalisch-Chemische Tabellen, 1923–1936; see also, Thacker, Folkins and Miller, *Ind. Eng. Chem.*, **33**, 584 (1941); Wagman, Kilpatrick, Taylor, Pitzer and Rossini, *J. Res. Nat. Bur. Stand.*, **34**, 143 (1945); Prosen and Rossini, *ibid.*, **34**, 263 (1945); Prosen, Pitzer and Rossini, *ibid.*, **34**, 403 (1945); Wagman, Kilpatrick, Pitzer and Rossini, *ibid.*, **35**, 467 (1945); Kilpatrick, Prosen, Pitzer and Rossini, *ibid.*, **36**, 559 (1946). For reviews of experimental methods, etc., see Rossini, *Chem. Rev.*, **18**, 233 (1936), **27**, 1 (1940); Roth, *Z. Elek.*, **38**, 94 (1932); **45**, 335 (1939).

information is obtained concerning the *actual* or absolute heat contents of substances, but only with regard to the *differences* of their heat contents. For this reason it is permissible to choose any arbitrary scale of reference for heat content; that is, a zero of heat content may be chosen by convention, just as the zero of the centigrade temperature scale is arbitrarily chosen as the ice point. Differences of temperature are the same, irrespective of whether the actual temperatures are recorded on the absolute or centigrade scales, provided they are expressed in the same units. A similar situation exists in connection with the heat content; the ΔH value will be the same, irrespective of whether the heat contents H are expressed on an absolute scale or an arbitrary, conventional scale. The convention usually employed in thermochemistry is arbitrarily to take *the heat contents of all elements as zero in their standard states at all temperatures*.* The heat of formation of a compound is the difference between the heat content of the compound and that of its elements, and since the latter are taken as zero, by convention, it follows that on the basis of this convention *the heat content of a compound is equal to its heat of formation.*

Utilizing the heat contents derived in this manner, the standard heat change of a reaction can be readily calculated from the known standard heat contents, i.e., the standard heats of formation, of the substances concerned. Consider, for example, the reaction, at 25° C,

$$CH_4(g) + 2O_2(g) = CO_2(g) + 2H_2O(l),$$
$$-17.89 \quad\quad 0 \quad\quad -94.05 \quad -2 \times 68.32$$

and writing the standard heat of formation at 25° C of each species below the formula, it is seen that the total heat content of the products is $-94.05 + (-2 \times 68.32)$ whereas that of the reactants is $-17.89 + 0$. The standard increase ΔH^0 of heat content in the reaction is thus given by the difference of these two quantities, viz.,

$$\Delta H^0_{298} = [-94.05 - (2 \times 68.32)] - (-17.89 + 0) = -212.80 \text{ kcal.}$$

Instead of using the heats of formation to calculate the heat of reaction, the procedure may be reversed, and the heat of formation of a compound derived from the heat of reaction, provided the heats of formation of all the other substances involved are known.

Problem: From the result of the problem in § 12c, determine the heat of formation of 1 mole of solid naphthalene from graphite and hydrogen gas at 1 atm. pressure at 25° C.

In this case,

$$C_{10}H_8(s) + 12O_2(g) = 10CO_2(g) + 4H_2O(l), \quad \Delta H^0_{298} = -1{,}228.2 \text{ kcal.,}$$
$$x \quad\quad\quad 0 \quad\quad -10 \times 94.05 \quad -4 \times 68.32$$

* For certain problems dealing with changes of a physical nature, which are based on variations of the heat content of a system with temperature, the convention is to take the heat content of the system, element or compound, to be zero at 0° C.

the standard heat of formation of each species being written below its formula, x being that of the naphthalene, which is required. Since ΔH^0 is $-1{,}228.2$ kcal., it follows that

$$-1228.2 = [-(10 \times 94.05) - (4 \times 68.32)] - (x + 0),$$

$$x = 14.4 \text{ kcal.}$$

The standard heat of formation ΔH^0 of naphthalene at 25° C is thus 14.4 kcal. (The result is not particularly accurate in this instance, because it represents the difference between two large numbers.)

12g. Heat of Combustion.—Organic compounds containing carbon and hydrogen alone, or together with oxygen, can be burnt in oxygen gas to yield carbon dioxide and (liquid) water as the sole products. The heat change accompanying the complete combustion of 1 mole of a compound, at a given temperature and 1 atm. pressure, is called the **heat of combustion**. From the data in the preceding section, it is seen, for example, that the heat of combustion of methane at 25° C is -212.80 kcal. The heats of combustion of solids and liquids are usually measured at constant volume in a "bomb calorimeter"; the results can then be used to calculate ΔH, as explained above. The standard heats of combustion of a number of familiar organic compounds are given in Table VI.[2] The data may be used to calculate

TABLE VI. HEATS OF COMBUSTION AT 25° C IN KCAL. PER MOLE

Substance	ΔH	Substance	ΔH
Methane (g), CH_4	-212.8	Benzene (l), C_6H_6	-781.0
Ethane (g), C_2H_6	-372.8	Toluene (l), C_7H_8	-934.5
Propane (g), C_3H_8	-530.6	Xylene (l), C_8H_{10}	$-1{,}088$
n-Butane (g), C_4H_{10}	-688.0	Benzoic acid (s), C_6H_5COOH	-771.4
n-Pentane (l), C_5H_{12}	-838.7	Phenol (s), C_6H_5OH	-732.0
n-Hexane (l), C_6H_{14}	-995.0	Naphthalene (s), $C_{10}H_8$	$-1{,}228.2$
Ethylene (g), C_2H_4	-337.3	Sucrose (s), $C_{12}H_{22}O_{11}$	$-1{,}348.9$
1,3-Butadiene (g), C_4H_8	-607.9	Urea (s), $CO(NH_2)_2$	-151.5
Methanol (l), CH_3OH	-173.6	Ethyl acetate (l), $CH_3COOC_2H_5$	-538.0
Ethanol (l), C_2H_5OH	-326.7	Aniline (l), $C_6H_5NH_2$	-811.9

heats of reaction and of formation which cannot be determined directly. Examples of such calculations are provided by the problems in §§ 12d, 12f.

12h. Heat of Hydrogenation.—The change of heat content accompanying the hydrogenation of 1 mole of an unsaturated compound has been determined in a number of instances. The results are of interest for certain theoretical purposes, but they can also be utilized for the derivation of heats of formation and heats of combustion.

Problem: The standard heat of hydrogenation of gaseous propylene to propane is -29.6 kcal., and the heat of combustion of propane is -530.6 kcal. at 25° C. (Unless otherwise stated, constant pressure is to be understood.) Utilizing the

[2] See ref. 1, also Rossini, *Ind. Eng. Chem.*, **29**, 1424 (1937); Prosen and Rossini, *J. Res. Nat. Bur. Stand.*, **27**, 289 (1941); **33**, 255 (1944); **36**, 269 (1946); Prosen, Johnson and Rossini, *ibid.*, **35**, 141 (1945); **36**, 455, 463 (1946).

known heats of formation of carbon dioxide (-94.0 kcal.) and of liquid water (-68.3 kcal.), determine the heat of combustion and the standard heat of formation of propylene.

The data are as follows:

(i) $C_3H_6(g) + H_2(g) = C_3H_8(g)$, $\Delta H^0 = -\ \ 29.6$ kcal.
(ii) $C_3H_8(g) + 5O_2(g) = 3CO_2(g) + 4H_2O(l)$, $\Delta H^0 = -530.6$
(iii) $C(s) + O_2(g) = CO_2(g)$, $\Delta H^0 = -\ \ 94.0$
(iv) $H_2(g) + \tfrac{1}{2}O_2(g) = H_2O(l)$, $\Delta H^0 = -\ \ 68.3$

To obtain the standard heat of formation of propylene, add equations (i) and (ii), and then subtract the sum from three times equation (iii) and four times (iv); the result is

$$3C(s) + 3H_2(g) = C_3H_6(g), \quad \Delta H^0 = 5.0 \text{ kcal.}$$

The heat of combustion can either be obtained from this result by reversing it and adding three times equations (iii) and (iv), or it may be derived by utilizing (i), (ii) and (iv); in any event, it is found that

$$C_3H_6(g) + 4\tfrac{1}{2}O_2(g) = 3CO_2(g) + 3H_2O(l), \quad \Delta H^0 = -491.9 \text{ kcal.}$$

12i. Phase Changes.—A phase change, such as vaporization of a liquid, fusion or sublimation of a solid, and transition from one crystalline modification to another, is invariably associated with a change of heat content. Such heat changes are often referred to as the "latent heat" of vaporization, fusion, etc., although the tendency at the present time is toward the omission of the term "latent." These quantities represent the difference in the heat contents of 1 gram, or 1 mole, of the two phases under consideration at the temperature and pressure at which the phase change takes place. Thus, the heat content of 1 gram of steam (water vapor) is 539.4 cal. greater than that of the same weight of liquid water at 100° C and 1 atm. pressure. The heat of vaporization of water at 100° C is thus 539.4 cal. g.$^{-1}$ and this quantity of heat must be supplied to 1 g. of liquid water at 100° C to convert it into vapor at the same temperature and 1 atm. pressure. Like other heat changes, latent heats of various types vary with the temperature; thus at 25° C, the heat of vaporization of water is 583.6 cal. g.$^{-1}$, the pressure being 23.76 mm., or 0.0313 atm. Molar heats of vaporization and fusion are frequently employed in thermodynamics, and these are the changes of heat content associated with the vaporization or fusion of 1 mole of the substance under consideration, at the given temperature and pressure. The results may be expressed in the form of equations similar to the thermochemical equations represented above; thus,

$$H_2O(s) = H_2O(l), \quad \Delta H_{273} = 1.438 \text{ kcal.}$$
$$H_2O(l) = H_2O(g, 1 \text{ atm.}), \quad \Delta H_{373} = 9.717 \text{ kcal.}$$

Heats of vaporization and fusion are utilized in the treatment of phase changes (Chapter XI), but for the present they will be employed in connection with thermochemical problems. For example, it is sometimes required to know the heat of a particular reaction when one (or more) of the substances taking part is in a different physical state than usual; the necessary adjustment can be readily made, as the following instance will show. The standard heat of formation of liquid water at 25° C is $-$ 68.32 kcal. per mole, i.e.,

$$H_2(g) + \tfrac{1}{2}O_2(g) = H_2O(l), \qquad \Delta H^0_{298} = -\,68.32 \text{ kcal.}$$

Suppose the standard heat of formation of water vapor at 25° C is required; use is made of the heat of vaporization of water, which is 583.6 × 18.016 cal. or 10.514 kcal. mole^{-1}, i.e.,

$$H_2O(l) = H_2O(g, 0.0313 \text{ atm.}), \qquad \Delta H_{298} = 10.514 \text{ kcal.}$$

For the present purpose it is necessary to know the value of ΔH for this process with water vapor at 1 atm. pressure. If the vapor behaved as an ideal gas, the result would be the same as that given above, since ΔH would be independent of the pressure (§ 9e). Although water vapor is not ideal, the difference in the heat content at 1 atm. pressure from that at 0.0313 atm. has been calculated (cf. § 20e) as only 0.005 kcal. at 25° C, i.e.,

$$H_2O(g, 0.0313 \text{ atm.}) = H_2O(g, 1 \text{ atm.}), \qquad \Delta H_{298} = 0.005 \text{ kcal.}$$

Combination with the preceding equation gives

$$H_2O(l) = H_2O(g, 1 \text{ atm.}), \qquad \Delta H_{298} = 10.52 \text{ kcal.},$$

and hence

$$H_2(g) + \tfrac{1}{2}O_2(g) = H_2O(g), \qquad \Delta H^0_{298} = -\,57.80 \text{ kcal.}$$

with each substance in its standard gaseous state.

When there is change in the crystalline form of any substance involved in a reaction there is a change in the heat content. If the data for a given reaction with two separate forms of a particular substance are available, the heat of transition of one form to the other can be evaluated. For example, the heats of combustion of the two allotropic forms of carbon, viz., diamond and graphite, are known to be $-$ 94.50 and $-$ 94.05 kcal. g. atom^{-1}, respectively, at 25° C, i.e.,

$$C(\text{diamond}) + O_2(g) = CO_2(g), \qquad \Delta H_{298} = -\,94.50 \text{ kcal.}$$
$$C(\text{graphite}) + O_2(g) = CO_2(g), \qquad \Delta H_{298} = -\,94.05$$

By subtracting these two thermochemical equations, it is immediately seen that

$$C(\text{diamond}) = C(\text{graphite}), \qquad \Delta H_{298} = -\,0.45 \text{ kcal.}$$

The transition of 1 g. atom of carbon as diamond to the stable form, i.e., graphite, at 25° C is associated with a decrease of 0.45 kcal. in the heat content.

12j. Effect of Temperature on Heat of Reaction: The Kirchhoff Equation.—An expression for the variation of the heat of reaction with temperature can be derived in a simple manner. If H_A is the total heat content of the reactants and H_B is that of the products, at the same temperature and pressure, then, as indicated in § 12a [cf. equation (12.1)], ΔH is given by

$$\Delta H = H_B - H_A, \tag{12.4}$$

all the quantities referring to the same pressure. If this equation is differentiated with respect to temperature, it is seen that

$$\left[\frac{\partial(\Delta H)}{\partial T}\right]_P = \left(\frac{\partial H_B}{\partial T}\right)_P - \left(\frac{\partial H_A}{\partial T}\right)_P. \tag{12.5}$$

According to equation (9.11), $(\partial H/\partial T)_P$ is equal to C_P, and hence

$$\left[\frac{\partial(\Delta H)}{\partial T}\right]_P = (C_P)_B - (C_P)_A, \tag{12.6}$$

where $(C_P)_A$ and $(C_P)_B$ are the total heat capacities, at the given (constant) pressure, of the reactants and the products, respectively. The right-hand side of equation (12.6) is the increase in the heat capacity of the system accompanying the chemical reaction, and so it may be represented by ΔC_P; thus, (12.6) takes the form

$$\left[\frac{\partial(\Delta H)}{\partial T}\right]_P = \Delta C_P, \tag{12.7}$$

which is the expression of what is generally know as the **Kirchhoff equation** (G. R. Kirchhoff, 1858), although a similar result was obtained earlier by C. C. Person (1851). *The rate of variation of the heat of reaction with temperature, at constant pressure, is thus equal to the increase in the heat capacity accompanying the reaction.*

The heat capacity is an extensive property, and so the heat capacity of the system in its initial state is the sum of the heat capacities of the reactants, and that in the final state is the sum of the heat capacities of the products of the reaction. For the general chemical reaction

$$aA + bB + \cdots = lL + mM + \cdots,$$

the increase of heat capacity ΔC_P is thus given by

$$\Delta C_P = [l(C_P)_L + m(C_P)_M + \cdots] - [a(C_P)_A + b(C_P)_B + \cdots], \tag{12.8}$$

where the C_P terms are here the *molar* heat capacities of the species indicated. The expression in the first set of brackets in equation (12.8) gives the total heat capacity of the products, and that in the second brackets is for the reactants, so that the difference is equal to the increase of heat capacity for the reaction. An alternative form of (12.8) is

$$\Delta C_P = \sum(nC_P)_f - \sum(nC_P)_i, \tag{12.9}$$

where n is the number of moles of each substance taking part in the reaction and C_P is its *molar* heat capacity. The subscripts i and f refer to the initial state (reactants) and final state (products), respectively, so that the first term on the right-hand side of equation (12.9) is the sum of all the nC_P terms for the products, whereas the second term is the corresponding sum for the reactants.

The Kirchhoff equation as derived above should be applicable to both chemical and physical processes, but one highly important limitation must be borne in mind. For a chemical reaction there is no difficulty concerning $(\partial \Delta H/\partial T)_P$, i.e., the variation of ΔH with temperature, *at constant pressure*, since the reaction can be carried out at two or more temperatures and ΔH determined at the same pressure, e.g., 1 atm., in each case. For a phase change, such as fusion or vaporization, however, the ordinary latent heat of fusion or vaporization (ΔH) is the value under equilibrium conditions, when a change of temperature is accompanied by a change of pressure. If equation (12.7) is to be applied to a phase change the ΔH's must refer to *the same pressure* at different temperatures; these are consequently not the ordinary latent heats. If the variation of the *equilibrium* heat of fusion, vaporization or transition with temperature is required, equation (12.7) must be modified, as will be seen in § 271.

12k. Application of the Kirchhoff Equation.—The application of the Kirchhoff equation to determine the heat of reaction at one temperature if that at another is known involves the integration of equation (12.7); thus, between the temperature limits of T_1 and T_2 the result is

$$\Delta H_2 - \Delta H_1 = \int_{T_1}^{T_2} \Delta C_P dT, \tag{12.10}$$

where ΔH_1 and ΔH_2 are the heats of reaction at the two temperatures. If, in the simplest case, ΔC_P may be taken as constant and independent of temperature over a small range or, better, if ΔC_P is taken as the mean value $\Delta \bar{C}_P$ in the temperature range from T_1 to T_2, it follows from equation (12.10) that

$$\Delta H_2 - \Delta H_1 = \Delta \bar{C}_P (T_2 - T_1). \tag{12.11}$$

Problem: The mean molar heat capacities, at constant pressure, of hydrogen, oxygen and water vapor in the temperature range from 25° C to 100° C are as follows: $H_2(g)$, 6.92; $O_2(g)$, 7.04; $H_2O(g)$, 8.03 cal. deg.$^{-1}$ mole^{-1}. Utilizing the heat of reaction at 25° C obtained in § 12i, calculate the standard heat of formation of water vapor at 100° C.

The reaction is
$$H_2(g) + \tfrac{1}{2}O_2(g) = H_2O(g),$$
so that,
$$\Delta \bar{C}_P = \bar{C}_P(H_2O, g) - [\bar{C}_P(H_2, g) + \tfrac{1}{2}\bar{C}_P(O_2, g)]$$
$$= 8.03 - (6.92 + 3.52) = -2.41 \text{ cal. deg.}^{-1}$$

In this case, T_1 is 25° C and T_2 is 100° C, so that $T_2 - T_1$ is 75°; ΔH_1 is known from § 12i to be -57.80 kcal. It is required to find ΔH_2 by using equation (12.11), and for this purpose the same heat units, e.g., kcal., must be used for the ΔH's and $\Delta \bar{C}_P$; hence,

$$\Delta H_2 + 57.80 = -2.41 \times 10^{-3} \times 75,$$
$$\Delta H_2 = -57.98 \text{ kcal. mole}^{-1}.$$

In the general case, integration of (12.10) is possible if ΔC_P is known as a function of temperature. Since the variation of the heat capacity of many substances can be expressed by means of equation (9.24), it follows that

$$\Delta C_P = \Delta\alpha + \Delta\beta T + \Delta\gamma T^2 + \cdots, \tag{12.12}$$

where, for the general reaction given in § 12j,

$$\begin{aligned}\Delta\alpha &= (l\alpha_L + m\alpha_M + \cdots) - (a\alpha_A + b\alpha_B + \cdots)\\ \Delta\beta &= (l\beta_L + m\beta_M + \cdots) - (a\beta_A + b\beta_B + \cdots)\\ \Delta\gamma &= (l\gamma_L + m\gamma_M + \cdots) - (a\gamma_A + b\gamma_B + \cdots),\end{aligned} \tag{12.13}$$

and so on. Inserting equation (12.12) into (12.10), and carrying out the integration, it is found that

$$\Delta H_2 - \Delta H_1 = \Delta\alpha(T_2 - T_1) + \tfrac{1}{2}\Delta\beta(T_2^2 - T_1^2)\\ + \tfrac{1}{3}\Delta\gamma(T_2^3 - T_1^3) + \cdots, \tag{12.14}$$

so that the change of heat content at one temperature can be calculated if that at another temperature is known, and the α, β, γ, ..., values for all the substances concerned are available from Table II.

Problem: Calculate the standard heat of formation of water vapor at 100° C allowing for the variation with temperature of the heat capacities of the reactants and the product, and taking ΔH^0 as -57.80 kcal. mole^{-1} at 25° C.

From Table II,

$$\begin{aligned}C_P(H_2O, g) &= 7.219 + 2.374 \times 10^{-3}T + 0.2670 \times 10^{-6}T^2\\ C_P(H_2, g) &= 6.947 - 0.200 \times 10^{-3}T + 0.4808 \times 10^{-6}T^2\\ C_P(O_2, g) &= 6.095 + 3.253 \times 10^{-3}T - 1.0170 \times 10^{-6}T^2,\end{aligned}$$

so that

$$\begin{aligned}\Delta C_P &= C_P(H_2O, g) - [C_P(H_2, g) + \tfrac{1}{2}C_P(O_2, g)]\\ &= -2.776 + 0.947 \times 10^{-3}T + 0.295 \times 10^{-6}T^2 \text{ cal. deg.}^{-1}\end{aligned}$$

Hence,

$$\Delta\alpha = -2.776, \qquad \Delta\beta = 0.947 \times 10^{-3}, \qquad \Delta\gamma = 0.295 \times 10^{-6},$$

and by equation (12.14), since T_1 is 25° C, i.e., 298° K, and T_2 is 100° C, i.e., 373° K,

$$\begin{aligned}\Delta H_2 - \Delta H_1 &= -2.776(373 - 298) + \tfrac{1}{2} \times 0.947 \times 10^{-3}[(373)^2 - (298)^2]\\ &\quad + \tfrac{1}{3} \times 0.295 \times 10^{-6}[(373)^3 - (298)^3] \text{ cal.}\\ &= -208.2 + 23.8 + 2.5 = -182 \text{ cal.}\end{aligned}$$

Since ΔH_1 is -57.80 kcal., and $\Delta H_2 - \Delta H_1$ is -0.182 kcal., it follows that ΔH_2 is -57.98 kcal. mole^{-1}, as obtained in the preceding problem.

In some connections it is useful to derive a general expression which will give the heat content change at any temperature; this may be done by the general integration of equation (12.7), which yields

$$\Delta H = \Delta H_0 + \int \Delta C_P dT, \tag{12.15}$$

where the integration constant ΔH_0 may be regarded as the difference in the heat contents of products and reactants if the substances could exist at the absolute zero, and if the expression for ΔC_P were valid down to that temperature. Upon inserting the general equation (12.12) into (12.15) and carrying out the integration, the result is

$$\Delta H = \Delta H_0 + \Delta\alpha T + \tfrac{1}{2}\Delta\beta T^2 + \tfrac{1}{3}\Delta\gamma T^3 + \cdots. \qquad (12.16)$$

The values of $\Delta\alpha$, $\Delta\beta$, $\Delta\gamma$, etc., are presumed to be known from heat capacity data, and so a knowledge of ΔH at any one temperature will permit the integration constant ΔH_0 to be calculated. Insertion of the results in (12.16) then gives an equation for the heat of reaction as a power series function of temperature. The result will be applicable over the same temperature range only as are the empirical heat capacity constants α, β, γ, etc.

Problem: The standard heat of formation ΔH^0 of ammonia gas is -11.03 kcal. mole^{-1} at 25° C; utilizing the data in Table II, derive a general expression for the heat of formation applicable in the temperature range from 273° to 1500° K.

The reaction is

$$\tfrac{1}{2}N_2(g) + \tfrac{3}{2}H_2(g) = NH_3(g),$$

and hence,

$\Delta C_P = C_P(NH_3, g) - [\tfrac{1}{2}C_P(N_2, g) + \tfrac{3}{2}C_P(H_2, g)]$.
$C_P(NH_3) = 6.189 + 7.787 \times 10^{-3}T - 0.728 \times 10^{-6}T^2$
$\tfrac{1}{2}C_P(N_2) = 3.225 + 0.707 \times 10^{-3}T - 0.0404 \times 10^{-6}T^2$
$\tfrac{3}{2}C_P(H_2) = 10.421 - 0.300 \times 10^{-3}T + 0.7212 \times 10^{-6}T^2$,
$\Delta C_P = -7.457 + 7.38 \times 10^{-3}T - 1.409 \times 10^{-6}T^2$ cal. deg.$^{-1}$
$\Delta H^0 = \Delta H_0^0 - 7.457T + \tfrac{1}{2} \times 7.38 \times 10^{-3}T^2 - \tfrac{1}{3} \times 1.41 \times 10^{-6}T^3$ cal.
$\quad = \Delta H_0^0 - 7.46 \times 10^{-3}T + 3.69 \times 10^{-6}T^2 - 0.47 \times 10^{-9}T^3$ kcal. mole^{-1}.

At 25° C, i.e., 298° K, the value of ΔH^0 is -11.03 kcal., and hence it is found from this equation that ΔH_0^0 is -9.13 kcal., so that

$$\Delta H^0 = -9.13 - 7.46 \times 10^{-3}T + 3.69 \times 10^{-6}T^2 - 0.47 \times 10^{-9}T^3 \text{ kcal. mole}^{-1}.$$

Although the examples given above have referred to reactions involving only gases, the equations derived, e.g., (12.7), (12.14), (12.15), etc., can be applied to any chemical reaction. In the evaluation of ΔC_P by equation (12.8) or (12.9), the appropriate nC_P term must be included for every reactant and product, irrespective of its form. If the variation of the heat capacity with temperature of the solids or liquids concerned can be represented by an expression of the form of (9.24), then equations such as (12.14) and (12.16) can still be employed. Otherwise, the appropriate equations can be derived without difficulty.

Problem: The variation with temperature of the heat capacity of carbon (graphite), between 273° and 1373° K, is given by

$$C_P = 2.673 + 2.62 \times 10^{-3}T + 1.17 \times 10^5 T^{-2} \text{ cal. deg.}^{-1} \text{ g. atom}^{-1}.$$

Utilizing data already available derive a general expression for the heat of the reaction

$$C(s) + H_2O(g) = CO(g) + H_2(g)$$

as a function of the temperature, in the range specified above.

From Table II and the expression given above for the heat capacity of carbon,

$$\Delta C_P = [C_P(CO, g) + C_P(H_2, g)] - [C_P(C, s) + C_P(H_2O, g)]$$
$$= 3.397 - 3.36 \times 10^{-3}T - 0.066 \times 10^{-6}T^2 - 1.17 \times 10^5 T^{-2} \text{ cal. deg.}^{-1}$$

Hence, by equation (12.15), after converting ΔC_P into kcal. deg.$^{-1}$,

$$\Delta H = \Delta H_0 + \int \Delta C_P dT$$
$$= \Delta H_0 + 3.397 \times 10^{-3}T - 1.68 \times 10^{-6}T^2 - 0.022 \times 10^{-9}T^3 + 1.17 \times 10^2 T^{-1} \text{ kcal.}$$

The (standard) heats of formation of $H_2O(g)$ and $CO(g)$ at 25° C are -57.80 and -26.42 kcal. mole^{-1}, respectively, so that ΔH for the given reaction is 31.38 kcal. at 25° C, i.e., 298° K. Inserting these values in the result just obtained, ΔH_0 is found to be 30.13 kcal., and hence ΔH can now be calculated at any temperature in the specified range.

121. Heat Changes of Reactions in Solution.—When a reaction takes place in solution, or when one or more of the reactants or products are in solution instead of the pure state, the heat change is affected just as for a phase change. This is because the formation of a solution is almost invariably accompanied by a change of heat content, that is, heat is evolved or absorbed. The heat change per mole of solute dissolved, referred to as the **heat of solution**, is not a constant quantity, however, for it depends upon the amount of solvent. In other words, *the heat of solution at a given temperature varies with the concentration of the solution*. When a solute, e.g., a solid, gradually dissolves in a particular solvent, the composition of the solution changes from pure solvent to the final solution. The heat of solution *per mole* at any instant thus varies during the course of the solution process; this quantity, known as the "differential heat of solution" will be considered more fully later (Chapter XVIII).

At the present time, the matter of interest is the *total* heat change per mole of solute when solution is complete; this is the **integral heat of solution**. Thus, if ΔH is the total change of heat content when m moles of solute are dissolved in a definite quantity, e.g., 1,000 grams, of solvent, the integral heat of solution is equal to $\Delta H/m$. The integral heats of solution for various solutions of hydrochloric acid, of different molalities, at 25° C, are given below.

Molality (m)	0.139	0.278	0.555	1.11
Moles H_2O/Moles HCl	400	200	100	50
$\Delta H/m$	-17.70	-17.63	-17.54	-17.40 kcal.

If the heat changes are extrapolated to infinite dilution the integral heat of dilution is found to be -17.88 kcal. per mole; thus, when 1 mole of hydrogen chloride is dissolved in a large quantity of water, so as to form an extremely dilute solution, the total heat evolved is 17.88 kcal. at 25° C. Results analogous to those quoted above have been obtained with other solutes of various kinds, not necessarily gaseous in nature. Because of the interaction between hydrogen chloride and water, however, the heats of solution are unusually large. Further, as for gases in general, $\Delta H/m$ is negative in this case. For solids which do not interact with the solvent, as is the case with most hydrated salts in water, the integral heat of solution is usually positive.*

One consequence of the variation of the heat of solution with the composition of the solution is that dilution of a solution from one concentration to another is also accompanied by a heat change. Consideration of the change of heat content per mole at any instant in the course of the dilution process, known as the "differential heat of dilution," will be left to a subsequent chapter. The net change *per mole of solute* associated with the dilution of a solution from one concentration to another is the **integral heat of dilution**. For example, using the data for hydrochloric acid given above, it is possible to write

$$HCl(g) + 400H_2O(l) = HCl(400H_2O), \quad \Delta H = -17.70 \text{ kcal.}$$
$$HCl(g) + 50H_2O(l) = HCl(50H_2O), \quad \Delta H = -17.40$$

so that by subtraction,

$$HCl(50H_2O) + 350H_2O(l) = HCl(400H_2O), \quad \Delta H = -0.30 \text{ kcal.}$$

The dilution of the $HCl(50H_2O)$, i.e., 1.11 molal, to the $HCl(400H_2O)$, i.e., 0.139 molal, solution is thus accompanied by the evolution of a total amount of 0.30 kcal. per mole of hydrogen chloride, at 25° C. By utilizing the extrapolated heat of solution at infinite dilution, the integral heat of dilution of any solution to infinite dilution can be calculated in an analogous manner. Thus,

$$HCl(g) + 50H_2O(l) = HCl(50H_2O) \quad \Delta H = -17.40 \text{ kcal.}$$
$$HCl(g) + aq = HCl(aq) \quad \Delta H = -17.88$$

so that,

$$HCl(50H_2O) + aq = HCl(aq) \quad \Delta H = -0.48 \text{ kcal.}$$

As indicated in § 12b, the symbol aq implies a large amount of water, so that $HCl(aq)$ refers to an infinitely dilute solution of hydrochloric acid.

The consequence of heats of solution in connection with the heat changes in chemical reactions may be illustrated by reference to the reaction between ammonia and hydrogen chloride. If the gaseous reactants are employed and the product is solid ammonium chloride, the change may be represented by

$$NH_3(g) + HCl(g) = NH_4Cl(s) \quad \Delta H = -41.9 \text{ kcal.}$$

* A number of values of heats of solution are given in Table 6 at the end of the book.

If the ammonium chloride eventually occurs in an aqueous solution containing 1 mole of the salt to 200 moles of water, then use is made of the known integral heat of solution, viz.,

$$NH_4Cl(s) + 200H_2O(l) = NH_4Cl(200H_2O) \qquad \Delta H = 3.9 \text{ kcal.},$$

so that by addition

$$NH_3(g) + HCl(g) + 200H_2O(l) = NH_4Cl(200H_2O) \qquad \Delta H = -38.0 \text{ kcal.}$$

Since the integral heat of solution varies to some extent with the concentration of the ammonium chloride solution, this result is applicable in particular to the composition of solution specified.

If the reaction takes place between aqueous solutions of ammonia and hydrochloric acid, the thermochemical equations are as follows:

$$NH_3(g) + 100H_2O(l) = NH_3(100H_2O) \qquad \Delta H = -8.5 \text{ kcal.}$$
$$HCl(g) + 100H_2O(l) = HCl(100H_2O) \qquad \Delta H = -17.5$$

which, combined with the result obtained above, gives

$$NH_3(100H_2O) + HCl(100H_2O) = NH_4Cl(200H_2O) \qquad \Delta H = -12.0 \text{ kcal.}$$

Direct experimental measurement of the heat change of the reaction between aqueous solutions of ammonia and hydrochloric acid has confirmed the calculated result.

When applying the Kirchhoff equation, in order to determine the variation with temperature of the heat content change accompanying a reaction in solution, the heat capacity to be employed is a special quantity, called the "partial molar heat capacity." This quantity will be described in Chapter XVIII, in connection with a general discussion of the properties of substances in solution.

13. Flame and Explosion Temperatures

13a. Maximum Reaction Temperatures: Flame Temperatures.—In the foregoing treatment of thermochemical changes it has been supposed that the reaction takes place at constant temperature, and that the heat liberated (or absorbed) is removed (or supplied) by the surroundings. It is this quantity of heat which is recorded as the heat of reaction. It is possible, however, to conceive the reaction taking place under adiabatic conditions, so that no heat enters or leaves the system. For a constant pressure process, as is usually postulated, this means that ΔH will be zero. In a reaction performed adiabatically the temperature of the system will change, so that the products will be at a different temperature from that of the reactants. If ΔH at constant temperature is positive, i.e., heat is absorbed, the temperature of the adiabatic system will fall, but if ΔH is negative, the temperature will rise during the course of the reaction. From a knowledge of heats of reaction and of the variation with temperature of the heat capacities of the reactants and products, it is possible to calculate the final temperature

FLAME AND EXPLOSION TEMPERATURES

of the system. This temperature is of particular interest in connection with the combustion of gaseous hydrocarbons in oxygen or air. In such cases it gives the **maximum flame temperature**, the actual temperature being somewhat lower because of various disturbing factors.

Several different procedures, all based on the same fundamental principles, are available for calculating the maximum temperature attainable in a given reaction. The simplest is to imagine the reaction taking place at ordinary temperature (25° C), assuming this to be the initial temperature of the reactants, and then to find to what temperature the products can be raised by means of the heat evolved in the reaction.

Problem: The heat of combustion of methane is -212.80 kcal. at 25° C; the difference in the heat contents of liquid water and vapor at 1 atm. pressure at 25° C is 10.52 kcal. Using the data in Table II calculate the maximum temperature of the flame when methane is burnt in the theoretical amount of air ($1O_2$ to $4N_2$) at 25° C and constant pressure, assuming combustion to be complete.

Since the water will ultimately be in the form of vapor, at the high temperature, the value of ΔH at ordinary temperatures is required with water vapor as the product; thus,

$$CH_4(g) + 2O_2(g) = CO_2(g) + 2H_2O(l), \qquad \Delta H_{298} = -212.80 \text{ kcal.}$$
$$2H_2O(l) = 2H_2O(g), \qquad \Delta H_{298} = 2 \times 10.52$$

so that

$$CH_4(g) + 2O_2(g) = CO_2(g) + 2H_2O(g), \qquad \Delta H_{298} = -191.76 \text{ kcal.}$$

It may now be supposed that this quantity of heat is utilized to raise the temperature of the products, consisting of 1 mole CO_2, 2 moles H_2O and 8 moles of N_2, which were associated in the air with the 2 moles of O_2 used in the combustion of 1 mole of CH_4. If T_2 is the maximum temperature attained in the combustion, then the heat required to raise the temperature of the products from 25° C, i.e., 298° K, to T_2 must be equal in magnitude to the heat of reaction, but opposite in sign. The sum of ΔH for the temperature increase and ΔH for the reaction must be zero, so that ΔH for the whole (adiabatic) process is zero, as postulated above. If $\Sigma(nC_P)_f$ is the total heat capacity of the products, then the heat required to raise the temperature from 298° K to T_2, i.e., ΔH(heating), is given by

$$\Delta H(\text{heating}) = -\Delta H_{298}(\text{reaction}) = \int_{298}^{T_2} \Sigma(nC_P)_f dT. \qquad (13.1)$$

In the present case,

$$\Sigma(nC_P)_f = C_P(CO_2) + 2 \times C_P(H_2O, g) + 8 \times C_P(N_2),$$

and from Table II,

$$C_P(CO_2) = 6.396 + 10.100 \times 10^{-3}T - 3.405 \times 10^{-6}T^2$$
$$2 \times C_P(H_2O, g) = 14.438 + 4.748 \times 10^{-3}T + 0.534 \times 10^{-6}T^2$$
$$8 \times C_P(N_2) = 51.592 + 11.300 \times 10^{-3}T - 0.646 \times 10^{-6}T^2,$$

so that

$$\Sigma(nC_P)_f = 72.43 + 26.15 \times 10^{-3}T - 3.517 \times 10^{-6}T^2 \text{ cal. deg.}^{-1}$$

Upon inserting this result into equation (13.1), with ΔH(heating) equal to $-\Delta H_{298}$(reaction), i.e., 191.76 kcal. or 191,760 cal., it follows that

$$191,760 = [72.43T + \tfrac{1}{2} \times 26.15 \times 10^{-3}T^2 - \tfrac{1}{3} \times 3.517 \times 10^{-6}T^3]_{298}^{T_2}$$
$$= 72.43(T_2 - 298) + 13.08 \times 10^{-3}[T_2^2 - (298)^2]$$
$$- 1.172 \times 10^{-6}[T_2^3 - (298)^3],$$
$$214,470 = 72.43T_2 + 13.08 \times 10^{-3}T_2^2 - 1.172 \times 10^{-6}T_2^3.$$

Solving by the method of successive approximations, it is found that T_2 is about 2250° K or 1980° C.

13b. Calculated and Actual Flame Temperatures.—The calculated maximum flame temperature for the combustion of methane in the theoretical amount of air is seen to be nearly 2000° C. Similar results, approximately 2000° C, have been estimated for several hydrocarbons, and also for carbon monoxide and hydrogen. Actual flame temperatures have been determined in a number of cases, the values being about 100° below those calculated.[3]

There are several reasons why the results obtained in the manner described are higher than the experimental flame temperatures. In the first place, it is unlikely that the reaction can be carried out under such conditions that the process is adiabatic, and no heat is lost to the surroundings. Further, it is improbable that the theoretical quantity of air will be sufficient to cause complete combustion of the hydrocarbon. In practice, excess air must be used, and since the temperature of the additional oxygen, as well as that of the large amount of accompanying nitrogen, must be raised by the heat of the reaction, the temperature attained will be lower than if combustion were complete in the theoretical amount of air. The effect of any "inert" gas not utilized in the reaction can be readily seen by performing the calculation in the problem given above on the basis of the assumption that the methane is completely burnt in the theoretical amount of pure oxygen. The maximum temperature is found to be over 4000° K, so that the presence of nitrogen in the air lowers the theoretical maximum by nearly 2000°. By the same procedure it is also possible to calculate the maximum flame temperature when a hydrocarbon is burnt in excess of air; the "products" will then include the oxygen remaining when the combustion is complete, in addition to the nitrogen and the actual reaction products. The value for methane, for example, will be found to be less than 2250° K.

Another reason why observed flame temperatures are lower than the calculated values is that at the high temperature attained in a burning hydrocarbon, dissociation of the water vapor into hydrogen and oxygen, or hydrogen and hydroxyl, and, especially, of the carbon dioxide into carbon monoxide and oxygen is very appreciable. These reactions involve the absorption of considerable amounts of heat, allowance for which should be included in the calculation. From a knowledge of the equilibrium constants of the dissociation reactions and their variation with temperature it is possible, by a series of approximations, to obtain a more accurate estimate of the flame temperature. For most common hydrocarbons the results derived in this manner, for combustion in air, are of the order of 100° lower than those which neglect dissociation. For combustion in oxygen, when the temperatures are much higher, the discrepancy is greater.

Finally, it may be mentioned that the combustion of a hydrocarbon is not the relatively simple process represented by the usual chemical equation. Various

---

[3] Jones, Lewis, Friauf and Perrott, *J. Am. Chem. Soc.*, **53**, 869 (1931).

compounds other than carbon dioxide and water are frequently formed, and so the heat of reaction is not equal to the value determined from measurements in an explosion bomb, in the presence of a large excess of oxygen, at ordinary temperatures.

13c. Influence of Preheating Reacting Gases.—Apart from the use of oxygen, the temperature of a hydrocarbon flame may be increased by preheating the reacting gases. In the calculation of the maximum temperature it is necessary to obtain, first, the value of, or an expression for, ΔH for the reaction at the temperature to which the gases are heated; the method described above may then be used. An alternative procedure is, however, as follows. The heat of reaction at any temperature T_1 is given by equation (12.15), i.e.,

$$\Delta H(\text{reaction}) = \Delta H_0(\text{reaction}) + \int_0^{T_1} \Delta C_P dT. \qquad (13.2)$$

The increase in the heat capacity ΔC_P may be represented by

$$\Delta C_P = \Sigma(nC_P)_f - \Sigma(nC_P)_i, \qquad (13.3)$$

as in equation (12.9), so that (13.2) becomes

$$\Delta H(\text{reaction}) = \Delta H_0(\text{reaction}) + \int_0^{T_1} \Sigma(nC_P)_f dT - \int_0^{T_1} \Sigma(nC_P)_i dT. \qquad (13.4)$$

On the other hand, the expression for ΔH for the heating of the products, from the initial temperature T_1 to the final (maximum) temperature T_2 is

$$\Delta H(\text{heating}) = \int_{T_1}^{T_2} \Sigma(nC_P)_f dT. \qquad (13.5)$$

Since the sum of the two ΔH values given by equations (13.4) and (13.5) must be zero, it follows that

$$\Delta H_0(\text{reaction}) + \int_0^{T_2} \Sigma(nC_P)_f dT - \int_0^{T_1} \Sigma(nC_P)_i dT = 0. \qquad (13.6)$$

The value of $\Delta H_0(\text{reaction})$ and the heat capacities of the reactants and products may be presumed to be known, and so it is possible to calculate the maximum temperature T_2 for any given initial temperature T_1 of the reactants.

An alternative treatment makes direct use of $\Delta H_T(\text{reaction})$ at any temperature T, e.g., 25° C, thus avoiding the necessity of first calculating $\Delta H_0(\text{reaction})$; in this case equation (13.4) takes the form

$$\Delta H_T(\text{reaction}) = \Delta H_0(\text{reaction}) + \int_0^T \Sigma(nC_P)_f dT - \int_0^T \Sigma(nC_P)_i dT,$$

and combination with (13.6) gives

$$\Delta H_T(\text{reaction}) + \int_T^{T_2} \Sigma(nC_P)_f dT - \int_T^{T_1} \Sigma(nC_P)_i dT = 0, \qquad (13.7)$$

where, as before, T_1 is the temperature to which the reactants are preheated, and T_2 is the maximum temperature attainable in the reaction under adiabatic conditions at constant pressure. It is of interest to note that if T_1 is made identical with T, so that the heat of reaction ΔH_T is given at the initial temperature of the reacting gases, equation (13.7) becomes identical with (13.1). Further, if T is taken as 0° K, ΔH_T becomes equivalent to ΔH_0 and then equation (13.7) reduces to (13.6). Hence, equation (13.7) may be taken as the general expression, applicable to all cases, which may be modified according to circumstances.

13d. Maximum Explosion Temperatures and Pressures.—The methods described in the preceding section, e.g., equation (13.7), may be used to calculate the maximum temperature, often referred to as the **explosion temperature**, attained in a combustion reaction at constant volume, instead of at constant pressure. The only changes necessary are (a) the replacement of the ΔH value at constant pressure by ΔE at constant volume, and (b) the use of the heat capacities at constant volume, i.e., C_V, instead of those for constant pressure. The conversion of ΔH_P to ΔE_V, i.e., of Q_P to Q_V, can be made with sufficient accuracy by means of equation (12.3), and failing other information the heat capacity C_V of a gas may be taken as 2 cal. deg.$^{-1}$ mole^{-1} less than the value of C_P at the same temperature (cf., however, § 21a). The explosion temperatures calculated in this manner represent maximum values, for the reasons given in connection with flame temperatures.

From the maximum temperature, the pressure attained in an explosion at constant volume can be estimated, utilizing the approximation that the gas laws are applicable. The additional information required is either the original pressure or the volume of the vessel and the quantities of the gases involved. The pressures calculated in this manner are, of course, maximum values, and are based on the supposition that the system is in a state of thermodynamic equilibrium when the rapid combustion is completed. This condition is unlikely to be satisfied by the system; nevertheless, the results give a good indication of the maximum pressure accompanying an explosive reaction.

Problem: A mixture of hydrogen gas and the theoretical amount of air, at 25° C and a total pressure of 1 atm., is exploded in a closed vessel. Estimate the maximum explosion temperature and pressure, assuming adiabatic conditions. In order to simplify the calculation, the mean heat capacities $\bar{C}_P$ of nitrogen (8.3 cal. deg.$^{-1}$ mole^{-1}) and of water vapor (11.3 cal. deg.$^{-1}$ mole^{-1}), for the temperature range from 25° to 3000° C, may be used; they may be regarded as independent of the (moderate) pressure.

The value of ΔH for the reaction

$$H_2(g) + \tfrac{1}{2}O_2(g) = H_2O(g),$$

is known to be -57.8 kcal. at 25° C, and since Δn is here $-\tfrac{1}{2}$, it follows from equation (12.3) that Q_V (or ΔE) is $-57,500$ cal., since $RT\Delta n$ is equal to -300 cal.

For the present purpose, equation (13.7) or (13.1) takes the form

$$\Delta E + \int_{298}^{T_2} \sum (nC_V)_f dT = 0,$$

and since the mean heat capacities are to be used,

$$\Delta E + \sum (n\bar{C}_V)_f (T_2 - 298) = 0.$$

In the present case $\sum (n\bar{C}_V)_f$ is equal to $\bar{C}_V(H_2O, g) + 2 \times \bar{C}_V(N_2)$, since 2 moles of nitrogen, associated with the $\tfrac{1}{2}$ mole of oxygen employed in the combustion of the mole of hydrogen, are included in the "products." Assuming $\bar{C}_V$ to be less than $\bar{C}_P$ by 2 cal. deg.$^{-1}$ mole^{-1}, it follows from the data, that $\sum (n\bar{C}_V)_f$ is 9.3 + 2

× 6.3 = 21.9 cal. deg.$^{-1}$. Hence,
$$-57,500 + 21.9(T_2 - 298) = 0,$$
$$T_2 = 2920° \text{ K}.$$

The maximum explosion temperature is thus about 2900° K, which is considerably higher than the corresponding flame temperature; the latter is found to be about 2400° K.

The maximum pressure is derived from the ideal gas equation. The original gas, at 1 atm. and 298° K, consisted of $1H_2$, $\tfrac{1}{2}O_2$ and $2N_2$, whereas the final gas at 2920° K consists of $1H_2O$ and $2N_2$. Since $PV = nRT$, with V and R constant, it is easily shown that the maximum explosion pressure is

$$P_2 = \frac{3}{3.5} \times \frac{2920}{298} = 8.4 \text{ atm}.$$

In the foregoing treatment the calculation of the maximum pressure involves a knowledge of the heat capacities of the products. This procedure has been reversed in a method for determining heat capacities, first used by R. Bunsen (1850) and improved by M. Pier (1908). A known amount of a mixture of two volumes of hydrogen and one of oxygen and a definite quantity of the gas whose heat capacity is to be measured, are placed in a steel bomb. The gases are then exploded by a spark, and the maximum pressure attained is measured. From this the corresponding temperature can be calculated by using the ideal gas law. Since the heat capacity of the water vapor is known, the mean heat capacity of the experimental gas, which must be inert in character, can be calculated. Data extending to very high temperatures have been obtained in this manner.

14. Calculation of Heat of Reaction

14a. Heats of Combustion.—A number of methods have been proposed for the calculation of heats of formation and combustion from a knowledge of the formula of the substance concerned. Although the results obtained are not always accurate, they are useful when experimental data are not available. An examination of the heats of combustion of organic compounds shows that isomeric substances have almost the same values, and that in any homologous series there is a change of 150 to 160 kcal. per mole for each CH_2 group. These results suggest that each carbon and hydrogen atom that is burnt to carbon dioxide and water, respectively, contributes a more or less definite amount to the heat of combustion. There is a possibility, therefore, of developing approximate rules relating the composition of the given substance to its heat of combustion.

One such rule (W. M. Thornton, 1917) [4] is that the heat of combustion is about $-52.5n$ kcal. per mole, where n is the number of atoms of oxygen required to burn a molecule of the compound. Heats of combustion calculated in this manner are in satisfactory agreement with observation for hydrocarbons, e.g., octane (C_8H_{18}): calculated, -1312 kcal., observed,

---

[4] Thornton, *Phil. Mag.*, **33**, 196 (1917); for more exact rules applicable to paraffin hydrocarbons, see Rossini, *Ind. Eng. Chem.*, **29**, 1424 (1937); Ewell, *ibid.*, **32**, 778 (1940); Prosen and Rossini, *J. Res. Nat. Bur. Stand.*, **34**, 263 (1945).

− 1307 kcal.; benzene (C_6H_6): calculated, − 787 kcal., observed, − 782 kcal., but are less satisfactory for compounds containing oxygen, e.g., succinic acid ($C_4H_6O_4$): calculated, − 367 kcal., observed, − 356 kcal.; sucrose ($C_{12}H_{22}O_{11}$): calculated, − 1260 kcal., observed, − 1350 kcal. Although the results were meant to apply to heats of combustion at constant volume, i.e., ΔE, they are certainly not sufficiently accurate to permit of a distinction between ΔE and ΔH. Further, no allowance is made for the state, e.g., solid, liquid or gaseous, of the compound under consideration, for the errors are probably greater than the latent heats.

In its simplest form, the method proposed by M. S. Kharasch (1929) [5] is virtually the same as that just described. The molar heat of combustion of a liquid compound at constant pressure is equal to $-26.05x$ kcal., where x is the number of valence electrons of carbon not shared with oxygen in the original substance, but which are shared with oxygen, i.e., in carbon dioxide, when combustion is complete. In general, x is equal to twice the number n of oxygen atoms utilized in the combustion of a molecule, so that this rule is equivalent to stating that the heat of combustion is $-52.1n$ kcal. per mole. However, Kharasch has realized the necessity for including allowances for various types of structure in the compound, and by the use of these correction factors results have been obtained which are within one per cent, or less, of the experimental heats of combustion.

It is perhaps unnecessary to mention that from a knowledge of the heat of combustion, the heat of formation of the compound from its elements can be calculated (§ 12g). The results will not be very accurate for, as indicated earlier, they usually involve the difference between two relatively large numbers, one of which, namely the estimated heat of combustion, may be appreciably in error.

14b. Bond Energies and Heat of Reaction.—A more fundamental approach to the problem of calculating heats of formation and reaction is by the use of **bond energies**. By the bond energy is meant the *average* amount of energy, per mole, required to break a particular bond in a molecule and separate the resulting atoms or radicals from one another. Thus, the C—H bond energy is one-fourth of the amount of energy required to break up a mole of methane molecules into separate, i.e., gaseous, carbon atoms and hydrogen atoms. There are good reasons for believing that different energies are required to remove the successive hydrogen atoms, one at a time, from a methane molecule, but the so-called bond energy is the mean value. From a knowledge of heats of dissociation of various molecules into atoms and of the standard heats of formation of others, it has been possible to derive the mean energies of a number of different bonds. Some of these values, as calculated by L. Pauling (1940), are given in Table VII;[6] they are based on 125 kcal. per g. atom as the heat of vaporization of carbon. For the C=O and C≡N bonds the energies vary to some extent with the nature of the compounds; thus for C=O, it is 142 kcal. in formaldehyde,

[5] Kharasch, *J. Res. Nat. Bur. Stand.*, **2**, 359 (1929).
[6] L. Pauling, "The Nature of the Chemical Bond," 2nd ed., 1940, pp. 53, 131.

149 kcal. in other aldehydes, and 152 kcal. in ketones, acids and esters. The energy of the C≡N bond is 150 kcal. in most cyanides, but 144 kcal. in hydrogen cyanide.

TABLE VII. BOND ENERGIES IN KCAL. PER MOLE

Bond	Energy	Bond	Energy
H—H	103.4 kcal.	H—I	71.4 kcal.
C—C	58.6	C—N	48.6
Cl—Cl	57.8	C—O	70.0
Br—Br	46.1	C—Cl	66.5
I—I	36.2	C—Br	54.0
C—H	87.3	C—I	45.5
N—H	83.7	O=O	118
O—H	110.2	N≡N	170
H—Cl	102.7	C=C	100
H—Br	87.3	C≡C	123

For C=O and C≡N, see text.

By the use of the bond energies in Table VII it is possible to derive satisfactory heats of formation and reaction in many cases, provided the substances involved do not contain certain double-bonded compounds. Whenever wave-mechanical resonance is possible, the energy required to dissociate the molecule, as calculated from the results given above, is too small. It is necessary in such instances to make allowance for the "resonance energy."[7] Although this varies from one compound to another, its value is approximately 38 kcal. mole^{-1} for benzene and its simple derivatives, 75 kcal. mole^{-1} for naphthalene compounds, 28 kcal. per mole for carboxylic acids, and 24 kcal. mole^{-1} for esters.

Suppose it is required to determine the standard heat of formation of ethane, that is, the heat of the reaction

$$2C(s) + 3H_2(g) = C_2H_6(g).$$

This may be regarded as equivalent to the vaporization of 2 g. atoms of solid carbon, requiring 2×125 kcal., the breaking of 3 moles of H—H bonds, requiring 3×103.4 kcal., the resulting atoms are then combined to form one C—C bond, yielding 58.6 kcal., and six C—H bonds, yielding 6×87.3 kcal. in ethane. The gain in energy, which may be taken as ΔH, accompanying the formation of ethane at ordinary temperatures is thus given by

$$\Delta H = [(2 \times 125) + (3 \times 103.4)] - [58.6 + (6 \times 87.3)] = -22.2 \text{ kcal.}$$

The value determined experimentally from the heat of combustion is -20.24 kcal. at 25° C. From the heat of formation, it is possible to calculate the heat of combustion, using the known heats of formation of carbon dioxide and water. Theoretically, it should be possible to evaluate the heat of

---

[7] See, for example, L. Pauling, "The Nature of the Chemical Bond," 2nd ed., 1940, Chap. IV; S. Glasstone, "Theoretical Chemistry," 1944, Chap. III.

combustion directly, but in doing so allowance should be made for the resonance energy of carbon dioxide, 33 kcal. per mole.

Problem: Utilize bond and resonance energies to evaluate the heat of combustion of benzoic acid.

The reaction is
$$C_6H_5COOH + 7\tfrac{1}{2}O_2 = 7CO_2 + 3H_2O,$$
and the bond energies are as follows:

Reactants			Products		
Bonds:			Bonds:		
	4C—C	234.4 kcal.		14C=O	2128 kcal.
	3C=C	300		6O—H	661
	5C—H	436.5	Resonance:		
	C=O	152		7CO$_2$	231
	C—O	70			
	O—H	110.2			3020 kcal.
	7½O=O	885			
Resonance:					
	Benzene ring	38			
	Carboxyl	28			
		2254 kcal.			

The increase of heat content for the combustion is thus $2254 - 3020 = -766$ kcal. (The experimental value is -771 kcal. at 25° C.)

The bond energies recorded in Table VII are based on data for substances in the gaseous state; strictly, therefore, they should be used for reactions involving gases only. However, molar heats of fusion and vaporization are usually of the order of 1 to 10 kcal.; hence, provided equal numbers of molecules of solids and liquids appear on both sides of the chemical equation, the conventional bond energies may be employed to yield results of a fair degree of accuracy.

EXERCISES *

1. The heat of hydrogenation of ethylene (C_2H_4) to ethane (C_2H_6) is -32.6 kcal at 25° C. Utilizing the heat of combustion data in Table VI, determine the change of heat content accompanying the cracking reaction n-butane (C_4H_{10}) → $2C_2H_4 + H_2$ at 25° C.

2. The heats of combustion of n-butane and isobutane are -688.0 and -686.3 kcal., respectively, at 25° C. Calculate the heat of formation of each of these isomers from its elements, and also the heat of isomerization, i.e., n-butane → isobutane, at 25° C.

3. Calculate ΔH_P and ΔE_V at 25° C for the reactions
$$C_6H_6(g) + 7\tfrac{1}{2}O_2 = 3H_2O(l) + 6CO_2$$
$$C_6H_6(l) + 7\tfrac{1}{2}O_2 = 3H_2O(g) + 6CO_2,$$

* Unless otherwise specified, heat capacity and heat content data will be found in Tables II and V, respectively, or at the end of the book.

using heat of formation data. The heat of vaporization of benzene is 103 cal. g.$^{-1}$ and that of water is 583.6 cal. g.$^{-1}$ at 25° C.

4. Calculate the change in heat content for the reaction occurring in the lead storage battery, viz.,

$$Pb + PbO_2 + 2H_2SO_4 \text{ (in 20\% aq. soln.)} = 2PbSO_4 + 2H_2O \text{ (in 20\% } H_2SO_4 \text{ soln.)},$$

assuming the volume of solution to be so large that its concentration does not change appreciably. The integral heat of solution (ΔH) of sulfuric acid in water to form a 20% solution is $- 17.3$ kcal. mole^{-1} at 25° C. The heat of formation of water in the acid solution may be taken as the same as for pure water.

5. The change of heat content for the reaction

$$SO_2(aq) + Cl_2(g) + 2H_2O(l) = H_2SO_4(aq) + 2HCl(aq)$$

is $- 74.1$ kcal. at 25° C. The integral heats of solution (ΔH) at infinite dilution of sulfur dioxide, sulfuric acid and hydrochloric acid are $- 8.5$, $- 23.5$ and $- 17.9$ kcal., respectively. Determine the heat of formation of pure sulfuric acid from its elements.

6. Taking the mean molar heat capacities of gaseous hydrogen, iodine and hydrogen iodide as 6.95, 8.02 and 7.14 cal. deg.$^{-1}$, respectively, calculate the heat of formation of hydrogen iodide from the gaseous elements at 225° C. The heat of sublimation of solid iodine is 58.5 cal. g.$^{-1}$ at 25° C.

7. The heat of solution of zinc in very dilute hydrochloric acid solution is $- 36.17$ kcal. per g. atom, but in a solution consisting of HCl.100H$_2$O, the heat of solution is $- 36.19$ kcal. The integral heat of solution of 1 mole zinc chloride in 200 moles water is $- 15.30$ kcal., and in an infinite amount of water it is $- 16.00$ kcal. What is the integral heat of infinite dilution of the HCl.100H$_2$O solution per mole of HCl?

8. Derive a general expression for the variation with temperature of the change of heat content for the reaction

$$H_2S(g) + 1\tfrac{1}{2}O_2(g) = H_2O(g) + SO_2(g).$$

Calculate the value of ΔH at 800° K.

9. Derive a general expression for the variation with temperature of the standard heat change of the reaction

$$ZnO(s) + C(s) = Zn(s) + CO(g).$$

Calculate the value of ΔH^0 at 600° K.

10. Determine the maximum flame temperature when methane is burnt with the theoretical amount of air (1O$_2$ to 4N$_2$) at 25° C and constant pressure, assuming combustion to be 80% complete. Dissociation of the products at high temperatures may be neglected.

11. Carbon monoxide is mixed with 25% more than the amount of air (1O$_2$ to 4N$_2$) required theoretically for complete combustion, and the mixture is preheated to 500° C. Determine the maximum flame temperature, assuming the carbon monoxide to be completely converted into the dioxide.

12. A small quantity of liquid ethanol is placed in an explosion bomb together with twice the theoretical amount of oxygen at 25° C and 1 atm. pressure. Taking the heat of vaporization of the alcohol as 9.5 kcal. mole^{-1} at 25° C, calculate the maximum explosion temperature and the maximum explosion pressure, assuming

adiabatic conditions and ideal behavior. The combustion of the ethanol is to be taken as complete, with water vapor and carbon dioxide as the sole products.

13. Compare the maximum flame temperatures for the complete combustion of (i) acetylene, (ii) hydrogen gas containing 2% by weight of atomic hydrogen, in the theoretical amount of oxygen, originally at 25° C, at 1 atm. pressure. For the reaction $2H = H_2$, the value of ΔH is -103.4 kcal. at 25° C; the heat capacity of atomic hydrogen may be taken as independent of temperature, with C_P equal to 5.0 cal. deg.$^{-1}$ mole^{-1}.

14. A mixture of equimolecular amounts of hydrogen and carbon monoxide, together with the theoretical quantity of air ($1O_2$ to $4N_2$) for the combustion, at a total pressure of 5 atm. at 25° C, is exploded in a closed vessel. Estimate the maximum temperature and pressure that could be attained, assuming combustion to be complete. The heat capacity data in Table II may be regarded as being applicable in the required pressure range, and the gases may be treated as behaving ideally.

15. Sulfur dioxide at 550° C is mixed with the theoretical amount of air, assumed to consist of 21 moles O_2 to 79 moles N_2, at 25° C, and the mixture is passed into a converter in which the gases react at 450° C. Assuming virtually complete conversion, how much over-all heat is liberated for each mole of sulfur trioxide formed?

16. Steam at a temperature of 150° C is passed over coke at 1000° C, so that the reaction $C(s) + H_2O(g) = CO(g) + H_2(g)$ takes place with an efficiency of 80%, i.e., 20% of the steam remains unreacted; the gases emerge at 700° C. Calculate the amount of heat which must be supplied per kg. of steam passing over the coke; the heat capacity of the latter may be taken as equal to that of graphite.

17. The reaction $C(s) + \frac{1}{2}O_2(g) = CO(g)$ is exothermic, whereas $C(s) + H_2O(g) = CO(g) + H_2(g)$ is endothermic; theoretically, therefore, it should be possible to pass a mixture of steam and air ($1O_2$ to $4N_2$) over heated coke so that its temperature remains constant. Assuming virtually complete reaction in each case, estimate the proportion of steam to air, both preheated to 100° C, which should maintain the temperature of the coke at 1000° K.

18. According to Prosen and Rossini (see ref. 4), the heat of formation of a gaseous normal paraffin containing more than five carbon atoms at 25° C is given by $-10.408 - 4.926n$ kcal. mole^{-1}, where n is the number of carbon atoms in the molecule. Calculate the heat of formation of gaseous n-hexane (C_6H_{14}) and compare the result with those given by the rules of Thornton and of Kharasch, and with the experimental value derived from the heat of combustion.

19. Estimate from the bond energies the approximate heat changes for the following gaseous reactions: (i) $H_2 + \frac{1}{2}O_2 = H_2O$, (ii) $C_2H_4 + Br_2 = C_2H_4Br_2$, (iii) $CH_3I + H_2O = CH_3OH + HI$, (iv) $3CH_3COCH_3$ (acetone) $= C_6H_3(CH_3)_3$ (mesitylene) $+ 3H_2O$.

20. Determine ΔH for the reaction

$$CH_3COCH_3(l) + 2O_2(g) = CH_3COOH(l) + CO_2(g) + H_2O(g),$$

taking the heats of combustion of liquid acetone and acetic acid as -427 and -209 kcal., respectively. Compare the result with that derived from bond energies allowing for the resonance energy in carbon dioxide and acetic acid.

CHAPTER VI

CALCULATION OF ENERGY AND HEAT CAPACITY

15. Classical Theory

15a. The Kinetic Theory of Gases.—The treatment of heat capacities given in the preceding chapters has been based partly on the use of thermodynamic relationships, and partly on experimental data relating to the actual heat capacities and their variation with temperature. The results and conclusions are, therefore, independent of any theories concerning the existence and behavior of molecules. However, by means of such theories valuable information has been obtained of a practical character, having a definite bearing on thermodynamic problems. For this reason, consideration will be given to certain theories of heat capacity, especially as some of the principles involved will be required in another connection at a later stage.

If it is postulated that an ideal gas consists of rapidly moving, perfectly elastic particles, i.e., the molecules, which do not attract each other, and which have dimensions negligible in comparison with the total volume of the gas, it is possible by the application of the laws of classical mechanics to derive an expression for the gas pressure. This approach to the study of gases, known as the **kinetic theory of gases**, leads to the result

$$PV = \tfrac{1}{3} N m \overline{c^2}, \tag{15.1}$$

where P is the pressure of the gas, V is the volume containing N molecules of mass m, and $\overline{c^2}$ is the mean square of the velocities of the molecules at the experimental temperature. If this result is combined with the equation of state for 1 mole of an ideal gas, viz., $PV = RT$, is found that

$$\tfrac{1}{3} N m \overline{c^2} = RT,$$
$$\tfrac{1}{2} N m \overline{c^2} = \tfrac{3}{2} RT, \tag{15.2}$$

where N is now the Avogadro number, i.e., the number of individual molecules in one mole. Since the mean kinetic energy of translation per molecule is $\tfrac{1}{2} m \overline{c^2}$, it follows that $\tfrac{1}{2} N m \overline{c^2}$ is equal to the total translational energy of the mole of gas; if this is represented by $E_{\text{tr.}}$, equation (15.2) gives

$$E_{\text{tr.}} = \tfrac{3}{2} RT. \tag{15.3}$$

15b. Kinetic Theory and Heat Capacity of Monatomic Gases.—If the energy of a molecule is supposed to be entirely translational,* then $E_{\text{tr.}}$ as

---
* Monatomic molecules undoubtedly possess other forms of energy, e.g., electronic and nuclear, but these may be regarded as being independent of temperature. It will be seen from equations (9.9) and (9.11) that only energy which varies with temperature can affect

given by equation (15.3) may be identified with the energy content E of 1 mole of gas, so that by equation (9.9) the molar heat capacity at constant volume is *

$$C_V = \left(\frac{\partial E}{\partial T}\right)_V = \tfrac{3}{2}R. \tag{15.4}$$

The molar heat capacity of an ideal gas, whose energy is that of translational motion only, should thus have a constant value, independent of temperature as well as of pressure (§ 9e), namely $\tfrac{3}{2}R$. Since R is 1.987 cal. deg.$^{-1}$ mole^{-1}, it follows that

$$C_V = 2.980 \text{ cal. deg.}^{-1} \text{ mole}^{-1}.$$

It was seen in § 9g that for an ideal gas $C_P - C_V = R$, so that for the gas under consideration,

$$C_P = C_V + R = \tfrac{5}{2}R$$
$$= 4.967 \text{ cal. deg.}^{-1} \text{ mole}^{-1}. \tag{15.5}$$

The ratio of the heat capacities of the gas at constant pressure and volume, respectively, is given by equations (15.4) and (15.5) as

$$\frac{C_P}{C_V} = \tfrac{5}{3} = 1.667. \tag{15.6}$$

For certain monatomic gases, such as helium, neon, argon, and mercury and sodium vapors, the ratio of the heat capacities at moderate temperatures has been found to be very close to 1.67, as required by equation (15.6). The values of the individual heat capacities at constant pressure and constant volume are 5.0 and 3.0 cal. deg.$^{-1}$ mole^{-1}, respectively, in agreement with equations (15.5) and (15.4). It appears, therefore, that for a number of monatomic gases the energy of the molecules, at least that part which varies with temperature and so affects the heat capacity, is entirely, or almost entirely, translational in character (see, however, § 16f).

15c. Polyatomic Molecules: Rotational and Vibrational Energy.—For gases consisting of molecules containing two or more atoms, the ratio of the heat capacities is less than 1.67, under ordinary conditions, and the values of C_P and C_V are larger than those given above. The only exceptions are the diatomic molecules hydrogen and deuterium, for which equations (15.4) and (15.5) have been found to apply at very low temperatures, about 50° K. However, the values are not constant, as they often are for monatomic molecules, and at ordinary temperatures C_P and C_V have increased to approximately 6.9 and 4.9 cal. deg.$^{-1}$ mole^{-1}, respectively; the ratio is then about 1.4, as it is for most other diatomic gases, e.g., nitrogen, oxygen, carbon monoxide, hydrogen chloride, etc., (§ 10e). In each case, further

the heat capacity, and so these other forms of energy can be neglected for the present. Polyatomic molecules also possess rotational and vibrational energy; these vary with temperature and so contribute to the heat capacity (§ 15c).

* Since E_{tr} is independent of the volume, the partial differential notation in equation (15.4) is unnecessary, but it is retained here and subsequently for the sake of consistency.

rise of temperature causes the heat capacities to increase, although the difference between C_P and C_V always remains equal to about 2 cal. deg.$^{-1}$ mole^{-1}, provided the pressure is not greatly in excess of atmospheric.

The reason why the heat capacities of polyatomic molecules are larger than the values given by equations (15.4) and (15.5) is that molecules of this type possess rotational energy and vibrational energy, in addition to the energy of translation. The rotational energy of a polyatomic molecule is due to the rotation of the molecule as a whole about three (or two) axes at right angles to one another, and vibrational energy is associated with the oscillations of the atoms within the molecule. The inclusion of these energies in the energy content of the molecule, and their increase with rising temperature, accounts for the discrepancy between the actual behavior of polyatomic molecules and the theoretical behavior described in § 15b. At sufficiently low temperatures the effect of the rotational energy and, especially, of the vibrational energy becomes inappreciable; this explains why hydrogen and deuterium behave like monatomic gases in the vicinity of 50° K, i.e., − 220° C. Other polyatomic gases would, no doubt, exhibit a similar behavior, but they liquefy before the rotational energy contribution to the heat capacity has become negligible. The decrease of C_P and C_V to 5 and 3 cal. deg.$^{-1}$ mole^{-1}, respectively, although theoretically possible, can thus not be observed.

15d. Principle of Equipartition of Energy.—The contributions of the vibrational and rotational motions to the energy and heat capacity of a molecule can be calculated by classical methods, and although the results are not correct, except at sufficiently high temperatures, for reasons to be explained later, the procedure is, nevertheless, instructive. According to the **principle of the equipartition of energy**,[1] each kind of energy of a molecule that can be expressed in the general form ax^2, referred to as a "square term" or "quadratic term," where x is a coordinate or a momentum, contributes an amount $\frac{1}{2}RT$ to the average energy of 1 mole. The translational (kinetic) energy of a molecule is equal to $\frac{1}{2}m(\dot{x}^2 + \dot{y}^2 + \dot{z}^2)$, where $\dot{x}$, $\dot{y}$ and $\dot{z}$ represent the components of its velocity in three directions at right angles. The momentum p_x, for example, is equal to $m\dot{x}$, and so the energy is given by $(p_x^2 + p_y^2 + p_z^2)/2m$, thus consisting of three quadratic terms. By the principle of the equipartition of energy, therefore, the translational energy should be $3 \times \frac{1}{2}RT$, i.e., $\frac{3}{2}RT$, in agreement with equation (15.3).

The rotational energy of a molecule, assumed to be of constant dimensions, is proportional to the square of the angular momentum, so that each type of rotation is represented by one square term, and thus should contribute $\frac{1}{2}RT$ per mole to the energy content. A diatomic molecule, or any linear molecule, exhibits rotation about two axes perpendicular to the line joining the nuclei. The rotational energy of a diatomic, or any linear,

[1] R. C. Tolman, "The Principles of Statistical Mechanics," 1938, pp. 93 *et seq.*; R. H. Fowler and E. A. Guggenheim, "Statistical Thermodynamics," 1939, pp. 121 *et seq.*; J. E. Mayer and M. G. Mayer, "Statistical Mechanics," 1940, pp. 144 *et seq.*; S. Glasstone, "Theoretical Chemistry," 1944, pp. 300 *et seq.*

molecule should thus be $2 \times \frac{1}{2}RT$, i.e., RT, per mole. A nonlinear molecule, on the other hand, can rotate about three axes, so that the rotation should contribute $3 \times \frac{1}{2}RT$, i.e., $\frac{3}{2}RT$, per mole to the energy of the system.

The energy of a harmonic oscillator is given by the sum of two quadratic terms, one representing the kinetic energy and the other the potential energy of the vibration. By the equipartition principle, the energy is thus expected to be $2 \times \frac{1}{2}RT$, i.e., RT, per mole for each possible mode of vibration. In general, a nonlinear molecule containing n atoms possesses $3n - 6$ modes of vibration, so that the vibrational energy contribution should be $(3n - 6)RT$ per mole. Diatomic and linear molecules, however, have $3n - 5$ vibrational modes, and the vibrational energy is expected to be $(3n - 5)RT$ per mole.

15e. Classical Calculation of Heat Capacities.—For a diatomic molecule two types of rotation are possible, as seen above, contributing RT per mole to the energy. Since there are two atoms in the molecule, i.e., n is 2, there is only one mode of vibration, and the vibrational energy should be RT per mole. If the diatomic molecules rotate, but the atoms do not vibrate, the total energy content E will be the sum of the translational and rotational energies, i.e., $\frac{3}{2}RT + RT = \frac{5}{2}RT$, per mole; hence,

$$C_V(\text{tr.} + \text{rot.}) = \left(\frac{\partial E}{\partial T}\right)_V = \tfrac{5}{2}R,$$

so that the heat capacity at constant volume should be about 5 cal. deg.$^{-1}$ mole^{-1}. The same result would be obtained if there is vibration of the atoms in the diatomic molecule, but no rotational motion. If, however, the molecule rotates and the atoms also vibrate, the energy content should be $\frac{3}{2}RT + RT + RT$ i.e., $\frac{7}{2}RT$, per mole; then

$$C_V(\text{tr.} + \text{rot.} + \text{vib.}) = \left(\frac{\partial E}{\partial T}\right)_V = \tfrac{7}{2}R,$$

that is, about 7 cal. deg.$^{-1}$ mole^{-1}.

The heat capacities of the diatomic gases, hydrogen, oxygen, carbon monoxide and hydrogen chloride, are all very close to 5 cal. deg.$^{-1}$ mole^{-1} at ordinary temperatures. It appears, therefore, that in these diatomic molecules there is either vibration or rotation, but not both, the rotational motion being the more probable. According to the classical equipartition principle, the value of C_V should remain at 5 cal. deg.$^{-1}$ mole^{-1} with increasing temperature, until the other type of energy, i.e., rotation or vibration, respectively, becomes excited; the heat capacity at constant volume should then increase sharply to 7 cal. deg.$^{-1}$ mole^{-1}. This is, however, quite contrary to experience: the heat capacity increases gradually, and not suddenly, as the temperature is raised. Even at 2000° C, the heat capacities C_V of hydrogen, nitrogen, oxygen and carbon monoxide are only 6.3, although that of hydrogen chloride is 6.9 cal. deg.$^{-1}$ mole^{-1}. For chlorine gas the value of C_V is already 6.0 at ordinary temperatures, and becomes equal to 7.0 cal. deg.$^{-1}$ mole^{-1} at about 500° C; it then increases, very slowly, to 7.2 cal. deg.$^{-1}$ mole^{-1} at 2000° C.

16. Quantum Statistical Theory of Heat Capacity

16a. Quantum Theory of Energy.—It is clear that as far as diatomic molecules are concerned the experimental heat capacities and their variation with temperature are not entirely consistent with the equipartition principle; the same is found to be true for other polyatomic molecules. In every case, however, the theoretical heat capacity is approached, or slightly exceeded, as the temperature is raised. Nevertheless, the approach is gradual, whereas the classical treatment would imply a sudden change from one value to another. An attempt to overcome this discrepancy was made by suggesting that the development of a new type of motion, with increasing temperature, did not occur with all the molecules at the same time. Thus, at any temperature a gas might consist of a mixture of molecules, some of which were rotating and not vibrating, but others were rotating as well as vibrating. However, no adequate theoretical treatment could be developed to account quantitatively for the experimental facts.

A much more satisfactory and complete interpretation of the observations is provided by the **quantum theory**, according to which *a molecule acquires its energy in the form of definite amounts or quanta*. As far as translational energy is concerned, the quanta are so small that the energy is taken up in a virtually continuous manner. As a result, the behavior corresponds to that required by the classical theory. Rotational quanta are also small, although considerably larger than those of translational energy; hence, at ordinary temperatures nearly all the molecules possess appreciable amounts (quanta) of rotational energy. The energy contribution is again in good agreement with that required by the equipartition principle, i.e., $\frac{1}{2}RT$ per mole, for each type of rotation.

The vibrational quanta, on the other hand, are much larger, and at ordinary temperatures the vibrational energy of most molecules is that of the lowest quantum level. In this event, the vibrational energy does not affect the heat capacity, as is the case for the majority of diatomic molecules. As the temperature is raised the molecules acquire increasing numbers of quanta of vibrational energy, with the result that the contribution to the heat capacity increases toward the classical value of R, i.e., about 2 cal. deg.$^{-1}$ mole^{-1}, for each mode of vibration. This accounts for the steady increase of the heat capacities at constant volume of several diatomic molecules from 5 at ordinary temperatures to 7 cal. deg.$^{-1}$ mole^{-1} as the temperature is raised. With chlorine, the vibrational energy quanta are not too large, because the binding energy of two chlorine atoms is smaller than that for most other diatomic molecules (cf. Table VII). Consequently, the number of molecules possessing one or more quanta of vibrational energy increases with temperature, even at normal temperatures; hence, the molar heat capacity C_V of chlorine already exceeds $\frac{5}{2}R$ at 0° C, and reaches $\frac{7}{2}R$ at about 500° C. The quantum theory thus permits of a qualitative explanation of the heat capacity observations for diatomic molecules, and also of other molecules. It will now be shown that a quantitative interpretation is also possible.

16b. The Partition Function.—The molecules of a gas do not all have the same energy; in fact any value of the energy, consistent with the requirement that it shall be made up of whole numbers of the various quanta, e.g., of translational, rotational and vibrational energy, is possible. However, it can be shown by the methods of statistical mechanics that at each temperature there is a particular distribution of the total energy of a gas among its constituent molecules which is much more probable than any other distribution.[2]

Consider a system, such as an ideal gas, consisting of a total number of N molecules *which do not attract or repel each other*. According to the quantum theory, the energy of each molecule at any instant must have a definite value involving a specific number of quanta. Let ϵ_i represent the energy, in excess of the lowest possible value, of one of these permitted states of the molecules. It has been shown by the methods of wave mechanics that even at the absolute zero a molecule would still possess a definite amount of energy, particularly vibrational energy; this quantity, referred to as the **zero-point energy** of the molecule, is the lowest value which acts as the reference level postulated above. Suppose N_0 is the number of molecules in this lowest energy state, and N_i is the number in the level in which the energy is ϵ_i. The most probable energy distribution among the molecules, contained in a vessel of *constant volume*, at the absolute temperature T, is then given by statistical mechanics as

$$N_i = N_0 e^{-\epsilon_i/kT}, \qquad (16.1)$$

where k, known as the **Boltzmann constant**, is equal to the molar gas constant R divided by the Avogadro number.* The foregoing result is an expression of the **Maxwell-Boltzmann law** of the distribution of energy. Although the law was derived from classical mechanics, it has been found that, at all temperatures above the lowest, quantum considerations lead to a result which is almost identical with that given by equation (16.1). One modification is, however, necessary: this is the introduction of a **statistical weight factor** g_i representing the number of possible quantum states having the same, or almost the same, energy ϵ_i. The appropriate form of the energy distribution law is then

$$N_i = \frac{N_0}{g_0} g_i e^{-\epsilon_i/kT}, \qquad (16.2)$$

where g_0 is the statistical weight of the lowest energy level.

If the various energy values, represented by the general term ϵ_i, are ϵ_0, ϵ_1, ϵ_2, etc., and the numbers of molecules possessing these energies are N_0,

---

[2] For a fuller treatment of this subject and of other matters considered in this chapter, see the works mentioned in ref. 1.

* The Boltzmann constant k is usually expressed in ergs deg.$^{-1}$; thus R is 8.314×10^7 ergs deg.$^{-1}$ mole^{-1} and the Avogadro number is 6.023×10^{23} mole^{-1}, so that k is $8.314 \times 10^7/6.023 \times 10^{23} = 1.380_5 \times 10^{-16}$ erg deg.$^{-1}$.

N_1, N_2, etc., or N_i, in general, then the total number of molecules N is equal to the sum of these individual numbers; that is,

$$N = N_0 + N_1 + N_2 + \cdots + N_i + \cdots$$
$$= \frac{N_0}{g_0}(g_0 e^{-\epsilon_0/kT} + g_1 e^{-\epsilon_1/kT} + \cdots + g_i e^{-\epsilon_i/kT} + \cdots),$$

where g_0, g_1, g_2, etc., are the statistical weights of the respective levels. This result may be put in the form

$$N = \frac{N_0}{g_0} \sum_{i=0}^{\infty} g_i e^{-\epsilon_i/kT}, \qquad (16.3)$$

where the summation is taken over all integral values of i, from zero to infinity, corresponding to all possible energy states of the molecules. The summation term in equation (16.3) is known as the **partition function** of the molecule, and is represented by the symbol Q *; thus,

$$Q = g_0 e^{-\epsilon_0/kT} + g_1 e^{-\epsilon_1/kT} + g_2 e^{-\epsilon_2/kT} \cdots + g_i e^{-\epsilon_i/kT} + \cdots$$
$$= \sum_{i=0}^{\infty} g_i e^{-\epsilon_i/kT}. \qquad (16.4)$$

It follows, therefore, from equations (16.3) and (16.4) that $N_0/g_0 = N/Q$; substituting this result in (16.2) it is seen that the distribution law may be written in the form

$$N_i = \frac{N}{Q} g_i e^{-\epsilon_i/kT}. \qquad (16.5)$$

16c. Energy, Heat Capacity and the Partition Function.—Since, in general, N_i molecules each possess energy ϵ_i, the total energy content, in excess of the zero-level value, of the molecules in the ith level is $N_i \epsilon_i$; the total energy, in excess of the zero level, of the whole system of N molecules is then given by the sum of all such terms. If E is the actual energy content of the N molecules and E_0 is the total energy when they are all in the lowest possible level, then

$$E - E_0 = N_0 \epsilon_0 + N_1 \epsilon_1 + N_2 \epsilon_2 + \cdots + N_i \epsilon_i + \cdots, \qquad (16.6)$$

and hence, by equation (16.5)

$$E - E_0 = \frac{N}{Q}(g_0 \epsilon_0 e^{-\epsilon_0/kT} + g_1 \epsilon_1 e^{-\epsilon_1/kT} + \cdots + g_i \epsilon_i e^{-\epsilon_i/kT} + \cdots). \qquad (16.7)$$

Upon differentiating equation (16.4), which defines the partition function, with respect to temperature, the volume being constant, as postulated above, and multiplying the result by kT^2, it is found that

$$kT^2 \left(\frac{\partial Q}{\partial T}\right)_V = g_0 \epsilon_0 e^{-\epsilon_0/kT} + g_1 \epsilon_1 e^{-\epsilon_1/kT} + \cdots + g_i \epsilon_i e^{-\epsilon_i/kT} + \cdots.$$

* This should not be confused with the heat quantity for which the same symbol is used.

It is evident, therefore, that equation (16.7) may be written as

$$E - E_0 = \frac{NkT^2}{Q}\left(\frac{\partial Q}{\partial T}\right)_V = NkT^2\left(\frac{\partial \ln Q}{\partial T}\right)_V.$$

If the system under consideration consists of 1 mole of gas, N is the Avogadro number and Nk is equal to R, the molar gas constant; the expression for the energy content per mole is then

$$E - E_0 = RT^2\left(\frac{\partial \ln Q}{\partial T}\right)_V. \qquad (16.8)$$

Finally, differentiation of equation (16.8) with respect to temperature, at constant volume, remembering that E_0 is constant, gives

$$C_V = \left(\frac{\partial E}{\partial T}\right)_V = \left[\frac{\partial}{\partial T}\left(RT^2\frac{\partial \ln Q}{\partial T}\right)\right]_V, \qquad (16.9)$$

for the molar heat capacity of an ideal gas system at constant volume. The value of C_P can then be obtained by the addition of R, in accordance with equation (9.29). The result expressed by equation (16.9) is applicable to any gas, provided the forces between the molecules are small, that is to say, provided the gas is almost ideal. Corrections for deviations from ideal behavior can be made if necessary (§ 21d).

In order to simplify the computation of heat capacities from partition functions it is often convenient to utilize the fact that equation (16.9) can be converted into the form

$$C_V = \frac{R}{T^2}\left[\frac{Q''}{Q} - \left(\frac{Q'}{Q}\right)^2\right], \qquad (16.10)*$$

where

$$Q' = -\sum \frac{\epsilon_i}{k} g_i e^{-\epsilon_i/kT} \quad \text{and} \quad Q'' = \sum \left(\frac{\epsilon_i}{k}\right)^2 g_i e^{-\epsilon_i/kT}.$$

By means of the equations derived above it should be possible to calculate the heat capacity of a gas at any temperature provided information concerning the partition function is available. The problem is thus reduced to a study of the evaluation of this property of a molecular species.

16d. Separation of Energy Contributions.—The energy values $\epsilon_0, \epsilon_1, \ldots, \epsilon_i, \ldots$, used in the definition of the partition function [equation (16.4)] refer to the total energy of a single molecule, including the translational, rotational and vibrational contributions; allowance must also frequently be made for the electronic energy because the molecules are not necessarily all in a single electronic energy level. This is the case, for example, for a number of monatomic substances, such as atomic oxygen and the halogens, which have multiplet electronic levels at ordinary temperatures, and excited electronic states for which allowance must be made at higher temperatures.

* It should be noted that $Q' = \partial Q/\partial(1/T)$ and $Q'' = \partial^2 Q/\partial(1/T)^2$.

For most common polyatomic molecules, with the exception of oxygen and nitric oxide, the electronic contribution to the partition function is virtually a factor of unity at ordinary temperatures, so that it can be ignored. At high temperatures, however, it becomes important and must be taken into consideration. There is also a nuclear effect on the partition function, but this may be neglected for the present (cf., § 24j).

If the various forms of energy of a molecule may be regarded as completely separable from one another, which is probably justifiable, at least as far as electronic, translational and combined rotational and vibrational energies are concerned, it is possible to write

$$\epsilon = \epsilon_e + \epsilon_t + \epsilon_r + \epsilon_v$$

or

$$e^{-\epsilon/kT} = e^{-\epsilon_e/kT} \times e^{-\epsilon_t/kT} \times e^{-\epsilon_r/kT} \times e^{-\epsilon_v/kT}, \tag{16.11}$$

where ϵ is the total energy per molecule, and ϵ_e, ϵ_t, ϵ_r and ϵ_v represent the separate electronic, translational, rotational and vibrational contributions, respectively. It is evident from equation (16.11) that each exponential term in (16.4), which defines the partition function, may be taken as the product of a number of terms of the same type, one for each kind of energy. Further, the statistical weight factor g is equal also to the product of the separate factors for the various forms of energy. As a result, the complete partition function Q may be divided into a number of factors,* viz.,

$$Q = Q_e \times Q_t \times Q_r \times Q_v, \tag{16.12}$$

where Q_e, Q_t, Q_r and Q_v are called the electronic, translational, rotational and vibrational partition functions, respectively. Each of these is defined by an expression identical in form with equation (16.4), but in which g and ϵ refer to the particular type of energy under consideration. The subdivision of the total partition function into a number of products, each characteristic of one type of energy, greatly simplifies its evaluation.

It will be observed from equations (16.8) and (16.9) that the energy and heat capacity of a gas depend on the *logarithm* of the complete partition function. Since the latter is equal to the *product* of the factors for the several forms of energy, the total energy content and heat capacity of the molecules will be equal to the *sum* of the contributions obtained by inserting the appropriate partition functions in the aforementioned equations. Thus, from equation (16.12),

$$\ln Q = \ln Q_e + \ln Q_t + \cdots,$$

so that

$$\frac{\partial \ln Q}{\partial T} = \frac{\partial \ln Q_e}{\partial T} + \frac{\partial \ln Q_t}{\partial T} + \cdots.$$

* This result follows from the mathematical fact that the product of sums is equal to the sum of the products.

Consequently, by equation (16.8),

$$E - E_0 = RT^2 \left(\frac{\partial \ln Q_e}{\partial T}\right)_V + RT^2 \left(\frac{\partial \ln Q_t}{\partial T}\right)_V + \cdots,$$

and similarly for the heat capacity. This result is the basis of the procedure which will now be adopted for the derivation of expressions for the energy and heat capacity of various types of molecules. It may be mentioned, in anticipation, that the only partition function factor which is dependent on the volume is the translational contribution; hence, in all other cases the partial differential notation, with the constant volume restriction, need not be used.

16e. Translational Partition Function.—Since all molecules, monatomic or polyatomic, possess translational energy, the corresponding contribution to the partition function will be considered first. This is determined by utilizing the expression

$$\epsilon_{t(1)} = \frac{n^2 h^2}{8ml^2}, \tag{16.13}$$

for the component *in one direction* of the translational energy of a molecule, derived by means of wave mechanics; n is a quantum number which may have any integral (or zero) value, m is the mass of the molecule which is confined in a box of length l in the direction parallel to the given energy component, and h is the Planck (quantum theory) constant, i.e., 6.624×10^{-27} erg sec. The statistical weight of each translational level is unity, and so the partition function for translational motion in one direction is given by equation (16.4), after inserting the energy expression from (16.13), as

$$Q_{t(1)} = \sum e^{-\epsilon_{t(1)}/kT} = \sum_{n=0}^{\infty} e^{-n^2 h^2 / 8ml^2 kT}, \tag{16.14}$$

the summation being carried over all values of the quantum numbers from zero to infinity. Since translational levels are very closely spaced,* the variation of energy may be regarded as being virtually continuous, instead of stepwise, as it actually is. The summation in equation (16.14) may thus be replaced by integration, so that

$$Q_{t(1)} = \int_0^\infty e^{-n^2 h^2 / 8ml^2 kT}\, dn$$

$$= \frac{(2\pi mkT)^{1/2}}{h} l. \tag{16.15}$$

---

* This can be shown by inserting the value for m, for any molecular species, in equation (16.13), together with the Planck constant; taking l as 1 cm., for example, the energy separation of successive quantum levels can then be found by setting $n = 1, 2$, etc. The result may thus be compared with $\frac{1}{2}kT$, the average translational energy of a single molecule in one direction. The latter will be found to be of the order of 10^{17} times the former, at ordinary temperatures, showing that the translational quantum levels must be closely spaced.

The contribution of each of the components of the translational energy, in three directions at right angles, is represented by equation (16.15), and the complete translational partition function is obtained by multiplication of the three identical expressions; thus,

$$Q_t = \frac{(2\pi mkT)^{3/2}}{h^3} V, \qquad (16.16)$$

where l^3 has been replaced by V, the volume of the gas.

Problem: Calculate the translational partition function for 1 mole of oxygen at 1 atm. pressure at 25° C, assuming the gas to behave ideally.

Since the pressure is given, it is convenient to replace V in equation (16.16) by RT/P, so that

$$Q_t = \frac{(2\pi mkT)^{3/2}}{h^3} \cdot \frac{RT}{P}.$$

It can be readily shown that Q_t is dimensionless, so that all that is necessary is to see that numerator and denominator are expressed in the same units. In this case, m, k and h are known in c.g.s. units, P is 1 atm., so that R should be in cc.-atm. deg.$^{-1}$ mole^{-1}. The mass m of the oxygen molecule is equal to the molecular weight 32.00 divided by the Avogadro number, i.e., 5.313×10^{-23} g. The temperature T is 298.2°; hence,

$$Q_t = \frac{(2 \times 3.1416 \times 5.313 \times 10^{-23} \times 1.380 \times 10^{-16} \times 298.2)^{3/2}}{(6.624 \times 10^{-27})^3} \cdot \frac{82.06 \times 298.2}{1.000}$$
$$= 4.28 \times 10^{30}.$$

If the expression for the translational partition function is inserted into equation (16.8), it is readily found, since π, m, k, h and V are all constant, that the translational contribution E_t to the energy, in excess of the zero-point value, is equal to $\tfrac{3}{2}RT$ per mole, which is precisely the classical value. The corresponding molar heat capacity at constant volume is thus $\tfrac{3}{2}R$. As stated earlier, therefore, translational energy may be treated as essentially classical in behavior, since the quantum theory leads to the same results as does the classical treatment. Nevertheless, the partition function derived above [equation (16.16)] is of the greatest importance in connection with other thermodynamic properties, as will be seen in Chapter IX.

16f. Electronic States: Monatomic Gases.—Many monatomic substances, as well as a few polyatomic molecules, e.g., oxygen, nitric oxide and nitrogen dioxide, have multiplet electronic ground states. That is to say, in their normal states there are two or more different electronic levels with energies so close together that they may be considered as a single level with a statistical weight factor greater than unity. In addition to this possibility, there may be excited electronic states, whose energy is appreciably greater than that of the normal (ground) states. Such excited states become increasingly occupied as the temperature is raised. In cases of this kind the electronic partition function is greater than unity and varies with temperature; its value must be determined for the calculation of energies and heat capacities.

The statistical weight factor of each electronic level, normal or excited, is equal to $2j + 1$, where j is the so-called "resultant" quantum number of the atom in the given state. The expression for the electronic factor in the partition function is then given by

$$Q_e = \sum (2j + 1)e^{-\epsilon_e/kT}, \qquad (16.17)$$

where ϵ_e is the energy of the electronic state in excess of the lowest state, i.e., the ground state. In the ground state, the energy ϵ_e is zero, so that the exponential factor $e^{-\epsilon_e/kT}$ is unity. The contribution of this state to the electronic partition function is thus $2j + 1$. For helium, neon, etc., and mercury, the value of j in the lowest energy state is found to be zero, from the spectra of these atoms; hence, $2j + 1$ is unity, and since no higher (excited) level need be considered, the electronic partition function factor is also unity, and so can be disregarded. Even if $2j + 1$ were not unity, the effect on the energy and heat capacity would still be zero, for these quantities are dependent on derivatives of the logarithm of the partition function with respect to temperature [cf. equations (16.8) and (16.9)]. Provided the ground state of the atom is the only state which need be considered, Q_e is $2j + 1$, where j refers to the ground state, and hence it is an integer which is independent of temperature; the contribution to E and C_V will consequently be zero.

It appears from spectroscopic studies that helium, neon, argon and mercury atoms are almost exclusively in a single (ground) state and that higher (excited) electronic energy states are not occupied to any appreciable extent, except at very high temperatures. These gases, therefore, behave in the manner required by the classical treatment. The heat capacities of these monatomic substances are thus equal to the translational value, i.e., $\frac{3}{2}R$, at all reasonable temperatures. If accurate observations could be made at high temperatures, however, a small increase would be observed, because of the occupation of higher electronic levels.

For some atoms one or more electronic states above the ground state are appreciably occupied even at moderate temperatures, and hence the appropriate terms must be included in the partition function. For example, in the lowest state of the chlorine atom, i.e., when ϵ_e is zero, the value of j is $\frac{3}{2}$; not very far above this is another state in which j is $\frac{1}{2}$. The electronic partition function for atomic chlorine at ordinary temperatures is therefore given by equation (16.17) as

$$\begin{aligned}Q_e &= (2 \times \tfrac{3}{2} + 1)e^{-0/kT} + (2 \times \tfrac{1}{2} + 1)e^{-\epsilon_1/kT} \\ &= 4 + 2e^{-\epsilon_1/kT},\end{aligned} \qquad (16.18)$$

where ϵ_1 is the electronic energy of the upper level, in excess of the value in the ground state. At higher temperatures other terms would have to be included for electronic states of higher energy, but these may be neglected here. The value of the energy ϵ_1 is found from the spectrum of atomic chlorine; it is derived from the separation of two lines whose frequencies, in wave numbers,* differ by 881 cm.$^{-1}$. According to the quantum theory, the energy ϵ corresponding to a frequency of ν cm.$^{-1}$ is represented by

$$\epsilon = \nu h c, \qquad (16.19)$$

where h is the Planck constant and c is the velocity of light. Upon inserting the values of h, c and k (see Table 1, Appendix), it is found that hc/k is 1.438_5 cm.

---

* The frequency in "wave numbers," or cm.$^{-1}$ units, is equal to the frequency expressed in vibrations per second, i.e., sec.$^{-1}$ units, divided by the velocity of light, 2.9977×10^{10} cm. sec.$^{-1}$.

so that from equations (16.18) and (16.19) the electronic partition function for atomic chlorine is found to be

$$Q_e = 4 + 2e^{-881hc/kT} = 4 + 2e^{-1268/T}. \qquad (16.20)$$

Hence the electronic contribution to the energy content and the heat capacity at any (moderate) temperature can be determined by means of equations (16.8) and (16.9), respectively.

Problem: Calculate the electronic contribution to the heat capacity of atomic chlorine at (i) 300° K, (ii) 500° K.

The calculation is most conveniently made by means of equation (16.10).
(i) At 300° K,
$$Q_e = 4 + 2e^{-1268/300} = 4.029.$$

Q'_e, as defined in connection with equation (16.10), contains one term only, since ϵ/k is zero for the lowest level; hence,

$$Q'_e = -1268 \times 2e^{-1268/300} = -37.02,$$

since ϵ/k is 1268 for the higher level, as seen in equation (16.20). Q''_e also consists of one term; thus,

$$Q''_e = (1268)^2 \times 2e^{-1268/300} = 4.695 \times 10^4.$$

Hence, by equation (16.10),

$$C_e = \frac{R}{(300)^2}\left[\frac{4.695 \times 10^4}{4.029} - \left(\frac{37.02}{4.029}\right)^2\right]$$
$$= 0.128R = 0.254 \text{ cal. deg.}^{-1} \text{ g. atom}^{-1}.$$

(ii) At 500° K,
$$Q_e = 4 + 2e^{-1268/500} = 4.158$$
$$Q'_e = -1268 \times 2e^{-1268/500} = -200.8$$
$$Q''_e = (1268)^2 \times 2e^{-1268/500} = 2.546 \times 10^5$$
$$C_e = \frac{R}{(500)^2}\left[\frac{2.546 \times 10^5}{4.158} - \left(\frac{200.8}{4.158}\right)^2\right]$$
$$= 0.236R = 0.469 \text{ cal. deg.}^{-1} \text{ g. atom}^{-1}.$$

The normal heat capacity of atomic chlorine, due to translational energy only, is $\frac{3}{2}R$, i.e., 2.98 cal. deg.$^{-1}$ g. atom^{-1}, and so the electronic contribution would be appreciable even at 300° K.

In general, an energy level, electronic or otherwise, can contribute to the partition function if the exponential term $e^{-\epsilon/kT}$ has an appreciable magnitude. As a very rough rule, it may be stated that if for any level ϵ/k is greater than about $5T$, then the contribution of that particular level may be neglected at the temperature T. As seen above, ϵ/k is 1268 deg. for the first excited level of atomic chlorine, and so this will affect the energy and heat capacity at temperatures exceeding 1268/5, i.e., about 250° K. The results in the problem given above show that at 300° K the electronic contribution to the molar heat capacity of atomic chlorine is $0.128R$, which is quite appreciable. Since $e^{-\epsilon/kT}$ increases rapidly with increasing temperature, the effect on the energy and heat capacity also becomes much more apparent.

The foregoing calculations show how the classical treatment, which does not take into account the possibility of electronic states, would fail if applied to atomic chlorine at temperatures exceeding about 250° K. The higher the temperature the greater the discrepancy. It was perhaps fortunate that the original prediction, based on the kinetic theory, that the molar heat capacity of a monatomic gas should be $\frac{3}{2}R$ at constant volume, independent of the temperature, was confirmed first by measurements with mercury vapor (A. Kundt and E. Warburg, 1876), and later with sodium vapor and the inert gases of the atmosphere. Spectroscopic measurements indicate that for the atoms of all these elements the energy difference between the lowest and next electronic level is so great that temperatures of several thousand degrees would be required for the second level to make an appreciable contribution to the energy and heat capacity. If, like atomic chlorine and, more particularly, atomic oxygen, the aforementioned substances had possessed low-lying electronic levels, discrepancies would have been observed which would have proved inexplicable until the modern development of the quantum theory.

16g. Diatomic Molecules: Electronic Partition Function.—In general, for diatomic molecules the effect of electronic levels above the ground state can be neglected, since the energy of the next state is usually so high as to be of little practical interest. The energy separation between the lowest and the next (excited) electronic level of the oxygen molecule is probably one of the smallest for any diatomic substance, yet this excited level begins to make a detectable contribution only at temperatures exceeding 2000° K. For other molecules, therefore, still higher temperatures are necessary before electronic states above the ground level have any noticeable effect on the partition function. Consequently, it is evident that, for all ordinary purposes, it is not necessary to consider any state other than the ground state of a diatomic molecule. Nearly all such stable molecules, with the notable exceptions of oxygen and nitric oxide, occur in a single electronic level, i.e., in a singlet state, and so the electronic partition function at all reasonable temperatures is unity. In molecular oxygen, however, there are three very closely spaced levels, i.e., a triplet state, and the electronic partition function is 3.0. Similarly, nitric oxide has two such levels, i.e., a doublet state, and so the electronic factor is virtually 2.0.* In neither case, however, is there any detectable variation with temperature in the vicinity of 300° K, because the energies of the excited electronic levels are so much higher than that of the normal state. Hence, at temperatures below at least 2000° K the electronic contribution to the energy content or heat capacity is negligible for all diatomic molecules. As mentioned earlier, the actual value of the electronic partition function is important in another thermodynamic connection to be discussed in a subsequent chapter.

* In nitric oxide, which is an exception among stable diatomic molecules, each level has a multiplicity of two (Λ-type doubling), so that the electronic partition function is actually 4.0.

16h. Diatomic Molecules: Rotational Partition Function.

—The value of the energy of a molecule in any rotational level can be expressed in terms of an integral (or zero) quantum number J, so that the rotational energy of the Jth level may be represented by ϵ_J. The corresponding statistical weight is $2J + 1$, and the rotational factor in the partition function, apart from nuclear spin effects, is given by

$$Q_r = \sum (2J + 1)e^{-\epsilon_J/kT}. \tag{16.21}$$

The rotational energies ϵ_J can be derived from a study of the spectrum of the molecule, and the value of each $(2J + 1)e^{-\epsilon_J/kT}$ term can be calculated as J takes on a series of integral values. It may seem, at first sight, that it would be necessary to include a very large number of terms in the summation of equation (16.21). However, this is not the case, for as J increases ϵ_J also increases, and hence the exponential factor $e^{-\epsilon_J/kT}$ decreases even more rapidly. The number of terms which contribute appreciably to the summation is therefore not so large as would at first appear.

For a diatomic molecule having two different nuclei, e.g., NO, HCl, OH, ICl, and even HD, where the nuclei are of different isotopes, the rotational quantum number J in equation (16.21) can have the successive values of 0, 1, 2, 3, etc., but for molecules possessing two identical atomic nuclei, e.g., H_2, D_2, O_2 and Cl_2, this is not strictly true. In principle, all molecules of the latter type can exist in "ortho" and "para" states, the statistical weights, $(i + 1)(2i + 1)$ and $i(2i + 1)$, respectively, depending on the spin quantum number i of the nucleus. If this quantum number is zero, the para states have a statistical weight of zero, and hence have no actual existence; such is the case, for example, with the most abundant isotopic form of molecular oxygen, in which both nuclei have a mass number of 16. In one of the states, either ortho or para, depending on various circumstances, the rotational quantum number J can have even values only, i.e., 0, 2, 4, etc., and in the other state it can be odd only, i.e., 1, 3, 5, etc. In order to obtain the complete rotational partition function, it is necessary, therefore, to use the correct nuclear spin statistical weight and the proper rotational quantum numbers for each state when making the summation indicated in equation (16.21). It may be pointed out that a nuclear spin factor should also be included for molecules with dissimilar nuclei, but this has the same value for all rotational levels, viz., $(2i + 1)(2i' + 1)$, where i and i' are the spin quantum numbers of the two nuclei.

For all diatomic molecules, with the exception of hydrogen below 300° K and of deuterium below 200° K, a considerable simplification is possible for temperatures above the very lowest. In the first place, the nuclear spin factor may be ignored for the present (see, however, § 24j), since it is independent of temperature and makes no contribution to the heat capacity. The consequence of the nuclei being identical is then allowed for by introducing a **symmetry number** σ, giving *the number of equivalent spatial orientations that a molecule can occupy as a result of simple rotation.* The value of σ is 2 for symmetrical diatomic molecules, and for unsymmetrical molecules it is unity. The rotational partition function is then given by

$$Q_r = \frac{1}{\sigma} \sum_{J=0,1,2,3,\cdots}^{\infty} (2J+1)e^{-\epsilon_J/kT}, \qquad (16.22)$$

where J can have all integral values from zero to infinity. Since the symmetry number is constant for any given molecule, it does not actually affect the heat capacity, but it is inserted here for the sake of completeness (see, however, § 24k).

At temperatures at which the modification described above is permissible, a further simplification is possible. The rotational energy of a rigid diatomic molecule, i.e., one of fixed length, in any level of quantum number J, is given by the expression

$$\epsilon_J = J(J+1)\frac{h^2}{8\pi^2 I}, \qquad (16.23)$$

where h is Planck's constant, and I is the moment of inertia * of the molecule. In the lowest energy level J is zero, so that ϵ_J is zero; hence ϵ_J as given by equation (16.23) represents the rotational energy, in the Jth level, in excess of the lowest, i.e., zero-point, value. This is the quantity, therefore, which is to be used in the determination of the partition function. Inserting the expression for ϵ_J given by equation (16.23) in (16.22), it is seen that

$$Q_r = \frac{1}{\sigma} \sum_{J=0,1,2,\cdots}^{\infty} (2J+1)e^{-J(J+1)h^2/8\pi^2 IkT}.$$

Provided the moment of inertia is moderately large and the temperature not too low, i.e., for diatomic molecules, other than hydrogen and deuterium, virtually all temperatures at which they are gaseous, the summation may be replaced by integration and, after neglecting certain small quantities, it is found that

$$Q_r = \frac{8\pi^2 IkT}{\sigma h^2}. \qquad (16.24)$$

TABLE VIII. MOMENTS OF INERTIA OF DIATOMIC MOLECULES

Molecule	I g. cm.^{2}	Molecule	I g. cm.^{2}
H_2	0.459×10^{-40}	N_2	13.9×10^{-40}
HD	0.612	CO	14.5
D_2	0.920	NO	16.4
HF	1.34	O_2	19.3
OH	1.48	Cl_2	114.6
HCl	2.66	ICl	245
HBr	3.31	Br_2	345
HI	4.31	I_2	743

* The moment of inertia of a diatomic molecule AB, consisting of two atoms whose actual masses are m_A and m_B, is given by $[m_A m_B/(m_A + m_B)]r^2$, where r is the distance between the centers of the atoms.

The moment of inertia can be derived from spectroscopic data or it can be calculated from the dimensions of the molecule, so that the rotational partition function can be determined. The values of the moments of inertia of a number of diatomic molecules in their ground states are given in Table VIII.[3]

Problem: Calculate the rotational partition function of (i) hydrogen gas, (ii) iodine chloride gas, at 300° K.

(i) From Table VIII, the moment of inertia of molecular hydrogen is 0.459×10^{-40} g. cm.^{2}, and its symmetry number σ is 2; hence utilizing the known values of k and h, it is seen from equation (16.24) that at 300° K,

$$Q_r = \frac{8 \times (3.14)^2 \times 0.459 \times 10^{-40} \times 1.38 \times 10^{-16} \times 300}{2 \times (6.624 \times 10^{-27})^2}$$
$$= 1.71.$$

(ii) For iodine chloride, I is 245×10^{-40} g. cm.^{2}, and since the symmetry number is unity,

$$Q_r = \frac{8 \times (3.14)^2 \times 245 \times 10^{-40} \times 1.38 \times 10^{-16} \times 300}{(6.624 \times 10^{-27})^2}$$
$$= 1820.$$

A consideration of the units will show that Q_r is dimensionless.

For the present purpose, it is not necessary to know the actual partition function, but only its variation with temperature. For a rigid molecule I is constant, and since π, k and h are also constant, it is readily seen from equation (16.24), using equations (16.8) and (16.9), that

$$E_r = RT^2 \frac{d \ln Q_r}{dT} = RT$$

and

$$C_r = \left(\frac{\partial E_r}{\partial T}\right)_V = R.$$

These results are identical with those obtained from the equipartition principle (§ 15d), so that, as a good approximation, classical methods can be used for the evaluation of the rotational energy and heat capacity of diatomic gases, except hydrogen and deuterium, at all temperatures above the very lowest.

16i. Rotational Heat Capacity of Molecular Hydrogen.—The spin quantum number of the hydrogen (H) nucleus is $\frac{1}{2}$, and at ordinary (and higher) temperatures molecular hydrogen, in which the ortho and para forms have attained equilibrium, consists of $(i + 1)(2i + 1)$, i.e., three, parts of the former to $i(2i + 1)$, i.e., one, part of the latter. If the gas is cooled in the absence of a catalyst, the relative amounts of ortho and para molecules remain unchanged. The so-called

[3] Data mainly adapted from G. Herzberg, "Molecular Spectra and Molecular Structure: Diatomic Molecules," 1939.

"normal" hydrogen, which is the gas commonly used in experimental work, may thus be regarded as consisting of a mixture of three parts of orthohydrogen and one part of parahydrogen.* In order to determine the heat capacity of this gas at any temperature it is necessary to evaluate the separate partition functions of the ortho and para forms, and then to calculate the corresponding heat capacities; the rotational heat capacity of "normal" hydrogen is then $\frac{3}{4}C_r$(ortho) + $\frac{1}{4}C_r$(para), for the given temperature. The rotational partition function of orthohydrogen is

$$Q_r(\text{ortho}) = (i+1)(2i+1) \sum_{J=1,3,5,\cdots}^{\infty} (2J+1)e^{-\epsilon_J/kT},$$

since only odd values of the rotational quantum number are associated with the ortho states, and that of the para form is

$$Q_r(\text{para}) = i(2i+1) \sum_{J=0,2,4,\cdots}^{\infty} (2J+1)e^{-\epsilon_J/kT},$$

even values only of the rotational quantum number being permitted; the nuclear spin factor has been included in each case.

It can be seen from equation (16.23) that the energy separation between successive rotational levels is related in an inverse manner to the moment of inertia of the molecule. For all gases, except H_2, D_2 and HD, the rotational energy separations are small enough for a large proportion of the molecules to occupy the higher levels even at low temperatures. The behavior in respect to rotational energy is therefore virtually classical under all reasonable conditions. With hydrogen, however, the moment of inertia is so small (Table VIII), and the separation between successive rotational energy levels so large, that a temperature of about 300° K has to be reached before the gas can be treated in a classical manner. In the vicinity of 50° K virtually all the molecules of hydrogen are in their lowest possible rotational states; the quantum number J is then unity for the ortho and zero for the para molecules. The respective partition functions are now

$$Q_r(\text{ortho}) = (i+1)(2i+1)e^{-h^2/4\pi^2 IkT} \quad \text{and} \quad Q_r(\text{para}) = i(2i+1),$$

using the values of ϵ_J given by equation (16.23). It is evident from the appropriate form of equation (16.9) that C_r(ortho) and C_r(para) will both be zero, and hence the rotational heat capacity of "normal" hydrogen will also be zero at temperatures of 50° K and below. As the temperature is raised additional terms contribute to the rotational partition function, and the heat capacity increases. By utilizing the ϵ_J values derived from the spectrum of molecular hydrogen, the rotational partition functions of the ortho and para forms have been determined at a number of temperatures by summing the appropriate $ge^{-\epsilon/kT}$ terms in each case. From these results the heat capacities of "normal" hydrogen have been calculated, and the values have been found to be in good agreement with measurements made from 80° to 300° K. In fact, the heat capacities obtained from the partition functions, after making allowance for departure from ideal behavior, are probably more accurate than those derived from experiments at low temperatures.

* If ortho-para equilibrium is attained at every temperature, the proportion of the para form will increase as the temperature is lowered from about 300° K. The system is then known as "equilibrium" hydrogen. At 20° K it consists of almost 100 per cent parahydrogen.

At temperatures of about 300° K and above, a sufficient number of rotational energy levels contribute to the partition functions for the hydrogen gas to be treated in a classical manner. The rotational partition function will then be given by equation (16.24), and the corresponding heat capacity will thus be R, as for other diatomic gases. Since deuterium has a higher moment of inertia than ordinary hydrogen, a somewhat lower temperature, namely, about 200° K, is sufficient for virtually classical behavior to be attained.[4]

16j. Diatomic Molecules: Vibrational Partition Function.—The vibrational factor in the partition function may be evaluated by using vibrational energy values ϵ_v derived from spectroscopic measurements. The statistical weight of each vibrational level is unity, and so the partition function is merely equal to the sum of the $e^{-\epsilon_v/kT}$ terms. Except at high temperatures the number of such terms having appreciable magnitudes is not large, and the summation can be made without difficulty, if required. However, for most purposes a simpler procedure is possible. The values of the energy in the various quantum levels of a harmonic oscillator are given by the expression

$$\epsilon_v = (v + \tfrac{1}{2})hc\omega, \tag{16.25}$$

where v is the vibrational quantum number which can be zero or integral; h and c are, as usual, the Planck constant and the velocity of light, respectively, and ω cm.$^{-1}$ is the vibration frequency, in wave numbers, of the given oscillator. In the lowest energy level v is zero, and the vibrational energy, i.e., the zero-point energy, is then given by equation (16.25) as $\tfrac{1}{2}hc\omega$. The vibrational energy of any level referred to the lowest energy state, which is the value required for the partition function, is thus

$$\epsilon_v = (v + \tfrac{1}{2})hc\omega - \tfrac{1}{2}hc\omega = vhc\omega. \tag{16.26}$$

The vibrational partition function is then determined by

$$Q_v = \sum e^{-\epsilon_v/kT} = \sum_{v=0}^{\infty} e^{-vhc\omega/kT}. \tag{16.27}$$

The exponential factor is of the form e^{-vx}, where x is $hc\omega/kT$, so that

$$Q_v = \sum_{v=0}^{\infty} e^{-vx} = 1 + e^{-x} + e^{-2x} + e^{-3x} + \cdots$$
$$= (1 - e^{-x})^{-1}, \tag{16.28}$$

where

$$x = \frac{hc\omega}{kT} = 1.439\frac{\omega}{T}, \tag{16.29}$$

and consequently,

$$Q_v = (1 - e^{-hc\omega/kT})^{-1} = (1 - e^{-1.439\omega/T})^{-1}. \tag{16.30}$$

[4] Dennison, *Proc. Roy. Soc.*, A115, 483 (1927); Giauque, *J. Am. Chem. Soc.*, 52, 4816 (1930); Davis and Johnston, *ibid.*, 56, 1045 (1934); A. Farkas, "Orthohydrogen, Parahydrogen, etc.," 1935.

This expression may be used for the vibrational partition function of a diatomic molecule at all temperatures; the only approximation involved is that the oscillations are supposed to be harmonic in nature. The anharmonicity correction must be made for precision calculations, but its effect is not large. The only property of the molecule required for the evaluation of Q_v by equation (16.30) is the vibration frequency, which can be obtained from a study of its spectrum. The values of this frequency for a number of diatomic molecules are given in Table IX.[5]

TABLE IX. VIBRATION FREQUENCIES OF DIATOMIC MOLECULES

Molecule	ω cm.$^{-1}$	Molecule	ω cm.$^{-1}$
H_2	4405	N_2	2360
HD	3817	CO	2168
D_2	3119	NO	1907
HF	4141	O_2	1580
OH	3728	Cl_2	565
HCl	2989	ICl	384
HBr	2650	Br_2	323
HI	2309	I_2	214

Problem: Calculate the vibrational partition function of (i) molecular hydrogen, (ii) molecular chlorine, at 300° K, assuming them to be harmonic oscillators.

(i) For hydrogen, ω is 4405 cm.$^{-1}$, and since $hc\omega/kT$ is equal to $1.439\omega/T$, it follows that

$$Q_v = (1 - e^{-1.439 \times 4405/300})^{-1}$$
$$= (1 - e^{-21.13})^{-1} = 1.000 \text{ (to several significant figures).}$$

(ii) For chlorine, ω is 565 cm.$^{-1}$, and hence

$$Q_v = (1 - e^{-1.439 \times 565/300})^{-1}$$
$$= (1 - e^{-2.710})^{-1} = 1.072.$$

The vibrational contribution E_v to the energy content, in excess of the zero-level value, is obtained by inserting Q_v as represented by equation (16.30) into (16.8); the result, omitting the partial differential notation which is here unnecessary (§ 16d), is

$$E_v = RT^2 \frac{d \ln Q_v}{dT}$$
$$= \frac{RT}{e^{hc\omega/kT} - 1} \cdot \frac{hc\omega}{kT} = RT \frac{x}{e^x - 1}, \qquad (16.31)$$

where, as defined by (16.29), x is equal to $hc\omega/kT$. Differentiation of equation (16.31) with respect to temperature then gives the vibrational heat capacity; thus,*

$$C_v = R \frac{e^{hc\omega/kT}}{(e^{hc\omega/kT} - 1)^2} \left(\frac{hc\omega}{kT}\right)^2 = R \frac{e^x x^2}{(e^x - 1)^2}. \qquad (16.32)$$

[5] See ref. 3.

* The symbol C_v for the vibrational contribution to the heat capacity should not be confused with C_V which is the *total* heat capacity at constant volume.

The evaluation of the quantities in equations (16.31) and (16.32), referred to as **Einstein functions** (cf. § 17b), is simplified by tables which give E_v and C_v directly from x, i.e., from $hc\omega/kT$.[6]

At moderate temperatures $hc\omega/kT$ is relatively large for many diatomic molecules; the vibrational partition function is then close to unity, and E_v and C_v are very small. This means that virtually all the molecules are in their lowest, i.e., $v = 0$, vibrational level, and the vibrational contribution to the energy and heat capacity will be zero. Such is found to be the case for hydrogen, oxygen, nitrogen and carbon monoxide up to about 370° K, as indicated earlier (§ 15e). As the temperature is raised $hc\omega/kT$ decreases, and the partition function increases accordingly. This means that increasing numbers of molecules now occupy the higher, i.e., $v = 1, 2$, etc., vibrational energy levels. The vibrational contributions to the energy and heat capacity of the gas increase at the same time. At sufficiently high temperatures $hc\omega/kT$ becomes small enough for $e^{hc\omega/kT} - 1$ to be virtually equal to $hc\omega/kT$.* Upon making this substitution in equation (16.31) it is seen that

$$E_v = RT,$$

and hence

$$C_v = R. \qquad (16.33)$$

The energy and heat capacity at high temperatures should consequently be identical, as a first approximation, with the values given by the classical equipartition principle.

The temperature at which $hc\omega/kT$ becomes small enough for the behavior to be classical depends on the vibration frequency ω, which varies from one molecular species to another (Table IX). There is, of course, no exact temperature at which this condition arises, for the approach to classical behavior, i.e., the transition from equation (16.32) to (16.33), must be gradual. However, there is a rough temperature at which the vibrational contributions to the energy and heat capacity become equal to the classical values to a certain degree of accuracy, e.g., three significant figures. For molecular hydrogen, ω is large, viz., 4405 cm.$^{-1}$, and the vibrational heat capacity does not attain the classical value of R per mole until a temperature of about 3500° K. The vibration frequencies of nitrogen, oxygen and carbon monoxide are somewhat less than for hydrogen, but their behavior is nevertheless not completely classical below 3000° K. With molecular chlorine, on the other hand, the situation is changed, for the vibration frequency is only 565 cm.$^{-1}$, and the vibrational contribution to the heat capacity becomes R per mole at about 1000° K; it is, in fact, quite considerable, more than 1 cal. deg.$^{-1}$ mole^{-1}, even at 300° K, thus accounting for the heat capacity of chlorine being greater than for other diatomic molecules (§ 15e).

[6] Sherman and Ewell, *J. Phys. Chem.*, **46**, 641 (1942); see also, Wilson, *Chem. Rev.*, **27**, 17 (1940); Stull and Mayfield, *Ind. Eng. Chem.*, **35**, 639 (1943); H. S. Taylor and S. Glasstone, "Treatise on Physical Chemistry," 3rd ed., 1942, Chapter IV (J. G. Aston), Appendix I.

* In the exponential expansion $e^x = 1 + x + x^2/2! + \cdots$, all terms beyond the second may be neglected when x is small, so that $e^x \approx 1 + x$, and hence $e^x - 1 \approx x$.

Problem: Calculate the vibrational contribution to the molar heat capacity of chlorine gas at 300° K, taking the vibration frequency as 565 cm.$^{-1}$. Estimate the total C_V for chlorine gas at 300° K.

In this case, since hc/k is 1.439 cm. deg.,

$$\frac{hc\omega}{kT} = \frac{1.439 \times 565}{300} = 2.71.$$

Either by insertion of this value in equation (16.32), or by the use of the tables of Einstein functions, it is found that

$$C_v = 0.56R = 1.12 \text{ cal. deg.}^{-1} \text{ mole}^{-1}.$$

At 300° K, the translational and rotational contributions to the heat capacity will be classical, i.e., $\frac{3}{2}R$ and R, respectively, making a total of $\frac{5}{2}R$ or 4.97 cal. deg.$^{-1}$ mole^{-1}. If the vibrational contribution 1.12 is added, the total is 6.09 cal. deg.$^{-1}$ mole^{-1}. (The experimental value which is not very accurate, is close to this result; some difference is to be expected, in any case, because the calculations given above are based on ideal behavior of the gas. The necessary corrections can be made by means of a suitable equation of state, § 21d.)

16k. Diatomic Molecules: Combined Partition Functions.—The treatment of the preceding sections has been based on the assumption that vibrational and rotational energies are independent; in other words, the approximation has been made of taking the molecules to be rigid, in spite of the fact that the atoms vibrate. In addition, the vibration has been treated as perfectly harmonic in nature. For the accurate evaluation of partition functions, especially at moderate and high temperatures, it is necessary to take into account the interaction of the vibrational and rotational motions of the molecule, and to allow for the anharmonicity of the atomic oscillations. It is sufficient to state here that the information required for this purpose can be obtained from spectroscopic measurements, although the treatment of the data is not simple. Nevertheless, the calculations have been carried out in a number of cases and the results are recorded in the literature.

16l. Polyatomic Molecules.—The general principles involved in the evaluation of the partition function, and hence the energy and heat capacity, of a molecule containing more than two atoms are quite similar to those described for diatomic molecules. Unless there is definite evidence to the contrary, as there is for nitrogen dioxide, which has an odd number of electrons, it is supposed that the ground state of the molecule consists of a single electronic level, and that excited states make no contribution to the total partition function. The electronic contribution to the energy in excess of the lowest level, and to the heat capacity may thus be taken as zero. Further, as with monatomic and diatomic molecules, the translational energy may be treated as classical at all feasible temperatures, so that the translational motion contributes $\frac{3}{2}RT$ per mole to the energy content and $\frac{3}{2}R$ per mole to the heat capacity.

At all reasonable temperatures the rotational levels of a molecule containing more than two atoms, like those of diatomic molecules, are occupied sufficiently for the behavior to be virtually classical in character. Assuming that the molecule can be represented as a rigid rotator, the rotational partition function, excluding the nuclear spin factor, for a *nonlinear* molecule is given by

$$Q_r = \frac{8\pi^2(8\pi^3 ABC)^{1/2}(kT)^{3/2}}{\sigma h^3}, \quad (16.34)$$

where A, B and C are the moments of inertia of the molecule with respect to three perpendicular axes, and σ is the symmetry number (§ 16h). For some molecules, e.g., NH_3, PCl_3, $CHCl_3$ and CH_3Cl, two of the three moments of inertia, e.g., A and B, are equal, so that the product of inertia ABC becomes A^2C. For spherically symmetrical molecules, such as CH_4 and CCl_4, A, B and C are equal, so that the product is A^3. If the nonlinear molecule is planar, e.g., benzene and water, the sum of two of the moments of inertia is equal to the third, i.e., $A + B = C$. The symmetry numbers of some polyatomic molecules are as follows: CO_2, H_2O, $SO_2(2)$; $NH_3(3)$; CH_4, $C_6H_6(12)$. For HCN, N_2O (i.e., NNO), COS, etc., σ is unity.

A *linear* molecule containing more than two atoms is analogous to a diatomic molecule. It has two identical moments of inertia, and the rotational partition function is given by the same equation (16.24) as for a diatomic molecule.

The moments of inertia of a molecule can be derived from spectroscopic data, or they may be calculated from the interatomic distances obtained by electron diffraction methods. The values for a number of simple molecules are given in Table X.[7]

TABLE X. MOMENTS OF INERTIA OF POLYATOMIC MOLECULES

Molecule	Moment of Inertia	Molecule	Moments of Inertia
HCN*	18.9×10^{-40} g. cm.^{2}	H_2O	1.02, 1.92, 2.94 $\times 10^{-40}$
N_2O*	66.9	H_2S	2.68, 3.08, 5.76
CO_2*	71.9	SO_2	12.3, 73.2, 85.5
C_2H_2*	23.8	NH_3	2.78, 2.78, 4.33
CH_4†	5.27	CH_3Cl	5.46, 61.4, 61.4
CCl_4†	520	CH_3Br	5.36, 85.3, 85.3

* Linear molecules.
† Spherically symmetrical molecules.

By differentiation with respect to temperature of Q_r, as given by equation (16.34), remembering that all the quantities except T are constant, it is readily found that E_r is $\frac{3}{2}RT$ and C_r is $\frac{3}{2}R$. These are the results to be expected from the equipartition principle for energy expressible in three square terms, as would be the case for a nonlinear molecule containing more than

---

[7] Data mainly adapted from Landolt-Börnstein, Physikalisch-Chemische Tabellen, 5th ed., 3rd Supplement; G. Herzberg. "Infra-Red and Raman Spectra of Polyatomic Molecules," 1945.

two atoms. Just as with diatomic molecules, excepting hydrogen and deuterium, the rotational contributions to the energy and heat capacity may be taken as the classical values at all temperatures.

As seen in § 15d, a molecule containing n atoms has $3n - 6$ modes of vibration if nonlinear, and $3n - 5$ modes if linear. Each of these modes of vibration contributes a factor of exactly the same form as equation (16.30) to the over-all vibrational partition function; the latter is thus the product of $3n - 6$ (or $3n - 5$) factors, each of the form $(1 - e^{-hc\omega/kT})^{-1}$. The values of the $3n - 6$ (or $3n - 5$) vibration frequencies are obtained from a study of the various spectra of the molecule or by a comparison with the results for related compounds. The vibration frequencies of some familiar molecules containing more than two atoms are recorded in Table XI. On account of molecular symmetry, two or more vibrations may have the same frequency; this is indicated by the numbers in parentheses in Table XI.[8]

TABLE XI. VIBRATION FREQUENCIES OF POLYATOMIC MOLECULES

Molecule	Frequencies (cm.$^{-1}$)	Molecule	Frequencies (cm.$^{-1}$)
HCN	729(2), 2001, 3451	H_2O	1654, 3825, 3935
N_2O	596(2), 1300, 2276	H_2S	1290, 2611, 2684
CO_2	667(2), 1340, 2349	SO_2	519, 1151, 1361
C_2H_2	612(2), 729(2), 1974, 3287, 3374	NH_3	950, 1627(2), 3334, 3414(2)
CH_4	1358(3), 1390(2), 3030, 3157(3)	CH_3Cl	732, 1020(2), 1355, 1460(2), 2900, 3047(2)
CCl_4	218(2), 314(3), 461, 776(3)	CH_3Br	610, 957(2), 1305, 1450(2), 2900, 3061(2)

For the determination of the vibrational contribution to the energy content or heat capacity it is not necessary actually to evaluate the product of the $3n - 6$ (or $3n - 5$) factors in the partition function. The expressions for the energy and heat content [equations (16.8) and (16.9)] involve the logarithm of the partition function, which is equal to the *sum* of the logarithms of the component factors. Thus, it is usually simpler to determine the contributions of each of the $3n - 6$ (or $3n - 5$) modes of vibration separately by means of equations (16.31) and (16.32), and then to add the results to obtain the total value for the molecule.

Problem: Calculate the heat capacity C_V of water vapor at (i) 500° K, (ii) 1000° K, assuming ideal behavior.

The translational and rotational contributions may be taken as classical, each being $\frac{3}{2}R$, so that the total is $3R$ per mole, i.e., 5.96 cal. deg.$^{-1}$ mole^{-1}, at each temperature.

(i) At 500° K, the values of $hc\omega/kT$ corresponding to the frequencies of (a) 1654, (b) 3825, (c) 3935 cm.$^{-1}$ are 4.77, 11.0 and 11.35, respectively. Setting these values of x into equation (16.32), the results for C_v are

(a) 0.392,   (b) 0.004,   and   (c) 0.003 cal. deg.$^{-1}$ mole^{-1},

[8] See ref. 7.

the total vibrational contribution being the sum of these quantities, viz., 0.399 ≈ 0.40 cal. The total heat capacity at constant volume is thus $5.96 + 0.40 = 6.36$ cal. deg.$^{-1}$ mole^{-1}.

(ii) At 1000° K, the values of $hc\omega/kT$ are 2.38, 5.50 and 5.67 respectively, so that the contributions to C_v are

(a) 1.265,   (b) 0.247,   and   (c) 0.225 cal. deg.$^{-1}$ mole^{-1},

the sum being 1.74 cal. The total heat capacity of water vapor at constant volume at 1000° K should thus be $5.96 + 1.74 = 7.70$ cal. deg.$^{-1}$ mole^{-1}.

The increase of the heat capacity of water vapor with temperature is in agreement with experiment.

The procedure described above, as for diatomic molecules, is based on the approximation that rotational and vibrational energies are separable, and that the oscillations are simple harmonic in character. The allowance for interaction, etc., has been made in a number of cases by utilizing actual energy levels derived from spectroscopic measurements. The results are, however, not greatly different from those obtained by the approximate method that has been given here.

16m. Internal Rotation.—There are certain molecules for which the foregoing procedure for calculating partition functions, and related properties, is not completely satisfactory; these are molecules containing groups which are apparently capable of free rotation about a single bond. One of the simplest examples is ethane, in which the two —CH$_3$ groups might be expected to rotate with respect to each other. Most aliphatic hydrocarbons and alkyl derivatives of benzene, water, hydrogen sulfide, ammonia and formaldehyde fall into the same category. For every type of internal rotation of a group within the molecule, the molecule as a whole possesses one less vibrational mode; the total number of internal rotations and vibrations is thus $3n - 6$ for a nonlinear molecule. If the internal rotation is completely unrestricted and classical in behavior, the energy can be represented by one quadratic term, so that the corresponding contributions to the energy content and heat capacity are $\frac{1}{2}RT$ and $\frac{1}{2}R$ per mole, respectively. Provided the vibration frequencies of the molecule are available from spectroscopic measurements, the evaluation of the energy content and the heat capacity would not be a difficult matter.

A comparison of thermodynamic properties obtained experimentally with those derived from the partition functions has revealed the fact that in many molecules, such as those mentioned above, in which free internal rotation might have been expected, the rotation is actually restricted. The contributions to the energy content, etc., are then appreciably different from those calculated on the assumption of free internal rotation. The results indicate that before one group can rotate freely past another, as in ethane, the molecules must acquire a certain amount of the appropriate energy. If this amount were very small, the internal rotation would be virtually free and unrestricted, but such appears not to be the case. Various lines of evidence show that the necessary energy is appreciable, and this results in a restriction of the rotation. If the magnitude of the required energy were known it would be possible to calculate, with a fair degree of accuracy, the contribution of the restricted internal rotation to the partition function and the properties derived from it. However, no satisfactory independent method is yet available, and the procedure adopted is to estimate the value of the energy by utilizing an accurate experimental thermodynamic quantity. Once this restricting energy is known, other properties can be calculated by making appropriate allowances. The details of the treatment are not simple, and as they lie outside the scope of this book it is not necessary to consider them further.[9]

17. Heat Capacity of Solids

17a. Atomic Heat Capacity: Classical Theory.—An ideal elementary solid may be regarded as consisting of a space lattice of independent atoms vibrating about their respective equilibrium positions, but not interacting with each other in any way. If the vibrations are simple harmonic in character, the energy can be expressed as the sum of two quadratic terms; hence, according to the principle of the equipartition of energy, the contribution to the energy content will be RT per g. atom for each vibrational mode. Since the atoms would be free to oscillate in all directions in space, each atom may be supposed to have three independent modes of vibration. The energy content E of the ideal solid element should thus be $3RT$ per g. atom, and hence the heat capacity at constant volume should be $3R$, i.e., 5.96 cal. deg.$^{-1}$ g. atom^{-1}. This theoretical conclusion is in general agreement with the familiar empirical rule of Dulong and Petit, which states that the atomic heat capacities of most solid elements, with the exception of carbon, boron, beryllium and silicon, measured at atmospheric pressure is about 6.4 cal. deg.$^{-1}$ at ordinary temperatures. The theoretical atomic heat capacity of $3R$ should apply to constant volume conditions, and by correcting the observed, i.e., constant pressure, measurements by means of equation (21.9), G. N. Lewis (1907) obtained the results given in Table XII.[10] It is

TABLE XII. HEAT CAPACITIES OF ELEMENTS AT CONSTANT VOLUME IN CAL. DEG.$^{-1}$ G. ATOM^{-1}

Element	C_V	Element	C_V	Element	C_V
Aluminum	5.7	Iodine	6.0	Platinum	5.9
Antimony	5.9	Iron	5.9	Silver	5.8
Cadmium	5.9	Lead	5.9	Thallium	6.1
Copper	5.6	Nickel	5.9	Tin	6.1
Gold	5.9	Palladium	5.9	Zinc	5.6

seen that at ordinary temperatures (about 20° C) the values of C_V are approximately constant, the mean being 5.9, with a variation of ± 0.2, cal. deg.$^{-1}$ g. atom^{-1}.

[9] For empirical and semiempirical rules for calculating the heat capacities of hydrocarbons and other organic compounds, see Bennewitz and Rossner, *Z. phys. Chem.*, **B39**, 126 (1938); Fugassi and Rudy, *Ind. Eng. Chem.*, **30**, 1029 (1938); Edmister, *ibid.*, **30**, 352 (1938); Dobratz, *ibid.*, **33**, 759 (1941); Glockler and Edgell, *ibid.*, **34**, 582 (1942); Stull and Mayfield, *ibid.*, **35**, 639, 1303 (1943); Pitzer, *J. Am. Chem. Soc.*, **63**, 2413 (1941); Spencer, *ibid.*, **67**, 1859 (1945).

[10] Lewis, *J. Am. Chem. Soc.*, **29**, 1165, 1516 (1907); Lewis and Gibson, *ibid.*, **39**, 2554 (1917).

In spite of the apparent agreement between the experimental data and the theoretical prediction based on the equipartition principle, there are nevertheless significant discrepancies. In the first place, the heat capacity of carbon, e.g., diamond, is only 1.45 cal. deg.$^{-1}$ g. atom^{-1} at 293°K, and it increases with increasing temperature, attaining a value of 5.14 cal. deg.$^{-1}$ g. atom^{-1} at 1080° K. Somewhat similar results have been obtained with boron, beryllium and silicon. Further, although the atomic heat capacities of most solid elements are about 6 cal. deg.$^{-1}$ g. atom^{-1} at ordinary temperatures, and do not increase markedly as the temperature is raised, a striking decrease is always observed at sufficiently low temperatures. In fact, it appears that the heat capacities of all solids approach zero at 0° K. Such a variation of the heat capacity of a solid with temperature is not compatible with the simple equipartition principle, and so other interpretations have been proposed.[11]

17b. The Einstein Heat Capacity Equation.—The first step in the improvement of the theory of the heat capacity of solid elements was made by A. Einstein (1907), who applied the quantum theory in place of the classical equipartition principle to calculate the energy of the atomic oscillators. The expression obtained for the atomic heat capacity of a solid at constant volume may be written in the form

$$C_V = 3R \frac{e^{hc\omega/kT}}{(e^{hc\omega/kT} - 1)^2} \left(\frac{hc\omega}{kT}\right)^2, \qquad (17.1)$$

where ω is the oscillation frequency, in wave numbers, of the atoms in the crystal lattice. Comparison of this result with the contribution made by a single oscillator to the heat capacity, as given by equation (16.32), shows that the former differs from the latter by a factor of three, as is to be expected.

According to equation (17.1), the atomic heat capacity of a solid element should approach zero at very low temperatures, but at high temperatures, when $hc\omega/kT$ is small in comparison with unity, the expression reduces to $3R$, in agreement with the result of the classical treatment. This is in general accord with the experimental behavior. The physical significance of the variation of the heat capacity of a solid with temperature is then similar to that given in connection with the vibrational contribution to the heat capacity of a gas (§ 16j). At very low temperatures, all the atoms in the solid are in the lowest vibrational level, and they then contribute nothing to the heat capacity. With increasing temperature the energy of the crystal increases and the higher levels are increasingly occupied; hence the heat capacity becomes appreciable. At sufficiently high temperatures a considerable number of atoms possess fairly large numbers of quanta of vibrational energy, and the behavior can then be expressed with reasonable accuracy by the classical treatment. In spite of the fact that the Einstein equation represents qualitatively, at least, the variation of heat capacities with temperature, it does not completely solve the problem of the heat

---

[11] See the works mentioned in ref. 1.

capacities of solid elements, for at low temperatures the calculated values fall off more rapidly than do the experimental heat capacities.

17c. The Debye Heat Capacity Equation.—There is little doubt that in its essentials the Einstein theory of the heat capacity of solid elements is correct, but it requires some modification in detail. Owing to the proximity of the atoms in a crystal, it is very improbable that they will act as independent units oscillating with a uniform frequency. As a result of interactions, the atoms will execute complex motions, but these may be regarded as being made up of a series of simple harmonic vibrations with various frequencies. A solid containing N atoms will thus behave as a system of $3N$ coupled oscillators, and there will be a total of $3N$ independent frequencies. The lowest frequency will be zero, but there is a definite limit to the highest frequency; this maximum, designated by ω_m, arises when the wave length of the oscillations is of the same order as the interatomic distances. In order to determine the distribution of frequencies, P. Debye (1912) disregarded the atomic structure of the solid, and treated it as a homogeneous, elastic medium. The vibrations of the atoms could then be considered as equivalent to elastic waves, similar to sound waves, propagated through this continuous medium. In this way an expression for C_V containing a complicated integral was obtained; the integral can, however, be evaluated in two special cases, (a) moderate and high temperatures, and (b) very low temperatures. In the former case, the atomic heat capacity equation becomes

$$C_V = 3R\left[1 - \frac{1}{20}\left(\frac{\theta}{T}\right)^2 + \frac{1}{560}\left(\frac{\theta}{T}\right)^4 - \cdots\right], \qquad (17.2)$$

where θ, known as the **characteristic temperature** of the element, is defined by

$$\theta = \frac{hc\omega_m}{k} = 1.439\omega_m \text{ deg.} \qquad (17.3)^*$$

At sufficiently high temperatures θ/T becomes small enough for all the terms in the brackets of equation (17.2), other than the first, to be neglected; C_V then becomes equal to the classical value $3R$.

It will be seen from the Debye equation (17.2) that C_V is a function of θ/T only, and hence the plot of C_V against T/θ (or log T/θ) should yield a curve that is the same for all solid elements.† The nature of the curve is shown in Fig. 9, and it is an experimental fact that the heat capacities of many elements, and even of a few simple compounds, e.g., ionic crystals such

---

* Since $hc\omega_m$ has the dimensions of energy and k is energy per degree, θ has the dimensions of a temperature.

† The same is true for the Einstein equation (17.1) which may be written as

$$C_V = 3R\frac{e^{\theta_E/T}}{(e^{\theta_E/T} - 1)^2}\left(\frac{\theta_E}{T}\right)^2.$$

The Einstein characteristic temperature θ_E, equal to $hc\omega/k$, where ω is the *mean* frequency, is smaller than the Debye θ which involves the *maximum* frequency.

as sodium and potassium chlorides, which crystallize in the cubic system, have been found to fall on or very close to this universal curve. The slight discrepancies that have been observed are probably to be attributed to the approximation of treating a crystalline solid as a continuous medium in the determination of the distribution of the vibration frequencies.

It is evident from Fig. 9 that the heat capacity of an element attains its classical value of $3R$ when T/θ is approximately unity (see problem given below). If the characteristic temperature of an element is relatively small, e.g., less than about 300, the value of T/θ will be equal to unity at temperatures below 300° K. For such elements the law of Dulong and Petit will evidently hold at ordinary temperatures. This is the case for the majority of the solid elements.

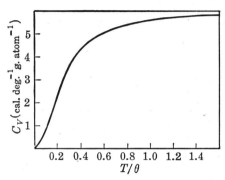

FIG. 9. Debye atomic heat capacity curve

On the other hand, if θ is large, as it is for carbon and some other light elements, a temperature of 300° K will coincide with the rising portion of the Debye curve. The atomic heat capacity will thus be well below $3R$, and it will increase rapidly with temperature, the limiting value being reached only at high temperatures. By giving θ a value of 1860, the variation of the heat capacity of diamond with temperature can be expressed with considerable accuracy by the Debye heat capacity equation. According to equation (17.3) the characteristic temperature θ will be large if the frequency ω_m is large; this frequency is a rough measure of the binding energy between the atoms in the crystal. The internal structures and high melting points of carbon, boron and silicon are compatible with exceptionally high values of the binding energy, and hence with the large values of the characteristic temperature of these elements.

Problem: Calculate the value of C_V for any element when its temperature is equal to the Debye characteristic temperature θ.

For such a temperature θ/T is unity, and hence by equation (17.2)

$$C_V = 3R \left[1 - \frac{1}{20} + \frac{1}{560} - \cdots \right] = 2.856R$$
$$= 5.68 \text{ cal. deg.}^{-1} \text{ g. atom}^{-1}.$$

Hence, when the temperature is equal to the characteristic temperature, i.e., when T/θ is unity, the heat capacity is just less than the classical value $3R$.

If the characteristic temperature of any solid element is known, the complete variation of the heat capacity at all moderate and reasonably high temperatures can be obtained from the Debye equation (17.2) or the equivalent curve (Fig. 9). The value of θ can be derived from a single experimental determination of the heat capacity, preferably made somewhere just above the middle of the rising part of the curve, that is, when C_V is about 3 cal. deg.$^{-1}$ g. atom^{-1}. The measurement should not be made in the region where C_V approaches $3R$, because of the flatness of the curve, nor at too low temperatures, for then equation (17.2) does not hold. Some values of the characteristic temperatures of a number of elements are quoted in Table XIII.[12]

TABLE XIII. DEBYE CHARACTERISTIC TEMPERATURES OF ELEMENTS

Element	θ	Element	θ	Element	θ
Aluminum	390	Copper	315	Nickel	375
Antimony	140	Gold	170	Potassium	100
Bismuth	104	Iron	450	Platinum	225
Calcium	225	Lead	90	Sodium	150
Carbon (Diamond)	1860	Magnesium	290	Silver	212
Cobalt	385	Manganese	350	Zinc	210

Problem: Calculate the heat capacity of diamond at 1080° K.

From Table XIII, the characteristic temperature θ is 1860; hence, by equation (17.2),

$$C_V = 3R \left[1 - \frac{1}{20}\left(\frac{1860}{1080}\right)^2 + \frac{1}{560}\left(\frac{1860}{1080}\right)^4 - \cdots \right]$$
$$= 3R \times 0.8673 = 5.17 \text{ cal. deg.}^{-1} \text{ g. atom}^{-1}.$$

(The experimental value, given earlier, is 5.14 cal. deg.$^{-1}$ g. atom^{-1}.)

It may be mentioned that the Debye characteristic temperature can be derived from other properties of the element, particularly from the compressibility and Poisson's (elasticity) ratio. Where such data are available it is thus possible to obtain reasonably accurate heat capacities, at moderate and high temperatures, from elasticity measurements.

17d. The Debye Equation at Low Temperatures.—At low temperatures the solution of the complete Debye equation leads to the result

$$C_V = 3R \left[\tfrac{4}{5}\pi^4 \left(\frac{T}{\theta}\right)^3 - \cdots \right]$$
$$= 464.4 \left(\frac{T}{\theta}\right)^3 - \cdots \text{ cal. deg.}^{-1} \text{ g. atom}^{-1}. \qquad (17.4)$$

The important conclusions, therefore, to be drawn from the Debye theory are that at low temperatures the atomic heat capacity of an element should be proportional to T^3, and that it should become zero at the absolute zero of temperature. In order for equation (17.4) to hold, it is necessary that the temperature should be less than about $\theta/10$; this means that for most

[12] Data adapted from various sources, e.g., R. H. Fowler and E. A. Guggenheim, "Statistical Thermodynamics," 1939; F. Seitz, "The Modern Theory of Solids," 1940; J. C. Slater, "Introduction to Chemical Physics," 1939.

substances the temperature must be below 30° K. Heat capacity measurements made at sufficiently low temperatures have served to confirm the reliability of the T^3 relationship for a number of elements, and even for some simple compounds. The proportionality of heat capacity to the third power of the absolute temperature has been found to be of great value for the purpose of extrapolating heat capacities to the absolute zero. Such extrapolations are necessary in connection with the determination of an important thermodynamic property (Chapter IX).

Since C_V is equal to 464.4 $(T/\theta)^3$ at very low temperatures, one heat capacity value under these conditions can be used to derive the characteristic temperature. With this known, the variation of C_V with temperature at higher temperatures can be obtained from equation (17.2). Alternatively, if θ is found, as described in § 17c, from heat capacity measurements at moderate temperatures, the values at low temperatures, i.e., less than $\theta/10$, can be estimated from equation (17.4). However, where the characteristic temperature θ has been determined by two methods, that is, from low temperature and high temperature measurements, the agreement is not exact, showing, as is to be expected, that the Debye theory is not perfect.

Problem: The atomic heat capacity of copper is 0.1155 cal. deg.$^{-1}$ at 20.20° K; calculate the value at 223° K.

From equation (17.4)

$$0.1155 = 464.4 \left(\frac{20.2}{\theta}\right)^3$$

$$\theta = 321.$$

Upon inserting this value into equation (17.2) it is found that at 223° K,

$$C_V = 3R\left[1 - \frac{1}{20}\left(\frac{321}{223}\right)^2 + \frac{1}{560}\left(\frac{321}{223}\right)^4 - \cdots\right]$$

$$= 3R \times 0.904 = 5.39 \text{ cal. deg.}^{-1} \text{ g. atom}^{-1}.$$

(The experimental result is about 5.5.)

17e. Heat Capacities at High Temperatures.—Although the theoretical treatment of heat capacities requires the limiting high temperature value to be $3R$, i.e., 5.96 cal. deg.$^{-1}$ g. atom^{-1}, experimental determinations have shown that with increasing temperature C_V increases still further. The increase is, however, gradual; for example, the heat capacity of silver is 5.85 cal. deg.$^{-1}$ g. atom^{-1} at 300° K and about 6.5 cal. deg.$^{-1}$ g. atom^{-1} at 1300° K. This increase is attributed mainly to the relatively free electrons of the metal behaving as an "electron gas." By the use of the special form of quantum statistics, viz., Fermi-Dirac statistics, applicable to electrons, the relationship

$$C_{\text{el.}} = aT,$$

where a is a constant for each element, has been derived for the contribution of the electron gas to the heat capacity. For most metals a lies between 10^{-4} and 2.5×10^{-4} in calorie units, and so the electron heat capacity is 0.03 to 0.075 at 300° K and 0.13 to 0.32 cal. deg.$^{-1}$ g. atom^{-1} at 1300° K. The influence of the free electrons is thus not large, although it becomes appreciable as the temperature is raised.*

17f. Heat Capacities of Compounds.—As indicated above, the Debye equation represents the variation with temperature of the heat capacities, at constant volume, of a number of simple compounds. In these cases equation (17.2) gives the heat capacity *per gram atom*, so that it must be multiplied by the number of atoms in the molecule to obtain the molar heat capacity. The compounds are, in general, substances crystallizing in the cubic system, to which the Debye treatment is particularly applicable. Among these, mention may be made of sodium and potassium chloride, potassium bromide, calcium fluoride and magnesium oxide. For certain other compounds, especially metallic halides, and also for some elements which do not form cubic crystals, e.g., rhombic sulfur, graphite and iodine, the heat capacity per g. atom is given by a Debye function, as in equation (17.2), but with $(T/\theta)^n$ in place of T/θ. The value of n is usually less than unity, and it must be determined empirically; thus the heat capacities must be known at two temperatures in order that θ and n may be obtained. Once these are available, the variation of heat capacity over a range of temperatures can be represented.[13]

If the Debye function in some form were applicable to all compounds it would mean that at sufficiently high temperatures the molar heat capacity at constant volume would be equal to $3R \times n$, where n is the number of atoms in the molecule; thus C_V should be approximately $6n$ cal. deg.$^{-1}$ mole^{-1}. This conclusion is analogous to the rule proposed by H. Kopp (1865), who suggested that the molar heat capacity of a compound is approximately equal to the sum of the atomic heat capacities of its constituent elements. For most elements, particularly those of higher atomic weight, the heat capacity at ordinary temperatures may be taken as about 6 cal. deg.$^{-1}$, but for the lighter elements Kopp suggested somewhat smaller values, as follows: carbon (1.8), hydrogen (2.3), boron (2.7), silicon (3.8), oxygen (4.0), fluorine (5.0), phosphorus (5.4), and sulfur (5.4). The results given by Kopp's rule are very approximate, but they may be useful when experimental data are not available.

EXERCISES

1. Assuming classical behavior, and no internal rotation, what would be the maximum value of C_P for (i) a linear molecule, (ii) a nonlinear molecule, containing n atoms?

2. Show that the partition function for 1 mole must be a dimensionless quantity. Verify by reference to the expressions for the translational and rotational (diatomic) partition functions.

3. Show that the translational partition function for 1 mole of an ideal gas is given by $Q_t = 1.879 \times 10^{20} M^{3/2} T^{3/2} V$, where M is the molecular (or atomic) weight and V is the molar (or atomic) volume in cc.

* The effect described here is quite distinct from that due to the occupation of the higher electronic levels (§ 16f); the contribution of the electron gas will apply even if all the atoms are in their lowest electronic states.

[13] Lewis and Gibson, *J. Am. Chem. Soc.*, **39**, 2554 (1917).

EXERCISES

4. Calculate the translational partition function for 1 mole of oxygen at 1 atm. pressure and 25° C, assuming ideal behavior.

5. The normal state of atomic oxygen is an "inverted triplet" consisting of three levels with j values of 2, 1 and 0. The frequency separation between the $j = 2$ (lowest) and the $j = 1$ (second) levels is 157.4 cm.$^{-1}$ and that between the $j = 2$ and the $j = 0$ (third) levels is 226 cm.$^{-1}$. Calculate the electronic partition function of atomic oxygen at 300° K and the corresponding contribution to the atomic heat capacity.

6. The frequency separation of the first excited electronic level of atomic oxygen from the lowest level is 15807 cm.$^{-1}$; at what temperature might the excited level be expected to affect the heat capacity?

7. Derive the value of the universal constant a in the expression $Q_r = a\sigma^{-1}IT$ for the rotational partition function of any diatomic (or any linear) molecule; I is the moment of inertia in c.g.s. units and T is the absolute temperature. Calculate the rotational partition function of carbon dioxide (a linear symmetrical molecule) at 25° C.

8. Evaluate the universal constant b in the expression $Q_r = b\sigma^{-1}(ABC)^{1/2}T^{3/2}$ for the rotational partition of a nonlinear polyatomic molecule; A, B and C are the three moments of inertia of the molecule in c.g.s. units. Determine the rotational partition function of the water molecule at 25° C.

9. Calculate the value of $e^{-\epsilon/kT}$ when ϵ/k is equal to $5T$. Justify the statement in the text that if for any energy level $\epsilon/k > 5T$, the contribution of that level to the partition function, and consequently to the heat capacity, is negligible.

10. Determine the vibrational partition function of carbon dioxide at 25° C.

11. Calculate the molar heat capacity at constant pressure of carbon dioxide at (i) 300° K, (ii) 500° K, (iii) 1000° K, assuming ideal behavior.

12. Calculate the molar heat capacity at constant pressure of methane at (i) 300° K, (ii) 500° K, (iii) 1000° K, assuming ideal behavior.

13. Plot the Debye curve, according to equation (17.2), for C_V against T/θ, using values of 0.2, 0.4, 0.6, $\cdots$, 1.2 for the latter.

14. Using the Debye curve obtained in the preceding exercise, and the values of C_V for aluminum given below, determine the (mean) Debye characteristic temperature for this element.

T	84.0°	112.4°	141.0° K
C_V	2.45	3.50	4.18 cal. deg.$^{-1}$ g. atom^{-1}.

Calculate the value of C_V at 25° C.

15. Plot the following values of C_V for aluminum, obtained at low temperatures, against T^3, and hence determine the (mean) Debye temperature; compare the result with that obtained in Exercise 14.

T	19.1°	23.6°	27.2°	32.4°	35.1° K
C_V	0.066	0.110	0.162	0.301	0.330 cal. deg.$^{-1}$ g. atom^{-1}.

16. The Debye characteristic temperature of silver is 212. Calculate the atomic heat capacity C_V of this metal at 20.0° K and 300° K.

17. For the heat capacities of gaseous paraffins and olefins, with more than three carbon atoms, Edmister (see ref. 9) suggested the empirical formula

$$C_P = 2.56 + 0.51n + 10^{-3}T(1.3n^2 + 4.4n - 0.65mn$$
$$+ 4.95m - 5.7) \text{ cal. deg.}^{-1} \text{ mole}^{-1},$$

where n is the number of carbon atoms and m the number of hydrogen atoms in the hydrocarbon molecule; T is the absolute temperature.

For paraffins, Pitzer (see ref. 9) proposed the formula

$$C_P = 5.65n - 0.62 + 10^{-2}t(1.11n + 1.58) \text{ cal. deg.}^{-1} \text{ mole}^{-1},$$

where n is the number of carbon atoms; t is the centigrade temperature. Compare the molar heat capacities of n-butane gas (C_4H_{10}) at 500° K, as given by the two expressions. According to Spencer [*J. Am. Chem. Soc.*, **67**, 1859 (1945)], the heat capacity of this gas is given by

$$C_P = -0.012 + 92.506 \times 10^{-3}T - 47.998 \times 10^{-6}T^2 + 9.706 \times 10^{-9}T^3 \text{ cal. deg.}^{-1} \text{ mole}^{-1}.$$

Calculate the value at 500° K.

CHAPTER VII

THE SECOND LAW OF THERMODYNAMICS

18. Conversion of Heat Into Work

18a. Scope of First and Second Laws.—To the chemist the essential interest of the **second law of thermodynamics** lies in the fact that it provides a means of predicting whether a particular reaction can occur under specified conditions. The first law of thermodynamics merely indicates that in any process there is an exact equivalence between the various forms of energy involved, but it provides no information concerning the feasibility of the process. In general, however, the second law supplies an answer to the question of whether a specified thermodynamic process is or is not possible. For example, the first law does not indicate whether water can spontaneously run uphill or not; all it states is that *if* water does run uphill, unless heat is supplied from outside, there will be a fall of temperature, the resulting decrease of energy content being equivalent to the work done against gravity. Similarly, there is nothing in the first law of thermodynamics to indicate whether a bar of metal of uniform temperature can spontaneously become warmer at one end and cooler at the other. All that the law can state is that *if* this process occurred, the heat energy gained by one end would be exactly equal to that lost by the other. It is the second law of the thermodynamics which provides the criterion as to the possibility, or rather the probability, of various processes.

Another important aspect of the second law, which is really fundamental to the problem enunciated above, deals with the conversion into work of energy absorbed as heat. The first law states that when heat is converted into work, the work obtained is equivalent to the heat absorbed, but it gives no information concerning the conditions under which the conversion is possible. It will be seen shortly that the heat absorbed at any one temperature cannot be completely transformed into work without leaving some change in the system or its surroundings; this fact is embodied in the second law of thermodynamics, and its consequences are of great significance.

18b. Spontaneous and Irreversible Processes.—In order to understand something of the conditions which determine whether a particular process will occur or not, it is of interest to examine certain processes which are known to be spontaneous, that is, processes which take place without external intervention of any kind. The expansion of a gas into an evacuated space, or from any region of higher into one of lower pressure, takes place spontaneously, until the pressure distribution is uniform throughout. Similarly, one gas will diffuse spontaneously into another until the mixing is complete and the system has the same composition in all parts. Diffusion of a solute from a concentrated solution into pure solvent, or into a dilute solution, will similarly take place without external intervention. Finally, reference may be made to the spontaneous conduction of heat along a bar of metal which is hot at one end and cold at the other, and to the spontaneous transfer of heat by radiation from a hotter to a colder body. These processes will continue until the temperature of the bar is uniform, in the former case, and until the two bodies attain the same temperature, in the latter instance. It will be observed that in every case the spontaneous process represents a tendency for the system to approach a state of thermodynamic equilibrium (§ 4c).

A fundamental characteristic of the processes described, and in fact of all spontaneous processes, is that they have never been observed to reverse themselves without the intervention of an external agency. A system which is in equilibrium under a given set of conditions will undergo no detectable change if the conditions are not altered. In other words, *spontaneous processes are not thermodynamically reversible* (cf. § 8a).* This fact, founded upon experience, is the basis of the second law of thermodynamics. Such processes as the spontaneous concentration of a gas at one end of a vessel, leaving a lower pressure at the other end, the spontaneous unmixing of a uniform gas mixture, or a bar of metal becoming spontaneously hot at one end and cold at the other end, have never been observed. It may be remarked, incidentally, that it is not altogether justifiable to say that these processes are impossible. It is *possible* for a gas to concentrate spontaneously in one part of a vessel, but the *probability* of this occurring, to judge from actual experience, is extremely small.

18c. Reversal of Spontaneous Processes.—By the use of an external agency, it is possible to bring about the reversal of a spontaneous process. For example, by introducing a piston into the vessel, the gas which has expanded into a vacuum could be restored to its original volume by compression. Work would have to be done on the gas, and at the same time an equivalent amount of heat would be produced, and the temperature of the gas would rise. If this heat could be completely reconverted into work by means of a hypothetical machine, then the original state of the gas would have been restored, and there would be no change in external bodies. It is a fundamental fact of experience, however, that *the complete conversion of heat into work is impossible, without leaving some effect elsewhere*. This result is in accord with the statement made above that spontaneous processes are not reversible in the thermodynamic sense.

Before stating the second law, other attempts to reverse spontaneous processes will be considered. When a bar of metal, which was originally hotter at one end, has attained a uniform temperature, the initial state

* In order to avoid the possibility of misunderstanding, it may be pointed out here that many changes which occur spontaneously in nature, e.g., expansion of a gas, evaporation of a liquid and even chemical reactions, can be carried out reversibly, at least in principle, as described in § 8a. However, when they do occur spontaneously, without external intervention, they are thermodynamically irreversible.

might conceivably be restored in the following manner. Heat is withdrawn from one end of the bar, completely converted into work, and then the work could be utilized to heat the other end of the bar, e.g., electrically or by friction. Actually, it is impossible to carry out this series of processes without leaving some changes for, as already stated, the complete conversion of heat into work without such changes is impossible.

It is evident that certain spontaneous physical processes could be reversed if the complete conversion of heat into work could be achieved; it will now be shown that similar considerations apply to chemical reactions. A piece of zinc, for example, will dissolve spontaneously in an aqueous solution of copper sulfate, according to the equation

$$Zn + CuSO_4 = ZnSO_4 + Cu,$$

with the evolution of a definite amount of heat. This reaction could be reversed by passing an electric current between the metallic copper and the solution of zinc sulfate in an appropriate manner, thus regenerating metallic zinc and copper sulfate. In order that the reversal might not leave changes elsewhere, it would be necessary for the heat evolved in the original reaction to be completely converted into electrical work. Once again, experience shows that a complete conversion of this kind, without producing other changes, is not possible. It is seen, therefore, that the spontaneous chemical process, like the physical changes considered above, is not thermodynamically reversible.

18d. The Second Law of Thermodynamics.—The second law of thermodynamics has been stated in various forms, one of which, concerning the irreversibility of spontaneous processes, has been already given. For subsequent purposes, however, a more useful form is that based on the inability to convert heat completely into work; thus, *it is impossible to construct a machine, operating in cycles, which will produce no effect other than the absorption of heat from a reservoir and its conversion into an equivalent amount of work.* The term "operating in cycles" is introduced to indicate that the machine must return to its initial state at regular stages (cf. § 7c), so that it can function continuously.

It will be recalled from the statements in § 9d that in an isothermal, reversible expansion of an ideal gas the work done is exactly equal to the heat absorbed by the system. In other words, in this process the heat is completely converted into work. However, it is important to observe that this conversion is accompanied by an increase in the volume of the gas, so that the system has undergone a change. If the gas is to be restored to its original volume by reversible compression, work will have to be done on the system, and an equivalent amount of heat will be liberated. The work and heat quantities involved in the process are exactly the same as those concerned in the original expansion. Hence, the net result of the isothermal expansion and compression is that the system is restored to its original state, but there is no net absorption of heat and no work is done. The foregoing is an illustration of the universal experience, that *it is not possible to convert heat into work by means of an isothermal, i.e., constant temperature, cycle.* This may be regarded as another aspect of the second law of thermodynamics.

A consequence of the impossibility of converting heat isothermally into work in a continuous manner is the impracticability of what is called "perpetual motion of the second kind," that is, the utilization of the vast stores of energy in the ocean and in the earth. There is nothing contrary to the first law of thermodynamics in this concept, but the fact that it has not been found feasible provides support for the second law. The ocean, for example, may be regarded as a heat reservoir of constant temperature,* and the law states that it is not possible to convert the heat continuously into work without producing changes elsewhere.

18e. Macroscopic Nature of the Second Law.—An insight into the fundamental basis of the second law of thermodynamics may be obtained by utilizing the concepts of the kinetic theory of matter. According to this theory an increase of temperature, resulting from the absorption of heat by a body, represents an increase in the kinetic energy of the random motion of the molecules. Hence, when the energy of a moving body is converted into heat by friction, the directed motion of the body as a whole is transformed into the chaotic motion of individual molecules. The reversal of the process, that is, the spontaneous conversion of heat into work, would require that all the molecules should spontaneously acquire a component of motion in one preferred direction. The probability of this occurring in a system consisting of a large number of molecules is very small. As indicated earlier, it cannot be stated that it is impossible for a spontaneous process to reverse itself spontaneously; there is merely a very great probability against it.

If it were possible to deal with systems involving only a few molecules, spontaneous processes could be reversed. Imagine a system consisting of five molecules possessing random motion. It is not improbable that at some instant all five molecules will have a preferred component of motion in one direction. The system as a whole will then have directed motion at that instant. The chaotic movement of the molecules, i.e., heat, will thus have been converted into directed motion, i.e., work.

Similarly, if a vessel contained only five or six molecules uniformly distributed, there would be a considerable probability that at some instant there will be a larger number of molecules at one end of the vessel than at the other. That is to say, a pressure difference would arise spontaneously within the vessel, provided it is permissible to speak of pressures when a few molecules are concerned. If the vessel contained a large number of molecules, it is highly improbable that any appreciable unequal distribution would arise spontaneously (cf. § 24a). The second law of thermodynamics is then to be regarded as applicable to macroscopic systems, and since it is

---

* Where there are temperature *differences* in the ocean, e.g., at different levels, the (partial) conversion of the heat of the ocean into work is possible; this is not contrary to the second law of thermodynamics.

systems of this type which are the basis of human observation and experience, no exception to the law has yet been observed.

It has long been realized that the impossibility, or rather the improbability, of reversing spontaneous processes is based on the inability to deal with individual molecules or small groups of molecules. If a device were available which could distinguish between fast- ("hot") and slow-moving ("cold") molecules, it would be possible to produce spontaneously a temperature gradient in a gas. Similarly, if a device could discriminate between two different types of gas molecules, a partial unmixing, which is the reverse of diffusion, could be achieved. However, the fact that no such devices are known is in accordance with the second law of thermodynamics.

18f. Conversion of Heat into Work.—In order for any form of energy to be available for the performance of work, it must be associated with a difference of potential or, in other words, with a directive influence. The work that can be done by falling water, for example, is due to the difference in potential energy at the upper and lower levels; similarly, electrical work is associated with a difference of electrical potential, generally known as an electromotive force. In a heat reservoir at constant temperature there is no directive influence, but two such reservoirs, *at different temperatures*, provide the difference of energy potential that is necessary for the conversion of heat into work. In order to carry out this conversion, heat is absorbed from the reservoir at the higher temperature, often referred to as the "source"; part of this heat is converted into work, and the remainder is returned to the heat reservoir at the lower temperature, referred to as the "sink". It is seen, therefore, that *a portion only of the heat taken in from the reservoir at the higher temperature can be converted into work*. The fraction of the heat absorbed by a machine that it can transform into work is called the **efficiency** of the machine; thus, if heat Q is taken up from the source, and W is the work done, the efficiency is equal to W/Q. Experience shows, in agreement with the statement of the second law of thermodynamics, that W is invariably less than Q in a continuous conversion process. The efficiency of a machine for the continuous conversion of heat into work is thus always less than unity. It will be understood, of course, that the first law of thermodynamics will still be applicable, for the energy difference between Q and W is returned to the lower temperature reservoir.

18g. The Carnot Theorem.—A highly important discovery in connection with the problem of the efficiency of **heat engines,** i.e., machines for the conversion of heat into work, was made by S. Carnot (1824). The principle he enunciated, which is derived from the second law of thermodynamics, may be stated in the following form: *All reversible heat engines operating between two given temperatures have the same efficiency.* In other words, provided the machines function in a thermodynamically reversible manner, the efficiency is independent of the working substance, or substances, or mode of operation; it depends only on the temperatures of the source and sink. Incidentally, it will be shown that this particular efficiency, namely that of a reversible engine, is the maximum possible for the given temperatures.

In order to prove the Carnot theorem, it will be assumed that there exist two reversible heat engines I and II, working between the same two temperatures, but having different efficiencies. Suppose that in each cycle the machine I takes in heat Q_2 from the source at T_2, converts an amount W into work, and gives up the remainder $Q_2 - W = Q_1$ to the sink at T_1. The machine II, on the other hand, is supposed to convert a smaller amount W' of the heat Q_2 taken in at T_2 into work, returning a quantity $Q_2 - W' = Q_1'$, which is greater than Q_1, to the sink at T_1. Let the machines be coupled together so that I operates in a direct manner, i.e., taking up heat at T_2, doing work and giving up the remainder at T_1, whereas II functions in the reverse manner, i.e., taking in heat at T_1, having work done upon it, and giving up heat at T_2.* This is permissible since the machines are assumed to be reversible. The various heat and work changes in each complete cycle of the combined machines are then as indicated in Fig. 10 and represented below. In accordance with the convention earlier (§ 7b), heat absorbed by the system is taken as positive and heat liberated as negative; similarly, work done by the system is positive and that done on the system is negative.

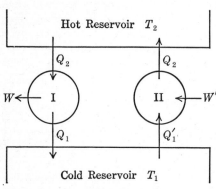

FIG. 10. Proof of the Carnot theorem

I	II
Heat transfer at $T_2 = Q_2$	Heat transfer at $T_2 = -Q_2$
Work done $= W$	Work done $= -W'$
Heat transfer at $T_1 = -Q_1$	Heat transfer at $T_1 = Q_1'$.

The net result of the complete cycle by the two reversible engines, bringing them both back to their initial states, without producing any external changes, is given by

Heat transfer at $T_1 = Q_1' - Q_1$
Work done $= W - W'$.

Since Q_1 is equal to $Q_2 - W$, and Q_1' is equal to $Q_2 - W'$ it follows that $Q_1' - Q_1$ is equal to $W - W'$, so that the heat absorbed at T_1 is equal to the work done by the engine. That is to say, the combined hypothetical, reversible machine, functioning in cycles, is able to convert completely into work the whole of the heat taken up from a reservoir at the temperature T_1, without leaving changes elsewhere. This is contrary to the second law of thermodynamics, and so it must be concluded that the two reversible machines I and II cannot have different efficiencies. The Carnot principle

* In other words, machine II functions as a refrigerator (see § 18j).

of the equality of efficiency for all reversible cycles working between the same two temperatures is thus a direct consequence of the second law of thermodynamics.

18h. The Carnot Cycle.—Since all reversible heat engines operating between the same two temperatures have equal efficiencies, it is sufficient to consider any convenient machine of this type, for all others will have the same efficiency. The one which lends itself to simple thermodynamic treatment makes use of the cycle described by S. Carnot (1824). In this hypothetical heat engine the working substance is 1 mole of an ideal gas; it is contained in a cylinder fitted with a weightless and frictionless piston, thus permitting reversible processes to be performed. It is supposed that there are available two large heat reservoirs which remain at constant temperatures, viz., T_2 (upper) and T_1 (lower), respectively. Further, it is assumed that completely adiabatic processes can be carried out when required, by surrounding the cylinder with a perfect nonconducting jacket so that no heat enters or leaves the system (§ 10a). The Carnot cycle consists of four stages which can be represented on a pressure-volume diagram, sometimes referred to as an "indicator diagram," as in Fig. 11.

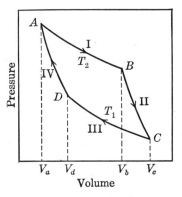

Fig. 11. Pressure-volume changes in Carnot cycle

I. The cylinder containing the mole of ideal gas, occupying a volume V_a, is placed in the heat reservoir at the higher temperature. The external pressure is adjusted so that it is always infinitesimally less than the gas pressure, and the temperature of the gas is infinitesimally less than that of the reservoir. In this manner, the gas is expanded isothermally and reversibly until its volume has increased to V_b. The path of the process is represented by the isothermal curve AB in Fig. 11. Since the gas is ideal, the work done W_I is given by equation (8.4); thus, for 1 mole of gas,

$$W_\mathrm{I} = RT_2 \ln \frac{V_b}{V_a}. \tag{18.1}$$

The heat Q_2 taken up from the reservoir must be equal to the work done (cf. § 9d), and hence by equation (9.21),

$$Q_2 = RT_2 \ln \frac{V_b}{V_a}. \tag{18.2}$$

II. The cylinder of gas is removed from the reservoir at T_2 and is surrounded by the nonconducting jacket, so that the gas can be expanded reversibly, i.e., infinitesimally slowly, and adiabatically. Work is done in the expansion, but since no heat enters or leaves the system, the temperature must fall (§ 10a). The reversible expansion is continued until the temperature has fallen to T_1, which is that of the lower temperature heat reservoir. The path is indicated by the adiabatic curve BC, the final volume being V_c. The work done is given by equation (10.8); thus,

$$W_{II} = - C_V(T_1 - T_2) \\ = C_V(T_2 - T_1), \qquad (18.3)$$

where C_V is the heat capacity of the ideal gas, assumed constant in the given temperature range. If the heat capacity were not constant, C_V would represent the mean value.

III. The nonconducting jacket is now removed and the cylinder is placed in the heat reservoir at T_1. The gas is compressed isothermally and reversibly, the external pressure being maintained infinitesimally greater than the gas pressure, and the temperature of the gas infinitesimally greater than that of the reservoir. The process is represented by the isothermal path CD in Fig. 11, the final volume being V_d. The work done is given by

$$W_{III} = RT_1 \ln \frac{V_d}{V_c}. \qquad (18.4)$$

Since V_c is greater than V_d, the value of W_{III} will be negative; this is, of course, because work is done *on* the gas in the compression. At the same time the quantity of heat Q_1, exactly equivalent to W_{III}, will be returned to the heat reservoir at T_1.

IV. The cylinder is removed from the heat reservoir and the nonconducting jacket is replaced. The gas is then compressed adiabatically and reversibly along DA until the initial state A is regained, the temperature of the gas rising from T_1 to T_2. The state D in stage III is deliberately chosen so that it lies on the same adiabatic as A. The work done is given by

$$W_{IV} = C_V(T_1 - T_2), \qquad (18.5)$$

where C_V has the same value as in equation (18.3), since for an ideal gas it must be independent of the volume or pressure.

As a result of the four stages just described the system has returned to its original state, so that a reversible cycle has been completed. The total work done W is the sum of the four work terms W_I, W_{II}, W_{III} and W_{IV}, but since W_{II} and W_{IV} are seen, by equations (18.3) and (18.5), to be equal but of opposite sign, it follows that

$$W = W_I + W_{III} = RT_2 \ln \frac{V_b}{V_a} + RT_1 \ln \frac{V_d}{V_c}. \qquad (18.6)$$

Since A and D lie in one adiabatic curve, while C and B lie on another, it follows from equation (10.5) that

$$\left(\frac{V_d}{V_a}\right)^{\gamma-1} = \frac{T_2}{T_1} \quad \text{and} \quad \left(\frac{V_c}{V_b}\right)^{\gamma-1} = \frac{T_2}{T_1},$$

and consequently

$$\frac{V_d}{V_a} = \frac{V_c}{V_b} \quad \text{or} \quad \frac{V_b}{V_a} = \frac{V_c}{V_d}.$$

Upon substitution of this result into equation (18.6) it is found that

$$W = RT_2 \ln \frac{V_b}{V_a} - RT_1 \ln \frac{V_b}{V_a}$$
$$= R(T_2 - T_1) \ln \frac{V_b}{V_a}. \qquad (18.7)$$

By definition, the efficiency of a heat engine is equal to the ratio of the total work W done in the cycle to the heat Q_2 taken in at the upper temperature; hence, by equations (18.2) and (18.7), the efficiency of the hypothetical Carnot engine is

$$\frac{W}{Q_2} = \frac{T_2 - T_1}{T_2}. \qquad (18.8)$$

In accordance with the Carnot theorem (§ 18g), *this expression gives the efficiency of any reversible heat engine operating between the temperatures* T_1 *and* T_2. As is to be expected, the efficiency is determined only by the temperatures of the two heat reservoirs acting as source (T_2) and sink (T_1), and is independent of the nature of the working substance. The lower the temperature of the sink, for a given temperature of the source, the greater will be the efficiency of the machine. Similarly, for a given temperature of the sink, the efficiency will be increased by using a high temperature source. In practice it is not convenient for the sink to be below atmospheric temperature, and so it is desirable that the upper temperature should be high. This fact underlies the use of high pressure steam or of mercury in boilers for power production.

Problem: The boiling point of water at a pressure of 50 atm. is 265° C. Compare the theoretical efficiencies of a steam engine operating between the boiling point of water at (i) 1 atm., (ii) 50 atm., assuming the temperature of the sink to be 35° C in each case.

(i) At 1 atm. pressure, the boiling point of water is 100° C, i.e., 373° K, and this represents the upper temperature T_2; the lower temperature T_1 is 35° C, i.e., 308° K, so that

$$\text{Efficiency} = \frac{T_2 - T_1}{T_2} = \frac{373 - 308}{373} = 0.174.$$

(ii) At 50 atm. pressure, T_2 is 265° C, i.e., 538° K, and T_1 is 308° K as in (i); hence

$$\text{Efficiency} = \frac{T_2 - T_1}{T_2} = \frac{538 - 308}{538} = 0.428.$$

The possible increase of efficiency is very marked.

Two special cases of equation (18.8) are of interest. First, if the efficiency of the reversible heat engine is to be unity, T_1 must be zero. Hence, the whole of the heat taken in at the higher temperature can be converted into work in a cycle, only if the lower temperature is the absolute zero. The second case is that in which T_1 and T_2 are equal, that is to say, the cycle is an isothermal one; in this event equation (18.8) shows the efficiency to be zero. This is in agreement with the conclusion reached earlier (§ 18d) that there can be no conversion of heat into work in an isothermal cycle.

18i. Maximum Efficiency of Heat Engine.—One of the essential properties of a reversible cycle is that *it has the maximum efficiency of any cycle operating between the same two temperatures.* That this is the case may be understood from a consideration of the Carnot engine in § 18h. Since the work terms in the adiabatic stages II and IV cancel one another, the work done in the cycle is that involved in the isothermal stages. It was seen earlier (§ 8c) that in a given isothermal expansion the work done by the system is a maximum when the expansion is carried out reversibly; similarly, the work done on the system in an isothermal compression is a minimum when performed reversibly. It follows, therefore, that in the Carnot reversible cycle the work done by the system at T_2 is the maximum possible for given expansion from A to B (Fig. 11), whereas the work done on the system at T_1 is a minimum for the compression from C to D. It is evident, therefore, that the total work done by the system is the maximum for the specified conditions. Since all reversible engines have the same efficiency as the Carnot cycle, it follows that the efficiency of a reversible heat engine operating between two given temperatures is the maximum possible for those temperatures.

18j. Refrigeration Engine.—In the foregoing treatment the Carnot cycle has been used as a heat engine, taking up heat at a higher temperature, giving some out at a lower temperature, and doing work in the process. Since the cycle is reversible, it is possible to operate it in the reverse direction, so that by doing work on the machine it can be made to take in heat at the lower temperature and give out heat at the upper temperature. In other words, the machine is functioning as a **refrigeration engine**, for by continually absorbing heat from the vessel at the lower temperature its temperature can be maintained at this low level, or lowered further. The work done on the refrigeration engine, using a Carnot cycle, must be equal, but of opposite sign, to the work done by the heat engine. The work done on the refrigeration engine is given by equation (18.7) with the sign reversed, and this may be written in the form

$$-W = R(T_2 - T_1) \ln \frac{V_b}{V_a} = R(T_2 - T_1) \ln \frac{V_c}{V_d}. \tag{18.9}$$

The heat Q_1 taken in at the lower temperature T_1 is equal, but opposite in sign, to that involved in stage III of the Carnot engine; hence, by equation (18.4),

$$Q_1 = -RT_1 \ln \frac{V_d}{V_c} = RT_1 \ln \frac{V_c}{V_d}. \tag{18.10}$$

The ratio of the work done on the machine to the heat absorbed at the lower temperature, that is, the **coefficient of performance** of the refrigeration engine, is given by

$$-\frac{W}{Q_1} = \frac{T_2 - T_1}{T_1}, \qquad (18.11)$$

and consequently,

$$-W = \frac{T_2 - T_1}{T_1} Q_1. \qquad (18.12)$$

Just as the Carnot (reversible) cycle gives the maximum proportion of work which can be obtained by a machine operating between two given temperatures, so equation (18.12) represents the minimum amount of work necessary for removing a quantity of heat Q_1 from a reservoir at T_1 and transferring it to a reservoir at the higher temperature T_2. If any stage of a refrigeration cycle were irreversible, more work than that represented by equation (18.12) would have to be done to transfer the given amount of heat between the same two temperatures.

Problem: Calculate the minimum amount of work in ergs required to freeze 1 g. of water at 0° C by means of a refrigeration engine which operates in surroundings at 25° C. How much heat, in calories, is given up to the surroundings?

The heat of fusion of ice is 79.8 cal. g.$^{-1}$ at 0° C, and so this quantity of heat must be transferred from 0° C, i.e., 273° K ($= T_1$), to 25° C, i.e., 298° K ($= T_2$); hence, by equation (18.12),

$$-W = \frac{298 - 273}{273} \times 79.8 = 7.30 \text{ cal.}$$

To convert into ergs it is necessary to multiply by 4.18×10^7, so that the work required is

$$7.30 \times 4.18 \times 10^7 = 3.05 \times 10^8 \text{ ergs.}$$

The heat given up at the higher temperature is the sum of the heat absorbed at the lower temperature, i.e., 79.8 cal., and of the work done on the engine, i.e., 7.30 cal.; the total is 87.1 cal.

The maximum work obtainable from a heat engine increases as the lower temperature is decreased, or the upper increased; similarly, it can be seen from equation (18.12) that the minimum amount of work which must be done in a given refrigeration process increases as the refrigeration temperature T_1 is lowered. Since $T_2 - T_1$ increases at the same time as T_1 is decreased, the ratio $(T_2 - T_1)/T_1$, in equation (18.12), increases rapidly as the temperature T_1 is diminished. If the latter temperature were to be the absolute zero, it is evident from equation (18.12) that an infinite amount of work would be necessary to transfer heat to an upper temperature even if this is only very slightly above 0° K. It follows, therefore, that *as the temperature of a system is lowered the amount of work required to lower the temperature further increases rapidly, and approaches infinity as the absolute zero is attained*. This fact has sometimes been expressed in the phrase "the unattainability of the absolute zero of temperature".

18k. The Thermodynamic (Kelvin) Temperature Scale.—The possibility of utilizing the efficiency of a reversible engine as the basis of a temperature scale was suggested by William Thomson (Lord Kelvin) in 1848. Suppose a reversible machine operates between two given temperature reservoirs; the temperature of each reservoir on the thermodynamic (Kelvin) scale is then defined as *proportional to the quantity of heat transferred to or from it in a reversible cycle*. Disregarding for the moment the signs of the heat quantities, if Q_2 is the heat transfer for the reservoir at the higher temperature and Q_1 is the amount of heat transferred at the lower temperature, then the respective temperatures on the thermodynamic (Kelvin) scale are θ_2 and θ_1, given by

$$\frac{\theta_2}{\theta_1} = \frac{Q_2}{Q_1}. \qquad (18.13)$$

In this way the ratio of the two temperatures is defined in a manner independent of any particular thermometric substance (cf. § 2b).

By inverting each side of equation (18.13) and subtracting the result from unity, it follows that

$$\frac{Q_2 - Q_1}{Q_2} = \frac{\theta_2 - \theta_1}{\theta_2}. \qquad (18.14)$$

When referred to a heat engine, Q_2 is the heat taken up at the higher temperature and Q_1 is the amount returned at the lower temperature; hence $Q_2 - Q_1$ is the quantity of heat converted into work, i.e., W, so that equation (18.14) may be written as

$$\frac{W}{Q_2} = \frac{\theta_2 - \theta_1}{\theta_2}. \qquad (18.15)$$

This expression defines the efficiency of the reversible heat engine in terms of the Kelvin temperatures.

The condition for the zero of the Kelvin scale may be derived by setting θ_1 in equation (18.15) equal to zero, the result is seen to be

$$\frac{W}{Q_2} = 1 \qquad (\text{for } \theta_1 = 0),$$

so that the zero of the thermodynamic scale is the lower temperature of a reversible cycle with an efficiency of unity, that is, one capable of converting heat completely into work. As seen in § 18h, this result is only possible if the lower temperature is the absolute zero on the ideal gas scale of temperature. From this fact, and the identity of equations (18.15) and (18.8), it follows that the Kelvin scale and the ideal gas scale are really the same. In order that temperatures on the two scales may coincide exactly it is only necessary to define the size of the degree so as to be the same on both scales, that is, one hundredth part of the range between the ice point and the steam point at 1 atm. pressure (§ 2b). In view of the identity of the ideal gas scale and the thermodynamic scale defined in this manner, temperatures on the former scale, like those on the latter, may be regarded as absolute, and independent of the thermometric substance. This is the justification for the use of the symbol "° K" (degrees Kelvin) for the so-called absolute temperatures based on the hypothetical ideal gas thermometer (§ 2c).

19. ENTROPY

19a. Combination of Carnot Cycles.—Although in § 18k, for convenience in deriving the Kelvin scale of temperature, the numerical values only of Q_1 and Q_2 were considered, it should be recalled that by convention (§ 7b) Q is the heat *taken up by* the system. In a Carnot cycle, therefore, the total heat absorbed is $Q_2 + Q_1$, where Q_2 has a positive value and Q_1 has a negative value, since the former is taken up at the higher temperature and the latter is given out at the lower. The work W done in the cycle must be equal to the total heat absorbed (§ 7c), so that

$$W = Q_2 + Q_1.$$

If this expression for W is substituted in equation (18.8), it is seen that

$$\frac{Q_2 + Q_1}{Q_2} = \frac{T_2 - T_1}{T_2}, \tag{19.1}$$

and consequently,

$$\frac{Q_2}{T_2} + \frac{Q_1}{T_1} = 0. \tag{19.2}$$

Any reversible cycle may be regarded as being made up of a number of Carnot cycles. Consider, for example, the cycle represented in Fig. 12 by the closed curve ABA; imagine a series of isothermal and adiabatic curves drawn across the diagram, so that a number of Carnot cycles are indicated. Starting from A, and following all the cycles down successively to B, and back again to A, it can be seen that all the paths inside the area enclosed by the curves ABA cancel each other, leaving only the path indicated by the zigzag outline. The larger the number of cycles taken in this manner the closer will the resultant path correspond to ABA, which represents the reversible cycle under consideration. The latter may thus be regarded as equivalent to the contribution of an infinite number of small Carnot cycles. For each of these cycles equation (19.2) shows that the sum of the two Q/T terms involved is zero; hence, for all the Carnot cycles equivalent to the general reversible cycle ABA, it follows that

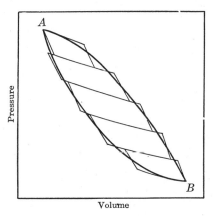

FIG. 12. Cyclic process as succession of Carnot cycles

$$\sum_{\text{cycle}} \frac{Q_{\text{rev.}}}{T} = 0, \tag{19.3}$$

where the summation includes two terms for each of the individual Carnot cycles. Since an infinite number of small Carnot cycles are required to duplicate the process ABA, it is convenient to write equation (19.3) in the form

$$\sum_{\text{cycle}} \frac{q_{\text{rev.}}}{T} = 0, \tag{19.4}$$

where $q_{\text{rev.}}$ represents the infinitesimally small quantity of heat absorbed at the temperature T in each of the small isothermal, reversible changes which make up the reversible cycle ABA.*

19b. Definition of Entropy.—The summation in equation (19.4), applicable to the complete (reversible) cycle ABA, may be divided into two parts, one for the path from A to B, and the other back from B to A; thus,

$$\sum_{\text{cycle}} \frac{q_{\text{rev.}}}{T} = \sum_{A \to B} \frac{q_{\text{rev.}}}{T} + \sum_{B \to A} \frac{q_{\text{rev.}}}{T} = 0. \tag{19.5}$$

It may be possible to go from A to B by a number of *different* reversible paths, always returning to A by the *same* reversible path BA. In every case the result represented by equation (19.5) must hold, and since the second summation on the right of this equation will always be the same, since the path $B \to A$ is the same, it follows that the value of the first summation must be independent of the path from A to B, provided only that it is reversible. In general, therefore, it can be seen that the summation of the $q_{\text{rev.}}/T$ terms between A and B, or of the corresponding summation from B to A, must be independent of the reversible path. The values of these summations are thus determined by the states A and B, that is, by the pressure, volume and temperature, and are independent of the manner in which the system was brought to these states. It is thus possible to express the value of each summation in terms of a function S, which depends only on the state of the system; thus,

$$\sum_{A \to B} \frac{q_{\text{rev.}}}{T} = S_B - S_A = \Delta S, \tag{19.6}$$

where S_A is the value of the function in the state A, and S_B in the state B.

The increase of the function S accompanying the change in state from A to B is ΔS, and its value, as seen by equation (19.6), is given by the sum of the $q_{\text{rev.}}/T$ terms between A and B, where $q_{\text{rev.}}$ is the heat absorbed by the system in an infinitesimal, *reversible, isothermal* stage occurring at the temperature T. For each of these small stages it is therefore possible to write

$$dS = \frac{q_{\text{rev.}}}{T}, \tag{19.7}$$

where dS represents the accompanying increase of the function S. This function or property of the system is called its **entropy** (Greek: *change*) in

---

* Since the paths within the area ABA cancel, the $q_{\text{rev.}}$ values are effectively those for the isothermal stages of the path ABA itself.

the given state; the entropy is not easily defined directly, and so it is best described in terms of the *entropy increase* accompanying a particular process. In an infinitesimal stage of an appreciable process the entropy increase dS is given by equation (19.7) as *the heat q_{rev} taken up isothermally and reversibly divided by the absolute temperature T at which it is absorbed*, i.e., q_{rev}/T. For an appreciable change the entropy increase is defined by equation (19.6) as the (algebraic) sum of all the q_{rev}/T terms between the initial and final states of the system. The only condition applicable to the path is that it shall be reversible.

Since the entropy in any state depends only on that state, it may be regarded as a thermodynamic property in the sense considered in § 4d. Hence, the increase of entropy of the system accompanying the change from state A to state B has the definite value $S_B - S_A$, and this quantity is completely independent of the path from A to B; it may be reversible or irreversible. However, it is important to remember that if a system changes from A to B in an irreversible manner, the increase of entropy is given by the summation of the q_{rev}/T terms between A and B, where the q_{rev}'s refer to the succession of isothermal changes *when the process is performed reversibly*. The sum of the q/T terms for an irreversible process is an indefinite quantity, depending on the path taken from A to B, and having no special thermodynamic significance as far as the system is concerned.

Because the entropy, like the energy, is a single-valued function of the state of the system, dS, like dE, is a complete differential. This fact adds considerably to the thermodynamic usefulness of the entropy function. The entropy of a system, like the energy, is an extensive property, dependent upon the amount of matter in the system. For example, if the amount of matter is doubled, the heat quantities required for the same change of state will also be doubled, and the entropy will clearly increase in the same proportion. Another consequence of the entropy being an extensive property is that when a system consists of several parts, the total entropy change is the sum of the entropy changes of the individual portions.

19c. Entropy Change and Unavailable Heat.—There is a simple relationship between entropy change and the heat which is rejected at the lower temperature in a heat engine that is of practical significance. It can be readily seen from equation (19.2) that for a *reversible cycle*

$$- Q_1 = T_1 \frac{Q_2}{T_2}, \qquad (19.8)$$

where $- Q_1$ is the heat returned to the reservoir at the lower temperature T_1, and hence is not available for conversion into work. Since the heat is taken up reversibly, Q_2/T_2 is the increase of entropy of the system in the heat absorption stage, i.e., ΔS_2. It follows, therefore, from equation (19.8) that

$$- Q_1 = T_1 \Delta S_2. \qquad (19.9)$$

Incidentally, since $- Q_1/T_1$ is equal to Q_2/T_2, by equation (19.2) or (19.8), the entropy decrease $- \Delta S_1$ at the lower temperature is equal to ΔS_2, so that Q_1 may equally be represented by $T_1 \Delta S_1$. Either of these relationships gives the quantity of heat that is returned to the reservoir at the lower temperature in a reversible cycle in terms of the entropy change. For a nonreversible cycle the proportion of heat that is unavailable for work is, of course, greater since the efficiency is less.

In actual practice, the heat absorbed by the working substance, e.g., water, is not all taken up at the one temperature T_2 but over a range of temperatures, e.g., from T_2 to T_2'; the appropriate form of equation (19.2) is then

$$\sum_{T_2 \to T_2'} \frac{q_{\text{rev.}}}{T} + \frac{Q_1}{T_1} = 0,$$

the heat Q_1 being rejected at the (approximately) constant temperature T_1. In this case,

$$-Q_1 = T_1 \sum_{T_2 \to T_2'} \frac{q_{\text{rev.}}}{T} = T_1 \Delta S_2, \qquad (19.10)$$

where ΔS_2 is now the increase of entropy accompanying the reversible absorption of heat in the temperature range from T_2 to T_2', instead of at T_2 alone, as in equation (19.9). Incidentally, provided T_1 is constant, the alternative form $Q_1 = T_1 \Delta S_1$ could be used, as in the previous case.

19d. Entropy Change in Reversible Process.—In a complete cycle the total entropy change of a system must be zero, since it has returned exactly to its original thermodynamic state; hence, as expressed by equation (19.4),

$$\sum_{\text{cycle}} \frac{q_{\text{rev.}}}{T} = 0$$

for the system. This result refers only to the substance or substances included in the system, sometimes called the "working substance," and it is necessary now to consider the "surroundings." In a *reversible* process, the heat quantity $q_{\text{rev.}}$ taken up by the system at any stage is supplied reversibly by the heat reservoir, i.e., the surroundings, the temperature of which differs only infinitesimally from that of the system. Hence, at every stage, the change in entropy of the surroundings will be equal to the entropy change of the system, but of opposite sign, since the latter takes in heat when the former gives it out, and vice versa. In every reversible process, therefore, the sum of the entropy changes of the system and its surroundings will be zero. In a complete reversible cycle, therefore, neither will undergo any resultant change of entropy.

19e. Entropy Change in Irreversible Process.—It has been shown in § 18i that the efficiency of a reversible engine is a maximum for the given working temperatures. Hence, the efficiency of a cycle involving an irreversible stage must be less than that of a Carnot cycle. Suppose, for example, that the higher temperature (T_2) stage, in which the heat $Q_{2(\text{irr.})}$ is absorbed, takes place in an irreversible manner. The remainder of the cycle may be supposed to be the same as for a Carnot cycle, the heat $Q_{1(\text{rev.})}$ being given out reversibly. The work done is given by $Q_{2(\text{irr.})} + Q_{1(\text{rev.})}$, and since the efficiency is less than for a completely reversible cycle, it follows that [cf. equation (19.1)]

$$\frac{Q_{2(\text{irr.})} + Q_{1(\text{rev.})}}{Q_{2(\text{irr.})}} < \frac{T_2 - T_1}{T_2} \tag{19.11}$$

and hence,

$$\frac{Q_{2(\text{irr.})}}{T_2} + \frac{Q_{1(\text{rev.})}}{T_1} < 0. \tag{19.12}$$

In general, therefore, for a cycle which is not completely reversible, the sum, over the whole cycle, of all the Q/T terms, or the q/T terms for a series of infinitesimal stages, would be less than zero.

Consider a perfectly general cycle ABA made up of a path $A \to B$, which involves one or more irreversible stages, and the completely reversible path $B \to A$; according to the arguments presented above, therefore,

$$\sum_{A \to B} \frac{q_{\text{irr.}}}{T} + \sum_{B \to A} \frac{q_{\text{rev.}}}{T} < 0. \tag{19.13}$$

Since the heat is absorbed in an irreversible manner, the first summation in equation (19.13) will not be definite, but will depend on the particular path taken. By the definition of the entropy change of the system [cf. equation (19.6)], it is seen that

$$\sum_{B \to A} \frac{q_{\text{rev.}}}{T} = S_A - S_B,$$

where $q_{\text{rev.}}$ is taken up reversibly, and hence equation (19.13) becomes

$$\sum_{A \to B} \frac{q_{\text{irr.}}}{T} + S_A - S_B < 0,$$

or, reversing the signs throughout,

$$S_B - S_A - \sum_{A \to B} \frac{q_{\text{irr.}}}{T} > 0. \tag{19.14}$$

In an irreversible process $A \to B$, therefore, the summation of the $q_{\text{irr.}}/T$ terms is actually *less* than the increase of entropy of the system.

The entropy change of the *surroundings* in the irreversible stage $A \to B$ must now be considered. This can best be ascertained from the change of entropy when the surroundings are restored to their original state; the required entropy change must then be equal in magnitude to this quantity but of opposite sign. The initial state of the surroundings can be restored by adding the various amounts of heat involved in the $q_{\text{irr.}}$ terms at the appropriate temperatures. In order to obtain the entropy change, these heat quantities are added reversibly (cf. § 19b), irrespective of the fact that they were not supplied reversibly to the system during the stage $A \to B$. The increase of entropy when the surroundings are brought back to their original state, after the process $A \to B$, is the sum of the $q_{\text{irr.}}/T$ terms, and hence the entropy change during this process is equal to $-\sum_{A \to B} q_{\text{irr.}}/T$.

The total entropy change of the system and its surroundings in the irreversible process $A \rightarrow B$ may thus be summarized as follows:

Increase of entropy of system $= S_B - S_A$

Increase of entropy of surroundings $= - \sum_{A \rightarrow B} \dfrac{q_{\text{irr.}}}{T}$

Net increase of entropy of system
and surroundings $(\Delta S_{\text{net}}) = S_B - S_A - \sum_{A \rightarrow B} \dfrac{q_{\text{irr.}}}{T}$.

As seen by equation (19.14), this quantity is greater than zero, and hence it follows that *in any irreversible process there is a net gain of entropy of the system and its surroundings.* If the cycle ABA is completed by the reversible path $B \rightarrow A$, as suggested above, the net entropy change in this process is zero, and hence the whole cycle must be accompanied by a gain of entropy. It follows, therefore, that *an irreversible process, or a cycle of which any part is irreversible, is accompanied by a gain of entropy of the combined system and its surroundings.*

Since natural, or spontaneously occurring, processes are irreversible (§ 18b), it must be concluded that all such processes are associated with a net increase of entropy.* From some points of view this is one of the most important consequences of the second law of thermodynamics; the law may in fact be stated in the form that *all processes occurring in nature are associated with a gain of entropy of the system and its surroundings.*†

It should be remembered that the net gain of entropy accompanying an irreversible process refers to the combination of the system and its surroundings, that is to say, to an *isolated system of constant energy* (cf. § 6d). It will be seen later that if the energy (and volume) of the system remained constant in an irreversible process, its entropy would increase, quite apart from that of its surroundings. It often happens that the entropy of the system, i.e., the working substance, actually decreases in a spontaneous, irreversible process because of the associated energy change, e.g., in the solidification of a supercooled liquid, but the entropy of the surroundings simultaneously increases by a greater amount, so that there is a net increase of entropy, as required by the second law of thermodynamics.

19f. Irreversible Processes and Degradation of Energy.—In a complete cycle, reversible or irreversible, the system returns to its original state, and hence it undergoes no resultant change of entropy. Any net increase in the entropy of the system and its surroundings, as in an irreversible cycle, will then be an increase

* Many changes which are naturally spontaneous ,e.g., expansion of a gas, solution of zinc in copper sulfate, etc., can be carried out, actually or in principle, in a reversible manner. It should be clearly understood that in the latter event the total entropy of the system and its surroundings remains unchanged. There is an increase of entropy only when the change occurs spontaneously and hence irreversibly.

† It is sometimes stated that the entropy of the "universe" is increasing; this, however, implies a knowledge of processes occurring outside the earth and so cannot be justified.

in the entropy of the surroundings; thus, in a complete cycle,

$$\Delta S_{net} = \Delta S_{surr.}$$

For a reversible cycle, ΔS_{net}, and hence $\Delta S_{surr.}$, is zero (§ 19d), whereas for an irreversible process it is positive, as seen above. It follows, therefore, that in an irreversible cycle an amount of heat $Q_{ex.}$ has been returned to the surroundings *in excess of that which is transferred in a reversible cycle*, where $Q_{ex.}$ is defined by

$$\frac{Q_{ex.}}{T} = \Delta S_{surr.} = \Delta S_{net},$$

that is,

$$Q_{ex.} = T \Delta S_{net}, \qquad (19.15)$$

the temperature T being that of the surroundings, i.e., the sink. This result gives a fundamental significance to the net entropy increase ΔS_{net} of the constant energy system: the product of ΔS_{net} and the temperature at which heat is rejected is equal to the quantity of heat that is "wasted" or "degraded" in an irreversible cycle. It represents heat taken in at the higher temperature which would have been available for work if the process had been carried out reversibly. As a result of the irreversible nature of the process, however, it has been transferred or "degraded" to a lower temperature where its availability for work is diminished. If, in any reversible cycle, a quantity of heat Q_2 is taken into the system at the temperature T_2, and Q_1 is rejected at the lower temperature T_1, then in an irreversible cycle the heat rejected, for the same quantity Q_2 of heat absorbed at T_2, is equal to $Q_1 + T_1 \Delta S_{net}$. For a given absorption of heat, therefore, the work done in the irreversible cycle is $T_1 \Delta S_{net}$ less than that done in a reversible cycle operating between the same temperatures. This result gives something of a physical significance to the net entropy increase as a measure of the irreversibility of any process.[1]

19g. Temperature-Entropy Diagrams.—A convenient form of diagram for representing thermodynamic changes in state, often used in engineering problems, is one in which the two axes indicate the temperature and entropy, respectively. On such a diagram an isothermal path is obviously a straight line parallel to the S-axis. Further, since in an adiabatic process no heat enters or leaves the system, the entropy change accompanying any such reversible process must be zero. A reversible, adiabatic change is, therefore, represented by a straight line parallel to the T-axis. It is because of the constancy of the entropy in a reversible, adiabatic process, the entropy change being zero, that the term **isentropic** (Greek: *same entropy*) is often used when referring to a process of this nature.

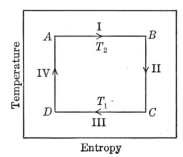

Fig. 13. Temperature-entropy diagram of Carnot cycle

[1] Cf. B. F. Dodge, "Chemical Engineering Thermodynamics," 1944, p. 71.

A Carnot cycle is very simply represented by a rectangle on a temperature-entropy diagram, as in Fig. 13; the isothermal stages at T_2 and T_1 are indicated by I and III, respectively, and the adiabatic paths are II and IV. The stages I, II, III and IV correspond exactly to those described in § 18h. It is at once obvious from the diagram that the entropy change Q_2/T_2 in stage I, where Q_2 is the heat transferred reversibly at the temperature T_2, is exactly equal to the entropy change $-Q_1/T_1$, in stage III, where $-Q_1$ is the heat transferred at T_1; thus,

$$\frac{Q_2}{T_2} = -\frac{Q_1}{T_1} \quad \text{or} \quad \frac{Q_2}{T_2} + \frac{Q_1}{T_1} = 0,$$

as found earlier [cf. equation (19.2)]. From this result the efficiency of the Carnot cycle can be readily derived, since the work done is equal to $Q_2 + Q_1$. It can be seen, therefore, that if entropy had been arbitrarily defined as $Q_{\text{rev.}}/T$, or as the sum of the $q_{\text{rev.}}/T$ terms, it would be a simple matter to derive the efficiency of a Carnot cycle. This procedure is sometimes adopted, but it has not been used here because in the treatment of §§ 19a, 19b the quantity $Q_{\text{rev.}}/T$, or $q_{\text{rev.}}/T$, appears as a logical development of the second law of thermodynamics, instead of being an apparently arbitrary function.

19h. Entropy Change and Phase Change.—In this and the following section the entropy changes accompanying certain simple processes will be evaluated. One case of interest for which the calculation can be made very readily is the increase of entropy associated with a phase change, e.g., solid to liquid (fusion), liquid to vapor (vaporization), or transition from one crystalline form to another. These changes can be carried out reversibly at a definite temperature, the system remaining in equilibrium throughout (cf. § 8a). The heat supplied under these conditions is the so-called "latent heat" accompanying the phase change. In the case of fusion, for example, $Q_{\text{rev.}}$ is equivalent to ΔH_f, the heat of fusion, and if the process has been carried out at the temperature T, the entropy increase, referred to as the **entropy of fusion**, is simply $\Delta H_f/T$. The **entropy of vaporization** can be determined in a similar manner. For example, the heat of vaporization of 1 mole of water at 25° C, in equilibrium with its vapor at a pressure of 0.0313 atm., is 10,514 cal. mole^{-1}. Since the temperature is $273.16 + 25.0 = 298.16°$ K, the entropy of vaporization of water at 25° C is $10,514/298.16 = 35.26$ cal. deg.$^{-1}$ mole^{-1}. It will be seen that the dimensions of entropy are heat/temperature, and hence it is usually expressed in terms of calories per degree, i.e., cal. deg.$^{-1}$. However, since the entropy is an extensive property the quantity of material constituting the system under consideration must be stated; consequently, the entropy of vaporization determined above was given as 35.26 cal. deg.$^{-1}$ mole^{-1}.

19i. Entropy Changes of Ideal Gas.—For an infinitesimal, isothermal process, the first law equation (7.5) may be written as

$$q = dE + w. \tag{19.16}$$

ENTROPY

If *the process is reversible,* and the *work is restricted to work of expansion,* as is almost invariably the case for processes considered in chemical thermodynamics, w may be replaced by PdV, where P is the pressure of the system, so that

$$q_{\text{rev.}} = dE + PdV. \tag{19.17}$$

Further, since the quantity of heat $q_{\text{rev.}}$ is transferred reversibly, at a constant temperature T, it follows that $q_{\text{rev.}}/T$ is equal to the entropy change dS accompanying the given infinitesimal change in state; hence, from equation (19.17), it is evident that

$$dS = \frac{dE + PdV}{T}. \tag{19.18}$$

For the special case of 1 mole of *an ideal gas*, dE may be replaced by $C_V dT$ [cf. equation (9.22)], where C_V is the molar heat capacity at constant volume, and P may be replaced by RT/V; equation (19.18) thus becomes

$$dS = C_V \frac{dT}{T} + R \frac{dV}{V}. \tag{19.19}$$

Making the *assumption* that C_V is independent of temperature (cf. § 9e), general integration of equation (19.19) gives an expression for the entropy of 1 mole of an ideal gas; thus,

$$S = C_V \ln T + R \ln V + s_0, \tag{19.20}$$

where s_0 is the integration constant. The value of this constant cannot be derived by purely thermodynamic methods, although it is possible to determine it by means of statistical mechanics, as will be explained in Chapter IX.

For an appreciable change with 1 mole of an ideal gas, between an initial thermodynamic state, indicated by the subscript 1, and the final state, indicated by the subscript 2, it follows from equation (19.20) that

$$\Delta S = S_2 - S_1 = C_V \ln \frac{T_2}{T_1} + R \ln \frac{V_2}{V_1}. \tag{19.21}$$

From this expression the entropy increase accompanying any given change in state of an ideal gas may be calculated.

An alternative form of (19.20) may be obtained by replacing V by RT/P, and utilizing the fact that for 1 mole of an ideal gas $C_P - C_V$ is equal to R; it is readily found that

$$S = C_P \ln T - R \ln P + s_0', \tag{19.22}$$

where s_0' is equal to $s_0 + R \ln R$. For a change in thermodynamic state, therefore,

$$\Delta S = S_2 - S_1 = C_P \ln \frac{T_2}{T_1} - R \ln \frac{P_2}{P_1}, \tag{19.23}$$

which is analogous to equation (19.21).

For a change of temperature *at constant volume*, the entropy increase of an ideal gas is seen from equation (19.21) to be

$$\Delta S_V = C_V \ln \frac{T_2}{T_1}, \tag{19.24}$$

whereas, for the same temperature change *at constant pressure*, (19.23) gives

$$\Delta S_P = C_P \ln \frac{T_2}{T_1}. \tag{19.25}$$

It should be noted that although the foregoing equations are ultimately based on (19.18), which applies to a reversible process, *the results are applicable to any change in thermodynamic state*, irrespective of whether it is carried out reversibly or not. This is because the entropy change depends only on the initial and final states, and not on the path between them.

For an *isothermal process*, T_1 and T_2 are identical, so that equations (19.21) and (19.23) reduce to

$$\Delta S_T = R \ln \frac{V_2}{V_1} = R \ln \frac{P_1}{P_2}. \tag{19.26}$$

The same result could have been derived directly from equation (9.21), which gives the heat absorbed in an isothermal, reversible process. If this quantity is divided by the constant temperature T, an expression for the entropy change, identical with equation (19.26), is obtained. It must not be forgotten, however, that this equation, like the others for entropy change derived in this section, is independent of the manner in which the process is carried out.

In the isothermal expansion of a gas, the final volume V_2 is greater than the initial volume V_1 and, consequently, by equation (19.26), ΔS is positive; that is to say, the expansion is accompanied by an increase of entropy of the system. Incidentally, when an ideal gas expands (irreversibly) into a vacuum, no heat is taken up from the surroundings (§ 9d), and so the entropy of the latter remains unchanged. In this case the net entropy increase is equal to the increase in entropy of the system, i.e., of the gas, alone.

In a reversible, adiabatic, i.e., isentropic, process, the entropy remains constant, and hence ΔS should be zero; the condition for *a reversible, adiabatic process* can thus be obtained by setting equation (19.21) equal to zero. The result is

$$C_V \ln \frac{T_2}{T_1} = - R \ln \frac{V_2}{V_1},$$

which is identical with equation (10.3). The characteristic equations for a reversible, adiabatic process, e.g., (10.5), (10.6) and (10.7), can thus be derived from entropy considerations.

19j. Entropy of Mixing.—Consider a number of ideal gases, incapable of interacting with one another in any way, placed in a vessel in which they are separated by partitions. If n_i is, in general, the number of moles of any gas, and v_i is the volume it occupies, the total entropy S_1 of the system, which is the sum of the entropies of the gases in the separate compartments, is given by equation (19.20) as

$$S_1 = \sum n_i (C_V \ln T + R \ln v_i + s_i), \tag{19.27}$$

the temperature being the same throughout. Suppose that all the partitions are removed, so that the gases become mixed. Each gas now occupies the total volume V of the system, and the entropy S_2 is then represented by

$$S_2 = \sum n_i (C_V \ln T + R \ln V + s_i). \tag{19.28}$$

If the separate gases were each at the same pressure before mixing,* the ratio of the initial volume v_i of any gas to the total volume V is equal to n_i/n, where n_i is the number of moles of that gas and n is the total number of moles in the system; thus,

$$\frac{v_i}{V} = \frac{n_i}{n} = \mathrm{N}_i, \tag{19.29}$$

where N_i is the mole fraction (§ 5b) of the given gas in the mixture. Replacing v_i in (19.27) by $\mathrm{N}_i V$, in accordance with equation (19.29), the result is

$$S_1 = \sum n_i (C_V \ln T + R \ln \mathrm{N}_i + R \ln V + s_i). \tag{19.30}$$

The increase of entropy resulting from the removal of the partitions and the mixing of the gases, known as the **entropy of mixing** ΔS_m, is equal to $S_2 - S_1$; hence, by equations (19.28) and (19.30),

$$\Delta S_m = - R \sum n_i \ln \mathrm{N}_i. \tag{19.31}$$

The entropy of mixing for a total of 1 mole of the mixture of ideal gases is obtained upon dividing equation (19.31) by the total number of moles n; the result is

$$\Delta S_m = - R \sum \frac{n_i}{n} \ln \mathrm{N}_i = - R \sum \mathrm{N}_i \ln \mathrm{N}_i, \tag{19.32}$$

assuming no change of temperature or total volume upon mixing.

It is of interest to note that since the mole fraction N_i of any gas in a mixture must be less than unity, its logarithm is negative; hence ΔS_m as defined by equation (19.32) is always positive. In other words, the mixing of two or more gases, e.g., by diffusion, is accompanied by an increase of entropy. Although equation (19.32) has been derived here for a mixture of ideal gases, it can be shown that it applies equally to an ideal mixture of liquids or an ideal solid solution.

Problem: Molecular hydrogen normally consists of three parts of orthohydrogen and one part of parahydrogen; at low temperatures the molecules of the former occupy nine closely spaced rotational levels, while the latter occupy only one. Calculate the entropy of mixing of the ten different kinds of hydrogen molecules.

There are nine kinds of ortho molecules, and since the total constitutes three-fourths of the hydrogen, each kind is present to the extent of $\frac{1}{9} \times \frac{3}{4} = \frac{1}{12}$. The mole fraction of each of the nine forms of orthohydrogen is thus $\frac{1}{12}$. Since there

* Under these conditions, the total pressure of the mixture of ideal gases will be the same as that of the individual gases before mixing.

is only one kind of parahydrogen, its mole fraction is just $\frac{1}{4}$. The entropy of mixing is obtained from equation (19.32), there being ten terms in the summation; for the nine identical ortho terms the mole fraction is $\frac{1}{12}$, and for the one para term it is $\frac{1}{4}$, viz.,

$$\Delta S_m = - R[9(\tfrac{1}{12} \ln \tfrac{1}{12}) + \tfrac{1}{4} \ln \tfrac{1}{4}]$$
$$= 2.208R = 4.39 \text{ cal. deg.}^{-1} \text{ mole}^{-1}.$$

This result will be utilized in § 24n.

19k. Entropy and Disorder.—Before closing this chapter, it is appropriate to refer to an interesting aspect of entropy which throws some light on the physical significance of this apparently theoretical property. The subject will be taken up in greater detail, from a somewhat different point of view, in Chapter IX, but a general indication may be given here. An examination of the various processes which take place spontaneously, and which are accompanied by a net increase of entropy, shows that they are associated with an increased randomness of distribution. For example, the diffusion of one gas into another means that the molecules of the two gases, which were initially separated, have become mixed in a random manner. Similarly, the spontaneous conduction of heat along a bar of metal means a more random distribution of the kinetic energies of the molecules. The conversion of mechanical work into heat, as seen earlier (§ 18e), is associated with the change from ordered motion of a body as a whole to the disordered or random motion of the molecules. It seems reasonable, therefore, to postulate a relationship between the entropy of a system and the randomness or degree of disorder in the given state.

The concept of *entropy as a measure of randomness*, or vice versa, is one of great value in many cases. Apart from its quantitative aspect, which will be considered later, it is also useful from the qualitative standpoint, for it is frequently possible to estimate whether a given process is accompanied by an increase or decrease of entropy from a consideration of the randomness or disorder in the initial and final states. Similarly, a knowledge of the entropy change often provides information concerning structural changes accompanying a given process. A simple illustration is provided by the melting of a solid. There is obviously an increase of disorder in passing from the solid to the liquid state, and hence an increase of entropy upon fusion is to be expected. As seen in § 19h, this is equal to the heat of fusion divided by the temperature at which melting occurs. In general, the greater the increase of disorder accompanying fusion, the greater will be the entropy increase. Thus, the molar entropy of fusion of ice at 0° C is 5.26 cal. deg.$^{-1}$ mole^{-1}, while that of benzene is 8.27 cal. deg.$^{-1}$ mole^{-1} at 5.4° C; because of the relatively large extent of order still present in liquid water above its melting point the entropy increase is smaller in the former than in the latter case.

EXERCISES

1. The use of biphenyl (m.p. 70°, b.p. 254° C) alone, or mixed with biphenylene oxide (m.p. 87°, b.p. 288° C), has been suggested as the working substance in a

heat engine. Consider the advantages and disadvantages with respect to the use of (i) water, (ii) mercury (b.p. 357° C), for the same purpose.

2. A reversible Joule cycle consists of the following stages: (i) an expansion at the constant pressure P_2, (ii) an adiabatic expansion to a lower pressure P_1, (iii) a compression at the constant pressure P_1, (iv) an adiabatic compression which restores the system to its initial state. Draw the indicator (P-V) diagram for the cycle, and prove that with an ideal gas as the working substance the efficiency is given by

$$\frac{W}{Q} = 1 - \left(\frac{P_1}{P_2}\right)^{[(C_P-C_V)/C_P]}$$

where Q is the heat taken up in stage (i) of the cycle. Show that it is possible to calculate the temperature at the end of each stage.

3. A Carnot cycle, in which the initial system consists of 1 mole of an ideal gas of volume V, is carried out as follows: (i) isothermal expansion at 100° C to volume $3V$, (ii) adiabatic expansion to volume $6V$, (iii) isothermal compression, (iv) adiabatic compression to the initial state. Determine the work done in each isothermal stage and the efficiency of the cycle.

4. It has been suggested that a building could be heated by a refrigeration engine operating in a reversed Carnot cycle. Suppose the engine, takes up heat from the outside at 2° C, work is done upon it, and then heat is given up to the building at 22° C. Assuming reversible behavior, how much work in ergs would have to be done for every kcal. of heat liberated in the building?

5. Mercury vapor at 357° C and 1 atm. pressure is heated to 550° C and its pressure is increased to 5 atm. Calculate the entropy change in the conventional units, the vapor being treated as an ideal monatomic gas.

6. An ideal gas undergoes throttled expansion (§ 11a), the pressure being 200 atm. on one side and 20 atm. on the other side of the throttle. The process is irreversible, and so the entropy change must be calculated by imagining the same change in thermodynamic state to be carried out reversibly. Determine the net change in entropy of the system and its surroundings. Is the sign in accordance with expectation?

7. Justify the statement that the solidification of a supercooled liquid is accompanied by a decrease in the entropy of the liquid but a greater increase in the entropy of the surroundings. Suggest how the change from supercooled water at $-10°$ C to ice at the same temperature and pressure could be carried out reversibly, so that the entropy change of the system could be evaluated.

8. Show that the entropy of mixing of a number of ideal gases, each at pressure p, to form a mixture at the total pressure p, at constant temperature, is $- R \sum N_i \ln N_i$ per mole of mixture. [Use equation (19.22).]

9. If air, consisting of 21 mole % oxygen and 79 mole % nitrogen, at 1 atm. pressure, could be separated into its pure constituent gases, each at 1 atm. pressure, at the same temperature, what would be the entropy change per mole of air? Ideal behavior may be postulated.

10. Show that equations (19.21) and (19.23) lead to the various conditions for an adiabatic change derived in § 10b.

11. The entropy of liquid ethanol is 38.4 cal. deg.$^{-1}$ mole^{-1} at 25° C. At this temperature the vapor pressure is 59.0 mm. and the heat of vaporization is $-$ 10.19 kcal. mole^{-1}. Assuming the vapor to behave ideally, calculate the entropy of ethanol vapor at 1 atm. pressure at 25° C.

CHAPTER VIII

ENTROPY RELATIONSHIPS AND APPLICATIONS

20. Temperature and Pressure Relationships

20a. Variation of Entropy with Temperature.—The equation (19.18), i.e.,

$$dS = \frac{dE + PdV}{T} \tag{20.1}$$

is of general applicability, *provided only that the work involved is work of expansion*. For an infinitesimal change at constant volume, when dV is zero, this becomes $dS_V = dE_V/T$, or

$$\left(\frac{\partial S}{\partial E}\right)_V = \frac{1}{T}. \tag{20.2}$$

By the general rules of partial differentiation [cf. equation (4.9)], both S and E being functions of T and V,

$$\left(\frac{\partial S}{\partial E}\right)_V = \left(\frac{\partial S}{\partial T}\right)_V \Big/ \left(\frac{\partial E}{\partial T}\right)_V,$$

and since $(\partial E/\partial T)_V$ is equal to C_V, the heat capacity at constant volume, it follows from equation (20.2) that

$$\left(\frac{\partial S}{\partial T}\right)_V = \frac{C_V}{T}. \tag{20.3}$$

If *constant volume* conditions are understood, the result may be stated in the form

$$dS_V = \frac{C_V}{T} dT = C_V d\ln T,$$

and integration consequently gives

$$(S - S_0)_V = \int_0^T \frac{C_V}{T} dT = \int_0^T C_V d\ln T, \tag{20.4}$$

where S is the entropy of the system at the temperature T, and S_0 is the hypothetical value at the absolute zero, the volume being the same in each case. For the entropy increase accompanying a change in temperature from T_1 to T_2 at constant volume, it is seen that

$$(S_2 - S_1)_V = \int_{T_1}^{T_2} \frac{C_V}{T} dT = \int_{T_1}^{T_2} C_V d\ln T. \tag{20.5}$$

If C_V is independent of temperature this expression becomes identical with equation (19.24).

Since the heat content H is equal to $E + PV$ by definition [equation (9.5)], it follows that at *constant pressure*

$$dH_P = dE_P + PdV,$$

and hence from (20.1), it is evident that dH_P is equal to TdS_P, or

$$\left(\frac{\partial S}{\partial H}\right)_P = \frac{1}{T}. \tag{20.6}$$

By means of arguments exactly similar to those used in deriving equation (20.3), and utilizing the fact that $(\partial H/\partial T)_P$ is equal to C_P, the heat capacity at constant pressure, it is found that

$$\left(\frac{\partial S}{\partial T}\right)_P = \frac{C_P}{T}. \tag{20.7}$$

Hence, for temperature changes at constant pressure,

$$(S - S_0)_P = \int_0^T \frac{C_P}{T} dT = \int_0^T C_P d\ln T, \tag{20.8}$$

and

$$(S_2 - S_1)_P = \int_{T_1}^{T_2} \frac{C_P}{T} dT = \int_{T_1}^{T_2} C_P d\ln T. \tag{20.9}$$

As before, if C_P is independent of temperature, equation (20.9) becomes identical with (19.25).

An important use of equation (20.9) is to determine the increase of entropy of a system for a specified change of temperature at constant pressure. Two procedures are possible, viz., graphical and analytical. In the former, the values of C_P/T are plotted against T, or C_P is plotted against $\ln T$, and the area under the curve between the ordinates representing T_1 and T_2 is measured. This gives the value of the integral in equation (20.9), and hence the increase of entropy for the change of temperature from T_1 to T_2. Alternatively, if C_P can be expressed as a function of the temperature, for example, by means of equation (9.24), the required integration can be carried out readily.

Problem: The heat capacity at 1 atm. pressure of solid magnesium, in the temperature range from 0° to 560° C, is given by the expression

$$C_P = 6.20 + 1.33 \times 10^{-3}T + 6.78 \times 10^4 T^{-2} \text{ cal. deg.}^{-1} \text{ g. atom}^{-1}.$$

Determine the increase of entropy, per g. atom, for an increase of temperature from 300° K to 800° K at 1 atm. pressure.

Dividing through the expression for C_P by T, it is found that

$$\frac{C_P}{T} = \frac{6.20}{T} + 1.33 \times 10^{-3} + 6.78 \times 10^4 T^{-3},$$

and hence, by equation (20.9),

$$(S_{800} - S_{300})_P = \int_{300}^{800} \left(\frac{6.20}{T} + 1.33 \times 10^{-3} + 6.78 \times 10^4 T^{-3}\right) dT$$

$$= 6.20 \ln \frac{800}{300} + 1.33 \times 10^{-3} \times (800 - 300) - \tfrac{1}{2} \times 6.78 \times 10^4$$
$$\times [(800)^{-2} - (300)^{-2}]$$
$$= 6.083 + 0.665 + 0.324 = 7.07 \text{ cal. deg.}^{-1} \text{ g. atom}^{-1}.$$

The entropy increase is thus 7.07 cal. deg.$^{-1}$ g. atom^{-1}.

Attention should be called to the fact that the equations derived in this section are quite general in the respect that they are not restricted to gaseous systems. *They are applicable to liquids and solids, as well as to gases.* However, it should be noted that equations (20.4) and (20.5), at constant volume, and (20.8) and (20.9), at constant pressure, in particular, require modification if there is a phase change, e.g., fusion, vaporization or change of crystalline form, within the given temperature range. In cases of this kind, the increase of entropy accompanying the phase change, as derived in § 19h, must be added to the values given by the foregoing equations. Some illustrations of this type will be considered in § 23b.

20b. Variation of Entropy with Pressure and Volume.—Upon rearrangement of equation (20.1), it is seen that

$$PdV = TdS - dE,$$

and by applying the condition of constant temperature, this equation may be expressed in the form

$$P = T\left(\frac{\partial S}{\partial V}\right)_T - \left(\frac{\partial E}{\partial V}\right)_T. \tag{20.10}$$

Upon differentiation with respect to temperature, at constant volume, the result is

$$\left(\frac{\partial P}{\partial T}\right)_V = T\frac{\partial^2 S}{\partial V \partial T} + \left(\frac{\partial S}{\partial V}\right)_T - \frac{\partial^2 E}{\partial V \partial T}. \tag{20.11}$$

Bearing in mind that C_V is equal to $(\partial E/\partial T)_V$, it is readily found, by differentiation of equation (20.3) with respect to volume, that

$$\frac{\partial^2 S}{\partial T \partial V} = \frac{1}{T} \cdot \frac{\partial^2 E}{\partial T \partial V}.$$

For complete differentials, such as dS and dE, the order of differentiation is immaterial, so that $\partial^2 S/\partial T \partial V$ is identical with $\partial^2 S/\partial V \partial T$ and $\partial^2 E/\partial V \partial T$ with $\partial^2 E/\partial T \partial V$; combination of the result just obtained with equation (20.11) thus leads to

$$\left(\frac{\partial S}{\partial V}\right)_T = \left(\frac{\partial P}{\partial T}\right)_V, \tag{20.12}$$

so that the variation of the entropy with volume at constant temperature can be derived from the easily accessible quantity $(\partial P/\partial T)_V$. For an *ideal gas*, for example, the latter is equal to R/V, so that

$$\left(\frac{\partial S}{\partial V}\right)_T = \frac{R}{V}.$$

This is identical with the form taken by equation (19.19) at constant temperature.

General differentiation of the relationship $H = E + PV$, which defines the heat content, gives

$$dH = dE + PdV + VdP,$$

and if this is combined with equation (20.1), the result is

$$dH = TdS + VdP. \tag{20.13}$$

At constant temperature this takes the form, analogous to equation (20.10),

$$V = -T\left(\frac{\partial S}{\partial P}\right)_T + \left(\frac{\partial H}{\partial P}\right)_T, \tag{20.14}$$

and differentiation with respect to temperature at constant pressure gives

$$\left(\frac{\partial V}{\partial T}\right)_P = -T\frac{\partial^2 S}{\partial P \partial T} - \left(\frac{\partial S}{\partial P}\right)_T + \frac{\partial^2 H}{\partial P \partial T}.$$

From equation (20.7), recalling that C_P is equal to $(\partial H/\partial T)_P$, it is found upon differentiating with respect to pressure that

$$\frac{\partial^2 S}{\partial T \partial P} = \frac{1}{T} \cdot \frac{\partial^2 H}{\partial T \partial P},$$

and comparison with the preceding equation leads to the result

$$\left(\frac{\partial S}{\partial P}\right)_T = -\left(\frac{\partial V}{\partial T}\right)_P. \tag{20.15}$$

Since $(\partial V/\partial T)_P$ can be obtained either from *P-V-T* data, or from a suitable equation of state, it is possible to determine the variation of entropy with pressure at constant temperature. Upon rearrangement of equation (20.15) and integration between the pressure limits of P_1 and P_2, at *constant temperature*, the result is

$$S_2 - S_1 = -\int_{P_1}^{P_2} \left(\frac{\partial V}{\partial T}\right)_P dP, \tag{20.16}$$

and hence the entropy change $S_2 - S_1$ may be obtained by plotting the values of $(\partial V/\partial T)_P$ at various pressures against the pressure, and thus evaluating the integral graphically. Alternatively, an expression for $(\partial V/\partial T)_P$ may be obtained as a function of the pressure by means of a suitable equation of state, and then equation (20.16) can be solved analytically.

20c. Entropy Corrections for Deviation from Ideal Behavior.—One of the most useful applications of equation (20.16) is to determine the correction which must be made to entropy values obtained for real gases to allow for departure from ideal behavior. As will be explained in Chapter IX it is possible to determine the entropy of a gas at 1 atm. pressure, but it is desirable to express the result in terms of an ideal gas at the same pressure; this is referred to as the **standard entropy** of the gas. The correction is obtained in the following manner.

The increase of entropy of the actual gas from 1 atm. pressure to a very low pressure P^*, where it behaves ideally, is given by equation (20.16) as

$$S(P^* \text{ atm.}) - S(1 \text{ atm.}) = -\int_1^{P^*} \left(\frac{\partial V}{\partial T}\right)_P dP,$$

constant temperature being understood. For an ideal gas, it is evident from the equation of state $PV = RT$ that $(\partial V/\partial T)_P$ is equal to R/P, and hence the entropy increase from the very low pressure P^* to 1 atm. is given by *

$$S^*(1 \text{ atm.}) - S^*(P^* \text{ atm.}) = -\int_{P^*}^1 \frac{R}{P} dP.$$

Since the actual gas may be regarded as behaving ideally at the low pressure P^*, the quantities $S(P^* \text{ atm.})$ and $S^*(P^* \text{ atm.})$ may be taken as identical; the required entropy correction to be added to the observed entropy value $S(1 \text{ atm.})$ is then obtained by adding the two equations given above, viz.,

$$S^*(1 \text{ atm.}) - S(1 \text{ atm.}) = S^0 - S = \int_{P^*}^1 \left[\left(\frac{\partial V}{\partial T}\right)_P - \frac{R}{P}\right] dP, \quad (20.17)$$

where S^0, the usual symbol for the standard entropy, has been written in place of $S^*(1 \text{ atm.})$, and S is the observed entropy of the actual gas at 1 atm. pressure.

The integral in equation (20.17) could be evaluated graphically, but in practice it is more convenient to use an equation of state, and in this connection the Berthelot equation (§ 5f) has been employed. From this equation it is found that

$$\left(\frac{\partial V}{\partial T}\right)_P = \frac{R}{P}\left(1 + \frac{27}{32} \cdot \frac{PT_c^3}{P_c T^3}\right),$$

and hence equation (20.17) becomes

$$S^0 - S = \frac{27}{32} \cdot \frac{RT_c^3}{P_c T^3} \int_{P^*}^1 dP,$$

$$S^0 = S + \frac{27}{32} \cdot \frac{RT_c^3}{P_c T^3}, \quad (20.18)$$

* An asterisk is used, in general, to represent a system in the postulated ideal state.

the very small pressure P^* being neglected in comparison with unity. The second term on the right-hand side thus gives the correction which must be added to the observed entropy S of the gas at 1 atm. pressure in order to obtain the standard entropy S^0, for the gas behaving ideally at the same pressure.

Problem: What is the correction to be added to the observed entropy of 1 mole of nitrogen gas at 77.32° K and 1 atm. pressure to allow for departure from ideal behavior?

For nitrogen, T_c is 126.0° K and P_c is 33.5 atm. The value of P_c must be expressed in atm. since this unit has been used in the derivation of equation (20.18). When T is 77.32° K, this equation gives

$$S^0 = S + \frac{27}{32} \cdot \frac{R \times (126.0)^3}{33.5 \times (77.32)^3} = S + 0.109R.$$

Since the entropy is usually expressed in cal. deg.$^{-1}$ mole^{-1}, the value of R is 1.987 in the same units, so that the correction term $0.109R$ is equal to 0.217 cal. deg.$^{-1}$ mole^{-1}. This result will be employed in § 23c.

20d. Thermodynamic Equations of State.—By combining equations (20.10) and (20.12), there is obtained what is called a **thermodynamic equation of state**, viz.,

$$P = T\left(\frac{\partial P}{\partial T}\right)_V - \left(\frac{\partial E}{\partial V}\right)_T, \qquad (20.19)$$

for it gives a relationship between pressure, volume and temperature which is applicable to all substances, solid, liquid or gaseous. An interesting consequence of this equation arises in connection with the ideal gas law $PV = RT$; for an ideal gas, it is seen that

$$\left(\frac{\partial P}{\partial T}\right)_V = \frac{R}{V},$$

and introduction of this result into equation (20.19) gives

$$\left(\frac{\partial E}{\partial V}\right)_T = 0.$$

The constancy of the energy content of an ideal gas, irrespective of the volume, at constant temperature, which was postulated earlier (§ 9d), is thus a direct consequence of the application of the second law of thermodynamics to the equation of state for an ideal gas.

For a van der Waals gas, $(\partial P/\partial T)_V$ is equal to $R/(V - b)$, and hence it follows from equation (20.19) that

$$\left(\frac{\partial E}{\partial V}\right)_T = \frac{a}{V^2}.$$

This result gives a physical significance to the pressure correction term in the van der Waals equation (cf. § 11c, also Chapter IV, Exercise 11).

Another thermodynamic equation of state is obtained by the combination of equations (20.14) and (20.15); thus,

$$V = T\left(\frac{\partial V}{\partial T}\right)_P + \left(\frac{\partial H}{\partial P}\right)_T. \qquad (20.20)$$

For an ideal gas, $(\partial H/\partial P)_T$, like $(\partial E/\partial V)_T$, can be readily shown to be zero, so that the heat content is independent of the pressure, as stated earlier. For real gases, however, this is not the case, and equation (20.20) may be used, in conjunction with a conventional equation of state, or actual P-V-T data, to determine the change of heat content accompanying a given pressure change. This is of special value in connection with heats of reaction, when the changes of heat content are determined for actual gases at 1 atm., or other, pressure, but the data are required for ideal behavior, i.e., at very low pressures.[1]

20e. Variation of Heat Content with Pressure.—By rearrangement of equation (20.20) and integration between the pressure limits of P_1 and P_2, at constant temperature, it is found that

$$H_2 - H_1 = \int_{P_1}^{P_2}\left[V - T\left(\frac{\partial V}{\partial T}\right)_P\right]dP, \qquad (20.21)$$

where H_1 and H_2 are the heat contents of the given substance at the temperature T, and pressures P_1 and P_2, respectively. The integral in this equation may be solved graphically from P-V-T data, if available, but for gases an equation of state may be employed. Here again, as in § 20c, the Berthelot equation is useful for gaseous substances; if the values of V, derived from equation (5.19), and of $(\partial V/\partial T)_P$, given in a preceding section, are inserted in (20.21) it can be shown without difficulty that

$$\begin{aligned}H_2 - H_1 &= \frac{9}{128}\cdot\frac{RT_c}{P_c}\left(1 - 18\frac{T_c^2}{T^2}\right)\int_{P_1}^{P_2}dP \\ &= \frac{9}{128}\cdot\frac{RT_c}{P_c}\left(1 - 18\frac{T_c^2}{T^2}\right)(P_2 - P_1).\end{aligned} \qquad (20.22)$$

The change of heat content for a given pressure change can thus be evaluated from the critical constants. If heat content data are to be corrected for departure from ideal behavior, P_2 is set equal to zero; H_2 is then the heat content of the gas at this pressure when it behaves ideally. Since the heat content of an ideal gas is independent of the pressure, H_2 gives the required corrected value.

Problem: Calculate the difference in the heat content of 1 mole of oxygen gas at 1 atm. pressure and the same gas when behaving ideally at 25° C.

For oxygen, T_c is 154.3° K and P_c is 49.7 atm.; the pressures P_1 and P_2 must be expressed in the same units as P_c, i.e., in atm. In the present case P_2 is set equal to zero, and H_2 is then replaced by H^*, the ideal gas value; P_1 is 1 atm., and T is 25° C, i.e., 298.16° (or 298.2°) K. Hence, by equation (20.22), taking R as

---

[1] See, for example, Rossini, *J. Res. Nat. Bur. Stand.*, **22**, 407 (1939).

1.987 cal. deg.$^{-1}$ mole^{-1}, the result is obtained in cal. mole^{-1}; thus,

$$H^* - H_1 = \frac{9 \times 1.987 \times 154.3}{128 \times 49.7}\left(1 - 18 \times \frac{(154.3)^2}{(298.2)^2}\right) \times (-1)$$
$$= 1.66 \text{ cal. mole}^{-1}.$$

This difference is small in comparison with the experimental errors in measurements of heat of reaction; it can, therefore, be neglected in most cases, as stated in § 12b.

Because of the approximate nature of the Berthelot equation of state, it is probable that equation (20.22) is not very reliable when the pressure difference $P_2 - P_1$ is large. In connection with the study of high-pressure gas reactions, however, it is sometimes required to know the difference between the heat content of a gas at high pressure and at zero pressure. If P-V-T data are available, equation (20.21) may be used directly, but if they are lacking, or if approximate results are adequate, a generalized treatment, involving the use of reduced quantities and the compressibility factor (§ 5i), is simple and convenient.

Utilizing the definition of the compressibility factor κ, that is, $PV = \kappa RT$, it is readily found that

$$V - T\left(\frac{\partial V}{\partial T}\right)_P = -\frac{RT^2}{P}\left(\frac{\partial \kappa}{\partial T}\right)_P,$$

and hence, by equation (20.21), *at constant temperature*,

$$H_2 - H_1 = -RT^2 \int_{P_1}^{P_2} \left(\frac{\partial \kappa}{\partial T}\right)_P \frac{dP}{P}. \tag{20.23}$$

Dividing both sides by T, and replacing T on the right-hand side by θT_c and P by πP_c, where θ and π are the reduced temperature and pressure (cf. § 5e), respectively, equation (20.23) becomes

$$\frac{H_2 - H_1}{T} = -R\theta \int_{\pi_1}^{\pi_2} \left(\frac{\partial \kappa}{\partial \theta}\right)_\pi \frac{d\pi}{\pi}. \tag{20.24}$$

The values of $(\partial \kappa/\partial \theta)_\pi$ can be derived from the generalized compressibility chart (Fig. 4) and hence the integral can be evaluated graphically. If the pressure P_2, i.e., π_2, is taken as zero, H_2 may be replaced by the ideal gas value H^*, so that equation (20.24) becomes

$$\frac{H^* - H}{T} = R\theta \int_0^\pi \left(\frac{\partial \kappa}{\partial \theta}\right)_\pi \frac{d\pi}{\pi}. \tag{20.25}$$

As seen in § 5i, κ is, as a first approximation, a universal function of θ and π; it is evident, therefore, that for a definite reduced temperature and reduced pressure, the right-hand side of equation (20.25) has the same value for all gases. It is consequently, possible to construct a generalized diagram giving the (approximate) value of $(H^* - H)/T$ for any gas with the reduced pressure as coordinate and the reduced temperature as parameter. Such a diagram is represented in Fig. 14.[2]

[2] Watson and Nelson, *Ind. Eng. Chem.*, **25**, 880 (1933). For experimental study of the variation of heat content with pressure, see Gilliland and Lukes, *ibid.*, **32**, 957 (1940).

It may be remarked that a reduced form of equation (20.16) or (20.17) and a corresponding generalized diagram have been developed. As they are not widely used, however, they will not be given here.[3]

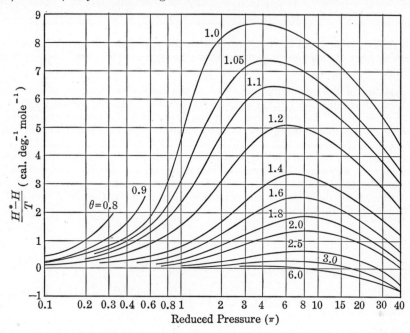

Fig. 14. Generalized $(H^* - H)/T$ curves

20f. Adiabatic Relationships.—Two relationships, analogous to (20.12) and (20.15), which are applicable to constant entropy, i.e., reversible adiabatic, processes can be derived in a similar manner. Writing equation (20.1) in the form

$$dE = TdS - PdV,$$

it is evident that at constant volume, i.e., when dV is zero,

$$\left(\frac{\partial E}{\partial S}\right)_V = T, \tag{20.26}$$

whereas at constant entropy, i.e., when dS is zero,

$$\left(\frac{\partial E}{\partial V}\right)_S = -P. \tag{20.27}$$

If equation (20.26) is differentiated with respect to volume, at constant entropy, and (20.27) with respect to entropy, at constant volume, and the

[3] Cf. Edmister, *Ind. Eng. Chem.*, **28**, 1112 (1936); York, *ibid.*, **32**, 54 (1940); Robinson and Bliss, *ibid.*, **32**, 396 (1940); Maron and Turnbull, *ibid.*, **34**, 544 (1942).

results equated, it is found that

$$\left(\frac{\partial T}{\partial V}\right)_S = -\left(\frac{\partial P}{\partial S}\right)_V. \tag{20.28}$$

From equation (20.13), that is,

$$dH = TdS + VdP,$$

the restrictions of constant pressure and constant entropy, respectively, lead to

$$\left(\frac{\partial H}{\partial S}\right)_P = T \quad \text{and} \quad \left(\frac{\partial H}{\partial P}\right)_S = V. \tag{20.29}$$

Upon differentiating with respect to pressure, at constant entropy, in the first case, and with respect to entropy, at constant pressure, in the second case, and equating the results, it is seen that

$$\left(\frac{\partial T}{\partial P}\right)_S = \left(\frac{\partial V}{\partial S}\right)_P. \tag{20.30}$$

20g. General Applicability of Results.—The four equations (20.12), (20.15), (20.28) and (20.30) are frequently referred to as the **Maxwell relations**. It is important to note that these results, as well as the general equations of state (20.19) and (20.20), are applicable to systems of all types, homogeneous or heterogeneous. Two conditions, however, must be borne in mind. The mass of the system is assumed to be constant, so that there is no loss or gain of matter in the course of any thermodynamic change. Systems of this kind, which may consist of one or more phases, are known as **closed systems**. The second condition is based on the postulate that the work done is work of expansion only, and that it is equal to PdV, where P is the pressure of the system (cf. §§ 19i, 20a). This means that the system must always remain in equilibrium with the external pressure; in other words, the pressure inside the system must either be equal to, or differ only by an infinitesimal amount from, the external pressure. An important application of equation (20.12) to heterogeneous systems will be given in § 27b.

21. Entropy and Heat Capacity Relationships

21a. Difference of Heat Capacities.—It was shown in § 9c that the difference between the heat capacities at constant pressure and constant volume, of any *homogeneous system* of constant composition, is given by

$$C_P - C_V = \left[P + \left(\frac{\partial E}{\partial V}\right)_T\right]\left(\frac{\partial V}{\partial T}\right)_P,$$

and hence, utilizing equation (20.19),

$$C_P - C_V = T\left(\frac{\partial P}{\partial T}\right)_V\left(\frac{\partial V}{\partial T}\right)_P. \tag{21.1}$$

This result is applicable to any single substance, either solid, liquid or gaseous, or to a homogeneous system containing definite amounts of two or more substances.

For 1 mole of *an ideal gas*, $(\partial P/\partial T)_V$ is equal to R/V, i.e., P/T, and $(\partial V/\partial T)_P$ is R/P; hence by equation (21.1),

$$C_P - C_V = R,$$

in agreement with equation (9.29). For a real gas, the values of $(\partial P/\partial T)_V$ and of $(\partial V/\partial T)_P$ can be obtained from P-V-T data, and hence $C_P - C_V$ can be determined. Even without these data, however, a qualitative indication of the results to be expected may be derived from an equation of state.

For a van der Waals gas, for example,

$$P = \frac{RT}{V-b} - \frac{a}{V^2},$$

and hence,

$$T\left(\frac{\partial P}{\partial T}\right)_V = \frac{RT}{V-b} = P + \frac{a}{V^2}. \tag{21.2}$$

In order to evaluate $(\partial V/\partial T)_P$ in a convenient form, the van der Waals equation is multiplied out to give

$$PV = RT - \frac{a}{V} + bP + \frac{ab}{V^2},$$

and, dividing through by P, the result is

$$V = \frac{RT}{P} - \frac{a}{PV} + b + \frac{ab}{PV^2}.$$

As a first approximation, V may be replaced by RT/P in the correction terms a/PV and ab/PV^2, so that

$$V = \frac{RT}{P} - \frac{a}{RT} + b + \frac{abP}{R^2T^2}.$$

Differentiation of this expression with respect to temperature, at constant pressure, yields

$$\left(\frac{\partial V}{\partial T}\right)_P = \frac{R}{P} + \frac{a}{RT^2} - \frac{2abP}{R^2T^3}. \tag{21.3}$$

It is readily found by arrangement of the previous equation that

$$\frac{R}{P} = \frac{V-b}{T} + \frac{a}{RT^2} - \frac{abP}{R^2T^3},$$

and combination with (21.3) gives

$$\left(\frac{\partial V}{\partial T}\right)_P = \frac{V-b}{T} + \frac{2a}{RT^2} - \frac{3abP}{R^2T^3}.$$

Introducing this result together with equation (21.2) into (21.1), and omitting some of the smaller terms, it is seen that

$$C_P - C_V = \left(P + \frac{a}{V^2}\right)\left(\frac{V-b}{T}\right) + P\left(\frac{2a}{RT^2} - \frac{3abP}{R^2T^3}\right)$$

$$= R + \frac{2a}{RT^2}P - \frac{3ab}{R^2T^3}P^2. \quad (21.4)$$

Except at low temperatures and high pressures, the last term may be neglected, so that

$$C_P - C_V \approx R + \frac{2a}{RT^2}P$$

$$\approx R\left(1 + \frac{2a}{R^2T^2}P\right). \quad (21.5)$$

It is evident, therefore, that for a real gas the value of $C_P - C_V$ is greater than R; the difference increases, as a first approximation, in a linear manner with the pressure, and is most marked at lower temperatures. The difference between $C_P - C_V$ and R is evidently greater for easily liquefiable gases, for these have, in general, higher a values.

Problem: Calculate $C_P - C_V$ for nitrogen at 25° C and 200 atm. pressure, the van der Waals a being 1.39 liter^2 atm. mole^{-2}.

Since a is given in liter^2 atm. mole^{-2}, P should be in atm., and R in liter-atm. deg.$^{-1}$ mole^{-1}, i.e., 0.0820; hence, equation (21.5) gives

$$C_P - C_V \approx R\left[1 + \frac{2 \times 1.39 \times 200}{(0.082)^2 \times (298)^2}\right]$$

$$\approx 1.93R.$$

(The actual value is approximately $1.9R$.)

At low temperatures and high pressures the value of $C_P - C_V$ can become very much larger than for an ideal gas. However, the difference between $C_P - C_V$ and R does not increase continuously with the pressure, as implied by equation (21.5). The reason for this is the $-3abP^2/R^2T^3$ term in equation (21.4). As the pressure is increased this becomes of increasing importance, and it is evident that at sufficiently high pressures $C_P - C_V$ should attain a maximum, and subsequently decrease with increasing pressure. The results obtained directly from actual P-V-T data are in qualitative agreement with this conclusion. At 20° C, the maximum, equal approximately to $2R$, is attained with nitrogen gas at about 300 atm. pressure.

Without going into details, it may be noted that a similar treatment based on the Berthelot equation of state gives the result

$$C_P - C_V \approx R\left(1 + \frac{27}{16} \cdot \frac{T_c^3}{P_c T^3}P\right), \quad (21.6)$$

applicable over a limited range of pressures.

Problem: Compare the value of $C_P - C_V$ for nitrogen at 25° C and 200 atm. pressure given by equation (21.6) with that obtained above from (21.5).

For nitrogen P_c is 33.5 atm. and T_c is 126.0° K; hence when T is 25° C, i.e., 298.2° K, and P is 200 atm., equation (21.6) gives

$$C_P - C_V \approx R\left[1 + \frac{27 \times (126.0)^3 \times 200}{16 \times 33.5 \times (298.2)^3}\right]$$
$$\approx 1.76R,$$

which may be compared with $1.93R$ from equation (21.5).

The thermodynamic equation (21.1) may also be expressed in terms of the reduced temperature, pressure and volume, and the compressibility factor. It is then possible to construct a generalized diagram for $C_P - C_V$ applicable to all gases (cf. § 20e).[4]

21b. Difference of Heat Capacities: Alternative Expression.—An alternative form of equation (21.1) has been used to determine the difference in the heat capacities of solids, liquids and gases. For a homogeneous system of constant mass the volume V is a single valued function of the temperature and pressure, and so it is possible to write

$$dV = \left(\frac{\partial V}{\partial T}\right)_P dT + \left(\frac{\partial V}{\partial P}\right)_T dP. \qquad (21.7)$$

For a process occurring at constant volume dV is zero, and hence equation (21.7) becomes

$$\left(\frac{\partial V}{\partial T}\right)_P dT_V = -\left(\frac{\partial V}{\partial P}\right)_T dP_V$$

or

$$\left(\frac{\partial P}{\partial T}\right)_V = -\left(\frac{\partial V}{\partial T}\right)_P \bigg/ \left(\frac{\partial V}{\partial P}\right)_T.$$

Upon introducing this result into equation (21.1) it follows that

$$C_P - C_V = -T\left(\frac{\partial V}{\partial T}\right)_P^2 \bigg/ \left(\frac{\partial V}{\partial P}\right)_T. \qquad (21.8)$$

This equation has been utilized in connection with P-V-T data, either in graphical or analytical form, to determine $C_P - C_V$ values for a number of gases over a considerable range of temperature and pressure.[5]

For solids and liquids another form of equation (21.8) is convenient. The quantity $\frac{1}{V}\left(\frac{\partial V}{\partial T}\right)_P$ is equal to the coefficient of (cubical) thermal expansion α of the substance constituting the system, and $-\frac{1}{V}\left(\frac{\partial V}{\partial P}\right)_T$ is the

---

[4] Schulze, Z. phys. Chem., **88**, 490 (1914); Edmister, ref. 3; Ind. Eng. Chem., **32**, 373 (1940).

[5] Deming and Shupe, Phys. Rev., **37**, 638 (1931); **38**, 2245 (1931); **40**, 848 (1932); Roper, J. Phys. Chem., **45**, 321 (1941).

compressibility coefficient β; hence,

$$C_P - C_V = \frac{\alpha^2 T V}{\beta}. \tag{21.9}$$

This equation has been found especially useful for the conversion of heat capacities of solids, in particular, measured at constant (atmospheric) pressure to the values at constant volume (§ 17a).

Problem: For metallic copper at 25° C, the coefficient of expansion α is 49.2×10^{-6} deg.$^{-1}$, and the compressibility coefficient β is 0.785×10^{-6} atm.$^{-1}$; the density is 8.93 g. cc.$^{-1}$ and atomic weight 63.57. Calculate the difference in the atomic heat capacities at constant pressure and constant volume.

In the present case V in equation (21.9) is the atomic volume, i.e., the atomic weight divided by the density; if this is expressed in cc. g. atom^{-1}, i.e., 63.57/8.93, it can be readily seen from equation (21.9) that using the values given for α and β, $C_P - C_V$ will be in cc.-atm. deg.$^{-1}$ g. atom^{-1}. Since 1 cc.-atm. is equivalent to 0.0242 cal. (Table 1, Appendix), it follows that at 25° C, i.e., 298.2° K,

$$C_P - C_V = \frac{(49.2 \times 10^{-6})^2 \times 298.2 \times 63.57 \times 0.0242}{0.785 \times 10^{-6} \times 8.93}$$
$$= 0.159 \text{ cal. deg.}^{-1} \text{ g. atom}^{-1}.$$

21c. Determination of Heat Capacity.—For a simple system, e.g., a single substance or a homogeneous mixture of constant composition, the entropy is dependent on two thermodynamic variables, e.g., temperature and pressure, only, so that

$$dS = \left(\frac{\partial S}{\partial T}\right)_P dT + \left(\frac{\partial S}{\partial P}\right)_T dP,$$

and hence, by the method used in § 21b, it follows that

$$\left(\frac{\partial T}{\partial P}\right)_S = -\left(\frac{\partial S}{\partial P}\right)_T \bigg/ \left(\frac{\partial S}{\partial T}\right)_P.$$

Utilizing equation (20.15) for $(\partial S/\partial P)_T$, and (20.7) for $(\partial S/\partial T)_P$, it is found upon rearrangement that

$$C_P = T \left(\frac{\partial V}{\partial T}\right)_P \bigg/ \left(\frac{\partial T}{\partial P}\right)_S. \tag{21.10}$$

This equation has been used for the determination of the heat capacities at constant pressure of both liquids and gases. The quantity $(\partial V/\partial T)_P$ is the rate of thermal expansion and this can be measured without difficulty. The other factor, $(\partial T/\partial P)_S$, is called the adiabatic temperature coefficient, since it applies to constant entropy, i.e., adiabatic, conditions. It can be determined by allowing the fluid to expand suddenly, and hence adiabatically, over a known pressure range, and observing the temperature change.[6]

[6] Joule, *Phil. Mag.*, **17**, 364 (1859); Lummer and Pringsheim, *Ann. Physik*, **64**, 555 (1898); Eucken and Mücke, *Z. phys. Chem.*, **B18**, 167 (1932); Dixon and Rodebush, *J. Am. Chem. Soc.*, **49**, 1162 (1927); Richards and Wallace, *ibid.*, **54**, 2705 (1932); Burlew, *ibid.*, **62**, 681, 690, 696 (1940).

21d. Variation of Heat Capacity at Constant Pressure with Pressure.—

It is a simple matter to derive an expression for the influence of pressure on heat capacity. Upon differentiating equation (20.7), i.e.,

$$\left(\frac{\partial S}{\partial T}\right)_P = \frac{C_P}{T},$$

with respect to pressure, at constant temperature, and equation (20.15), i.e.,

$$\left(\frac{\partial S}{\partial P}\right)_T = -\left(\frac{\partial V}{\partial T}\right)_P,$$

with respect to temperature, at constant pressure, and equating the results, it is found that

$$\frac{\partial^2 S}{\partial T \partial P} = \frac{1}{T}\left(\frac{\partial C_P}{\partial P}\right)_T = -\left(\frac{\partial^2 V}{\partial T^2}\right)_P,$$

$$\left(\frac{\partial C_P}{\partial P}\right)_T = -T\left(\frac{\partial^2 V}{\partial T^2}\right)_P. \quad (21.11)$$

This equation holds for any homogeneous substance, but it is usually applied to gases. For an ideal gas, it is evident from the equation $PV = RT$ that $(\partial^2 V/\partial T^2)_P$ is zero, and hence the heat capacity should be independent of the pressure (cf. § 9e). Real gases, however, exhibit marked variations of heat capacity with pressure, especially at low temperatures; at $-70°$ C, for example, the value of C_P for nitrogen increases from 6.8 at low pressures to 12.1 cal. deg.$^{-1}$ mole^{-1} at 200 atm. At ordinary temperatures, however, the heat capacity increases by about 2 cal. deg.$^{-1}$ mole^{-1} for the same increase of pressure.

The actual change of heat capacity with pressure is given by an expression obtained by the integration of equation (21.11). At a sufficiently low pressure, represented by P^*, where the gas behaves ideally, the heat capacity C_P^* may be regarded as virtually independent of pressure; this pressure may be taken as the lower limit of integration, and if the upper limit is any pressure P at which the heat capacity at constant pressure is C_P, then

$$C_P - C_P^* = -T \int_{P^*}^{P} \left(\frac{\partial^2 V}{\partial T^2}\right)_P dP, \quad (21.12)$$

at a constant temperature T. The values of $(\partial^2 V/\partial T^2)_P$ for a given gas at various pressures can be derived from P-V-T measurements, if available; by graphical integration $C_P - C_P^*$ can then be determined for any desired pressure.[7]

If the constants in a satisfactory equation of state, e.g., the Beattie-Bridgeman equation, were known, $(\partial^2 V/\partial T^2)_P$ could be expressed analytically as a function of the pressure, and then integrated in accordance with equation (21.12). The treatment may be illustrated in a simple manner by utilizing the van der Waals

---

[7] See ref. 5; also, Hoxton, *Phys. Rev.*, **36**, 1091 (1930).

equation. It was seen in § 21a that for a van der Waals gas [cf. equation (21.3)]

$$\left(\frac{\partial V}{\partial T}\right)_P = \frac{R}{P} + \frac{a}{RT^2} - \frac{2abP}{R^2T^3},$$

and hence

$$\left(\frac{\partial^2 V}{\partial T^2}\right)_P = -\frac{2a}{RT^3} + \frac{6abP}{R^2T^4},$$

so that from equation (21.12)

$$C_P - C_P^* = \int_{P^*}^{P} \left(\frac{2a}{RT^2} - \frac{6abP}{R^2T^3}\right) dP.$$

Assuming a and b to be independent of the pressure, and taking P^* as zero, it follows that

$$C_P - C_P^* = \frac{2a}{RT^2} P - \frac{3ab}{R^2T^3} P^2. \tag{21.13}$$

At moderate temperatures and pressures the second term on the right-hand side of equation (21.13) may be neglected; the value of C_P should thus increase in a linear manner with the pressure. The rate of increase should be less the higher the temperature. With increasing pressure the effect of the second term will become appreciable, especially at low temperatures, and at sufficiently high pressures C_P should reach a maximum and then decrease. These qualitative expectations are in agreement with the results obtained by the application of P-V-T data to equation (21.12), and also with the limited experimental determinations of heat capacities at high pressures.[8]

From the Berthelot equation of state in conjunction with (21.12) it is found that

$$C_P - C_P^* \approx \frac{81}{32} \cdot \frac{RT_c^3}{P_c T^3} P. \tag{21.14}$$

This result, as might be expected, is reliable at moderate temperatures and pressures, when $C_P - C_P^*$ is a linear function of the pressure.

Problem: Calculate the change of C_P for nitrogen when the pressure is increased to 100 atm. at 25° C, using (i) the van der Waals equation (ii) the Berthelot equation.

(i) For nitrogen, a is 1.39 and b is 3.92×10^{-2} in liter, atm. and mole units; R must therefore be in liter-atm. deg.$^{-1}$ mole^{-1}, and P in atm., i.e., 100 atm.; equation (21.13) then gives, with $T = 298°$ K,

$$C_P - C_P^* \approx \frac{2 \times 1.39 \times 100}{0.082 \times (298)^2}$$
$$\approx 0.038 \text{ liter-atm. deg.}^{-1} \text{ mole}^{-1},$$

the second term being negligible. Since 1 liter-atm. is equivalent to 24.2 cal., the change in heat capacity should be 0.93 cal. deg.$^{-1}$ mole^{-1}.

[8] Worthing, *Phys. Rev.* **33**, 217 (1911); Mackey and Krase, *Ind. Eng. Chem.*, **22**, 1060 (1930).

(ii) Taking T_c as 126.0° K and P_c as 33.5 atm., equation (21.14) gives

$$C_P - C_P^* \approx \frac{81 \times 0.082 \times (126)^3 \times 100}{32 \times 33.5 \times (298)^3}$$
$$\approx 0.047 \text{ liter-atm. deg.}^{-1} \text{ mole}^{-1}.$$

This is equivalent to 1.1 cal. deg.$^{-1}$ mole^{-1}. (The experimental value is about 1.0 cal. deg.$^{-1}$ mole^{-1}.)

As in the case of other thermodynamic properties, it is possible to derive a general relationship for $C_P - C_P^*$ involving the reduced temperature and pressure. The simplest method of approach to this problem is to utilize the familiar definition of C_P as $(\partial H/\partial T)_P$; hence,

$$C_P - C_P^* = \left(\frac{\partial H}{\partial T}\right)_P - \left(\frac{\partial H^*}{\partial T}\right)_P = \left[\frac{\partial (H - H^*)}{\partial T}\right]_P, \quad (21.15)$$

where, as before, H and H^* refer to the heat contents at an appreciable pressure P and at a very low pressure, respectively. Upon introducing the mathematical result

$$\frac{d[(H - H^*)/T]}{d \ln T} = T \frac{d[(H - H^*)/T]}{dT} = \frac{d(H - H^*)}{dT} - \frac{H - H^*}{T},$$

equation (21.15) gives

$$C_P - C_P^* = \frac{H - H^*}{T} + \left\{\frac{\partial [(H - H^*)/T]}{\partial \ln T}\right\}_P. \quad (21.16)$$

The temperature T may be expressed as θT_c, where θ is the reduced temperature; hence $d \ln T$ is equal to $d \ln \theta$, since T_c is constant, and equation (21.16) may consequently be written as

$$C_P - C_P^* = \frac{H - H^*}{T} + \left\{\frac{\partial [(H - H^*)/T]}{\partial \ln \theta}\right\}_P. \quad (21.17)$$

It was seen earlier (§ 20e) that $(H - H^*)/T$ is a function of the reduced temperature and pressure of a form applicable to all gases [cf. equation (20.25)], at least as a first approximation; hence the same must be true for $C_P - C_P^*$, in accordance with equation (21.17).† The first term on the right-hand side is obtained from equation (20.25) or its equivalent (Fig. 14), and the second term from the slope of the plot of $(H - H^*)/T$ against θ (or $\ln \theta$), all the values being at the reduced pressure corresponding to the pressure P at which $C_P - C_P^*$ is required.[9]

21e. Variation of Heat Capacity at Constant Volume with Volume.—By means of equations (20.3) and (20.12), a relationship analogous to (21.11) can be obtained for the variation of C_V with volume, which is related to the variation with pressure, at constant temperature; this is

$$\left(\frac{\partial C_V}{\partial V}\right)_T = T \left(\frac{\partial^2 P}{\partial T^2}\right)_V, \quad (21.18)$$

† Since a good approximation for a function does not, on differentiation, necessarily give a good approximation for the derivative of the function, this conclusion may be open to some objection.

[9] Dodge, *Ind. Eng. Chem.*, **24**, 1353 (1932); Watson and Nelson, *ibid.*, **25**, 880 (1933); see also, ref. 3.

which, upon integration, gives

$$C_V - C_V^* = T \int_{V^*}^{V} \left(\frac{\partial^2 P}{\partial T^2}\right)_V dV, \qquad (21.19)$$

where C_V^* is the constant heat capacity at low pressure, i.e., large volume V^*. The chief application of these equations, like (21.11) and (21.12), is to gases. For an ideal gas, it can be readily seen that since $(\partial^2 P/\partial T^2)_V$ is zero, C_V is independent of the volume (or pressure), as is to be expected, but for a real gas this is not necessarily the case. The values of $(\partial^2 P/\partial T^2)_V$ can be derived from P-V-T measurements, and hence $C_V - C_V^*$ can be obtained by graphical integration; alternatively, an analytical method, similar to that described above, may be used.

It is of interest to note that $(\partial^2 P/\partial T^2)_V$ is zero for a van der Waals gas, as well as for an ideal gas; hence, C_V should also be independent of the volume (or pressure) in the former case. In this event, the effect of pressure on C_P is equal to the variation of $C_P - C_V$ with pressure. Comparison of equations (21.4) and (21.13), both of which are based on the van der Waals equation, shows this to be true. For a gas obeying the Berthelot equation or the Beattie-Bridgeman equation $(\partial^2 P/\partial T^2)_V$ would not be zero, and hence some variation of C_V with pressure is to be expected. It is probable, however, that this variation is small, and so for most purposes the heat capacity of any gas at constant volume may be regarded as being independent of the volume or pressure. The maximum in the ratio γ of the heat capacities at constant pressure and volume, respectively, i.e., C_P/C_V, referred to earlier (§ 10e), should thus occur at about the same pressure as that for C_P, at any temperature.

22. The Joule-Thomson Effect

22a. The Joule-Thomson Coefficient.—Although the subject matter of this section has no direct connection with entropy, it may be considered here because the results are based on one of the thermodynamic equations of state derived in § 20d; to this extent the material may be regarded as a consequence of the entropy concept. In equation (20.20), viz.,

$$V = T\left(\frac{\partial V}{\partial T}\right)_P + \left(\frac{\partial H}{\partial P}\right)_T,$$

$(\partial H/\partial P)_T$ may be replaced by $-\mu_{\text{J.T.}} C_P$, as given by equation (11.5), where $\mu_{\text{J.T.}}$ is the Joule-Thomson coefficient, i.e., $(\partial T/\partial P)_H$; hence, for a fluid, i.e., liquid or gas, it is possible to write

$$V = T\left(\frac{\partial V}{\partial T}\right)_P - \mu_{\text{J.T.}} C_P \qquad (22.1)$$

or

$$\mu_{\text{J.T.}} = \frac{1}{C_P}\left[T\left(\frac{\partial V}{\partial T}\right)_P - V\right]. \qquad (22.2)$$

This expression is based upon thermodynamic considerations only, and hence is exact; the Joule-Thomson coefficient, at any temperature, may thus be determined by inserting experimental results for $(\partial V/\partial T)_P$, V and C_P under the given conditions. The values derived in this manner have been found to agree closely with those obtained by direct experiment.[10]

Problem: At 20° C, the value of the dimensionless quantity $(T/V)(\partial V/\partial T)_P$ for nitrogen was found to be 1.199 at 100 atm.; the specific volume was then 8.64 ml. g.$^{-1}$ and C_P was 8.21 cal. deg.$^{-1}$ mole^{-1}. Calculate the Joule-Thomson coefficient of nitrogen under these conditions.

Equation (22.2) may be put in the form

$$\mu_{\text{J.T.}} = \frac{V}{C_P}\left[\frac{T}{V}\left(\frac{\partial V}{\partial T}\right)_P - 1\right],$$

and if C_P is the molar heat capacity, V must be the molar volume. Since $\mu_{\text{J.T.}}$ is usually expressed in deg. atm.$^{-1}$, it will be convenient to have V in liter mole^{-1} and C_P in liter-atm. deg.$^{-1}$ mole^{-1}; the equation is then seen to be dimensionally correct. The molecular weight of nitrogen is 28.0, and the molar volume is $8.64 \times 28.0 \times 10^{-3}$ liter; C_P is 8.21×0.0413 liter-atm. deg.$^{-1}$ mole^{-1}, and hence

$$\mu_{\text{J.T.}} = \frac{8.64 \times 28.0 \times 10^{-3}}{8.21 \times 0.0413}(1.199 - 1) = 0.142° \text{ atm.}^{-1}$$

(The direct experimental value is 0.143° atm.$^{-1}$.)

For an ideal gas, satisfying the equation $PV = RT$ under all conditions, $(\partial V/\partial T)_P$ is equal to V/T; it follows, therefore, from equation (22.2), that the Joule-Thomson coefficient is always zero. For a real gas, however, this coefficient is usually not zero even at very low pressures, when ideal behavior is approached in other respects. That this is the case may be seen by making use of an equation of state for a real gas.

For a van der Waals gas, for example, it is seen from the results in § 21a that

$$T\left(\frac{\partial V}{\partial T}\right)_P = V - b + \frac{2a}{RT} - \frac{3abP}{R^2T^2},$$

and hence

$$T\left(\frac{\partial V}{\partial T}\right)_P - V = \frac{2a}{RT} - b - \frac{3abP}{R^2T^2}.$$

The equation (22.2) for the Joule-Thomson coefficient thus becomes

$$\mu_{\text{J.T.}} = \frac{1}{C_P}\left(\frac{2a}{RT} - b - \frac{3abP}{R^2T^2}\right). \quad (22.3)$$

At very low, or zero, pressure, the last term in the parentheses is negligible

---

[10] Deming and Shupe, *Phys. Rev.*, **37**, 638 (1931), **48**, 448 (1935); see also, Perry and Herrman, *J. Phys. Chem.*, **41**, 1189 (1935); Edmister, *Ind. Eng. Chem.*, **28**, 1112 (1936); Benedict, *J. Am. Chem. Soc.*, **59**, 1189 (1937); Maron and Turnbull, *Ind. Eng. Chem.*, **34**, 544 (1942).

and then

$$\mu^*_{\text{J.T.}} = \frac{1}{C^*_P}\left(\frac{2a}{RT} - b\right). \qquad (22.4)$$

Except in the special circumstances when $2a/RT$ is equal to b, the Joule-Thomson coefficient at low pressures, as given by equation (22.4), is not zero.

If the van der Waals a and b are known it is possible to obtain an indication of the value of the Joule-Thomson coefficient at any given temperature and pressure by means of equation (22.3). The results are approximate only, since a and b vary with temperature and pressure, the values generally employed (Table I) being based on critical data.

Problem: Calculate the Joule-Thomson coefficient of nitrogen gas at 20° C and 100 atm. pressure, taking C_P as 8.21 cal. deg.$^{-1}$ mole^{-1}.

Since a and b are usually given in liter, atm. and mole units, viz., 1.39 and 2.92×10^{-2}, R is 0.082 liter-atm. deg.$^{-1}$ mole^{-1} and C_P is 8.21×0.0413 liter-atm. deg.$^{-1}$ mole^{-1}; hence at 20° C (293° K), equation (22.3) gives, with P equal to 100 atm.,

$$\mu_{\text{J.T.}} = \frac{1}{8.21 \times 0.0413}$$

$$\times \left[\frac{2 \times 1.39}{0.082 \times 293} - 2.92 \times 10^{-2} - \frac{3 \times 1.39 \times 2.92 \times 10^{-2} \times 100}{(0.082)^2 \times (293)^2}\right]$$

$$= 0.19° \text{ atm.}^{-1}.$$

(This is somewhat larger than the experimental value given above. If equation (22.4) is employed, the result is still larger, viz., 0.25° atm.$^{-1}$.)

22b. The Joule-Thomson Inversion Temperature.—The condition for the Joule-Thomson inversion temperature, where the Joule-Thomson coefficient changes sign from positive to negative or vice versa (§ 11c), can be obtained by setting $\mu_{\text{J.T.}}$ equal to zero. *For a van der Waals gas*, equation (22.3) gives the condition as

$$\frac{2a}{RT_i} - b - \frac{3abP}{R^2 T_i^2} = 0 \qquad (22.5)$$

$$T_i^2 - \frac{2a}{Rb} T_i + \frac{3aP}{R^2} = 0. \qquad (22.6)$$

TABLE XIV. JOULE-THOMSON INVERSION TEMPERATURES FOR NITROGEN

Press.	Inversion Temperature	
	Upper	Lower
1 atm.	348° C	—
20	330.0°	− 167.0° C
60	299.6°	− 162.4°
100	277.2°	− 156.5°
180	235.0°	− 134.7°
220	212.5°	− 117.2°
300	158.7°	− 68.7°
376	40°	

This equation is a quadratic, and hence there will be, in general, two values of the inversion temperature T_i for every pressure, as stated in § 11c. The experimental values obtained for nitrogen gas are recorded in Table XIV, and are plotted (full line) in Fig. 15.[11] For all temperatures within the curve

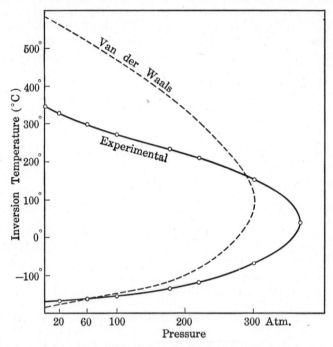

Fig. 15. Joule-Thomson inversion curve for nitrogen

the Joule-Thomson coefficient is positive, a throttled expansion being accompanied by a fall of temperature; outside the curve the coefficient is negative.

A generalized expression for the Joule-Thomson inversion temperature may be obtained by the use of a reduced equation of state. By setting equation (22.2) equal to zero, the condition for the inversion temperature T_i is then seen to be

$$T_i \left(\frac{\partial V}{\partial T} \right)_P - V = 0, \tag{22.7}$$

since C_P is not zero. If the pressure, volume and temperature are expressed in terms of the corresponding reduced quantities π, ϕ and θ, equation (22.7) becomes

$$\theta_i \left(\frac{\partial \phi}{\partial \theta} \right)_\pi - \phi = 0, \tag{22.8}$$

[11] Roebuck and Osterberg, *Phys. Rev.*, **48**, 450 (1935).

From the reduced form of the van der Waals equation (5.16), i.e.,

$$\left(\pi + \frac{3}{\phi^2}\right)(3\phi - 1) = 8\theta, \tag{22.9}$$

it follows upon differentiation, at constant reduced pressure, that

$$\left(\frac{\partial \phi}{\partial \theta}\right)_\pi = \frac{8\phi^3}{3\pi\phi^3 - 9\phi + 6}.$$

If this result is introduced into equation (22.8), and π is eliminated by means of (22.9), it is found that

$$\theta_i = \frac{3(3\phi - 1)^2}{4\phi^2}, \tag{22.10}$$

and substitution of this value for θ in (22.9) gives for the corresponding reduced pressure,

$$\pi_i = \frac{9(2\phi - 1)}{\phi^2}. \tag{22.11}$$

By combining equations (22.10) and (22.11), it is possible to eliminate ϕ and thus obtain a general relationship between the reduced inversion temperature and pressure applicable to all gases. Provided π_i is less than 9, solution of equation (22.11) gives two real values of ϕ for each value of π_i; insertion of these two ϕ's in equation (22.10) then gives the two reduced inversion temperatures θ_i for the particular reduced pressure π_i. By choosing various values of the latter from 0 to 9, the data can be obtained for a generalized, reduced inversion temperature-pressure curve, which should be applicable to any gas. The curve derived from equations (22.10) and (22.11) is shown by the broken line in Fig. 15.[12]

Comparison of the two curves in Fig. 15 indicates that the reduced equation *for a van der Waals gas* is qualitatively correct, but is not quantitatively accurate. This fact is brought out more clearly by considering some actual results. According to equation (22.11), or Fig. 15, the maximum value of the reduced pressure for which inversion is possible is 9; ϕ is then unity, and θ_i should be 3, by equation (22.10). For nitrogen, T_c is 126.0° K and P_c is 33.5 atm., so that the maximum pressure for the observation of a Joule-Thomson inversion should be $9P_c$, i.e., 9×33.5 or 301.5 atm., the temperature should then be $3T_c$, i.e., 3×126.0 or 378° K. The experimental values are 376 atm. and 313° K. Further, at very low, e.g., zero, pressure, ϕ can be either 0.5 or infinity, by equation (22.11); insertion of these values in (22.10) gives the two values of the reduced inversion temperature as 0.75 and 6.75, so that the actual (absolute) temperatures should be $0.75T_c$ and $6.75T_c$. For nitrogen, these would be 94.5° K and 850° K, compared with the (extrapolated) experimental values of about 103° K and 623° K.

Although equations (22.10) and (22.11) yield results that are not exact, they are, nevertheless, useful when direct measurements have not been made; they provide an indication of the range within which cooling of a given gas by the Joule-Thomson effect is possible. Such information would be of value in connection with the liquefaction of the particular gas.

[12] Porter, *Phil. Mag.*, **11**, 554 (1906); **19**, 888 (1910).

EXERCISES

1. Give the complete derivation of equation (20.7).

2. Give the complete derivation of equation (20.20), and show that the heat content of a gas which satisfies the equation $PV = RT$ at all temperatures and pressures is independent of pressure.

3. Derive an expression for $S^0 - S$ for a van der Waals gas, where S^0 is the standard entropy and S is the experimental entropy at 1 atm. at the same temperature, using equations (20.17) and (21.3). Calculate the value of $S^0 - S$ for nitrogen gas at 77.32° K (cf. problem in § 20c).

4. By means of the expression based on the Berthelot equation, determine the correction for the deviation from ideal behavior of the entropy of chlorine gas measured at its boiling point (239.0° K) at 1 atm.

5. Determine the increase in entropy of nitrogen gas when it is heated from 27° C to 1227° C at 1 atm. pressure. (Use C_P data in Table II.)

6. The heat capacity of solid iodine at any temperature $t°$ C, from 25° C to the melting point (113.6° C), at 1 atm. pressure, is given by

$$C_P = 13.07 + 3.21 \times 10^{-4}(t - 25) \text{ cal. deg.}^{-1} \text{ mole}^{-1}$$

[Frederick and Hildebrand, *J. Am. Chem. Soc.*, **60**, 1436 (1938)]. The heat of fusion at the melting point is 3,740 cal. mole^{-1}. The heat capacity of the liquid is approximately constant at 19.5 cal. deg.$^{-1}$ mole^{-1}, and the heat of vaporization at the normal boiling point (184° C) is 6,100 cal. mole^{-1}. Determine the increase of entropy accompanying the change of 1 mole of iodine from solid at 25° C to vapor at 184° C, at 1 atm. pressure.

7. Taking C_P for ice and water as 9.0 and 18.0 cal. deg.$^{-1}$ mole^{-1}, respectively, and the heat of fusion as 79.8 cal. g.$^{-1}$ at 0° C, determine the change of entropy accompanying the spontaneous solidification of supercooled water at $-10°$ C at 1 atm. pressure (cf. Exercise 7, Chapter VII).

8. Show that, in general, for any thermodynamic change in state, the corresponding entropy change is given by

$$\Delta S = S_2 - S_1 = \int_{T_1}^{T_2} C_P \frac{dT}{T} - \int_{P_1}^{P_2} \left(\frac{\partial V}{\partial T}\right)_P dP.$$

Suggest a method for determining ΔS for a gas.

9. Show that by the use of the compressibility factor κ, defined by $PV = \kappa RT$, it is possible to express equation (20.16) in the reduced form applicable to any gas. Describe the construction of a generalized chart for determining the (approximate) change of entropy of any gas with pressure at constant temperature.

10. Derive an expression for $H_2 - H_1$ for a van der Waals gas, where H_1 and H_2 are the heat contents at pressures P_1 and P_2, respectively, at the same temperature, using equation (20.21) and the results in § 21a. Calculate the difference in the heat content, in cal. per mole of oxygen, at 1 atm. and 100 atm. at 25° C.

11. By means of Fig. 14 estimate the change of heat content, in cal., accompanying the compression of 1 mole of ethane at 50° C from 30 atm. to 300 atm. pressure. What other information would be required to calculate the change of heat content if the temperature were changed, as well as the pressure?

12. Utilize Fig. 14 to determine the change in ΔH for the reaction $\frac{1}{2}N_2(g) + \frac{3}{2}H_2(g) = NH_3(g)$ as a result of increasing the pressure from a low value, virtually zero, to 200 atm. at 450° K.

EXERCISES

13. Estimate the difference between ΔH for the reaction $H_2(g) + \tfrac{1}{2}O_2(g) = H_2O(g)$ for the actual gases each at 1 atm. and 25° C, and the value that would be expected if the gases behaved ideally.

14. The coefficient of cubical expansion α of sodium at 20° C is 21.3×10^{-5} deg.$^{-1}$ and the compressibility coefficient β is 15.6×10^{-6} megabar^{-1} (1 megabar = 10^6 bar = 10^6 dynes cm.$^{-2}$); the density is 0.97 g. cc.$^{-1}$ [Eastman, et al., *J. Am. Chem. Soc.*, **46**, 1184 (1924)]. Calculate $C_P - C_V$ per g. atom of solid sodium at 20° C, and determine C_P at this temperature by using the Debye heat capacity equation.

15. Show that equation (21.5) can be derived directly from the simplified van der Waals equation applicable at moderate pressures (Exercise 7, Chapter II).

16. Verify equation (21.6).

17. Give the complete derivation of equation (21.18).

18. The simplified form of the van der Waals equation (Exercise 7, Chapter II) can also be written as $PV = RT + AP$, where A is a function of the temperature. Show that under these conditions $(\partial C_P/\partial P)_T = -T(\partial^2 A/\partial T^2)_P$ and hence derive an expression for $C_P - C_P^*$ in terms of the van der Waals constants.

19. By means of equation (21.4), derive an expression for the pressure at which $C_P - C_V$ for a van der Waals gas has a maximum value at any given temperature. Determine the pressure at which $C_P - C_V$ is a maximum for nitrogen gas at 25° C. What is the value of $C_P - C_V$ at this pressure? (The experimental pressure is about 300 atm. and $C_P - C_V$ is then approximately 4 cal. deg.$^{-1}$ mole^{-1}.)

20. Utilize equation (21.13) to find the pressure at which C_P is a maximum for a van der Waals gas. Why is this identical with the pressure at which $C_P - C_V$ is a maximum? If C_P^* for nitrogen is 6.96 cal. deg.$^{-1}$ mole^{-1}, calculate the maximum value of C_P/C_V for this gas at 25° C, making use of the result obtained in the preceding exercise.

21. Prove Reech's theorem, that for any homogeneous system,

$$C_P/C_V = (\partial P/\partial V)_S/(\partial P/\partial V)_T.$$

Show that this is essentially the basis of the Clément and Desormes method for determining γ for gases [equation (10.11)].

22. By means of equation (21.8) explain how a generalized (reduced) chart for $C_P - C_V$ could be constructed.

23. Calculate the Joule-Thomson coefficient of carbon monoxide at 25° C and 400 atm. pressure, given that $(T/V)(\partial V/\partial T)_P$ is 0.984, the molar volume is 76.25 cc. mole^{-1} and C_P is 8.91 cal. deg.$^{-1}$ mole^{-1} [Deming and Shupe, *Phys. Rev.*, **38**, 2245 (1931)].

24. Determine the value of the Joule-Thomson coefficient for the conditions in the preceding exercise, assuming the carbon monoxide to behave as a van der Waals gas.

25. By using equation (22.2) in conjunction with the compressibility factor (κ), derive a simple condition for the Joule-Thomson inversion temperature of a gas at any (reduced) pressure in terms of the variation of κ with the (reduced) temperature.

CHAPTER IX

ENTROPY: DETERMINATION AND SIGNIFICANCE

23. The Third Law of Thermodynamics

23a. Entropy at the Absolute Zero.—It was seen in § 20a that the entropy of any substance at the temperature T, and a given pressure, could be expressed by means of the relationship [cf. equation (20.8)]

$$S - S_0 = \int_0^T \frac{C_P}{T} dT = \int_0^T C_P d\ln T, \qquad (23.1)$$

where S_0 is the hypothetical entropy at the absolute zero. If the value of S_0 were known it would be possible to derive the entropy at any required temperature from heat capacity data. Following the development of the "heat theorem" of W. Nernst (1906), now chiefly of historical interest, M. Planck (1912) made a suggestion concerning the value of S_0 which has become known as the **third law of thermodynamics**. The first and second laws have led to the development of the concepts of energy content and entropy, respectively. The so-called third law differs, however, from these in the respect that it leads to no new concepts; it merely places a limitation upon the value of the entropy. For this reason, some writers hesitate to refer to it as a law of thermodynamics. Nevertheless, it is a generalization which leads to conclusions in agreement with experience, and hence is a "law" in the scientific sense of the term. In its broadest form the law may be stated as follows: *Every substance has a finite positive entropy, but at the absolute zero of temperature the entropy may become zero, and does so become in the case of a perfectly crystalline substance.* The exact significance of a "perfectly crystalline substance" will be more clearly understood later (§ 24m), but for the present it should be taken to represent substances in the pure solid state, to the exclusion of solid solutions and amorphous substances, such as glasses.[1]

23b. Experimental Determination of Entropy.—According to the third law of thermodynamics the entropy of a pure solid may be taken as zero at $0°$ K; hence S_0 in equation (23.1) is zero for a pure solid substance. The problem of evaluating the entropy of such a solid therefore resolves itself into the determination of the heat capacity at a series of temperatures right down to the absolute zero. The values of C_P/T are then plotted against T, or C_P is plotted against $\ln T$, i.e., 2.303 log T, for the whole range of tempera-

[1] Lewis and Gibson, *J. Am. Chem. Soc.*, **39**, 2554 (1917); **42**, 1529 (1920); Gibson, Parks and Latimer, *ibid.*, **42**, 1533, 1542 (1920); Gibson and Giauque, *ibid.*, **45**, 93 (1923); Pauling and Tolman, *ibid.*, **47**, 2148 (1925); Eastman, *Chem. Rev.*, **18**, 257 (1936); Eastman and Milner, *J. Chem. Phys.*, **1**, 445 (1933); see also, Cross and Eckstrom, *ibid.*, **10**, 287 (1942).

ture; since S_0 is zero, the area under the curve from $0°$ K to any temperature T then gives the entropy of the solid at this temperature, and at the constant pressure, e.g., 1 atm., at which the heat capacities were measured. Since the heat capacity determinations cannot be made right down to the absolute zero, the observations are carried to as low a temperature as possible, e.g., $10°$ of $15°$ K; the results are then extrapolated to $0°$ K. In many cases the data can be represented by a Debye function (§ 17c) and the characteristic temperature θ can be calculated; the values of C_V down to $0°$ K can then be obtained by equation (17.4) which is applicable at low temperatures. It will be noted that the Debye equation gives C_V whereas C_P is required for the evaluation of the entropy by equation (23.1). However, the difference between these two quantities is so small at temperatures from $0°$ to $10°$ or $15°$ K as to be quite negligible. This may be seen from equation (21.1), for example, since both T and $(\partial V/\partial T)_P$, as will be shown below (§ 23d), approach zero as the temperature becomes smaller. All the information is now available for the determination of the entropy of the solid at any temperature up to which the heat capacity is known.

If the solid undergoes a polymorphic change (or changes), as is frequently the case, the heat capacity curve for the solid will consist of two (or more) portions, one for each crystalline form. In cases of this type the entropy of transition $\Delta H_t/T_t$, where ΔH_t is the molar heat of transition and T_t is the transition temperature (§ 19h), must be included in the entropy calculations.

TABLE XV.* ENTROPIES OF SOLIDS AND LIQUIDS IN THEIR STANDARD STATES AT $25°$ C IN CALORIES PER DEGREE PER G. ATOM (OR PER MOLE)

Elements				Compounds			
C (diamond)	0.58	Fe(s)	6.5	NaF(s)	13.1	AgCl(s)	23.0
C (graphite)	1.36	Zn(s)	9.95	NaCl(s)	17.3	AgI(s)	27.6
Na(s)	12.2	Br$_2$(l)	36.7	NaBr(s)	20.1	Hg$_2$Cl$_2$(s)	47.0
Mg(s)	7.77	Ag(s)	10.2	KCl(s)	19.8	PbCl$_2$(s)	32.6
Al(s)	6.75	I$_2$(s)	27.9	KBr(s)	22.6	MgO(s)	6.55
K(s)	15.2	Hg(l)	18.5	KI(s)	24.1	Al$_2$O$_3$(s)	12.5
Cu(s)	7.97	Pb(s)	15.5	H$_2$O(l)	16.75	ZnO(s)	10.4

* For further data, see Table 5 at end of book.

If the substance whose entropy is to be determined is a liquid at ordinary temperatures, it is first necessary to make measurements on the solid form up to its melting point; the entropy of the solid at this temperature is then obtained by the procedure described above. To obtain the entropy of the liquid at the melting point it is necessary to add the entropy of fusion, which is equal to $\Delta H_f/T_m$, where ΔH_f is the molar heat of fusion and T_m is the melting point. The entropy of the liquid at any higher temperature may now be obtained by plotting C_P/T against T, or C_P against $\ln T$, for the liquid; the area under the curve between the melting point and any particular temperature, up to the boiling point, then gives the corresponding entropy increase which must be added to the value at the melting point.

Some actual experimental data will be quoted below (see Table XVI), but for the present sufficient indication has been given of the procedure used for determining the entropies of substances which are solid or liquid at ordinary temperatures. For such substances the standard states are the pure solid or pure liquid at 1 atm. pressure (cf. § 12e), and the standard entropies, per g. atom (for elements) or per mole (for compounds), at 25° C, derived from heat capacity measurements are recorded in Table XV.[2] As stated earlier (§ 19h), entropies are usually expressed in terms of calories per degree, and the quantity 1 cal. deg.$^{-1}$, the temperature being on the usual absolute scale, in terms of the centigrade degree, is often referred to as an **entropy unit** and abbreviated to E.U.

23c. Entropies of Gases.—For substances which are normally gaseous, it is necessary to follow the procedure outlined above for the solid and the liquid states up to the boiling point of the latter. The entropy of vaporization $\Delta H_v/T_b$, where ΔH_v is the heat of vaporization and T_b is the boiling point, at 1 atm. pressure, is then added to the entropy of the liquid to give that of the gas at the normal boiling point. The method may be illustrated by the experimental results obtained for nitrogen. The Debye characteristic temperature θ was found to be 68 from low temperature heat capacity

TABLE XVI. THE ENTROPY OF NITROGEN GAS AT ITS BOILING POINT

Process	E.U. mole^{-1}
0° to 10° K from Debye equation	0.458
10° to 35.61° K (transition point) by graphical integration	6.034
Transition, 54.71 cal. mole^{-1}/35.61°	1.536
35.61° to 63.14° K (melting point) by graphical integration	5.589
Fusion, 172.3 cal. mole^{-1}/63.14°	2.729
63.14° to 77.32° K (boiling point) by graphical integration	2.728
Vaporization, 1332.9 cal. mole^{-1}/77.32°	17.239
Total	36.31

measurements on solid nitrogen, and this was utilized to determine the entropy of the latter at 10° K. At 35.61° K the solid undergoes a change of crystalline form, the heat of transition being 54.71 cal. mole^{-1}; the entropy of transition is thus 54.71/35.61, i.e., 1.536 E.U. mole^{-1}. The melting point of the higher temperature crystalline form of nitrogen is 63.14° K, and the heat of fusion is 172.3 cal. mole^{-1}, giving an entropy of fusion of 172.3/63.14, i.e., 2.729 E.U. mole^{-1}. The boiling point of liquid nitrogen is 77.32° K, and the heat of vaporization is 1,332.9 cal. mole^{-1}, so that the entropy of vaporization is 1,332.9/77.32, i.e., 17.239 E.U. mole^{-1}. These values for the entropies of the phase changes are added to the entropy contributions of the two solid states and the liquid obtained by graphical integration of the experimental C_P/T values against T in the usual manner. The complete

[2] For entropies of inorganic substances, see Kelley, *U. S. Bur. Mines Bull.*, 394 (1936), 434 (1941), also 350 (1932); W. M. Latimer, "The Oxidation States of the Elements, etc.," 1938; for organic substances, see G. S. Parks and H. M. Huffman, "The Free Energies of Some Organic Compounds," 1932, and numerous papers in the *J. Am. Chem. Soc.*, etc.

results are recorded in Table XVI [3]; it is seen that the entropy of gaseous nitrogen at its boiling point at 1 atm. pressure is 36.31 E.U. mole^{-1}.

If precise heat capacity data were available for the gas, it would be possible to evaluate the increase in entropy between the boiling point and the standard temperature, e.g., 298.16° K, in the usual manner (§ 20a) by graphical integration over this temperature range. If T_1 is the normal boiling point and T_2 represents the standard temperature, the contribution of the gaseous state to the total entropy, at constant (atmospheric) pressure, is given by equation (20.9), viz.,

$$S_2 - S_1 = \int_{T_1}^{T_2} \frac{C_P}{T} dT = \int_{T_1}^{T_2} C_P d \ln T. \qquad (23.2)$$

However, sufficiently accurate measurements of gaseous heat capacities down to the boiling point have usually not been made, and so the entropy contribution of the gaseous state, assuming ideal behavior, is derived from heat capacities obtained from partition functions as described in Chapter VI. If the heat capacity can be expressed as a function of the temperature over the required range, the entropy change can be derived analytically, as described in § 20a.

For reasons which will become clear later, it is the practice to record the entropy of a gaseous substance in terms of an *ideal gas* at 1 atm. pressure; this is the standard state of the gas. The correction to be applied to the observed entropy at 1 atm. pressure to give the value for an ideal gas at the same pressure was calculated in § 20c. As seen in that section, the correction for nitrogen gas at its boiling point was found to be 0.217 cal. deg.$^{-1}$ mole^{-1}. Addition of this quantity to the observed entropy, i.e., 36.31 E.U., gives 36.53 E.U. mole^{-1} for the standard entropy of nitrogen gas at its boiling point. The additional entropy of nitrogen, as an ideal gas, from the boiling point to 298.16° K, i.e., 25° C, at 1 atm. pressure is found by statistical methods to be 9.36 E.U., so that the standard entropy of nitrogen gas at 298.16° K is 36.53 + 9.36, i.e., 45.89 E.U. mole^{-1}.

The standard entropy values of a number of gases will be given later, after the calculation of entropies by statistical methods, using partition functions, has been described.

23d. Tests of the Third Law of Thermodynamics.—The ultimate test of the postulate that the entropy of a perfect solid is zero at 0° K is that it leads to entropy values, such as those in Table XV, which, when combined with other data, such as heats of reaction, yield results, particularly equilibrium constants, which are in agreement with experiment. This aspect of the subject will be taken up later in the book (Chapter XIII). In the meantime other verifications of the third law of thermodynamics will be mentioned. One important fact is that entropies calculated on the basis of the third law, in the manner described above, are usually in complete agreement with those derived from statistical considerations. A few ap-

---

[3] Giauque and Clayton, *J. Am. Chem. Soc.*, **55**, 4875 (1933).

parent discrepancies have been observed, but these are readily explained by the "imperfect" nature of the solid (§ 24m).

An interesting test of the third law is possible when a solid is capable of existing in two or more modifications, i.e., enantiotropic forms, with a definite transition point. The entropy of the high temperature form (α) at some temperature above the transition point may, in some cases, be obtained in two independent ways. First, heat capacity measurements can be made on the form (β) stable below the transition point, and the entropy at this temperature may then be determined in the usual manner. To this is then added the entropy of transition, thus giving the entropy of the α-form at the transition point (cf. first three lines of Table XVI). The entropy contribution of the α-form from the transition temperature to the chosen temperature is then obtained from heat capacity measurements on the α-form. The second procedure is to cool the α-form rapidly below the transition point so that it remains in a metastable state. Its heat capacity can then be determined from very low temperatures up to temperatures above the normal transition point, and the entropy of the α-form is then obtained directly from these data. Measurements of this kind have been made with a number of substances, e.g., sulfur, tin, cyclohexanol and phosphine, and the entropies obtained by the two methods have been found to be in close agreement.[4]

The results with phosphine are particularly striking, for this substance exists in a high temperature (α) form, with two low temperature modifications (β and γ), both of which change into the α-form at the transition points of 49.43° and 30.29° K, respectively. The entropy of the α-form at 49.43° K has been obtained in two ways based on the third law of thermodynamics, a summary of the results being given in Table XVII.[5] In the first method (I),

TABLE XVII. THE ENTROPY OF SOLID PHOSPHINE AT 49.43° K

I	E.U. mole^{-1}	II	E.U. mole^{-1}
0° to 15° K, β-form (Debye)	0.338	0° to 15° K, γ-form (Debye)	0.495
15° to 49.43° K (graphical)	4.041	15° to 30.29° K (graphical)	2.185
Transition $\beta \to \alpha$ at 49.43° K	3.757	Transition $\gamma \to \alpha$ at 30.29° K	0.647
		30.29° to 49.43° K (graphical)	4.800
Total	8.136	Total	8.127

heat capacity measurements were made on the β-form up to the $\beta \to \alpha$ transition point (49.43° K); to the entropy obtained from these results was then added the entropy of the transition to the α-form, thus giving the entropy of the latter at 49.43° K. In the second method (II), measurements were made on the γ-form up to the $\gamma \to \alpha$ transition point (30.29° K); to this was added the entropy of the transition, so as to give the entropy of the α-form at 30.29° K. The increase of entropy of the latter from 30.29°

[4] Kelley, J. Am. Chem. Soc., 51, 1400 (1929); Eastman and McGavock, ibid., 59, 145 (1937); Stephenson and Giauque, J. Chem. Phys., 5, 149 (1937).

[5] Stephenson and Giauque, ref. 4.

to 49.43° K was then derived, in the usual manner, from heat capacity measurements on the α-form in this temperature range. It is evident from the data in Table XVII that the agreement between the entropies obtained by the two methods is excellent. Strictly speaking, this agreement proves only that the two crystalline modifications (β and γ) of the solid have the same entropy at 0° K. The value is not necessarily zero, as assumed in Table XVII, but in view of the fact that similar behavior has been observed with a variety of substances, both elements and compounds, it is probable that the entropy of each solid is actually zero at 0° K.

According to the third law of thermodynamics, the entropy of a perfect crystal should be zero at 0° K at all pressures; hence, it follows that

$$\left(\frac{\partial S}{\partial P}\right)_{T=0} = 0.$$

By equation (20.15), $(\partial S/\partial P)_T$ is equal to $-(\partial V/\partial T)_P$, and consequently

$$\left(\frac{\partial V}{\partial T}\right)_P = 0 \quad \text{(for } T = 0\text{)},$$

so that the rate of expansion of a solid with temperature should become zero at 0° K. Experimental observations have shown that the values of $(\partial V/\partial T)_P$ for a number of solids, e.g., copper, silver, aluminum, diamond, sodium chloride, silica, calcium fluoride and iron disulfide, do, in fact, approach zero as the temperature is lowered.[6] Incidentally, this provides the justification for the statement made earlier (§ 23b) that the difference between values of C_P and C_V for a solid becomes negligible at low temperatures. It should be pointed out that what has been proved above is that $(\partial S/\partial P)_T$ approaches zero at the absolute zero; in other words, the entropy of a solid becomes independent of the pressure. This does not prove that the entropy is actually zero, but it is probable that such is the case.

24. Statistical Treatment of Entropy

24a. Entropy and Probability.—It was seen in § 19k that a correlation was possible between the entropy of a system and the extent of order or disorder. An analogous relationship, which permits of quantitative development, will now be considered between the entropy and the "probability" of a system. Suppose, as in the experiments of Gay-Lussac and Joule (§ 9d), there are two similar globes, separated by a stopcock; one globe contains a gas, whereas the other is evacuated. If the stopcock is opened, then, according to the second law of thermodynamics, there is a very strong probability, amounting almost to a certainty, that provided there are present a large number of molecules, the gas will distribute itself uniformly through the globes. Some light may be shed on this matter by means of the theory of probability. Suppose that the whole system consists of a single molecule

[6] Buffington and Latimer, *J. Am. Chem. Soc.*, **48**, 2305 (1926).

only; there is an equal probability that it will be found in either of the two globes, assuming them to have equal volumes. The probability that the molecule will be found in one particular globe at any instant is thus $\frac{1}{2}$; that is to say, there is one chance in two of this condition being realized. If there are two similar molecules in the system, it can be readily shown that there is one chance in four that both will be in a given globe at the same time; thus, the probability of this distribution is $\frac{1}{4}$ or $(\frac{1}{2})^2$. In general, for a system containing N molecules the probability that all the molecules will remain in the original globe is $(\frac{1}{2})^N$. For a globe having a volume of one liter the value of N is about 10^{22} at ordinary temperature and pressure; even at such low pressures as 10^{-6} atm., it would still be of the order of 10^{16}. The probability that the molecules will remain in the original globe after the stopcock is opened is thus extremely small. This is also the probability that the molecules will return spontaneously to the one globe after having been distributed uniformly throughout the two globes.

Of course, the foregoing calculations deal with an extreme case of non-uniform distribution. Nevertheless, it is possible to show by means of the theory of probability that the chances of any appreciable spontaneous fluctuation from a uniform distribution of the gas throughout the whole available space is so extremely small that it is unlikely to be observed in millions of years, provided the system contains an appreciable number of molecules. It is possible to state, therefore, that the probability of the virtually uniform distribution of a considerable number of molecules in the space available is very large.

To sum up the situation, it may be concluded that the probability that all the molecules of a gas will remain in one part of the space available to them is extremely small under ordinary conditions. On the other hand, the probability of a virtually uniform distribution of the gas is large. The spontaneous process in which a gas, at constant temperature, fills uniformly the whole of the available volume is thus associated with a large increase in the probability of the system. In general, *all spontaneous processes represent changes from a less probable to a more probable state*, and since such processes are accompanied by an increase of entropy, it is to be expected that there may be a connection between the entropy of a system in a given state and the probability of that state.

24b. The Boltzmann-Planck Equation.—If S is the entropy and W is the probability of a particular state, then it should be possible to represent S as a function of W, i.e., $S = f(W)$. To ascertain the nature of this function, consider two systems having entropies S_A and S_B, and probabilities W_A and W_B, respectively. If the systems are combined the probability of the resulting system is the product $W_A \times W_B$, whereas the entropy, being additive (§ 19b), is the sum $S_A + S_B$; hence,

$$S_{AB} = S_A + S_B = f(W_A \times W_B).$$

Since S_A is equal to $f(W_A)$ and S_B is $f(W_B)$, it follows that

$$f(W_A) + f(W_B) = f(W_A \times W_B).$$

To satisfy this condition it is obvious that the function must be logarithmic, so that it is possible to write

$$S = k \ln W + \text{constant}, \qquad (24.1)$$

where k is a constant which must have the same dimensions as entropy, i.e., energy (heat) $\times$ deg.$^{-1}$. It will be seen shortly that k is, in fact, the Boltzmann constant, that is, the gas constant per single molecule, equal to R/N, where N is the Avogadro number. The value of the constant term in equation (24.1) is not obvious, but the considerations presented by L. Boltzmann (1890) and M. Planck (1912) have shown that it may be taken as zero, so that the **Boltzmann-Planck equation** takes the form

$$S = k \ln W. \qquad (24.2)$$

There is no complete proof of this expression, and consequently it is to be regarded as being in the nature of a reasonable postulate relating the entropy of a system to its probability.

24c. Significance of Thermodynamic Probability.—The problem that now arises is that of ascribing a precise significance to the "probability of a system in a given state," so that its value may be determined. This quantity, sometimes called the "thermodynamic probability," may be defined as *the total number of different ways in which the given system, in the specified thermodynamic state, may be realized.** The complete calculation of this probability can be carried out by the methods of statistical mechanics, to be described shortly, but in the meantime some general aspects of the subject will be considered. In a system consisting of a perfect solid at the absolute zero, all the molecules are in their lowest energy state, and they are arranged in a definite manner in the crystal. It would appear, therefore, that under such conditions there is only one way in which the system may be realized; thus W is unity, and hence by equation (24.2) the entropy should be zero. This conclusion is in agreement with the third law of thermodynamics. It will be seen later that there are certain solids in which the molecules may be arranged in different ways in the crystal in the vicinity of the absolute zero. Such solids are not perfect crystals in the sense of the third law of thermodynamics, and their entropies are not zero at 0° K. Similar considerations apply to solid solutions and to glasses.

24d. Entropy of Expansion of Ideal Gas.—A simple application of equation (24.2) in connection with the interpretation of thermodynamic probability, is to calculate the entropy of isothermal expansion of an ideal gas. For such a gas the energy content, at constant temperature, is independent of the volume, and so the thermodynamic probability of the system is determined by the number of ways, consistent with the given energy, in which the molecules can occur in the volume occupied by the gas.

Consider a single molecule contained in a vessel which can be divided into two equal parts by means of a shutter. If at any instant the shutter is closed, the

---

* It may be noted that "thermodynamic probability" is not a probability in the ordinary sense, but is proportional to the latter.

chances of the molecule being present in one part is one-half that of it being in the whole vessel. In general, the probability of a single molecule being found in any volume is proportional to that volume, at constant temperature and energy. If a is the thermodynamic probability, i.e., the number of ways in which a single molecule can occur, in unit volume, the probability for a volume V is then aV. Since the same probability is applicable to any one of the N molecules present in the given volume, the total probability of, or the number of ways of realizing, the system is $(aV)^N$.

Suppose an isothermal expansion is carried out in which the volume of the ideal gas is changed from V_1 to V_2. If the corresponding entropies are S_1 and S_2, and the probabilities are W_1 and W_2, it follows from equation (24.2) that the entropy change accompanying the process is given by

$$S_2 - S_1 = k \ln \frac{W_2}{W_1}.$$

For the isothermal expansion of the ideal gas the probabilities are seen to be proportional to V^N, so that

$$(S_2 - S_1)_T = k \ln \left(\frac{V_2}{V_1}\right)^N$$

$$= kN \ln \frac{V_2}{V_1}. \tag{24.3}$$

If the system consists of 1 mole of gas, N is the Avogadro number, and, assuming k to be the Boltzmann constant, as indicated above, it follows from equation (24.3) that

$$\Delta S_T = (S_2 - S_1)_T = R \ln \frac{V_2}{V_1},$$

for the isothermal volume change of an ideal gas. This result is seen to be identical with equation (19.26). The identification of k in equation (24.2) with the gas constant per single molecule, and the general form of the Boltzmann-Planck equation can thus be justified.

24e. Statistical Mechanics.—The evaluation of the total probability of a system can be carried out by the methods of statistical mechanics. The details of the arguments are somewhat involved, but it is sufficient to present them in general outline here.[7] The state of any single molecule can be completely described by specifying the values of f positional coordinates and f momenta, where f is the number of "degrees of freedom" of the molecule, equal to three times the number of atoms it contains. If a hypothetical space, known as **phase space**, having $2f$ axes could be imagined, the exact state of the molecule could be represented by a point in this imaginary phase space. For all molecules with external and internal configurations and energies which are virtually equal, within certain very narrow limits, the representative points will be found within a particular small volume, called a **unit cell**, in the phase space. By specifying the numbers of molecules

[7] For further discussion, see R. C. Tolman, "The Principles of Statistical Mechanics," 1938; R. H. Fowler and E. A. Guggenheim, "Statistical Thermodynamics," 1939; J. E. Mayer and G. M. Mayer, "Statistical Mechanics," 1940; S. Glasstone, "Theoretical Chemistry," 1944.

whose representative points are to be found in the different unit cells in the phase space, the condition of every molecule in the system is defined. This determines the **macroscopic state** of the system, as indicated by its observable properties, such as pressure, energy, etc.

For a system of identical molecules, an exchange of molecules between different unit cells will not affect the macroscopic state, i.e., the observable properties, of the system. Each distribution of the molecules among the permissible unit cells corresponding to *the same macroscopic state* of the system is known as a **microscopic state**. *The total number of microscopic states then represents the number of different ways in which the given system can be realized*, and this is its thermodynamic probability, in the sense defined above. If the different possible cells are indicated by the numerals 1, 2, ..., i, ..., the numbers of representative points in each of these cells are $N_1, N_2, ..., N_i, ...$, their sum being equal to N, the total number of molecules in the system. In classical statistics, the molecules, although identical, are treated as distinguishable, and so the total number of microscopic states, which is equal to the probability of the system, is given by the number of possible ways of arranging a total of N distinguishable articles in a number of groups, so that there are N_1 in the first, N_2 in the second, and so on, with, in general, N_i in the ith group. The evaluation of the number of ways of realizing this arrangement is a relatively simple mathematical problem; the result is

$$W = \frac{N!}{N_1! N_2! \ldots N_i! \ldots}. \qquad (24.4)$$

The development of quantum statistics has shown that this expression requires modification in two respects. The first arises because there are frequently a number of states whose energies are so close together that they behave classically as a single state. However, when determining the total number of ways a particular state can be realized, this multiplicity must be taken into account. Thus, to each energy level, or unit cell, there must be ascribed a statistical weight, as stated in § 16b. If g_i is, in general, the statistical weight corresponding to the ith cell, there will be a choice of g_i states for each of the N_i molecules, for every possible arrangement of the other molecules in this cell; the number of different microscopic states in the cell is thus increased by a factor of $g_i^{N_i}$. Applying this correction to each of the cells, equation (24.4) for the probability becomes

$$W = N! \frac{g_1^{N_1}}{N_1!} \cdot \frac{g_2^{N_2}}{N_2!} \ldots \frac{g_i^{N_i}}{N_i!} \ldots. \qquad (24.5)$$

The second change necessary in equation (24.4) is due to the fact that the classical concept of distinguishability of the molecules is found to be inconsistent with quantum mechanics. In order to correct for this, it is necessary to divide equation (24.5) by $N!$, so that the result is

$$W = \frac{g_1^{N_1}}{N_1!} \cdot \frac{g_2^{N_2}}{N_2!} \ldots \frac{g_i^{N_i}}{N_i!} \ldots. \qquad (24.6)$$

This expression may be regarded as a consequence of the classical treatment modified by the requirements of quantum statistics. However, two different forms of quantum statistics, namely the Bose-Einstein statistics, which are applicable to photons and to atoms and molecules containing an even number of elementary particles, i.e., electrons, protons and neutrons, and the Fermi-Dirac statistics, which apply to electrons and to atoms and molecules containing an odd number of elementary particles, have been developed. These give somewhat different results for the total number of ways in which a given state can be realized, but it is important to note that at all temperatures and pressures of chemical interest the final expressions may be reduced to a form equivalent to equation (24.6). The value of W given by the latter may thus be taken as representing the probability of a given state of a system of N molecules.

24f. Statistical Calculation of Entropy.—Since the evaluation of the entropy by equation (24.2) requires a knowledge of $\ln W$, this quantity will now be considered. Upon taking logarithms of equation (24.6), the result is

$$\ln W = (N_1 \ln g_1 + N_2 \ln g_2 + \cdots + N_i \ln g_i + \cdots)$$
$$- (\ln N_1! + \ln N_2! + \cdots + \ln N_i! + \cdots) = \sum N_i \ln g_i - \sum \ln N_i! \quad (24.7)$$

If the system consists of a very large number of molecules, as is the case under normal conditions, the various numbers $N_1, N_2, \ldots, N_i, \ldots$, are also very large. It is then possible to make use of the Stirling approximation for the factorials of large numbers; thus,

$$\ln N_1! = N_1 \ln N_1 - N_1$$
$$\ln N_2! = N_2 \ln N_2 - N_2$$
$$\cdots \quad \cdots \quad \cdots$$
$$\ln N_i! = N_i \ln N_i - N_i$$
$$\cdots \quad \cdots \quad \cdots$$

Upon adding these quantities, and recalling that $N_1 + N_2 + \cdots + N_i + \cdots$ is equal to the total number N of the molecules in the system, it is seen that

$$\sum \ln N_i! = \sum N_i \ln N_i - N.$$

This expression may be substituted in equation (24.7), so that

$$\ln W = \sum N_i \ln g_i - \sum N_i \ln N_i + N. \quad (24.8)$$

It was seen in § 16b that in an *ideal gas* the number of molecules N_i possessing energy ϵ_i at the temperature T was given by equation (16.5) as

$$N_i = \frac{N}{Q} g_i e^{-\epsilon_i/kT}, \quad (24.9)*$$

---

* It is of interest to note that this equation can itself be derived from equations (24.4), (24.5) or (24.6), so that the treatment does not involve any arguments other than those developed here and in the preceding section.

where Q is the partition function of the given molecular species. If this result is inserted for the second N_i in the term $-N_i \ln N_i$, the latter becomes

$$-N_i \ln N_i = -N_i \ln \left(\frac{N}{Q} g_i e^{-\epsilon_i/kT}\right)$$

$$= N_i \ln \frac{Q}{N} - N_i \ln g_i + \frac{N_i \epsilon_i}{kT},$$

and by combining this result with equation (24.8) it is seen that

$$\ln W = N \ln \frac{Q}{N} + \frac{\sum N_i \epsilon_i}{kT} + N.$$

The summation $\sum N_i \epsilon_i$ is given by equation (16.6); thus,

$$\sum N_i \epsilon_i = N_0 \epsilon_0 + N_1 \epsilon_1 + \cdots + N_i \epsilon_i + \cdots = E - E_0,$$

and hence, utilizing equation (16.8),

$$\frac{\sum N_i \epsilon_i}{kT} = \frac{E - E_0}{kT} = \frac{RT^2}{kT}\left(\frac{\partial \ln Q}{\partial T}\right)_V.$$

It follows, therefore, that

$$\ln W = N \ln \frac{Q}{N} + \frac{RT}{k}\left(\frac{\partial \ln Q}{\partial T}\right)_V + N. \tag{24.10}$$

The value for $\ln W$ from equation (24.10) may be substituted in the Boltzmann-Planck equation, so that the general expression for the entropy of an ideal gaseous system is found to be

$$S = k \ln W = kN \ln \frac{Q}{N} + RT \left(\frac{\partial \ln Q}{\partial T}\right)_V + kN.$$

If N is the Avogadro number, kN is equal to the molar gas constant R, and then S is the molar entropy; hence,

$$S = R \ln \frac{Q}{N} + RT \left(\frac{\partial \ln Q}{\partial T}\right)_V + R. \tag{24.11}$$

For certain later purposes, it is convenient to write this equation in a somewhat different form by utilizing the Stirling approximation

$$\ln N! = N \ln N - N,$$

so that

$$k \ln N! = kN \ln N - kN$$
$$= R \ln N - R.$$

Since

$$R \ln \frac{Q}{N} = R \ln Q - R \ln N,$$

it can be readily shown that equation (24.11) becomes

$$S = k \ln \frac{Q^N}{N!} + RT \left(\frac{\partial \ln Q}{\partial T} \right)_V. \qquad (24.12)$$

It is seen from the foregoing results, e.g., equations (24.11) and (24.12), that by combining statistical mechanics with the Boltzmann-Planck equation it is possible to derive a relationship between the molar entropy of any gas, assuming it to behave ideally, and the partition function of the given species. Since the partition function and its temperature coefficient may be regarded as known, from the discussion in Chapter VI, the problem of calculating entropies may be regarded as solved, in principle. In order to illustrate the procedure a number of cases will be considered.

24g. Entropy of Monatomic Molecules: The Sackur-Tetrode Equation.—A monatomic molecule has no vibrational or rotational energy, and so the only contributions to the partition functions are those for translational energy and for possible electronic states. The translational partition function is given by equation (16.16), and if the electronic factor is Q_e, it follows that for a monatomic gas

$$Q = Q_e \frac{(2\pi mkT)^{3/2}}{h^3} V,$$

where V is the volume occupied. Assuming Q_e to be independent of temperature, as will be the case for the majority of substances at not too high temperature (§ 16f), then

$$RT \left(\frac{\partial \ln Q}{\partial T} \right)_V = \tfrac{3}{2} R,$$

and hence the molar entropy of an ideal monatomic gas is given by equation (24.11) as

$$S = R \ln \left[Q_e \frac{(2\pi mkT)^{3/2}}{Nh^3} V \right] + \tfrac{5}{2} R, \qquad (24.13)$$

where V is now the molar volume. Since the gas is ideal, V may be replaced by RT/P, so that another form of equation (24.13) is

$$S = R \ln \left[Q_e \frac{(2\pi mkT)^{3/2}}{Nh^3} \cdot \frac{RT}{P} \right] + \tfrac{5}{2} R. \qquad (24.14)$$

These are alternative expressions of what is known as the **Sackur-Tetrode equation**, derived in a somewhat different manner by O. Sackur (1911–13) and H. Tetrode (1912).[8]

If the universal constants R, π, k, h and N are separated from m, T, P and Q_e, which are characteristic of the system, equation (24.14) becomes

$$S = R[\tfrac{3}{2} \ln m + \tfrac{5}{2} \ln T - \ln P + \ln Q_e + \ln R(2\pi k/h^2)^{3/2}/N + \tfrac{5}{2}].$$

[8] Sackur, *Ann. Physik*, **40**, 67 (1913); Tetrode, *ibid.*, **38**, 434 (1912); **39**, 255 (1913); Stern, *Physik. Z.*, **14**, 629 (1913).

The treatment of units is here the same as that employed in connection with the translational partition function in the problem in § 16e; m, k and h are in c.g.s. units, and if P is in atm., R in the $\ln R$ term should conveniently be in cc.-atm. deg.$^{-1}$ mole^{-1}. The actual weight m of the molecule may be replaced by M/N, where M is the ordinary molecular weight. Making these substitutions and converting the logarithms, it is found that

$$S = 2.303R(\tfrac{3}{2}\log M + \tfrac{5}{2}\log T - \log P + \log Q_e - 0.5055), \quad (24.15)$$

for the molar entropy, where P is the pressure in atm. It is of interest to compare equation (24.15) with (19.22), derived earlier for an ideal gas; since C_P for an ideal gas is equal to $\tfrac{5}{2}R$ these two expressions are identical in form, the constant s_0', which cannot be derived from purely thermodynamic considerations, involving the molecular weight of the gas, its electronic multiplicity and a numerical constant.

If the pressure P in equation (24.15) is taken as 1 atm., the entropy is that of the particular substance behaving as an ideal gas at this pressure, and hence represents the standard entropy S^0 (§ 20c). By taking R as 1.987 cal. deg.$^{-1}$ mole^{-1}, equation (24.15) then gives the standard molar entropy of the gas in these same units; thus,

$$S^0 = 4.576(\tfrac{3}{2}\log M + \tfrac{5}{2}\log T + \log Q_e - 0.5055), \quad (24.16)*$$

in cal. deg.$^{-1}$ mole^{-1}, i.e., E.U. mole^{-1}. It is thus a very simple matter to determine the standard entropy of a monatomic gas at reasonable temperatures. It should be noted that the electronic factor Q_e is always included in the expression for the entropy, but it was usually omitted from the corresponding equations in Chapter VI in which the partition function was related to the energy and the heat capacity. The reason for this is that the latter involve $(\partial \ln Q/\partial T)_V$ only, and if Q_e is constant, as it usually is, it makes no contribution to the energy or heat capacity. It will be seen from equation (24.11), however, that the entropy depends on $\ln Q$, as well as on the temperature coefficient; the actual value of Q_e, or rather of $\ln Q_e$, must consequently be included, even if it is constant.

Problem: Calculate the standard entropy of atomic chlorine at 25° C.

The atomic or molecular weight, actually the mean isotopic weight, is 35.46; this gives the value of M; T is 298.2° K, and Q_e may be taken as 4.03, as calculated in the problem in § 16f, the temperature variation being neglected. Hence, by equation (24.16),

$$S^0 = 4.576(\tfrac{3}{2}\log 35.46 + \tfrac{5}{2}\log 298.2 + \log 4.03 - 0.5055)$$
$$= 39.4 \text{ cal. deg.}^{-1} \text{ g. atom}^{-1}.$$

The calculated standard entropies for a number of monatomic gases at 25° C are recorded in Table XVIII, together with the entropies obtained

---

* The factor 4.576, which is frequently encountered in thermodynamic calculations, is equal to 2.303×1.987, where 2.303 is the factor for converting ordinary to natural logarithms, i.e., $\ln x = 2.303 \log x$, and 1.987 is the value of R in cal. deg.$^{-1}$ mole^{-1}.

from heat capacity measurements, as described earlier in this chapter.[9] For these particular elements the ground states are all singlet levels, so that Q_e is unity in each case.

TABLE XVIII. CALCULATED AND EXPERIMENTAL ENTROPIES OF MONATOMIC GASES AT 25° C

Gas	Calc.	Exp.
Helium	30.1 E.U. mole^{-1}	29.2 E.U. mole^{-1}
Argon	37.0	36.4
Cadmium	40.1	40.0
Zinc	38.5	38.4
Mercury	41.8	41.3
Lead	41.9	41.8

24h. Polyatomic Molecules.—Since the partition function appears only in the form of its logarithm in the general equation (24.11) or (24.12) for the entropy, it is permissible, as with the energy and heat capacity, to consider the total entropy as the sum of the contributions associated with the various forms of energy. As stated in § 16d, this procedure is approximate, although very little error can be involved in the separation of the translational contribution and also the electronic contribution if the molecules are almost entirely in the ground state, e.g., at normal temperatures. In this event, equation (24.14), (24.15) or (24.16) gives the sum of the translational and electronic entropies for any type of molecule, monatomic or polyatomic.

The complete partition function of a polyatomic molecule may now be represented by the product $Q_t \times Q_i$, where Q_t is the translational, including the electronic, factor, as derived above, and Q_i is the combined rotational and vibrational, i.e., internal, factor. Since $\ln Q$ is then equal to $\ln Q_t + \ln Q_i$, equation (24.12) may be written in the form

$$S = S_t + S_i = \left(k \ln \frac{Q_t^N}{N!} + k \ln Q_i^N \right) + RT \left[\left(\frac{\partial \ln Q_t}{\partial T} \right)_V + \frac{d \ln Q_i}{dT} \right]$$

$$= \left[k \ln \frac{Q_t^N}{N!} + RT \left(\frac{\partial \ln Q_t}{\partial T} \right)_V \right] + \left(k \ln Q_i^N + RT \frac{d \ln Q_i}{dT} \right). \quad (24.17)$$

The expression in the square brackets in equation (24.17), which is of the same form as (24.12), gives the combined translational and electronic entropies, and that in the parentheses is the internal contribution. The condition of constant volume has been omitted, since the rotational and vibrational partition functions are independent of the volume of the system (cf. §§ 16h, 16j, etc.). It will be noted that the $N!$ has been included in the translational term. As far as the final result is concerned, it is immaterial which term contains the $N!$, but there are theoretical reasons, which need not be considered here, for including it in the expression for the translational entropy. Since $k \ln Q_i^N$ is equivalent to $R \ln Q_i$, because kN is equal to R,

---

[9] Lewis, Gibson and Latimer, *J. Am. Chem. Soc.*, **44**, 1008 (1922); Rodebush and Dixon, *ibid.*, **47**, 1036 (1925).

it follows that the total rotational and vibrational contribution to the entropy is given by

$$S_i = R \ln Q_i + RT \frac{d \ln Q_i}{dT}. \quad (24.18)$$

For accurate results, Q_i should be the combined rotational and vibrational partition function derived from the actual energy levels of the molecule as obtained from spectroscopic measurements (§ 16k). For most purposes, at ordinary temperatures, very little error results from the separation of Q_i into the product of two independent factors, viz., Q_r and Q_v, representing the rotational and vibrational partition functions, respectively. Because equation (24.18) involves Q_i in logarithmic terms only, it follows that an expression of the same form can be used to give the separate rotational and vibrational entropies. Thus, if Q_i is replaced by Q_r, the result is S_r, the rotational contribution to the entropy, and similarly the vibrational contribution S_v is obtained by using Q_v for Q_i in equation (24.18). The sum of S_r and S_v derived in this manner represents S_i, which added to S_t, as given by equation (24.14), etc., yields the total entropy.

24i. The Vibrational Entropy.—It was seen in § 16l that the contribution to the partition function of each mode of vibration of frequency ω cm.$^{-1}$ is given by a factor $(1 - e^{-x})^{-1}$, where x is equal to $hc\omega/kT = 1.439\omega/T$. The vibrational entropy is represented by the appropriate form of equation (24.18) as

$$S_v = R \ln Q_v + RT \frac{d \ln Q_v}{dT},$$

and hence it is not a difficult matter to show that the contribution to the molar entropy of each mode of vibration is

$$S_v = \frac{Rx}{e^x - 1} - R \ln (1 - e^{-x}). \quad (24.19)$$

The terms in equation (24.19) can be readily obtained for any value of x, i.e., $hc\omega/kT$, by means of the tables of Einstein functions mentioned earlier (§ 16j).

If x is relatively large, as is the case at moderate temperatures for many stable diatomic molecules, which have large vibrational frequencies, e^x becomes very large and e^{-x} is very small. It is seen from equation (24.19) that in this event S_v is almost zero. This is true for hydrogen, deuterium, oxygen, nitrogen, carbon monoxide and hydrogen chloride, for example, at temperatures up to about 370° K. For these molecules, it has already been seen (§ 16j) that the vibrational contribution to the energy and heat capacity is also negligible at such temperatures. If the vibration frequency ω is relatively small, or the temperature fairly high, the value of S_v as given by equation (24.19) is not negligible; this is the case, for example, with chlorine even at ordinary temperatures. Molecules containing more than two atoms invariably have at least one vibration with a low frequency which makes an appreciable contribution to the entropy at all reasonable temperatures.

24j. Nuclear Spin Entropy.—Before considering the contribution to the entropy of a molecule associated with its rotational motion, it is necessary to refer to the subject of nuclear spin and its effect on entropy. When discussing the evaluation of energies and heat capacities from the partition function in Chapter VI, the nuclear spin was deliberately neglected. This procedure is justified by the fact that the nuclear spin contribution to the partition function is a constant factor, independent of temperature, except possibly in the vicinity of the absolute zero. However, as emphasized earlier in another connection, the expression for the entropy involves $\ln Q$ itself, in addition to its temperature coefficient, and consequently the nuclear spin factor must be included. For each atomic nucleus having a spin quantum number i, the contribution to the partition function is a factor $2i + 1$ at all reasonable temperatures. It follows, therefore, from equation (24.18) that the nuclear spin entropy is $R \ln (2i + 1)$ for every nucleus in the molecule. To obtain the total entropy of a molecule the appropriate $R \ln (2i + 1)$ terms must be added to the contribution of translation, vibration, rotation, etc.

Since atoms retain their nuclear spins unaltered in all processes except those involving ortho-para conversions, there is no change in the nuclear spin entropy. It is consequently the common practice to omit the nuclear spin contribution, leaving what is called the **practical entropy** or **virtual entropy**.*

It is of importance to note that, except for hydrogen and deuterium molecules, the entropy derived from heat capacity measurements, i.e., the **thermal entropy**, as it is frequently called, is equivalent to the practical entropy; in other words, the nuclear spin contribution is not included in the former. The reason for this is that, down to the lowest temperatures at which measurements have been made, the nuclear spin does not affect the experimental values of the heat capacity used in the determination of entropy by the procedure based on the third law of thermodynamics (§ 23b). Presumably if heat capacities could be measured right down to the absolute zero, a temperature would be reached at which the nuclear spin energy began to change and thus made a contribution to the heat capacity. The entropy derived from such data would presumably include the nuclear spin contribution of $R \ln (2i + 1)$ for each atom. Special circumstances arise with molecular hydrogen and deuterium to which reference will be made below (§ 24n).

24k. The Rotational Entropy.—At all temperatures above the lowest, the rotational partition function Q_r of a diatomic molecule, or of *any linear molecule*, is given to a good approximation by equation (16.24), i.e.,

$$Q_r = \frac{8\pi^2 I k T}{\sigma h^2}, \tag{24.20}$$

* It may be noted, incidentally, that the practical or virtual entropy also does not include the entropy of mixing of different isotopic forms of a given molecular species. This quantity is virtually unchanged in a chemical reaction, and so the entropy change of the process is unaffected by its complete omission.

where I is the moment of inertia of the molecule and σ is its symmetry number (§§ 16h, 16l). Insertion of this result in equation (24.18) gives the rotational contribution to the molar entropy of a linear molecule,

$$S_r = R \ln \frac{8\pi^2 I k T}{\sigma h^2} + R, \qquad (24.21)$$

since $d \ln Q_r/dT$ is equal to $1/T$. If the recognized values are inserted for the universal constants π, k and h, in c.g.s. units, and if R is taken as 1.987 cal. deg.$^{-1}$ mole^{-1}, it is found, after converting the logarithms, that

$$S_r = 4.576(\log I + \log T - \log \sigma + 38.82) \text{ cal. deg.}^{-1} \text{ mole}^{-1} \qquad (24.22)$$

the moment of inertia being still in c.g.s. units., i.e., g. cm.^{2}. This expression gives the molar rotational entropy in conventional E.U. mole^{-1} for any diatomic or linear molecule.

Problem: Calculate the total standard entropy of nitrogen gas at 25° C, using data in Tables VIII and IX.

Since the ground state of molecular nitrogen is a singlet level, Q_e is unity, and the molecular weight M being 28.00, the combined standard translational and electronic entropy, the latter being actually zero, is given by equation (24.16) as

$$S_t^0 = 4.576(\tfrac{3}{2} \log 28.00 + \tfrac{5}{2}\log 298.2 - 0.5055)$$
$$= 35.9 \text{ E.U. mole}^{-1}.$$

The vibrational frequency is 2360 cm.$^{-1}$ (Table IX), and hence x, i.e., $1.439\omega/T$, is 11.4. Insertion of this result into equation (24.19) gives a value for S_v that is negligibly small. The vibrational contribution to the entropy of molecular nitrogen at 25° C may thus be taken as zero.

The moment of inertia is 13.9×10^{-40} g. cm.^{2} (Table VIII), and σ is 2, since the nitrogen molecule is symmetrical; hence, by equation (24.22),

$$S_r = 4.576[\log (13.9 \times 10^{-40}) + \log 298.2 - \log 2 + 38.82]$$
$$= 9.8 \text{ E.U. mole}^{-1}.$$

The total entropy of nitrogen gas in its standard state at 25° C is thus the sum of 35.9 and 9.8 E.U., i.e., 45.7 E.U. mole^{-1}, which may be compared with the thermal, i.e., third law, value of 45.89 E.U. mole^{-1} given in § 23c.

For a *nonlinear molecule*, the rotational partition function [cf. equation (16.34)] may be taken with sufficient accuracy as

$$Q_r = \frac{8\pi^2(8\pi^3 ABC)^{1/2}(kT)^{3/2}}{\sigma h^3}, \qquad (24.23)$$

and since $d \ln Q_r/dT$ is $\tfrac{3}{2}/T$, it follows from equation (24.18) that

$$S_r = R \ln \frac{8\pi^2(8\pi^3 ABC)^{1/2}(kT)^{3/2}}{\sigma h^3} + \tfrac{3}{2}R, \qquad (24.24)$$

where A, B and C are the moments of inertia of the molecule. Extracting the constants and converting the logarithms, in the usual manner, it is found that

$$S_r = 4.576(\tfrac{1}{2} \log ABC + \tfrac{3}{2} \log T - \log \sigma + 58.51), \quad (24.25)$$

for the molar rotational entropy in cal. deg.$^{-1}$ mole^{-1}.

Problem: Calculate the rotational contribution to the molar entropy of ammonia at 25° C.

From Table X, the three moments of inertia are 2.78×10^{-40}, 2.78×10^{-40} and 4.33×10^{-40} g. cm.^{2}, σ is 3 and T is 298.2° K, so that by equation (24.25),

$$S_r = 4.576[\tfrac{1}{2} \log (2.78 \times 2.78 \times 4.33 \times 10^{-120}) + \tfrac{3}{2} \log 298.2 - \log 3 + 58.51]$$
$$= 11.5 \text{ cal. deg.}^{-1} \text{ mole}^{-1}.$$

It is seen that the rotational contribution to the entropy is quite appreciable, the amount increasing, in general, with the size of the molecule and the masses of the atoms, since these cause the moments of inertia to be relatively large.

241. Comparison of Third Law and Statistical Entropies.—If the vibration frequencies and moments of inertia of a molecule, which are often available from spectroscopic measurements, are known, the standard entropy of any gaseous substance can be evaluated with a fair degree of accuracy. By the use of combined rotational and vibrational partition functions, derived directly from the energies in the various levels (§ 16k), more precise entropy values are obtainable. In the great majority of cases the entropies obtained by the statistical method, based on the Boltzmann-Planck equation, are in complete agreement with those derived from heat capacity measurements, utilizing the third law postulate of zero entropy for the perfect solid at 0° K. There are a few cases, however, where some discrepancy has been observed; these are carbon monoxide, nitric oxide, nitrous oxide, hydrogen, deuterium, water and organic compounds, such as ethane, in which the internal rotation is restricted (§ 16m). The various types of discrepant behavior will now be considered.

24m. Random Orientation in the Solid.—For carbon monoxide, nitric oxide and nitrous oxide, the thermal entropy, based on the third law of thermodynamics, is found in each case to be about 1.1 cal. deg.$^{-1}$ mole^{-1} less than the value given by the statistical method. This result suggests that the entropies of the respective solids are not zero at 0° K, as required by the third law, but that they are actually about 1.1 E.U. mole^{-1}. A possible explanation of this fact is to be found in the similarity between the two ends of the respective molecules, so that the alternative arrangements

CO   OC   NO   ON   NNO   ONN

can occur in the crystal lattices. Instead of all the molecules being oriented in one direction in the crystal, two alternative arrangements are equally probable. The crystal is, therefore, not "perfect" in the sense required by the third law, and hence it is not correct to take the entropy as zero at 0° K. If the distribution of molecules between the two possible orientations were completely random, the probability of the state, as defined in § 24c, would be two, and the entropy should be $R \ln 2$, i.e., 1.38 cal. deg.$^{-1}$ mole^{-1}, instead of zero for the perfect crystal. The observed discrepancy of 1.1 E.U. mole^{-1} indicates that the arrangement of the two alternative orientations in the solid is not completely random at the low temperatures at which heat capacity measurements were made. In the three instances under consideration, the correct entropies are those obtained from the partition functions, for these are based on the properties of the gas, and do not involve extrapolation, through the liquid and solid states, down to the absolute zero as is necessary for the evaluation of thermal entropies.

Somewhat analogous considerations apply to the entropy of water vapor. The result derived from heat capacity measurements is again lower than the statistical value, and this can be accounted for by random orientation of the water molecules in the solid. The situation is complicated, however, by the distribution of hydrogen bonds in the ice crystal, and by other factors. In this instance, also, the crystal is not perfect, and so the entropy would not be zero at 0° K. The statistical value of the entropy is therefore the correct one to be used in thermodynamic calculations.

24n. Entropy of Hydrogen and Deuterium.—An entirely different phenomenon is responsible for the discrepancy between the third law and statistical entropies of hydrogen and deuterium. The thermal value for hydrogen is 29.64 cal. deg.$^{-1}$ mole^{-1} at 25° C and 1 atm. pressure, corrected to ideal behavior, but statistical calculations give 33.96 cal. deg.$^{-1}$ mole^{-1}, including the nuclear spin contribution. The latter is $R \ln (2i + 1)^2$, since the molecule consists of two atoms each with a nuclear spin i; for the hydrogen nucleus i is $\frac{1}{2}$, and so the nuclear spin entropy is $R \ln 4$, i.e., 2.75 cal. deg.$^{-1}$ mole^{-1}. If this is subtracted from 33.96 cal. deg.$^{-1}$ mole^{-1}, the result still differs from the thermal value. The responsible factor in this instance is the existence of hydrogen in ortho and para states. The statistical calculations are based on the assumption that ortho-para equilibrium is attained at all temperatures, but this condition is not realized in the hydrogen used for heat capacity measurements (cf. § 16i).

Because solid hydrogen at low temperatures still contains the ortho and para forms in the normal proportions of three to one, the entropy cannot become zero at 0° K. The actual value is equal to the entropy of mixing, as calculated in the problem in § 19j, viz., 4.39 E.U. mole^{-1}. If this is added to the apparent third law value of 29.64, the result is 34.03 cal. deg.$^{-1}$ mole^{-1}, which is in good agreement with the statistical entropy, 33.96 cal. deg.$^{-1}$ mole^{-1}. For use in connection with the entropies of other substances, when there are no changes in the ortho-para hydrogen ratio, the appropriate value of the practical entropy is obtained by subtracting the nuclear spin contribution of 2.75, given above, from the statistical value of 33.96 cal. deg.$^{-1}$ mole^{-1}; the entropy of molecular hydrogen is thus taken as 31.21 E.U. mole^{-1}. The data for molecular deuterium must be treated in an analogous manner.

24o. Restricted Internal Rotation.—When there is restriction to internal rotation of two parts of a molecule with respect to one another, as in ethane and other paraffin hydrocarbons, and in alcohols, amines, etc., the calculation of the entropy from the partition functions requires a knowledge of the energy which restricts rotation. As stated in § 16m, this can be obtained from a comparison of an experimentally determined thermodynamic property with that derived by means of partition functions. Actually, however, when there is a possibility of restricted rotation the entropy is usually obtained from heat capacity measurements.

Nevertheless, where the necessary data are not available, it is possible to make a very satisfactory estimate by the statistical method, the restricting energy being assumed to be equal to that in a related compound for which the value is known.[10]

24p. Standard Entropies of Gases.—The standard entropies for a number of gases, that is, corrected to ideal behavior at 1 atm. pressure, at 25° C are recorded in Table XIX.[11] They are based partly on statistical calcula-

TABLE XIX.* STANDARD ENTROPIES OF GASES AT 25° C IN CAL. PER DEGREE PER MOLE

H_2	31.21	HCl	44.66	CO_2	51.06
D_2	34.62	HBr	47.48	N_2O	52.58
N_2	45.77	HI	49.36	SO_2	59.24
O_2	49.00	CO	47.30	NH_3	46.03
Cl_2	53.31	NO	50.34	CH_4	44.50
$Br_2(g)$	58.63	$H_2O(g)$	45.11	C_2H_6	54.85
$I_2(g)$	62.29	H_2S	49.15	C_2H_4	52.48

* For further data see Table 5 at end of book.

tions and partly on thermal data, depending on which are considered more reliable in each case. The values are all practical entropies which may be used in conjunction with those given for solids and liquids in Table XV to calculate the entropy change in a chemical reaction. These entropy changes will be utilized for important thermodynamic purposes in Chapter XIII. Special problems associated with the determination of the entropies of ions in solution will be taken up in Chapter XIX.

Problem: Calculate the standard entropy change for the reaction

$$C(s) + H_2O(l) = CO(g) + H_2(g)$$

at 25° C.

From Table XIX, the total entropy of the products is 47.30 (for CO) + 31.21 (for H_2), i.e., 78.51 E.U. From Table XV, the entropy of the reactants is 1.36(C) + 16.75(H_2O, l), i.e., 18.11 E.U. The standard entropy change ΔS^0 for the reaction is thus 78.51 − 18.11 = 60.40 cal. deg.$^{-1}$ at 25° C.

EXERCISES

1. Since the atoms or molecules in a solid occupy fixed positions they may be treated as distinguishable; there is consequently only one way of realizing a perfect

---

[10] For reviews, see Pitzer, *Chem. Rev.*, **27**, 39 (1940); Aston, *ibid.*, **27**, 59 (1940); Wilson, *ibid.*, **27**, 17 (1940); also, H. S. Taylor and S. Glasstone, "Treatise on Physical Chemistry," 3rd ed., 1942, Chap. IV (J. G. Aston); for empirical methods of treating molecules with restricted internal rotation, see Pitzer, *J. Chem. Phys.*, **8**, 711 (1940); *Chem. Rev.*, **27**, 39 (1940); Pitzer and Scott, *J. Am. Chem. Soc.*, **63**, 2419 (1941); Ewell, *Ind. Eng. Chem.*, **32**, 778 (1940).

[11] See Kelley, ref. 2, Latimer, ref. 2; see also, Thacker, Folkins and Miller, *Ind. Eng. Chem.*, **33**, 584 (1941), Wagman, Kilpatrick, Taylor, Pitzer and Rossini, *J. Res. Nat. Bur. Stand.*, **34**, 143 (1945); Wagman, Kilpatrick, Pitzer and Rossini, *ibid.*, **35**, 467 (1945); Kilpatrick, Prosen, Pitzer and Rossini, *ibid.*, **36**, 559 (1946); for references, see Wilson, ref. 10.

solid, i.e., its thermodynamic probability is unity. In a solid solution consisting of N_1 molecules of one substance (or form) and N_2 molecules of another, the system can be realized in $N!/N_1!N_2!$ different ways, where N is equal to $N_1 + N_2$. (Note that for a substance in the pure state this reduces to unity.) Show that the entropy of formation of 1 mole of a solid solution from its pure solid constituents, i.e., the entropy of mixing, is $-R(\text{N}_1 \ln \text{N}_1 + \text{N}_2 \ln \text{N}_2)$, where N_1 and N_2 are the respective mole fractions [cf. equation (19.32)].

2. By assuming a solid solution of silver chloride ($\text{N}_1 = 0.272$) and silver bromide ($\text{N}_2 = 0.728$) to have an entropy of zero at 0° K, the thermal entropy at 25° C was found to be equal to the sum of the entropies of the pure constituents. Other measurements, not based on the third law, however, indicated that the entropy of the solid solution was greater than that of the constituents by about 1.1 cal. deg.$^{-1}$ mole^{-1} (Eastman and Milner, ref. 1). Account quantitatively for the discrepancy.

3. The heat of vaporization of mercury at its normal boiling point (357° C) is 13,600 cal. g. atom^{-1}, and the mean heat capacity of the liquid is 6.5 cal. deg.$^{-1}$ g. atom^{-1}. Assuming the vapor to be an ideal monatomic gas and the atoms to be in a singlet electronic state, calculate the entropy of (i) gaseous, (ii) liquid, mercury at 25° C. (The actual values are 41.8 and 18.5 E.U. g. atom^{-1}.)

4. Use the tables of entropy data to determine the entropy changes accompanying the following reactions: (i) $\text{Na}(s) + \text{KCl}(s) = \text{NaCl}(s) + \text{K}(s)$; (ii) $\text{AgCl}(s) + \frac{1}{2}\text{H}_2(g) = \text{HCl}(g) + \text{Ag}(s)$; (iii) $\text{Hg}(l) + \frac{1}{2}\text{O}_2(g) = \text{HgO}(s)$.

5. From the following data, evaluate the third law molar entropy of hydrogen chloride as an ideal gas at 1 atm. pressure and 25° C. The entropy of the solid at 98.36° K is 7.36 E.U. mole^{-1}; at this temperature transition to another solid form occurs, the heat of transition being 284.3 cal. mole^{-1}. The increase of entropy accompanying the heating of the second solid modification from 98.36° to 158.91° K, the melting point, is 5.05 E.U. The heat of fusion of the solid is 476.0 cal. mole^{-1}. From the melting point to the normal boiling point (188.07° K), the increase of entropy is 2.36 E.U. and the heat of vaporization is 3860 cal. mole^{-1}. The mean heat capacity of hydrogen chloride gas from the boiling point to 25° C may be taken as 6.98 cal. deg.$^{-1}$ mole^{-1} at 1 atm. pressure.

6. Use the moment of inertia and vibration frequency given in Chapter VI to calculate the standard (practical) entropy of hydrogen chloride at 25° C. Compare the result with that obtained in the preceding problem.

7. Calculate the standard (practical) entropy of hydrogen sulfide at 25° C, using data in Chapter VI.

8. Show that if the Debye heat capacity equation were applicable, the entropy of a perfect solid at very low temperatures should be equal to $\frac{1}{3}C_P$, where C_P is the heat capacity at the given temperature. What would be the value in terms of the Debye characteristic temperature?

9. By means of the following data for cyclohexane [Kelley, J. Am. Chem. Soc., 51, 1400 (1929)], which exists in two crystalline modifications, test the third law of thermodynamics. For form I, the Debye characteristic temperature is 112; the increase in entropy of the solid, from C_P measurements, from 13.5° to 263.5° K is 33.55 E.U. At 263.5° K there is a transition to form II, the heat change being 1,960 cal. mole^{-1}. The entropy change from 263.5° to the melting point (297.0° K) is 5.02 E.U., the heat of fusion is 406 cal. mole^{-1}. For form II, the Debye temperature is 84, and the entropy increase from 13.5° to 297.0° K, the melting point, is 45.34 E.U. mole^{-1}.

10. The following heat capacities in cal. deg.$^{-1}$ mole^{-1} have been recorded for solid chlorine [Giauque and Powell, *J. Am. Chem. Soc.*, **61**, 1970 (1939)]:

T	C_P	T	C_P	T	C_P	T	C_P
14.05° K	0.810	33.94° K	4.804	79.71° K	9.201	134.06° K	11.47
17.40	1.331	42.37	6.018	98.06	10.03	145.19	11.92
19.81	1.842	58.59	7.879	112.99	10.57	155.45	12.41
26.37	3.192	70.50	8.720	123.53	11.00	164.99	12.93

Determine the entropy of solid chlorine at its melting point, 172.12° K. The entropy contribution at temperatures below 14° K should be obtained by assuming the Debye equation to be applicable.

11. The thermal entropy of normal deuterium was found to be 33.90 E.U. mole^{-1}. Normal deuterium consists of two parts of ortho- to one part of para-molecules; at low temperatures the former occupy six and the latter nine closely spaced levels. The spin of each deuterium nucleus is 1 unit. Show that the practical standard entropy of deuterium gas at 25° C is 34.62 E.U. mole^{-1}. (Add the entropy of mixing to the thermal entropy and subtract the nuclear spin contribution.) Compare the result with the value which would be obtained from statistical calculations, using moment of inertia, etc. in Chapter VI.

12. Show that equations (19.25) and (19.26) for the entropy change of an ideal monatomic gas at constant pressure and constant temperature, respectively, follow from the Sackur-Tetrode equation (24.13) or (24.14).

CHAPTER X

FREE ENERGY

25. The Free Energy and Work Functions

25a. The Work Function.—Although the entropy concept is the fundamental consequence of the second law of thermodynamics, there are two other functions, which utilize the entropy in their derivation, that are more convenient for use in many instances. One of these, the free energy, will be employed extensively in subsequent portions of this book in connection with the study of equilibria, both chemical and physical, and the direction of chemical change.

The **work function**, represented by the symbol A, is defined by

$$A = E - TS, \tag{25.1}$$

where E is the energy content of the system, T is its temperature, and S its entropy. Since E, T and S are characteristic properties of the system, depending only on its thermodynamic state and not on its previous history, it is evident that the same considerations must apply to the work function. Hence A is to be regarded as *a single-valued function of the state of the system*, and dA can be treated as a complete differential (§ 4e). Further, since E and S are both extensive properties, A will also be extensive in character, its value being proportional to the quantity of matter constituting the system under consideration.

In order to obtain some understanding of the physical significance of the work function, consider an *isothermal change* from the initial state indicated by the subscript 1 to the final state indicated by 2; thus,

$$A_1 = E_1 - TS_1 \quad \text{and} \quad A_2 = E_2 - TS_2,$$

so that ΔA_T, the increase in the work function accompanying the process at constant temperature, is given by

$$A_2 - A_1 = (E_2 - E_1) - T(S_2 - S_1)$$

or

$$\Delta A_T = \Delta E_T - T\Delta S. \tag{25.2}$$

If ΔS in equation (25.2) is replaced by $Q_{\text{rev.}}/T$, where $Q_{\text{rev.}}$ is the heat taken up when the given change is carried out in a reversible manner, then

$$\Delta A_T = \Delta E_T - Q_{\text{rev.}}. \tag{25.3}$$

According to the first law of thermodynamics [equation (7.2)], assuming a

reversible, isothermal process,
$$-W_{\text{rev.}} = \Delta E_T - Q_{\text{rev.}}. \tag{25.4}$$
Comparison of equations (25.3) and (25.4) shows that
$$-\Delta A_T = W_{\text{rev.}}, \tag{25.5}$$
so that the decrease in the A function in any process at constant temperature is equal to the reversible work done by the system. Since the reversible work is, under these conditions, the maximum work that can be obtained from the given thermodynamic change in state, it follows that in an isothermal process the decrease of the work function is a measure of the maximum work obtainable from that change in state. It is this fact which justifies the use of the term "work function" for the quantity defined by equation (25.1), although it was at one time called the "free energy" (H. von Helmholtz). It should be noted that any given process, isothermal or otherwise, is accompanied by a definite change in the value of the work function A, but it is only for an isothermal process that this change is a measure of the maximum work available.

Since ΔA_T and ΔE_T are completely defined by the initial and final states of the system, the results obtained above, e.g., equations (25.3) and (25.5), provide proof of the statement made earlier (§ 8b), that in any *isothermal, reversible process* the values of W and Q are definite, depending only on the initial and final states. The work term W here includes all the forms of work performed on or by the system.

25b. The Free Energy.—The second, and more generally useful, function derived from the entropy is called the **free energy**, and is defined by
$$F = E - TS + PV, \tag{25.6}$$
where P and V refer, as usual, to the pressure and volume of the system. This definition may be written in two alternative forms which are frequently employed. First, by equation (25.1), A is equal to $E - TS$; hence,
$$F = A + PV. \tag{25.7}$$
Second, since by equation (9.5), H is equivalent to $E + PV$, it follows that
$$F = H - TS. \tag{25.8}$$
Like the work function, the free energy F is seen to be *a single-valued function of the thermodynamic state of the system*, so that dF is a complete differential. In addition, the free energy, like A, S, E and H, is an extensive property. Comparison of equations (25.1) and (25.8) reveals an interesting fact of general applicability: A is related to F in the same manner as E is to H. It will be seen later that there are many relationships involving A and E, with similar expressions relating F and H.

For a process taking place at *constant pressure*, it is evident from equation (25.7) that
$$\Delta F_P = \Delta A_P + P\Delta V. \tag{25.9}$$

If, in addition, the change is an *isothermal* one, i.e., the temperature is constant, ΔA is equal to $-W_{\text{rev.}}$, as seen above, so that equation (25.9) gives

$$-\Delta F_{P,T} = W_{\text{rev.}} - P\Delta V. \tag{25.10}$$

The quantity $W_{\text{rev.}}$ represents the total reversible work obtainable in the given change; this may include other forms of work, e.g., electrical or surface work, in addition to work of expansion. The latter is equal to $P\Delta V$ (§ 3g), and so $W_{\text{rev.}} - P\Delta V$ represents the reversible work, exclusive of work of expansion, that can be obtained from a given change in state. This quantity is sometimes referred to as the **net work**, and is represented by $W'_{\text{rev.}}$, so that by equation (25.10),

$$-\Delta F_{P,T} = W'_{\text{rev.}}. \tag{25.11}$$

The decrease of free energy accompanying a process taking place at constant temperature and pressure is thus equal to the reversible, i.e., maximum, work other than work of expansion, i.e., the maximum net work, obtainable from the process. It is because the change in F is a measure of the "useful" work that F has been called the free energy.[1] It has also been referred to as the "thermodynamic potential" (J. W. Gibbs), and as the "available energy" (Lord Kelvin). As mentioned in connection with the work function, the value of ΔF for any change is quite definite, no matter under what conditions the process is performed, but *only when the temperature and pressure are constant is the free energy change equal to the maximum net work available for the given change in state.*

25c. Work Function and Free Energy Relationships.—It was seen in § 19i, for an infinitesimal stage of an isothermal, reversible process, in which *the work done is restricted to work of expansion*, that [cf. equation (19.18)],

$$dS = \frac{dE + PdV}{T}, \tag{25.12}$$

and hence,

$$TdS = dE + PdV. \tag{25.13}$$

By differentiation of equation (25.1), i.e., $A = E - TS$, which defines the work function,

$$dA = dE - TdS - SdT, \tag{25.14}$$

so that from (25.13) and (25.14), it follows that

$$dA = -PdV - SdT. \tag{25.15}$$

At constant volume, therefore, $dA_V = -SdT_V$, or

$$\left(\frac{\partial A}{\partial T}\right)_V = -S, \tag{25.16}$$

[1] Lewis, *Proc. Am. Acad. Arts Sci.*, **35**, 3 (1899); *Z. phys. Chem.*, **32**, 364 (1900); see also, G. N. Lewis and M. Randall, "Thermodynamics and the Free Energy of Chemical Substances," 1923.

and at constant temperature, $dA_T = -\,PdV_T$, or

$$\left(\frac{\partial A}{\partial V}\right)_T = -\,P. \tag{25.17}$$

These relationships give the variation of the work function with temperature and volume, respectively

Differentiation of equation (25.6), i.e., $F = E - TS + PV$, yields

$$dF = dE - TdS - SdT + PdV + VdP, \tag{25.18}$$

and for an infinitesimal stage in a reversible process involving only work of expansion, use of equation (25.13) reduces this to

$$dF = VdP - SdT. \tag{25.19}$$

Hence, at constant pressure, $dF_P = -\,SdT_P$, or

$$\left(\frac{\partial F}{\partial T}\right)_P = -\,S, \tag{25.20}$$

and at constant temperature, $dF_T = VdP_T$, or

$$\left(\frac{\partial F}{\partial P}\right)_T = V, \tag{25.21}$$

giving the variation of the free energy with temperature and pressure.

Comparison of equation (25.16) with (25.20) and of (25.17) with (25.21) brings to light another useful generalization: A is related to V in the same manner as F is related to P. It is found that by exchanging A for F and V for P, equations involving A and V may be converted into analogous expressions relating F and P; however, since an increase of pressure corresponds to a decrease of volume, the change of variable from V to P, or vice versa, is accompanied by a change of sign [cf. equations (25.17) and (25.21)].

It must be emphasized that the results derived above, like those obtained in § 20b, *et seq.*, are applicable only to *closed systems*, as stated in § 20g. Such systems may be homogeneous or heterogeneous, and may consist of solid, liquid or gas, but the total mass must remain unchanged. It will be seen later that in some cases a system consists of several phases, and although the mass of the whole system is constant, changes may take place among the phases. In these circumstances the equations apply to the system as a whole but not to the individual phases. Since it has been postulated that the work done in a change in state is only work of expansion, equal to PdV, the second condition stated in § 20g, that the system is always in equilibrium with the external pressure, must also be operative.

25d. Isothermal Changes in the Work Function and Free Energy.—For an isothermal change dT is zero, and hence, as noted above, equation (25.15) becomes

$$dA_T = -\,PdV. \tag{25.22}$$

For an appreciable isothermal process the increase ΔA in the work function can be obtained by integration of equation (25.22) between the limits of the initial state 1 and the final state 2; thus,

$$\Delta A_T = (A_2 - A_1)_T = -\int_{V_1}^{V_2} P dV. \quad (25.23)$$

Comparison of this result with equation (8.2) shows it to be in agreement with the relationship derived in § 25a, between the change in the work function in an isothermal process and the reversible work obtainable.

The corresponding expressions involving the free energy could be written down by changing P for V, and altering the signs where necessary, but they may be derived in a simple manner. At constant temperature, equation (25.19) gives

$$dF_T = V dP, \quad (25.24)$$

and hence for any appreciable isothermal process,

$$\Delta F_T = (F_2 - F_1)_T = \int_{P_1}^{P_2} V dP.$$

In the special case of the system being 1 mole of an *ideal gas*, V may be replaced by RT/P, so that

$$\Delta F_T = RT \int_{P_1}^{P_2} \frac{dP}{P} = RT \ln \frac{P_2}{P_1}. \quad (25.25)$$

Since, for an ideal gas, P_2/P_1 is equal to V_1/V_2 at constant temperature, it follows that ΔF_T is equal to ΔA_T, as may be seen by replacing P in equation (25.23) by RT/V and integrating. It is important to point out, however, that this equality applies only to an isothermal process with an *ideal gas*, but it is not true generally. Considerable confusion was caused in the earlier development of chemical thermodynamics because of the failure to realize this limitation.

25e. The Gibbs-Helmholtz Equations.—If the value of S given by equation (25.20) is substituted in (25.8) the result is

$$F = H + T \left(\frac{\partial F}{\partial T}\right)_P, \quad (25.26)$$

and, similarly, combination of equation (25.16) with (25.1) gives

$$A = E + T \left(\frac{\partial A}{\partial T}\right)_V. \quad (25.27)$$

These two expressions are forms of the equation derived by J. W. Gibbs (1875) and H. von Helmholtz (1882), and usually referred to as the **Gibbs-Helmholtz equation**. Upon dividing equation (25.26) by T^2, and rearranging, it is readily found that an alternative form is

$$\left[\frac{\partial (F/T)}{\partial T}\right]_P = -\frac{H}{T^2}. \tag{25.28}$$

The analogous expression $[\partial(A/T)/\partial T]_V = -E/T^2$ can be derived from equation (25.27), but this is rarely used by chemists.

There are other forms of the Gibbs-Helmholtz equation which are more frequently employed; these deal with changes in the free energy, heat content, etc., accompanying an appreciable process. The process may be chemical or physical in nature; the only restriction is that it takes place in a closed system, i.e., one of constant mass, which is in equilibrium with the external pressure. For the initial and final states, indicated by the subscripts 1 and 2, respectively, of an *isothermal* process, equation (25.8) becomes

$$F_1 = H_1 - TS_1 \quad \text{and} \quad F_2 = H_2 - TS_2,$$

so that by subtraction,

$$F_2 - F_1 = (H_2 - H_1) - T(S_2 - S_1),$$
$$\Delta F = \Delta H - T\Delta S, \tag{25.29}$$

where ΔF, ΔH and ΔS represent the increase of free energy, heat content and entropy, respectively, for the given isothermal process.* Further, from equation (25.20),

$$-\Delta S = -(S_2 - S_1) = \left(\frac{\partial F_2}{\partial T}\right)_P - \left(\frac{\partial F_1}{\partial T}\right)_P$$
$$= \left[\frac{\partial (F_2 - F_1)}{\partial T}\right]_P = \left[\frac{\partial (\Delta F)}{\partial T}\right]_P. \tag{25.30}$$

If this result is inserted into equation (25.29) there is obtained

$$\Delta F = \Delta H + T\left[\frac{\partial (\Delta F)}{\partial T}\right]_P, \tag{25.31}$$

a very useful form of the Gibbs-Helmholtz equation.

Similarly, by means of equations (25.1) and (25.16) it is possible to derive the analogous expression

$$\Delta A = \Delta E + T\left[\frac{\partial (\Delta E)}{\partial T}\right]_V. \tag{25.32}$$

Still another form of the Gibbs-Helmholtz equation may be developed from (25.31) by utilizing the same procedure as was employed in converting equation (25.26) into (25.28); thus, dividing through equation (25.31) by

* Subscripts T might have been included to indicate that constant temperature is implied, but this condition is usually understood.

T^2 and rearranging, the resulting expression may be put in the form

$$\left[\frac{\partial(\Delta F/T)}{\partial T}\right]_P = -\frac{\Delta H}{T^2}. \tag{25.33}$$

This equation represents the variation of ΔF, or rather of $\Delta F/T$, with temperature at constant pressure. If ΔH is expressed as a function of the temperature (§ 12k), it is thus possible, upon integration, to derive an expression for ΔF in terms of the temperature. This matter, as well as other applications of the various forms of the Gibbs-Helmholtz equation, will be taken up in later sections.

Attention may be drawn to the fact that although certain restrictions were mentioned in the course of the foregoing deductions, the final results are of general applicability. The Gibbs-Helmholtz equations (25.31), (25.32) and (25.33), for example, will hold for any change in a closed system, irrespective of whether it is carried out reversibly or not. This is because the values of ΔF and ΔH (or ΔA and ΔE) are quite definite for a given change, and do not depend upon the path followed. The only condition that need be applied is the obvious one that the system must be in thermodynamic equilibrium in the initial and final states of the process, for only in these circumstances can the various thermodynamic functions have definite values (§ 4d).

25f. Conditions of Equilibrium.—An important use of the free energy and, to some extent, of the work function, is to obtain simple criteria of spontaneous processes and of thermodynamic equilibrium which lend themselves very readily to practical application. The result derived in § 19e, that for a spontaneous (irreversible) process there is a net increase of entropy of the system and its surroundings, is based on equation (19.14); in other words, the essential fact is that in an irreversible process $A \rightarrow B$, the sum of the q/T terms for all the isothermal stages is *less* than the increase in entropy of the system. Hence, in an infinitesimal stage of an irreversible process q/T is less than dS, i.e.,

$$dS > \frac{q}{T} \text{ (irreversible process)},$$

where dS refers to the *system alone*, and not to the net entropy of the system and surroundings. For a reversible change, on the other hand, dS, by definition, is equal to q/T, so that

$$dS = \frac{q}{T} \text{ (reversible process)}.$$

It is, therefore, possible to combine these two results in the general expression

$$dS \geqq \frac{q}{T}, \tag{25.34}$$

where the "greater than" sign refers to an irreversible process, while the "equal to" sign applies to a reversible process which, as seen in § 8a, is a succession of equilibrium states.

By the first law of thermodynamics, $q = dE + w$ [cf. equation (7.5)], so that (25.34) may be written in the form

$$dS \geqslant \frac{dE + w}{T}. \tag{25.35}$$

If, in any process, the energy content remains constant and no work is done against an external force, dE and w are both zero; hence,

$$dS \geqslant 0, \quad \text{if} \quad dE = 0 \quad \text{and} \quad w = 0. \tag{25.36}$$

In the event that the external pressure is the only "force," the work is entirely work of expansion, i.e., PdV; the condition that w is zero is then satisfied when dV is zero, i.e., the volume is constant. In these circumstances, equation (25.36) becomes

$$dS_{E,V} \geqslant 0, \tag{25.37}$$

where the subscripts E and V indicate constancy of these properties. Consequently, when the energy and volume are maintained constant, the entropy of a system increases in a spontaneous process, but remains unaltered for a small change in the system when it is in a state of thermodynamic equilibrium.* In other words, *the entropy of a system at equilibrium is a maximum at constant energy and volume,* since a spontaneous process always represents a closer approach to the equilibrium state under the given conditions.

By combining equation (25.14) with (25.35) it follows that

$$dA \leqslant - w - SdT,$$

where the "less than" sign now refers to the spontaneous (irreversible) process. If the temperature is maintained constant and no work is done, so that dT and w are zero,

$$dA \leqslant 0, \quad \text{if} \quad dT = 0 \quad \text{and} \quad w = 0.$$

Again, when the work is solely work of expansion, the result may be written in the form

$$dA_{T,V} \leqslant 0. \tag{25.38}$$

This means that *in a state of thermodynamic equilibrium, at constant temperature and volume, the work function is a minimum;* under the same conditions a spontaneous process is accompanied by a decrease in the work function.

---

* Since dS in equations (25. 34) to (25.37) refers to the gain in entropy of the *system alone*, it depends on the initial and final states only and not on the path between them; hence, $S_{E,V}$ will increase in any change that is potentially spontaneous, even if it is actually carried out in a reversible manner, provided E and V are constant. Compare in this connection the first footnote in § 19e, which refers to the entropy of *the system and its surroundings*.

When the work is only work of expansion, and the system is always in equilibrium with the external pressure, w in equation (25.35) may be replaced by PdV, where P is the pressure of the *system*; thus,

$$dS \geqslant \frac{dE + PdV}{T},$$

$$TdS \geqslant dE + PdV.$$

If this result is combined with equation (25.18), it is seen that

$$dF \leqslant VdP - SdT,$$

and hence at constant temperature and pressure,

$$dF_{T,P} \leqslant 0, \qquad (25.39)$$

where the "less than" sign refers to a spontaneous process. Since most chemical reactions and many physical changes are carried out under conditions of constant temperature and pressure, equation (25.39) is almost invariably used, rather than (25.37) or (25.38), to give the conditions of a spontaneous process or of thermodynamic equilibrium. Since $dF_{T,P}$ is either less than or equal to zero, according as the system changes spontaneously or is in equilibrium, it follows that *for a system in equilibrium, at a given temperature and pressure, the free energy must be a minimum*. Further, *all spontaneous processes taking place at constant temperature and pressure are accompanied by a decrease of free energy*. This conclusion is of fundamental significance, for it provides a very simple and convenient test of whether a given process is possible or not. For an appreciable process, such as a chemical reaction, carried out at constant temperature and pressure, equation (25.39) may be written as

$$\Delta F_{T,P} \leqslant 0, \qquad (25.40)$$

a form in which the condition for a spontaneous process or for equilibrium is frequently employed (Chapter XIII).

25g. Work Function and Free Energy from Partition Functions.—For 1 mole of a gas which behaves ideally, PV is equal to RT, and so equation (25.6), which defines the free energy, may be written

$$F = E - TS + RT. \qquad (25.41)$$

Utilizing equations (16.8) and (24.11), for E and S respectively, in terms of the partition functions, it follows that for 1 mole of an ideal gas,

$$F = E_0 - RT \ln \frac{Q}{N},$$

$$F - E_0 = -RT \ln \frac{Q}{N}. \qquad (25.42)*$$

* It should be noted that in this section Q refers to the partition function.

It is seen from equation (25.41) that at 0° K, when T is zero, F and E are identical, and hence equation (25.42) may also be expressed in the form

$$F - F_0 = - RT \ln \frac{Q}{N}. \qquad (25.43)$$

These equations permit the calculation of the molar free energy of an ideal gas, with reference to the value in the lowest energy state, i.e., at 0° K; they will be used in Chapter XIII in connection with the determination of equilibrium constants and the free energy changes accompanying chemical reactions.

Since the work content A is equal to $F - PV$ [equation (25.7)], and hence to $F - RT$ for 1 mole of an ideal gas, it is seen from equation (25.42) that

$$A - E_0 = - RT \ln \frac{Q}{N} - RT,$$

and this may be written in the equivalent form

$$A - A_0 = - RT \ln \frac{Q}{N} - RT, \qquad (25.44)$$

since A and E are equal at the absolute zero.

25h. Thermodynamic Formulae.—It is of interest at this point to refer to certain general procedures which may be used for the derivation of thermodynamic relationships. One of these, which has been already employed from time to time, is the following. If x is a single-valued function of the variables y and z, e.g., a thermodynamic property of a closed system, it is possible to write for the complete (exact) differential dx,

$$dx = M dy + N dz, \qquad (25.45)$$

where M and N are also functions of the variables. If z is constant, so that dz is zero, then equation (25.45) yields the result

$$\left(\frac{\partial x}{\partial y} \right)_z = M, \qquad (25.46)$$

whereas if y is maintained constant, so that dy is zero,

$$\left(\frac{\partial x}{\partial z} \right)_y = N. \qquad (25.47)$$

If equation (25.46) is differentiated with respect to z, with y constant, and (25.47) with respect to y, with z constant, the results must be identical, so that

$$\left(\frac{\partial M}{\partial z} \right)_y = \left(\frac{\partial N}{\partial y} \right)_z. \qquad (25.48)$$

This result, sometimes referred to as the Euler criterion, or the reciprocity relationship, will now be used to derive some thermodynamic expressions.

An examination of the present and preceding chapters will reveal four analogous equations which are of the form of (25.45); these are

(i) $dE = TdS - PdV$
(ii) $dH = TdS + VdP$
(iii) $dA = -SdT - PdV$
(iv) $dF = -SdT + VdP$

[cf. equations (20.1) or (25.13), (20.13), (25.15) and (25.19), respectively]. By the use of equation (25.48), there follow immediately the four Maxwell relationships, viz.,

$$\left(\frac{\partial T}{\partial V}\right)_S = -\left(\frac{\partial P}{\partial S}\right)_V \quad \text{from (i)}$$

$$\left(\frac{\partial T}{\partial P}\right)_S = \left(\frac{\partial V}{\partial S}\right)_P \quad \text{from (ii)}$$

$$\left(\frac{\partial S}{\partial V}\right)_T = \left(\frac{\partial P}{\partial T}\right)_V \quad \text{from (iii)}$$

$$\left(\frac{\partial S}{\partial P}\right)_T = -\left(\frac{\partial V}{\partial T}\right)_P \quad \text{from (iv)}.$$

If X and Y are functions of the variables x, y and z, such that

$$dX = Ldy + xdz$$
$$dY = Ldy + zdx,$$

then by equation (25.46) or (25.47),

$$L = \left(\frac{\partial X}{\partial y}\right)_z = \left(\frac{\partial Y}{\partial y}\right)_x.$$

By applying this result to the expressions (i), (ii), (iii) and (iv), a new set of relationships can be derived as follows:

$$\left(\frac{\partial E}{\partial S}\right)_V = \left(\frac{\partial H}{\partial S}\right)_P \quad \text{from (i) and (ii)}$$

$$\left(\frac{\partial E}{\partial V}\right)_S = \left(\frac{\partial A}{\partial V}\right)_T \quad \text{from (i) and (iii)}$$

$$\left(\frac{\partial H}{\partial P}\right)_S = \left(\frac{\partial F}{\partial P}\right)_T \quad \text{from (ii) and (iv)}$$

$$\left(\frac{\partial A}{\partial T}\right)_V = \left(\frac{\partial F}{\partial T}\right)_P \quad \text{from (iii) and (iv)}.$$

By the use of equations analogous to (4.9) and (4.10), and the familiar mathematical relationship of the form

$$\left(\frac{\partial x}{\partial y}\right)_z = -\left(\frac{\partial x}{\partial z}\right)_y \left(\frac{\partial z}{\partial y}\right)_x$$

(cf. § 21b), numerous thermodynamic equations can be derived.

Since there are eight common thermodynamic variables, viz., P, V, T, E, H, S, A and F, there are possible $8 \times 7 \times 6$, i.e., 336, first (partial) derivatives, and a large number of relationships among them exist. P. W. Bridgman (1914) has devised a system which permits the derivation of an expression for any of these first derivatives in terms of three quantities which are, in general, capable of experimental determination, viz., $(\partial V/\partial T)_P$, $(\partial V/\partial P)_T$ and $(\partial H/\partial T)_P$, i.e., C_P. The procedure adopted is to write in a purely formal manner

$$\left(\frac{\partial x}{\partial y}\right)_z = \frac{(\partial x)_z}{(\partial y)_z},$$

where x, y and z represent any of the eight variables, and then to make use of the Bridgman formulae to obtain the appropriate values of $(\partial x)_z$ and $(\partial y)_z$. There are actually fifty-six such formulae, but since $(\partial x)_z$ is equal to $-(\partial z)_x$ the number is effectively reduced to the twenty-eight results given below.

$(\partial T)_P = -(\partial P)_T = 1$
$(\partial V)_P = -(\partial P)_V = (\partial V/\partial T)_P$
$(\partial S)_P = -(\partial P)_S = C_P/T$
$(\partial E)_P = -(\partial P)_E = C_P - P(\partial V/\partial T)_P$
$(\partial H)_P = -(\partial P)_H = C_P$
$(\partial F)_P = -(\partial P)_F = -S$
$(\partial A)_P = -(\partial P)_A = -S - P(\partial V/\partial T)_P$
$(\partial V)_T = -(\partial T)_V = -(\partial V/\partial P)_T$
$(\partial S)_T = -(\partial T)_S = (\partial V/\partial T)_P$
$(\partial E)_T = -(\partial T)_E = T(\partial V/\partial T)_P + P(\partial V/\partial P)_T$
$(\partial H)_T = -(\partial T)_H = -V + T(\partial V/\partial T)_P$
$(\partial F)_T = -(\partial T)_F = -V$
$(\partial A)_T = -(\partial T)_A = P(\partial V/\partial P)_T$
$(\partial S)_V = -(\partial V)_S = C_P(\partial V/\partial T)_P/T + (\partial V/\partial T)_P^2$
$(\partial E)_V = -(\partial V)_E = C_P(\partial V/\partial P)_T + T(\partial V/\partial T)_P^2$
$(\partial H)_V = -(\partial V)_H = C_P(\partial V/\partial P)_T + T(\partial V/\partial T)_P^2 - V(\partial V/\partial T)_P$
$(\partial F)_V = -(\partial V)_F = -V(\partial V/\partial T)_P - S(\partial V/\partial P)_T$
$(\partial A)_V = -(\partial V)_A = -S(\partial V/\partial P)_T$
$(\partial E)_S = -(\partial S)_E = PC_P(\partial V/\partial T)_T/T + P(\partial V/\partial T)_P^2$
$(\partial H)_S = -(\partial S)_H = -VC_P/T$
$(\partial F)_S = -(\partial S)_F = -VC_P/T + S(\partial V/\partial T)_P$
$(\partial A)_S = -(\partial S)_A = PC_P(\partial V/\partial P)_T/T + P(\partial V/\partial T)_P^2 + S(\partial V/\partial T)_P$
$(\partial H)_E = -(\partial E)_H = -V[C_P - P(\partial V/\partial T)_P] - P[C_P(\partial V/\partial P)_T + T(\partial V/\partial T)_P^2]$
$(\partial F)_E = -(\partial E)_F = -V[C_P - P(\partial V/\partial T)_P] + S[T(\partial V/\partial T)_P + P(\partial V/\partial P)_T]$
$(\partial A)_E = -(\partial E)_A = P[C_P(\partial V/\partial P)_T + T(\partial V/\partial T)_P^2]$
$(\partial F)_H = -(\partial H)_F = -V(C_P + S) + TS(\partial V/\partial T)_P$
$(\partial A)_H = -(\partial H)_A = -[S + P(\partial V/\partial T)_P][V - T(\partial V/\partial T)_P] + P(\partial V/\partial P)_T$
$(\partial A)_F = -(\partial F)_A = -S[V + P(\partial V/\partial P)_T] - PV(\partial V/\partial T)_P.$

The use of the Bridgman formulae may be illustrated by employing them to derive an expression for $(\partial T/\partial P)_H$, which is the Joule-Thomson coefficient (§ 11b). The required results are

$(\partial T)_H = -(\partial H)_T = V - T(\partial V/\partial T)_P$
$(\partial P)_H = -(\partial H)_P = -C_P,$

so that
$$\left(\frac{\partial T}{\partial P}\right)_H = \frac{1}{C_P}\left[T\left(\frac{\partial V}{\partial T}\right)_P - V\right],$$
as in equation (22.2).[2]

26. Chemical Potential

26a. Partial Molar Properties.—Although the concept of partial molar quantities is employed in connection with thermodynamic properties other than the free energy (see Chapter XIX), the partial molar free energy will be used so frequently that the opportunity may be taken to introduce some of the general ideas here. It has been mentioned, from time to time, that the results obtained so far are based on the supposition that the system under consideration is a closed one, that is, one of constant mass. The change of any thermodynamic property is then due to a change in the state of the system, and not to the addition or removal of matter. In the study of systems consisting of two or more substances, i.e., solutions, and of heterogeneous systems containing two or more phases, it is necessary to consider **open systems**, where composition and mass may vary. In this connection the concept of **partial molar properties**, as developed by G. N. Lewis (1907), is of great value.[3]

Consider any thermodynamic extensive property, such as volume, free energy, entropy, energy content, etc., the value of which, for a homogeneous system, is completely determined by the state of the system, e.g., the temperature, pressure, and the amounts of the various constituents present; thus, G is a function represented by

$$G = f(T, P, n_1, n_2, \ldots, n_i, \ldots), \qquad (26.1)$$

where $n_1, n_2, \ldots, n_i, \ldots$, are the numbers of moles of the respective constituents, $1, 2, \ldots, i, \ldots$, of the system. If there is a small change in the temperature and pressure of the system, as well as in the amounts of its constituents, the change in the property G is given by

$$dG = \left(\frac{\partial G}{\partial T}\right)_{P, n_1, n_2, \ldots} dT + \left(\frac{\partial G}{\partial P}\right)_{T, n_1, n_2, \ldots} dP$$
$$+ \left(\frac{\partial G}{\partial n_1}\right)_{T, P, n_2, \ldots} dn_1 + \cdots \left(\frac{\partial G}{\partial n_i}\right)_{T, P, n_1, \ldots} dn_i + \cdots. \quad (26.2)$$

The derivative $(\partial G/\partial n_i)_{T, P, n_1, \ldots}$ is called the **partial molar property** for the constituent i, and it is represented by writing a bar over the symbol for

---

[2] Bridgman, *Phys. Rev.*, **2**, 3, 273 (1914); see also, P. W. Bridgman, "Condensed Collection of Thermodynamic Formulas," 1925; for other generalized treatments, see Shaw, *Phil. Trans. Roy. Soc.*, **A234**, 299 (1935), or summary in Margenau and Murphy, "The Mathematics of Physics and Chemistry," 1943, p. 18; Lerman, *J. Chem. Phys.*, **5**, 792 (1937); Tobolsky, *ibid.*, **10**, 644 (1942).

[3] Lewis, *Proc. Am. Acad. Arts Sci.*, **43**, 259 (1907); *Z. physik. Chem.*, **61**, 129 (1907); see also, G. N. Lewis and M. Randall, ref. 1; *J. Am. Chem. Soc.*, **43**, 233 (1921).

the particular property, i.e., $\bar{G}_i$, so that

$$\left(\frac{\partial G}{\partial n_1}\right)_{T,P,n_2,n_3,\ldots} = \bar{G}_1, \qquad \left(\frac{\partial G}{\partial n_2}\right)_{T,P,n_1,n_3,\ldots} = \bar{G}_2, \text{ etc.} \qquad (26.3)$$

It is thus possible to write equation (26.2) in the form

$$dG = \left(\frac{\partial G}{\partial T}\right)_{P,n_1,n_2,\ldots} dT + \left(\frac{\partial G}{\partial P}\right)_{T,n_1,n_2,\ldots} dP$$
$$+ \bar{G}_1 dn_1 + \bar{G}_2 dn_2 + \cdots \bar{G}_i dn_i + \cdots. \qquad (26.4)$$

If the temperature and pressure of the system are maintained constant, dT and dP are zero, so that

$$dG_{T,P} = \bar{G}_1 dn_1 + \bar{G}_2 dn_2 + \cdots \bar{G}_i dn_i + \cdots, \qquad (26.5)$$

which gives, upon integration, for a system of definite composition, represented by the numbers of moles $n_1, n_2, \ldots, n_i, \ldots,$*

$$G_{T,P,N} = n_1 \bar{G}_1 + n_2 \bar{G}_2 + \cdots + n_i \bar{G}_i + \cdots. \qquad (26.6)$$

By general differentiation of equation (26.6), at constant temperature and pressure but varying composition, it is seen that

$$dG_{T,P} = (n_1 d\bar{G}_1 + \bar{G}_1 dn_1) + (n_2 d\bar{G}_2 + \bar{G}_2 dn_2) + \cdots$$
$$+ (n_i d\bar{G}_i + \bar{G}_i dn_i) + \cdots$$
$$= (n_1 d\bar{G}_1 + n_2 d\bar{G}_2 + \cdots + n_i d\bar{G}_i + \cdots)$$
$$+ (\bar{G}_1 dn_1 + \bar{G}_2 dn_2 + \cdots + \bar{G}_i dn_i + \cdots). \qquad (26.7)$$

Comparing this result with equation (26.5), it follows that at a *given temperature and pressure*

$$n_1 d\bar{G}_1 + n_2 d\bar{G}_2 + \cdots + n_i d\bar{G}_i + \cdots = 0, \qquad (26.8)$$

which must obviously apply to a system of definite composition. This simple relationship is the basis of the important Gibbs-Duhem equation (§ 26c), first derived by J. W. Gibbs (1875) and later, independently, by P. Duhem (1886).

26b. Physical Significance of Partial Molar Property.—The physical significance of any partial molar property, such as the partial molar volume, partial molar free energy, etc., of a particular *constituent of a mixture* may now be considered. According to equation (26.3), it is the increase in the property G of the system resulting from the addition, *at constant temperature and pressure*, of 1 mole of that substance to such a large quantity of the system that its composition remains virtually unchanged. However, a more useful picture is obtained from equation (26.6): this states that the sum of the $n_i \bar{G}_i$ terms for all the constituents is equal to the total value G for the system of given composition at constant temperature and pressure. Hence, the partial molar property $\bar{G}_i$ of any constituent may be regarded as the

* In general, constant composition will be indicated by the subscript N, as in $G_{T,P,N}$.

contribution of 1 mole of that constituent to the total value of the property G of the system under the specified conditions. Upon first consideration, it might be imagined, in view of this interpretation, that the partial molar property $\bar{G}_i$ was equal to the value of G_i for 1 mole of the constituent i in the pure state. It will be seen later that this is only true in certain limited circumstances. In general, $\bar{G}_i$ in a solution is not equal to G_i for the pure substance, and further, the value of $\bar{G}_i$ varies as the composition of the system is changed.*

It may be pointed out, in view of equation (26.6), that the partial molar property $\bar{G}_i$ is an intensive property (§ 4d), and not an extensive one; that is to say, its value does not depend on the amount of material constituting the system, but only on the composition, at the given temperature and pressure. Allowance for the quantity of each constituent is made by the appropriate values of $n_1, n_2, \ldots, n_i, \ldots$, in equation (26.6), so that $n_1\bar{G}_1$, $n_2\bar{G}_2, \ldots, n_i\bar{G}_i, \ldots$, are the contributions to the total value of the property G. This same conclusion can be reached by precise mathematical arguments based on the use of Euler's theorem on homogeneous functions. At constant temperature and pressure the property G is a homogeneous function of the numbers of moles (n_i) of degree unity; hence, the derivative of G with respect to n_i, i.e., $\bar{G}_i$, in general, will be a homogeneous function of degree zero. In other words, the partial molar property is independent of the n_i's and hence of the amounts of material in the system. It should be remembered, however, that $\bar{G}_i$ is not independent of the composition of the system, that is, of the ratio of the various n's to one another.

26c. Partial Molar Free Energy: The Chemical Potential.—Although the partial molar quantities of various thermodynamic properties will be considered in the course of this book, the discussion at present will be restricted mainly to a consideration of the **partial molar free energy**, that is, $\bar{F}_i$ for the ith constituent. This quantity is, for present purposes, identical with the function described by J. W. Gibbs, known as the **molar chemical potential** or, in brief, the **chemical potential**, and which is represented by the symbol μ. Hence, by the definition given above, the partial molar free energy or chemical potential of a constituent of a mixture is

$$\left(\frac{\partial F}{\partial n_i}\right)_{T, P, n_1, \ldots} = \bar{F}_i = \mu_i. \tag{26.9}$$

It is thus possible to rewrite equation (26.4), replacing G by F, and using μ's for the partial molar quantities, to give

$$dF = \left(\frac{\partial F}{\partial T}\right)_{P, N} dT + \left(\frac{\partial F}{\partial P}\right)_{T, N} dP \\ + \mu_1 dn_1 + \mu_2 dn_2 + \cdots + \mu_i dn_i + \cdots. \tag{26.10}$$

---

* For a system consisting of a single (pure) substance, the partial molar property $\bar{G}_i$ has only a formal significance, for it is then identical with the molar property G_i.

In the special case when there is no change in the numbers of moles of the various substances present, that is to say, when the system is a closed one, dn_1, dn_2, etc., are all zero, so that equation (26.10) becomes

$$dF = \left(\frac{\partial F}{\partial T}\right)_{P,N} dT + \left(\frac{\partial F}{\partial P}\right)_{T,N} dP.$$

It has been shown previously (§ 25c) that for an infinitesimal change in a closed system, in equilibrium with the external pressure,

$$dF = VdP - SdT,$$

so that, equating coefficients,

$$\left(\frac{\partial F}{\partial T}\right)_{P,N} = -S, \qquad (26.11)$$

and

$$\left(\frac{\partial F}{\partial P}\right)_{T,N} = V. \qquad (26.12)$$

These equations are identical, as of course they should be, with equations (25.20) and (25.21) which are applicable to a closed system. If these results are now substituted in equation (26.10), it follows that for the *open system*

$$dF = -SdT + VdP + \mu_1 dn_1 + \mu_2 dn_2 + \cdots + \mu_i dn_i + \cdots. \qquad (26.13)$$

At constant temperature and pressure this becomes

$$dF_{T,P} = \mu_1 dn_1 + \mu_2 dn_2 + \cdots + \mu_i dn_i + \cdots, \qquad (26.14)$$

which is of the same form as equation (26.5).

When the partial molar property $\bar{G}_i$ is the partial molar free energy μ_i, equation (26.8) becomes

$$n_1 d\mu_1 + n_2 d\mu_2 + \cdots + n_i d\mu_i + \cdots = 0, \qquad (26.15)$$

which is one form of the **Gibbs-Duhem equation**, applicable to a system at *constant temperature and pressure*. This equation has many applications, especially in connection with the study of liquid-vapor equilibria, such as are involved in distillation.

26d. Equilibrium in Heterogeneous System.—The condition represented by equation (25.39) for a system at equilibrium, viz.,

$$dF_{T,P} = 0,$$

is applicable to a closed system. As indicated earlier, it is possible for a system consisting of several phases to be closed; nevertheless, one or more of the constituent phases may be open, in the sense that there may be exchanges of matter between them. In a case of this kind, the change of free energy of each phase (open system) for a small change at constant temperature and pressure is given by equation (26.14), and for the system as a whole (closed system) $dF_{T,P}$ is given by the sum of the changes for the individual phases. If the whole system is in equilibrium, at a given temperature and pressure, $dF_{T,P}$ is zero, as seen above, and hence, by equation (26.14),

$$\sum \mu_i dn_i = 0, \qquad (26.16)$$

where the summation includes all the μdn terms for *all the phases* constituting the system. This result forms the basis of the well known "phase rule", as will be seen in § 28a.

26e. Alternative Definitions of Chemical Potential.—Although the definition of the chemical potential as the partial molar free energy is the one which is most generally useful in chemical thermodynamics, it is of interest to show that it can be defined in other ways. This justifies the use of the general term "chemical potential", indicating that the property has a wider significance than the partial molar free energy.

The energy content of a system, like the free energy, may be expressed as a function of the thermodynamic coordinates and masses of the constituents; for the present purpose it is convenient to choose as the coordinates the entropy and the volume. Hence, it is possible to write

$$dE = \left(\frac{\partial E}{\partial S}\right)_{V,N} dS + \left(\frac{\partial E}{\partial V}\right)_{S,N} dV + \left(\frac{\partial E}{\partial n_1}\right)_{S,V,n_2,\ldots} dn_1 + \cdots + \left(\frac{\partial E}{\partial n_i}\right)_{S,V,n_1,\ldots} dn_i + \cdots. \qquad (26.17)$$

By utilizing equation (19.18) or (25.13), it is readily seen that $(\partial E/\partial S)_{V,N}$ and $(\partial E/\partial V)_{S,N}$, which refer to constant values of the n's and hence are effectively applicable to a closed system, are equal to T and $-P$, respectively. Consequently, it follows that

$$dE = TdS - PdV + \left(\frac{\partial E}{\partial n_1}\right)_{S,V,n_2,\ldots} dn_1 + \cdots + \left(\frac{\partial E}{\partial n_i}\right)_{S,V,n_1,\ldots} dn_i + \cdots. \qquad (26.18)$$

From the definition of F as $E - TS + PV$ [equation (25.6)], there is obtained upon differentiation equation (25.18), viz.,

$$dF = dE - TdS - SdT + PdV + VdP,$$

and hence at constant temperature and pressure, i.e., dT and dP are zero,

$$dF_{T,P} = dE - TdS + PdV.$$

Combination of this result with equation (26.18) then gives

$$dF_{T,P} = \left(\frac{\partial E}{\partial n_1}\right)_{S,V,n_2,\ldots} dn_1 + \left(\frac{\partial E}{\partial n_2}\right)_{S,V,n_1,\ldots} dn_2 + \cdots + \left(\frac{\partial E}{\partial n_i}\right)_{S,V,n_1,\ldots} dn_i + \cdots. \qquad (26.19)$$

Since the values of dn_1, dn_2, etc., are independent of one another, it follows from a comparison of equations (26.19) and (26.14), that

$$\mu_1 = \left(\frac{\partial E}{\partial n_1}\right)_{S,V,n_2,\ldots}, \qquad \mu_2 = \left(\frac{\partial E}{\partial n_2}\right)_{S,V,n_1,\ldots}, \qquad \text{etc.},$$

which are alternative definitions of the chemical potentials. It should be understood that $(\partial E/\partial n_1)_{S,V,n_2,\ldots}$ etc., are not partial molar energies, for they refer to constant entropy and volume, and not to constant temperature and pressure. The partial molar energy content would be defined by $(\partial E/\partial n_1)_{T,P,n_2,...}$, and this is, of course, not equal to the chemical potential.

By means of arguments similar to those employed above, it can be shown that the chemical potential may also be defined in other ways, viz.,

$$\mu_1 = \left(\frac{\partial H}{\partial n_1}\right)_{S,P,n_2,...} = \left(\frac{\partial A}{\partial n_1}\right)_{T,V,n_2,...} = -T\left(\frac{\partial S}{\partial n_1}\right)_{E,V,n_2,...}. \quad (26.20)$$

26f. Variation of Chemical Potential with Temperature and Pressure.— The variation of the chemical potential of any constituent of a system with temperature may be derived by differentiating equation (26.9) with respect to temperature, and equation (26.11) with respect to n_i; the results are

$$\frac{\partial^2 F}{\partial n_i \partial T} = \left(\frac{\partial \mu_i}{\partial T}\right)_{P,N} \quad (26.21)$$

and

$$\frac{\partial^2 F}{\partial T \partial n_i} = -\left(\frac{\partial S}{\partial n_i}\right)_{T,P,n_1,...} = -\bar{S}_i, \quad (26.22)$$

the latter being equal to the partial molar entropy, by definition. Since dF is a complete differential, the order of differentiation is immaterial, and hence equations (26.21) and (26.22) are equivalent, so that

$$\left(\frac{\partial \mu_i}{\partial T}\right)_{P,N} = -\bar{S}_i. \quad (26.23)$$

This result is analogous to equation (25.20) which is applicable in particular to a pure substance.

By equation (25.8), $F = H - TS$, and differentiation with respect to n_i, the temperature, pressure and the numbers of moles of the other constituents remaining constant, gives

$$\left(\frac{\partial F}{\partial n_i}\right)_{T,P,n_1,...} = \left(\frac{\partial H}{\partial n_i}\right)_{T,P,n_1,...} - T\left(\frac{\partial S}{\partial n_i}\right)_{T,P,n_1,...}$$

or, in alternative symbols,

$$\mu_i = \bar{F}_i = \bar{H}_i - T\bar{S}_i.$$

If the expression for the partial molar entropy given by equation (26.23) is introduced, it is seen that

$$\mu_i - T\left(\frac{\partial \mu_i}{\partial T}\right)_{P,N} = \bar{H}_i, \quad (26.24)$$

which is a form of the Gibbs-Helmholtz equation (§ 25e). Upon dividing through both sides of equation (26.24) by T^2, the result, analogous to (25.28), is

$$\left[\frac{\partial(\mu_i/T)}{\partial T}\right]_{P,N} = -\frac{\bar{H}_i}{T^2}. \quad (26.25)$$

This equation is particularly useful for expressing the variation of the chemical potential with temperature, at constant pressure and composition, of any constituent of a gaseous, liquid or solid solution.

The effect of pressure on chemical potential may be derived by differentiating equation (26.9) with respect to pressure and (26.12) with respect to n_i; it is seen that

$$\frac{\partial F}{\partial n_i \partial P} = \left(\frac{\partial \mu_i}{\partial P}\right)_{T,N}$$

$$\frac{\partial F}{\partial P \partial n_i} = \left(\frac{\partial V}{\partial n_i}\right)_{T,P,n_1,\ldots} = \bar{V}_i,$$

so that

$$\left(\frac{\partial \mu_i}{\partial P}\right)_{T,N} = \bar{V}_i, \qquad (26.26)$$

which may be compared with the analogous equation (25.21) for a closed system, e.g., a pure substance. The rate of change of chemical potential with pressure of a particular constituent of a system, at constant temperature, is thus equal to the partial molar volume of that constituent.

For a system of ideal gases, a further development of equation (26.26) is possible. For any constituent i of a mixture of ideal gases of total volume V, the equation of state is $p_i V = n_i RT$, where n_i is the number of moles of that constituent present in the mixture and p_i is its partial pressure; hence,

$$V = \frac{n_i RT}{p_i}.$$

The partial molar volume is then given by differentiating with respect to n_i, all the other n's remaining unchanged, at constant temperature and pressure; thus,

$$\bar{V}_i = \left(\frac{\partial V}{\partial n_i}\right)_{T,P,n_1,\ldots} = \frac{RT}{p_i}. \qquad (26.27)$$

If this result is substituted in equation (26.26), it is found that for any constituent of an ideal gas mixture,

$$\left(\frac{\partial \mu_i}{\partial P}\right)_{T,N} = \frac{RT}{p_i}. \qquad (26.28)$$

This result will be employed in connection with the thermodynamics of mixtures of gases in Chapter XII.

26g. Free Energy Change in Any Process.—A useful application of the chemical potential is to determine the free energy change accompanying any process. In a completely general case, which may refer to a chemical or a physical change, the system may consist initially of n_1, n_2, etc., moles of constituents with respective chemical potentials μ_1, μ_2, etc. The total free energy of the system is then given by equation (26.6) as $n_1\mu_1 + n_2\mu_2 + \cdots$. If, in the final state, the system consists of n_1', n_2', etc., moles of substances whose chemical potentials are μ_1', μ_2', etc., the total free energy in the final state is $n_1'\mu_1' + n_2'\mu_2' + \cdots$. The free energy change accompanying the process is thus

$$\Delta F = (n_1'u_1' + n_2'\mu_2' + \cdots) - (n_1\mu_1 + n_2\mu_2 + \cdots). \qquad (26.29)$$

A number of applications of this result will be found in later sections.

EXERCISES

1. Derive expressions for (i) the free energy change, (ii) the work function change, accompanying the appreciable isothermal expansion of a van der Waals gas. (Use the expression for V in terms of P derived in § 21a.)

2. Explain how the generalized compressibility diagram (Fig. 4) could be used to determine (i) the free energy change, (ii) the work function change, for any gas at constant temperature.

3. Utilize the expression for the partition function derived in Chapter VI to develop an equation for the free energy of an ideal, monatomic gas referred to the value F_0 in the lowest energy state.

4. The variation of the volume of a liquid with pressure is given approximately by $V = V_0(1 - \beta P)$ where β is the compressibility coefficient and V_0 is the volume at low (virtually zero) pressure. Derive an expression for the change of free energy accompanying the isothermal change of a liquid from pressure P_1 to P_2. For water at 25° C, β is 49×10^{-6} atm.$^{-1}$ at moderate pressures. Calculate the change of free energy in cal. accompanying the compression of 1 mole of water from 1 atm. to 10 atm. at 25° C. The specific volume of water at 1 atm. and 25° C is 1.00294 cc. g.$^{-1}$. Would there be any appreciable difference in the result if the water were taken to be incompressible, i.e., with $\beta = 0$?

5. For the reaction $2H_2S(g) + SO_2(g) = 2H_2O(l) + 3S(s)$ at 25° C and 1 atm. pressure of each gas, the change of heat content and the entropy change can be obtained from data in preceding chapters. What is the corresponding free energy change in calories? Is the reaction likely to occur spontaneously, with the substances in their standard states, under the given conditions?

6. Without the use of tables, prove that

$$\left(\frac{\partial F}{\partial T}\right)_V = -S + V\left(\frac{\partial P}{\partial T}\right)_V$$

which is equal to $-S + R$ for 1 mole of an ideal gas. Show that the same result can be obtained from the Bridgman table.

7. Show that for a given process the rate of variation of the free energy change ΔF with temperature at constant pressure is given by $[\partial(\Delta F)/\partial T]_P = -\Delta S$. Estimate the value of ΔF in Exercise 5 at a temperature of 30° C, assuming the rate of change of ΔF with temperature to remain constant between 25° and 30° C.

8. Show that the relationships $[\partial(\Delta F)/\partial T]_P = -\Delta S$ and $[\partial(\Delta F)/\partial P]_T = \Delta V$, which are applicable to any physical or chemical change, constitute the basis of the Le Chatelier principle of mobile equilibrium. [Make use of equation (25.40).]

9. Using the fact that at equilibrium $\Delta F_{T,P} = 0$, what conclusion can be drawn concerning the free energy change accompanying the transfer of water from liquid at 25° C to vapor at 23.76 mm., which is the equilibrium vapor pressure, at the same temperature? Hence, determine the free energy change in cal. for the

EXERCISES

transfer of 1 mole of liquid water to vapor at 1 atm. and 25° C. The vapor may be regarded as an ideal gas.

10. The equilibrium vapor pressure of ice at $-10°$ C is 1.950 mm. and that of supercooled water at the same temperature is 2.149 mm. Calculate the free energy change in cal. accompanying the change of 1 mole of supercooled water to ice at $-10°$ C, and the same total (atmospheric) pressure. Is the sign in agreement with expectation?

11. Combine the result of the preceding exercise with that obtained in Exercise 7, Chapter VIII to evaluate the change in heat content for the process $H_2O(l) = H_2O(s)$ at $-10°$ C and 1 atm.

12. Utilize the Bridgman table to verify the relationships

$$\left(\frac{\partial H}{\partial P}\right)_S = \left(\frac{\partial F}{\partial P}\right)_T \quad \text{and} \quad \left(\frac{\partial F}{\partial T}\right)_P = \left(\frac{\partial A}{\partial T}\right)_V.$$

13. Derive an expression for $(\partial F/\partial H)_S$ by means of the Bridgman table.

14. Show that equation (25.20) can be used to derive the expression

$$S = R \ln \frac{Q}{N} + RT \left(\frac{\partial \ln Q}{\partial T}\right)_P$$

for the entropy of a gas in terms of its partition function.

15. Prove the relationships in equation (26.20).

CHAPTER XI

PHASE EQUILIBRIA

27. Systems of One Component

27a. Equilibrium Between Phases of One Component.—Most of the discussion hitherto has been devoted to laying the foundations of thermodynamic theory; the time is now opportune to consider some applications of the results already derived to problems of physical and chemical importance. In the present chapter a number of subjects will be discussed which have a bearing on equilibria between two or more phases, e.g., liquid and vapor, solid and vapor, etc., of one or more constituents. Such systems remain of constant mass, *as a whole*, no matter what changes occur within them, and so they can be treated as closed.

Consider any system consisting of two phases, e.g., liquid and vapor, of a single substance in equilibrium at constant temperature and pressure. Suppose that a small amount of one phase is transferred to the other; it follows, therefore, from equation (25.39) or (25.40), that the corresponding free energy change is zero. As long as both phases are present, an appreciable transfer, e.g., of 1 mole, from one phase to the other, will not disturb the equilibrium at constant temperature and pressure. For example, if liquid water and its vapor are in equilibrium, a large amount of water can be transferred from one phase to the other, at constant temperature and pressure, without affecting the state of equilibrium. It is, therefore, possible to utilize equation (25.40), viz.,

$$\Delta F = 0, \qquad (27.1)$$

where ΔF is the free energy change accompanying the process under consideration. If F_A is the molar free energy of the substance in one phase, e.g., liquid, and F_B is that in the other, e.g., vapor, the transfer of 1 mole of liquid to the vapor state is accompanied by an increase F_B and a decrease F_A in the free energy; thus,

$$\Delta F = F_B - F_A.$$

Since this is zero, by equation (27.1), when the system is in equilibrium, it follows that

$$F_A = F_B. \qquad (27.2)$$

In other words, *whenever two phases of the same single substance are in equilibrium, at a given temperature and pressure, the molar free energy is the same in each phase*. This conclusion can be extended to three phases, which is the maximum number that can coexist in equilibrium for a system of one component.

It must be remembered that the treatment has been limited to a *system of one component*. If there are two or more components present, it will be seen later (§ 28a) that the chemical potential, in place of the molar free energy, of each component is the same in every phase when the system is at equilibrium.

27b. The Clapeyron Equation.—Since the molar free energy of a given substance is the same in two phases A and B of a one-component system at equilibrium, it follows that if the temperature and pressure are altered infinitesimally, the system remaining in equilibrium under the new conditions, the change in the free energy must be the same in each phase, i.e.,

$$dF_A = dF_B.$$

In a phase change there is no work done other than work of expansion, and so it is permissible to use equation (25.19), namely,

$$dF_A = V_A dP - S_A dT \quad \text{and} \quad dF_B = V_B dP - S_B dT.$$

Since dF_A is equal to dF_B,

$$V_A dP - S_A dT = V_B dP - S_B dT,$$

$$\frac{dP}{dT} = \frac{S_B - S_A}{V_B - V_A} = \frac{\Delta S}{\Delta V}. \tag{27.3}$$

The term ΔS is the entropy increase for the transfer of a specified quantity, e.g., 1 mole, of substance from phase A to phase B, and hence it is equal to $\Delta H/T$, where ΔH is here the molar latent heat of the phase change taking place at the temperature T; making this substitution, equation (27.3) becomes

$$\frac{dP}{dT} = \frac{\Delta H}{T \Delta V}, \tag{27.4}$$

where ΔV is the difference of the molar volumes in the two phases. This expression is a form of the equation first derived by B. P. E. Clapeyron (1834); it gives the variation of the equilibrium pressure with temperature for any two phases of a given substance.

An alternative derivation of the **Clapeyron equation** makes use of the Maxwell equation (20.12), viz.,

$$\left(\frac{\partial P}{\partial T}\right)_V = \left(\frac{\partial S}{\partial V}\right)_T, \tag{27.5}$$

which is applicable to any closed system, homogeneous or heterogeneous, in equilibrium with the external pressure. For a system consisting of two phases of the same substance and, in fact, for any univariant system of more than one component, the equilibrium pressure, e.g., the vapor pressure, is dependent on the temperature only, and is independent of the volume. It is thus possible, *for a univariant system*, to replace $(\partial P/\partial T)_V$ by dP/dT

without the constant volume restriction.* If ΔS is the entropy change when any given quantity of the substance is transferred from one phase to the other, at constant temperature, and ΔV is the accompanying increase of volume, then $\Delta S/\Delta V$ will be constant at a given temperature, for both ΔS and ΔV are extensive properties which are proportional to the quantity of material transferred. Thus, $(\partial S/\partial V)_T$ may be replaced by $\Delta S/\Delta V$ at the given temperature, so that equation (27.5) becomes

$$\frac{dP}{dT} = \frac{\Delta S}{\Delta V} = \frac{\Delta H}{T\Delta V},$$

utilizing the fact that ΔS is equal to $\Delta H/T$, where ΔH is the latent heat of the phase change. The result obtained in this manner is identical with equation (27.4). The quantities ΔH and ΔV must refer to the same amount of the substance under consideration; this is usually either 1 gram or 1 mole.

27c. Solid-Liquid (Fusion) Equilibria.—Solid and liquid phases of a given substance are in equilibrium at the melting (or freezing) point; hence, in the Clapeyron equation (27.4), T is the melting point when P is the external pressure exerted on the system. By writing equation (27.4) in the inverted form

$$\frac{dT}{dP} = \frac{T\Delta V}{\Delta H}, \qquad (27.6)$$

an expression is obtained which gives the variation of the melting point T with the external pressure P. If V_s is the molar volume of the solid phase and V_l is that of the liquid at the temperature T and pressure P, then ΔV may be taken as $V_l - V_s$, representing the increase of volume in transferring 1 mole from solid to liquid phase. The corresponding value of ΔH, i.e., the heat absorbed, in the same phase change is the molar heat of fusion, and this may be represented by ΔH_f, so that equation (27.6) becomes

$$\frac{dT}{dP} = \frac{T(V_l - V_s)}{\Delta H_f}. \qquad (27.7)$$

Alternatively, V_l and V_s may be taken as the respective specific volumes; ΔH_f is then the heat of fusion per gram.

From a knowledge of the volumes (or densities) of the liquid and solid phases, and of the heat of fusion, it is possible to determine quantitatively the variation of the melting point of the substance with pressure. Qualitatively, it may be observed that if V_l is greater than V_s, that is, the liquid has a smaller density than the solid, at the melting point, then dT/dP will be positive, and the melting point will increase with the applied pressure. This is the case for the majority of solids. However, if V_l is smaller than V_s, the liquid having the greater density, increase of pressure will cause the melting point to decrease. Very few substances, notably ice, bismuth and antimony, exhibit this type of behavior.

* It should be understood that a system containing an inert gas, in addition to a pure liquid (or solid) and its vapor, is not univariant.

Problem: The specific volume of liquid water is 1.0001 cc. g.$^{-1}$ and that of ice is 1.0907 cc. g.$^{-1}$ at 0° C; the heat of fusion of ice at this temperature is 79.8 cal. g.$^{-1}$. Calculate the rate of change of melting point of ice with pressure in deg. atm.$^{-1}$.

If the values of V_l, V_s and ΔH_f given above are inserted in equation (27.7), it is readily seen that dT/dP will then be in deg. cc. cal.$^{-1}$. In order to convert this into deg. atm.$^{-1}$, use may be made of the fact that 0.0242 cal. cc.$^{-1}$ atm.$^{-1}$ is equal to unity (cf. § 3h); hence, multiplication of the result obtained above by this figure will give dT/dP in deg. atm.$^{-1}$. Thus, since T is 273.2° K,

$$\frac{dT}{dP} = \frac{273.2 \times (1.0001 - 1.0907) \times 0.0242}{79.8}$$

$$= -0.0075° \text{ atm.}^{-1}.$$

Since dT/dP is small, it may be assumed to remain constant over an appreciable pressure range, so that the melting point of ice (or the freezing point of water) decreases by 0.0075° C for 1 atm. increase of the external pressure. It is because the specific volume of ice is greater than that of water at 0° C, that increase of pressure is accompanied by a decrease in the melting point.

Instead of utilizing the Clapeyron equation (27.7) to determine the variation of the melting point with pressure, it may be applied to calculate the heat of fusion from a knowledge of dT/dP, or rather of $\Delta T/\Delta P$, which is assumed to be constant; the specific volumes, or densities, of the solid and liquid phases must, of course, be known.

27d. Equilibrium Between Two Crystalline Forms.—The variation with pressure of the transition point at which two crystalline forms of a solid are in equilibrium is given by an equation of the same form as (27.7). Thus, if β represents the form stable below the transition point, and α the form stable above the transition point, then

$$\frac{dT}{dP} = \frac{T(V_\alpha - V_\beta)}{\Delta H_t}, \qquad (27.8)$$

where dT/dP is the rate of change of the transition temperature T with the external pressure P; V_α and V_β are the molar (or specific) volumes of the indicated forms, and ΔH_t is the molar (or specific) heat of transition, all determined at the temperature T. It is unnecessary to enter into a detailed consideration of equation (27.8), for the comments made in connection with (27.7) can be readily adapted to the present case.

Problem: The specific volume of monoclinic sulfur (stable above the transition point) is greater than that of rhombic sulfur by 0.0126 cc. g.$^{-1}$. The transition point at 1 atm. pressure is 95.5° C, and it increases at the rate of 0.035° atm.$^{-1}$. Calculate the heat of transition in cal. g.$^{-1}$.

If the values given for $\Delta T/\Delta P$ and $V_\alpha - V_\beta$ are inserted directly into equation (27.8), it is readily found that ΔH_t will be obtained in cc. atm.$^{-1}$ g.$^{-1}$. Since the result is required in cal. g.$^{-1}$, all that is necessary is to multiply by the conversion factor 0.0242 cal. cc.$^{-1}$ atm.$^{-1}$, which is equal to unity. Hence, at $273.2 + 95.5 = 368.7°$ K,

$$\Delta H_t = \frac{T(V_\alpha - V_\beta)}{(\Delta T/\Delta P)} = \frac{368.7 \times 0.0126 \times 0.0242}{0.035}$$

$$= 3.2 \text{ cal. g.}^{-1}.$$

27e. Liquid-Vapor (Vaporization) Equilibria.—As applied to the equilibrium between a liquid and its vapor at a given temperature and pressure, the Clapeyron equation is used in the forms of both equations (27.4) and (27.6). The increase of volume ΔV accompanying the transfer of 1 mole (or 1 gram) of liquid to the vapor state is equal to $V_v - V_l$, where V_v and V_l are the molar (or specific) volumes of the vapor and liquid, respectively; * ΔH_v is the molar (or specific) heat of vaporization, so that equation (27.6) becomes

$$\frac{dT}{dP} = \frac{T(V_v - V_l)}{\Delta H_v}. \tag{27.9}$$

The boiling point of a liquid is the temperature at which the pressure of the vapor in equilibrium with it is equal to the external pressure; hence, in the form of (27.9), the Clapeyron equation gives the variation of the boiling point T of a liquid with the external pressure P.

On the other hand, if the equation is inverted, it gives the rate of change of vapor pressure † of the liquid with the temperature; thus,

$$\frac{dp}{dT} = \frac{\Delta H_v}{T(V_v - V_l)}. \tag{27.10}$$

These equations may be utilized for various purposes; for example, if the variation of boiling point with pressure or, what is the same thing, the variation of vapor pressure with temperature, is known, it is possible to calculate the heat of vaporization. Alternatively, if the latter is available, it is possible to determine dT/dP or dp/dT, for the rate of change of boiling point or of vapor pressure, respectively.

Problem: Assuming the heat of vaporization of water to be constant at 539 cal. g.$^{-1}$, calculate the temperature at which water will boil under a pressure of 77.0 cm., the boiling point being 100.00° C at 76.0 cm. The specific volume of water vapor at 100° C and 76.5 cm. pressure is 1664 cc. g.$^{-1}$ and that of liquid water is (approximately) 1 cc. g.$^{-1}$.

Since ΔH_v, V_v and V_l should be mean values, assumed constant over a range of temperature, dT/dP in equation (27.9) may be replaced by $\Delta T/\Delta P$. This quantity is required in deg. cm.$^{-1}$ (of mercury); but it is convenient to obtain the result first in deg. atm.$^{-1}$. The procedure is similar to that described in the

---

* The volumes V_v and V_l measured at the same temperature and pressure, i.e., the vapor pressure, are sometimes called the "orthobaric volumes."

† The symbol p will be used for vapor pressure (and for partial pressure), and P for external (atmospheric) or total pressure.

problem in § 27c; the use of V_v, V_l and ΔH_v given above yields $\Delta T/\Delta P$ in deg. cc. cal.$^{-1}$, and multiplication by 0.0242 cal. cc.$^{-1}$ atm.$^{-1}$ converts this into deg. atm.$^{-1}$. Thus, by equation (27.9), taking T as 100.0° C, i.e., 373.2° K,

$$\frac{\Delta T}{\Delta P} = \frac{373.2 \times (1664 - 1) \times 0.0242}{539}$$

$$= 27.9° \text{ atm.}^{-1}.$$

If this is divided by 76.0, the number of cm. of mercury equivalent to 1 atm. pressure, the result is 0.37° cm.$^{-1}$, so that an increase of 1.0 cm. in the external pressure, in the vicinity of 76 cm., causes the boiling point to rise by 0.37° C. Hence the required boiling point at 77.0 cm. pressure is 100.37° C.

27f. The Clausius-Clapeyron Equation.—If the temperature is not too near the critical point, the volume of the liquid, i.e., V_l, is small in comparison with that of the vapor, i.e., V_v, at the same temperature and pressure; hence, $V_v - V_l$ may be replaced by V_v, and then equation (27.10) may be written as

$$\frac{dp}{dT} = \frac{\Delta H_v}{TV_v}. \tag{27.11}$$

Further, in regions well below the critical point, the vapor pressure is relatively small, so that the ideal gas law may be assumed to be applicable, i.e., $pV_v = RT$, where V_v is the molar volume of the vapor and p is its pressure at the temperature T. Substituting RT/p for V_v in equation (27.11), this becomes

$$\frac{1}{p} \cdot \frac{dp}{dT} = \frac{\Delta H_v}{RT^2},$$

$$\frac{d \ln p}{dT} = \frac{\Delta H_v}{RT^2}. \tag{27.12}$$

This expression is sometimes referred to as the **Clausius-Clapeyron equation**, for it was first derived by R. Clausius (1850), in the course of a comprehensive discussion of the Clapeyron equation. Although the Clausius-Clapeyron equation is approximate, for it neglects the volume of the liquid and supposes ideal behavior of the vapor, it has the advantage of great simplicity. In the calculation of dp/dT (or dT/dP) from a knowledge of the heat of vaporization,* or vice versa, it is not necessary to use the volumes of the liquid and vapor, as is the case in connection with equations (27.9) and (27.10). However, as may be expected, the results are less accurate than those derived from the latter expressions.

27g. Integration of the Clausius-Clapeyron Equation.—A particular advantage of the Clausius-Clapeyron equation is the readiness with which it can be integrated; thus, if the heat of vaporization is assumed to be independent of temperature, integration of equation (27.12) between the temperature limits of T_1 and T_2 and the corresponding vapor pressures p_1 and

---

* For the heats of vaporization of a number of common liquids, see Table 2 at end of book.

p_2, gives

$$\ln \frac{p_2}{p_1} = -\frac{\Delta H_v}{R}\left(\frac{1}{T_2} - \frac{1}{T_1}\right) = \frac{\Delta H_v}{R}\left(\frac{T_2 - T_1}{T_1 T_2}\right). \tag{27.13}$$

If ΔH_v is expressed in cal. mole^{-1}, then R will be 1.987 cal. deg.$^{-1}$ mole^{-1}, and hence, after converting the logarithms, equation (27.13) becomes

$$\log \frac{p_2}{p_1} = \frac{\Delta H_v}{4.576}\left(\frac{T_2 - T_1}{T_1 T_2}\right), \tag{27.14}$$

recalling that 4.576 arises from the product of the logarithm conversion factor (2.303) and the value of R, i.e., 1.987, in cal. deg.$^{-1}$ mole^{-1}. This equation may be used to calculate the heat of vaporization if the vapor pressures of the liquid at two adjacent temperatures are known. Because ΔH_v is not really independent of temperature, as was assumed in the integration of equation (27.12), the value obtained is actually a mean for the given temperature range. Alternatively, if a mean heat of vaporization is available, the vapor pressure at one temperature (or boiling point at a given pressure) can be calculated (approximately) if that at another is known.

Problem: The mean heat of vaporization of water in the temperature range between 90° and 100° C is 542 cal. g.$^{-1}$. Calculate the vapor pressure of water at 90.0° C, the value at 100.0° C being 76.0 cm.

In equation (27.14), ΔH_v is the molar heat of vaporization of water in cal.; since the molecular weight is 18.02, ΔH_v is 542 × 18.02 cal. mole^{-1}. At 100.0° C, i.e., 373.2° K, which will be taken as T_2, the value of p_2 is 76.0 cm.; it is required to find p_1 at T_1 equal to 90° C or 363.2° K. Hence, by equation (27.14),

$$\log \frac{76.0}{p_1} = \frac{542 \times 18.02}{4.576}\left(\frac{373.2 - 363.2}{363.2 \times 373.2}\right).$$

Since p_2 has been expressed in cm., p_1 will be in the same units; therefore,

$$p_1 = 52.9 \text{ cm.}$$

(The experimental value is 52.6 cm.).

27h. Vapor Pressure-Temperature Relationships.—General integration of the Clausius-Clapeyron equation (27.12), assuming ΔH_v to be constant, gives

$$\ln p = -\frac{\Delta H_v}{RT} + \text{constant}, \tag{27.15}$$

or, converting the logarithms and expressing R in cal. deg.$^{-1}$ mole^{-1}, so that ΔH_v is the heat of vaporization in cal. mole^{-1}, this becomes

$$\log p = -\frac{\Delta H_v}{4.576 T} + C, \tag{27.16}$$

where C is a constant. It follows, therefore, that the plot of the logarithm of the vapor pressure, i.e., $\log p$, against the reciprocal of the absolute temperature, i.e., $1/T$, should be a straight line of slope $-\Delta H_v/4.576$. Thus, the slope of this plot can be used to obtain an approximate indication of the mean molar heat of vaporization over a specified temperature range.

Because ΔH_v is not constant, equation (27.16) is applicable over a restricted range of temperature; in order to extend the range, allowance should be made for the variation of the heat of vaporization with temperature. Thus, ΔH_v may probably be expressed as a power series function of the absolute temperature, viz.,

$$\Delta H_v = \Delta H_0 + \alpha T + \beta T^2 + \cdots, \qquad (27.17)$$

where ΔH_0, α, β, etc., are constants for the given liquid. As a first approximation all terms beyond the linear one may be neglected, so that

$$\Delta H_v \approx \Delta H_0 + \alpha T, \qquad (27.18)$$

and if this result is substituted in equation (27.12), integration gives

$$\ln p = -\frac{\Delta H_0}{RT} + \frac{\alpha}{R} \ln T + \text{constant}. \qquad (27.19)$$

This expression, which is of the form

$$\log p = \frac{A}{T} + B \log T + C, \qquad (27.20)$$

where A, B and C are constants, is similar to the equation proposed empirically by G. R. Kirchhoff (1858) and others. It is seen, therefore, that over an appreciable temperature range the plot of $\log p$ against $1/T$ should not be exactly linear, in agreement with observation on a number of liquids.

Several other vapor pressure-temperature relationships of a more complicated character have been proposed from time to time, but as these have no obvious thermodynamic basis or significance they will not be considered here. It may be mentioned, however, that if the experimental vapor pressure data can be expressed with some accuracy as an empirical function of the temperature, for example of the form

$$\log p = \frac{A}{T} + B + CT + DT^2 + \cdots, \qquad (27.21)$$

it is possible by differentiation to derive an expression for $d \ln p/dT$ in terms of the temperature. By the use of the Clausius-Clapeyron equation, the heat of vaporization at any temperature can then be calculated. This procedure, however, assumes the applicability of the ideal gas law to the vapor. To avoid the error introduced by this approximation, dp/dT, which is equal to $p \times d \ln p/dT$ is determined from (27.21), and the Clapeyron equation (27.10) is used; the orthobaric volumes V_v and V_l of vapor and liquid must then be known.

Problem: The vapor pressure of liquid chlorine, in cm. of mercury, can be represented by the expression

$$\log p = -\frac{1414.8}{T} + 9.91635 - 1.206 \times 10^{-2}T + 1.34 \times 10^{-5}T^2.$$

The specific volume of chlorine gas at its boiling point is 269.1 cc. g.$^{-1}$ and that of the liquid is (approximately) 0.7 cc. g.$^{-1}$. Calculate the heat of vaporization of liquid chlorine in cal. g.$^{-1}$ at its boiling point, 239.05° K.

Differentiation of the equation for $\log p$ with respect to temperature gives

$$\frac{d \log p}{dT} = \frac{1414.8}{T^2} - 1.206 \times 10^{-2} + 2.68 \times 10^{-5}T.$$

Further, since

$$\frac{d \log p}{dT} = \frac{1}{2.303} \cdot \frac{d \ln p}{dT} = \frac{1}{2.303 p} \cdot \frac{dp}{dT},$$

it is readily found that at the boiling point, 239.05° K, when p is 76.0 cm.,

$$\frac{dp}{dT} = 3.343 \text{ cm. (of mercury) deg.}^{-1}$$

$$= \frac{3.343}{76.00} = 0.04398 \text{ atm. deg.}^{-1}.$$

By equation (27.10),

$$\Delta H_v = T(V_v - V_l)\frac{dp}{dT},$$

and if V_v and V_l are 269.1 and 0.7 cc. g.$^{-1}$, respectively, and dp/dT is 0.04398 atm. deg.$^{-1}$, ΔH_v will be in cc.-atm. g.$^{-1}$; to convert this to cal. g.$^{-1}$, it is necessary to multiply by 0.0242 cal. cc.$^{-1}$ atm.$^{-1}$, which is equal to unity. Thus, at 239.05° K,

$$\Delta H_v = 239.05 \times 268.4 \times 0.04398 \times 0.0242$$
$$= 68.3 \text{ cal. g.}^{-1}.$$

27i. The Ramsay-Young and Dühring Rules.

—If the $B \log T$ term in equation (27.20) is neglected, that is to say, if the heat of vaporization is taken as constant, the vapor pressure variation with temperature of two liquids A and B can be represented by

$$\log p_A = \frac{A_A}{T} + C_A \quad \text{and} \quad \log p_B = \frac{A_B}{T} + C_B. \tag{27.22}$$

If these substances have the same vapor pressure p at the temperatures T_A and T_B, respectively, then

$$\frac{A_A}{T_A} + C_A = \frac{A_B}{T_B} + C_B,$$

so that,

$$\frac{T_A}{T_B} = \frac{T_A}{A_B}(C_A - C_B) + \frac{A_A}{A_B}. \tag{27.23}$$

Suppose that at the temperatures T'_A and T'_B, respectively, the two liquids have the same vapor pressure p'; hence,

$$\frac{T'_A}{T'_B} = \frac{T'_A}{A_B}(C_A - C_B) + \frac{A_A}{A_B}. \tag{27.24}$$

Subtraction of equation (27.24) from (27.23) then gives

$$\frac{T_A}{T_B} - \frac{T'_A}{T'_B} = C_1(T_A - T'_A), \qquad (27.25)$$

where C_1 is a constant. This result was obtained empirically by W. Ramsay and S. Young (1885), who found that C_1 was small; in fact, if the substances A and B are related chemically C_1 is almost zero, so that equation (27.25) becomes

$$\frac{T_A}{T_B} = \frac{T'_A}{T'_B}. \qquad (27.26)$$

It is to be expected, therefore, that for any two liquids which are related chemically, *the ratio of the temperatures at which they have the same vapor pressure should be constant.* The same result can be stated in another way: *the ratio of the boiling points of two similar liquids should have the same value at all pressures.*

The extent to which the Ramsay-Young rule, represented by equation (27.26), is applicable may be illustrated by reference to water and ethanol. The vapor pressure of water is 12.2 mm. of mercury at 287.5° K, and ethanol has the same vapor pressure at 273.2° K; the ratio of these temperatures is 287.5/273.2, i.e., 1.052. Since the normal boiling points are 373.2° K for water and 351.5° K for ethanol, these are the temperatures at which both have the same vapor pressure of 1 atm.; the ratio of these temperatures is 373.2/351.5, i.e., 1.062. The two ratios thus agree to within about one per cent.

Problem: The vapor pressures of molten silver and sodium are both 1 mm. of mercury at 1218° C and 441° C, respectively. The normal boiling point of sodium is 882° C, estimate that of silver.

Let A represent silver, and B sodium; T_A is $1218 + 273 = 1491°$ K, and T_B is $441 + 273 = 714°$ K; T'_B is $882 + 273 = 1155°$ K, and consequently by equation (27.26),

$$\frac{1491}{714} = \frac{T'_A}{1155},$$

$$T'_A = 2412° \text{ K.}$$

The boiling point of silver is thus calculated to be $2412 - 273 = 2139°$ C. (The experimental value is 1950° C; in view of the extrapolation from 1 mm. to 760 mm. involved in this calculation, the agreement is quite reasonable.)

By rearrangement of equation (27.26) it is found that

$$\frac{T'_A - T_A}{T'_B - T_B} = \frac{T_A}{T_B} = \frac{T'_A}{T'_B}$$

or

$$\frac{T'_A - T_A}{T'_B - T_B} = \text{constant}, \qquad (27.27)$$

the constant being independent of pressure; this constant should, in fact, be equal to the ratio of the normal boiling points, e.g., T'_A/T'_B, of the two liquids. This result is an expression of the rule discovered by U. Dühring (1878). Like the Ramsay-Young equation (27.26), the Dühring equation (27.27) is particularly applicable to pairs of similar liquids, although it has been found to hold with a moderate degree of accuracy for substances of different types, provided both are associated, e.g., hydroxylic compounds, or both are nonassociated.

Provided the vapor pressures of a reference liquid are known over a range of temperatures, it is possible, by means of the Ramsay-Young or Dühring rules, to establish the complete vapor pressure-temperature variation of another substance from one datum for the latter. It should be noted that the results obtained cannot be very precise, for the equations (27.26) and (27.27) can be exact only if the vapor pressures are represented by the linear equations (27.15), and if the constant C_1 in equation (27.25) is zero.[1]

27j. Trouton's Rule and Vapor Pressure Relationship.—It has been found experimentally that for a number of substances the molar entropy of vaporization ΔS_v at the normal boiling point has approximately the same value of 21 cal. deg.$^{-1}$ mole^{-1}. Thus, if ΔH_v is the molar heat of vaporization at the normal boiling point T_b, i.e., at 1 atm. pressure, then

$$\Delta S_v = \frac{\Delta H_v}{T_b} \approx 21 \text{ cal. deg.}^{-1} \text{ mole}^{-1},$$

which is a statement of the generalization known as **Trouton's rule** (F. Trouton, 1884). This rule holds for many familiar compounds with molecular weights in the region of 100, provided they are not associated in the liquid state. Various attempts have been made to modify the Trouton rule, so as to make it more widely applicable. According to J. H. Hildebrand (1915) the entropy of vaporization is more nearly constant if it is measured at the same concentration of the vapor in each case, instead of at the same pressure of the vapor, as in Trouton's rule.

Although these empirical rules, strictly speaking, lie outside the realm of thermodynamics, they have been mentioned because of their practical value in certain cases. At the normal boiling point the vapor pressure of a liquid is equal to 1 atm.; hence, equation (27.16) can be written as

$$\log 1 = -\frac{\Delta H_v}{4.576 T_b} + C.$$

As seen above, however, $\Delta H_v/T_b$ may be taken as equal to 21 for many nonassociated substances; hence, for such substances C should have a constant value, viz.,

$$C = \frac{21.0}{4.576} = 4.59.$$

The expression for the vapor pressure p, in atm., of any substance to which Trouton's law is applicable thus becomes

$$\log p(\text{atm.}) = -\frac{\Delta H_v}{4.576 T} + 4.59. \tag{27.28}$$

As a first approximation, the molar heat of vaporization at the boiling point may be used for ΔH_v; an expression for the variation of the vapor pressure with temperature is thus obtained from the one datum. Alternatively, if the vapor pres-

---

[1] Ramsay and Young, *Phil. Mag.*, **20**, 515 (1885); **21**, 33 (1886); **22**, 32, 37 (1887); Dühring, *Ann. Physik*, **11**, 163 (1880); Leslie and Carr, *Ind. Eng. Chem.*, **17**, 810 (1925); Perry and Smith, *ibid.*, **25**, 195 (1933); Carr and Murphy, *J. Am. Chem. Soc.*, **51**, 116 (1929); Lamb and Roper, *ibid.*, **62**, 806 (1940).

sure is known at one temperature, ΔH_v may be evaluated from equation (27.28) and the result may then be assumed to apply at other temperatures. Another possibility, which is equivalent to the use of Trouton's law, is to choose the known vapor pressure as 1 atm. at the boiling point in order to determine ΔH_v; this value can then be inserted in equation (27.28).[2]

Problem: The normal boiling point of benzene is 80.1° C; estimate its vapor pressure at 40° C.

Since T_b is $80.1 + 273.2 = 353.3°$ K, then by Trouton's law, ΔH_v is $21 \times 353.3 = 7419$ cal. mole^{-1}; hence by equation (27.28),

$$\log p(\text{atm.}) = -\frac{7419}{4.576 T} + 4.59.$$

At 40° C, the value of T is $40.0 + 273.2 = 313.2°$ K, so that

$$\log p(\text{atm.}) = -\frac{7419}{4.576 \times 313.2} + 4.59 = -0.587,$$

$$p = 0.259 \text{ atm.}$$
$$= 0.259 \times 76.0 = 19.6 \text{ cm.}$$

(The experimental value is 18.1 cm. The agreement is not too good, but the calculation may be used to give an approximate indication of the vapor pressure when experimental data are lacking.)

It is of interest to note that equation (27.28) is essentially equivalent to the Ramsay-Young rule. In the first place, it is based on equation (27.16), and in the second place, it supposes that the constant C has the same value for all liquids; if this were the case, $C_A - C_B$ in equation (27.24) and hence C_1 in equation (27.25) would have to be zero. Hence, (27.28) would lead directly to the Ramsay-Young equation.

27k. Solid-Vapor (Sublimation) Equilibria.—An equation of exactly the same form as (27.10) is applicable to solid-vapor equilibria; thus,

$$\frac{dp}{dT} = \frac{\Delta H_s}{T(V_v - V_s)}, \qquad (27.29)$$

where ΔH_s is the molar (or specific) heat of sublimation, and V_v and V_s are the molar (or specific) volumes of vapor and solid, respectively, at the equilibrium temperature and pressure. This equation gives the variation of the vapor (sublimation) pressure of the solid. The Clausius modification in which V_s is neglected and V_v is taken as equal to RT/p, as for an ideal gas, is permissible, and hence an equation similar to (27.12) can be used for the vapor pressure of the solid. By means of this equation, or by the integrated form (27.14), the various calculations referred to in connection with liquid-vapor systems can be made. Since the procedure is obvious it is not necessary to enter into details.

---

[2] For estimation of heats of vaporization, see Watson, *Ind. Eng. Chem.*, **23**, 360 (1931); Meissner, *ibid.*, **33**, 1440 (1941); Othmer, *ibid.*, **34**, 1072 (1942). For vapor pressure charts, see Germann and Knight, *ibid.*, **26**, 1226 (1934); Cox, *ibid.*, **28**, 613 (1936); Killefer, *ibid.*, **30**, 477, 565 (1938); Davis, *ibid.*, **32**, 226 (1940); **33**, 401 (1941).

It is well known that the vapor pressure curves of the solid and liquid phases of a given substance meet at the triple point; thus, in Fig. 16 the curve AO represents solid-vapor equilibria, OB is for liquid-vapor, and OC for solid-liquid equilibria. The three curves meet at the triple point O where solid, liquid and vapor can coexist in equilibrium. It will be observed that near the triple point, at least, the slope of the curve AO on the pressure-temperature diagram is greater than that of OB; in other words, near the point O, the value of dp/dT is greater along AO than along OB. That this must be the case can be readily shown by means of the Clausius-Clapeyron equation. For the solid-vapor system, this can be written in the form

$$\frac{dp_s}{dT} = p_s \frac{\Delta H_s}{RT^2},$$

whereas, for the liquid-vapor system,

$$\frac{dp_l}{dT} = p_l \frac{\Delta H_v}{RT^2}.$$

At the triple point p_s and p_l, the vapor pressures of solid and liquid, respectively, are equal, and so also are the temperatures T; the relative values of the slopes dp_s/dT and dp_l/dT are thus determined by the heats of sublimation (ΔH_s) and of vaporization (ΔH_v), respectively. By the first law of thermodynamics, the change of heat content in the transition solid → vapor, at a given temperature, must be the same whether it is carried out directly or through the intermediate form of liquid. Hence, ΔH_s must be equal to $\Delta H_v + \Delta H_f$, at the same temperature, so that ΔH_s is greater than ΔH_v; the slope dp_s/dT of the curve AO is thus greater than dp_l/dT of the curve OB, in the vicinity of the triple point.

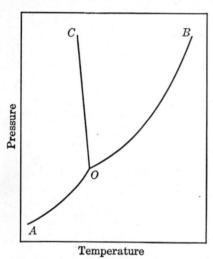

Fig. 16. Pressure-temperature equilibrium diagram

271. Variation of Equilibrium Latent Heat with Temperature.—If a given substance can occur in two phases, A and B, one of which changes into the other as the temperature is raised, then the value of the accompanying latent heat and its variation with temperature depend on whether the pressure is maintained constant, e.g., 1 atm., or whether it is the equilibrium value. In the former case the Kirchhoff equation (12.7) will apply, as stated in § 12j, but if equilibrium conditions are postulated allowance must be made for the change of pressure with temperature.

For any process, the change of ΔH with temperature and pressure can be represented by the general equation

$$d(\Delta H) = \left[\frac{\partial(\Delta H)}{\partial T}\right]_P dT + \left[\frac{\partial(\Delta H)}{\partial P}\right]_T dP,$$

and hence,

$$\frac{d(\Delta H)}{dT} = \left[\frac{\partial(\Delta H)}{\partial T}\right]_P + \left[\frac{\partial(\Delta H)}{\partial P}\right]_T \frac{dP}{dT}$$

$$= \Delta C_P + \left[\frac{\partial(\Delta H)}{\partial P}\right]_T \frac{dP}{dT}. \quad (27.30)$$

It can be seen from equation (20.20) that

$$\left[\frac{\partial(\Delta H)}{\partial P}\right]_T = \Delta V - T\left[\frac{\partial(\Delta V)}{\partial T}\right]_P,$$

and according to the Clapeyron equation (27.4), $dP/dT = \Delta H/T\Delta V$ for a phase change; insertion of these results into equation (27.30) then leads to

$$\frac{d(\Delta H)}{dT} = \Delta C_P + \frac{\Delta H}{T}\left\{1 - \frac{T}{\Delta V}\left[\frac{\partial(\Delta V)}{\partial T}\right]_P\right\} \quad (27.31)$$

for the variation of the *equilibrium* latent heat ΔH with temperature. In this expression ΔC_P is equal to $(C_P)_B - (C_P)_A$, where the constant pressure is the equilibrium pressure of the system at the temperature T.

For the liquid-vapor change the value of $d(\Delta H_v)/dT$ is not greatly different from ΔC_P, as may be seen from the following considerations. As shown in § 27f, ΔV may be taken as equal to V_v, the volume of the liquid being neglected in comparison with that of the vapor, and this may be replaced by RT/p, the vapor being assumed to behave as an ideal gas. In this event,

$$\left[\frac{\partial(\Delta V)}{\partial T}\right]_P = \frac{R}{p} = \frac{\Delta V}{T},$$

and hence the quantity in the braces in equation (27.31) is equal to zero. It follows, therefore, that provided the temperature is not too near the critical point, and the foregoing approximations may be made with some justification, equation (27.31) becomes

$$\frac{d(\Delta H_v)}{dT} \approx \Delta C_P = (C_P)_v - (C_P)_l, \quad (27.32)$$

where $(C_P)_v$ and $(C_P)_l$ are the heat capacities of vapor and liquid, respectively, at constant pressure equal to the vapor pressure at the given temperature.

In connection with the variation of the equilibrium heat of fusion with temperature, this simplification is not permissible. However, $[\partial(\Delta V)/\partial T]_P$ is usually small for the solid-liquid phase change, and so equation (27.31) reduces to

$$\frac{d(\Delta H_f)}{dT} \approx \Delta C_P + \frac{\Delta H_f}{T} = (C_P)_l - (C_P)_s + \frac{\Delta H_f}{T}. \quad (27.33)$$

27m. Dependence of Vapor Pressure on Total Pressure.—When a liquid or a solid vaporizes into a vacuum and a state of equilibrium is reached, the same pressure, i.e., the vapor pressure, is exerted on the two phases in equilibrium. If by some means, e.g., by introducing an insoluble, inert gas in the free space above it, the pressure on the liquid (or solid) is changed, the pressure of the vapor will be affected. The problem is treated below in a general manner, without restriction as to the nature of the two phases or the method of applying the pressure.

Consider any two phases A and B of a given substance in equilibrium at a specified temperature; the pressures on the respective phases are then P_A and P_B. Since the phases are in equilibrium, the molar free energies must be the same in each phase (§ 27a); hence, if a small change were made in the system in such a manner that the equilibrium was not disturbed, the free energy increase dF_A of one phase would be equal to the increase dF_B of the other phase. Suppose the pressure on the phase A is altered by an arbitrary amount dP_A; let the accompanying change in the pressure on B, which is required to maintain equilibrium, be dP_B. The molar free energy changes of the two phases are given by equation (25.24) as

$$dF_A = V_A dP_A \quad \text{and} \quad dF_B = V_B dP_B,$$

where V_A and V_B are the molar volumes of the phases A and B, respectively, under the equilibrium conditions. As seen above, dF_A must be equal to dF_B, so that

$$V_A dP_A = V_B dP_B.$$

Hence, the criterion of equilibrium, at constant temperature, is given by

$$\frac{dP_A}{dP_B} = \left(\frac{\partial P_A}{\partial P_B}\right)_T = \frac{V_B}{V_A}, \tag{27.34}$$

which is a generalized form of the equation first derived by J. H. Poynting (1881). For equilibrium to be retained, the pressure changes on the two phases must evidently be inversely proportional to the respective molar (or specific) volumes.

In the special case in which A is a liquid and B a vapor, the Poynting equation is

$$\frac{dp}{dP} = \frac{V_l}{V_v},$$

where P is the total pressure on the liquid and p is the vapor pressure. Assuming the vapor to behave as an ideal gas, V_v is equal to RT/p, so that

$$\frac{dp}{dP} = \frac{pV_l}{RT}. \tag{27.35}$$

The effect of external (total) pressure on the vapor pressure of a liquid is small, so that dp/dP may be replaced by $\Delta p/\Delta P$, where Δp is the increase of vapor pressure resulting from an appreciable increase ΔP in the total pressure exerted on the liquid. Hence, equation (27.35) may be written in the form

$$\frac{\Delta p}{p} = \frac{V_l}{RT} \Delta P, \tag{27.36}$$

from which the effect on the vapor pressure resulting from an increase of the external pressure, e.g., by introducing an inert gas, can be readily evaluated.

Problem: The true vapor pressure of water is 23.76 mm. at 25° C. Calculate the vapor pressure when water vaporizes into a space already containing an insoluble gas at 1 atm. pressure, assuming ideal behavior.

In using equation (27.36), it will be seen that Δp and p must be in the same units, the exact nature of which is immaterial; further, the units of R are determined by those of V_l and ΔP. Thus, if V_l is expressed in liters and ΔP in atm., R will be 0.082 liter-atm. deg.$^{-1}$ mole^{-1}. The specific volume of water may be taken, with sufficient accuracy, as 1.0, so that V_l is about 18 ml. or 0.018 liter mole^{-1}.

When the water vaporizes into a vacuum, the vapor pressure p is equal to the total pressure, i.e., 23.76 mm. at 25° C or 298° K; the external pressure is increased by approximately 1 atm., assuming the vapor pressure not to alter greatly, so that ΔP is 1.0. Hence, by equation (27.36),

$$\frac{\Delta p}{23.76} = \frac{0.018 \times 1.0}{0.082 \times 298}$$

$$\Delta p = \frac{0.018 \times 23.76}{0.082 \times 298} = 0.0175 \text{ mm.}$$

The vapor pressure is thus $23.76 + 0.018 \approx 27.78$ mm.

The effect of external pressure on the vapor pressure of a liquid is seen to be relatively small; nevertheless, the subject has some significance in connection with the theory of osmotic pressure.[3]

28. Systems of More Than One Component

28a. Conditions of Equilibrium.—If a system of several phases consists of more than one component, then the equilibrium condition of equal molar free energies in each phase requires some modification. Because each phase may contain two or more components in different proportions, it is necessary to introduce partial molar free energies, in place of the molar free energies. Consider a closed system of P phases, indicated by the letters $a, b, \ldots, P$, containing a total of C components, designated by $1, 2, \ldots, C$, in equilibrium at constant temperature and pressure which are the same for all the phases. The chemical potentials, or partial molar free energies (§ 26c), of the various components in the P phases may be represented by $\mu_{1(a)}, \mu_{2(a)}, \ldots, \mu_{C(a)}$; $\mu_{1(b)}, \mu_{2(b)}, \ldots, \mu_{C(b)}; \ldots; \mu_{1(P)}, \mu_{2(P)}, \ldots, \mu_{C(P)}$. Suppose various small amounts dn moles of the components are transferred from one phase to another, the temperature and pressure remaining constant; the whole closed system is in equilibrium, and so according to equation (26.16) the sum of all

---

[3] See, for example, S. Glasstone, "Textbook of Physical Chemistry," 2nd ed., 1946, Chap. IX.

the μdn terms for all the phases will be zero. It follows, therefore, that

$$\begin{aligned}
& \mu_{1(a)}dn_{1(a)} + \mu_{1(b)}dn_{1(b)} + \cdots + \mu_{1(P)}dn_{1(P)} \\
+ & \mu_{2(a)}dn_{2(a)} + \mu_{2(b)}dn_{2(b)} + \cdots + \mu_{2(P)}dn_{2(P)} \\
& \cdots \qquad \cdots \qquad \cdots \\
+ & \mu_{C(a)}dn_{C(a)} + \mu_{C(b)}dn_{C(b)} + \cdots + \mu_{C(P)}dn_{C(P)} = 0.
\end{aligned} \qquad (28.1)$$

At equilibrium, the total mass of each component will be constant, since the whole system is a closed one; hence,

$$\begin{aligned}
dn_{1(a)} + dn_{1(b)} + \cdots dn_{1(P)} &= 0 \\
dn_{2(a)} + dn_{2(b)} + \cdots dn_{2(P)} &= 0 \\
\cdots \qquad \cdots \qquad \cdots & \\
dn_{C(a)} + dn_{C(b)} + \cdots dn_{C(P)} &= 0.
\end{aligned} \qquad (28.2)$$

If the expression in equation (28.1) is to remain zero for all possible variations dn in the numbers of moles of the components, subject only to the restrictions represented by the equations (28.2), it is essential that

$$\begin{aligned}
\mu_{1(a)} &= \mu_{1(b)} = \cdots = \mu_{1(P)} \\
\mu_{2(a)} &= \mu_{2(b)} = \cdots = \mu_{2(P)} \\
&\cdots \\
\mu_{C(a)} &= \mu_{C(b)} = \cdots = \mu_{C(P)}.
\end{aligned} \qquad (28.3)$$

It is seen, therefore, that when a system consisting of a number of phases containing several components is in complete equilibrium, at a definite temperature and pressure which are uniform throughout, *the chemical potential of each component is the same in all the phases.* It may be noted that in the special case of a single component, the partial molar free energy (or chemical potential) is equal to the molar free energy (see footnote, § 26b), and the equations (28.3) become identical with (27.2).

If the phases of a system are not in equilibrium, the chemical potentials of the components will not be the same in each phase. There will then be a tendency for each component, for which such a difference exists, to pass spontaneously from the phase in which its chemical potential is higher to that in which it is lower, until the values become identical in the two phases. In other words, matter tends to flow spontaneously from a region of higher to one of lower chemical potential. There is thus seen to be an analogy between chemical potential and other forms of potential, e.g., electrical potential, energy potential, etc.

28b. The Phase Rule.—By means of the conclusion reached in the preceding section, it is possible to derive the familiar phase rule which gives the relationship between the number of components and phases in equilibrium in a system, and the number of variables, i.e., the degrees of freedom, which must be specified in order to define the system completely. The composition of a phase containing C components is given by $C - 1$ concentration terms, for if the concentrations of all but one of the components are known, that of the last component must be equal to the remainder. Hence, for the compositions of P phases to be defined it is necessary to state $P(C - 1)$ concentration terms. The total number of concentration variables of the system is thus $P(C - 1)$. In addition to the composition, the uniform temperature and pressure of the system must be specified, and assuming that no other factors, such as surface or electrical effects, influence the equilibrium, it follows that

$$\text{Total number of variables} = P(C - 1) + 2.$$

The fact of the closed system being in equilibrium, at a given temperature and pressure, leads to the result represented in the equations (28.3); this is equivalent to a set of $C(P - 1)$ independent equations which automatically fix $C(P - 1)$ of the possible variables. The number of variables remaining undetermined is then $[P(C - 1) + 2] - C(P - 1) = C - P + 2$. In order to define the system completely, therefore, this number of variables must be arbitrarily fixed, and hence must be equal to the number of degrees of freedom (F), or variance, of the system; hence,

$$F = C - P + 2, \qquad (28.4)$$

which is the **phase rule** derived by J. W. Gibbs (1875).

It may be noted that if a particular component is absent from any phase, the number of composition variables is reduced correspondingly. At the same time there will be a similar decrease in the number of independent equations determined by the equality of the chemical potentials, i.e., equations (28.3). The net effect will thus be to leave unchanged the number of degrees of freedom, as given by equation (28.4). The phase rule in its familiar form will then hold for the system even if all the components are not present in every phase.

28c. Univariant System of Several Components.—It was indicated in § 27b that an equation of the form of (27.4) is applicable to any univariant system, although its derivation was restricted to a system of one component. It is of interest to show, by a more detailed procedure, that the same result can be obtained if the univariant system consists of several components. However, to avoid too great a complication of symbols, etc., the discussion will be restricted to a system of two components. According to the phase rule, if such a system is to be univariant, i.e., $F = 1$, with $C = 2$, the value of P will be 3, so that there must be three phases in equilibrium.

A simple example of such a system would be a saturated solution of a non-volatile solid in equilibrium with vapor of the solvent; the three phases would then be (i) the solid component 1, e.g., a salt, (ii) a saturated solution of this substance in the liquid component 2, e.g., water, and (iii) the vapor of the latter. Suppose a small change is made in the system, which is maintained in equilibrium, involving changes of temperature dT and pressure dP and transfer of dn_1 moles of component 1 from the solid to the solution. The accompanying change in the partial molar free energy (chemical potential) of the substance in the solution is then given by

$$d\mu_1 = \left(\frac{\partial \mu_1}{\partial T}\right)_{P,N} dT + \left(\frac{\partial \mu_1}{\partial P}\right)_{T,N} dP + \left(\frac{\partial \mu_1}{\partial n_1}\right)_{T,P} dn_1, \quad (28.5)^*$$

whereas the corresponding change in the chemical potential of the pure solid, which is equivalent to the change in its molar free energy, is

$$d\mu_1' = dF_1' = \left(\frac{\partial F_1'}{\partial T}\right)_P dT + \left(\frac{\partial F_1'}{\partial P}\right)_T dP, \quad (28.6)$$

the primes being used to indicate the solid phase. Since the molar free energy of the solid is independent of the amount, there is no term involving dn_1 in equation (28.6). By utilizing equations (26.23) and (26.26), i.e., $(\partial \mu_i/\partial T)_{P,N} = -\bar{S}_i$, and $(\partial \mu_i/\partial P)_{T,N} = \bar{V}_i$, equation (28.5) can be written as

$$d\mu_1 = -\bar{S}_1 dT + \bar{V}_1 dP + \left(\frac{\partial \mu_1}{\partial n_1}\right)_{T,P} dn_1. \quad (28.7)$$

Similarly, by means of equations (25.20) and (25.21), i.e., $(\partial F/\partial T)_P = -S$, and $(\partial F/\partial P)_T = V$, equation (28.6) can be transformed into

$$d\mu_1' = -S_1' dT + V_1' dP. \quad (28.8)$$

Since the system has remained in equilibrium, the chemical potential of the given component must be the same in both phases, i.e., solid and solution; hence, the change of chemical potential in one phase must be equal to the accompanying change in the other phase. In other words, $d\mu_1$ and $d\mu_1'$, as given by equations (28.7) and (28.8), must be identical; it follows, therefore, by subtraction, that

$$-(\bar{S}_1 - S_1')dT + (\bar{V}_1 - V_1')dP + \left(\frac{\partial \mu_1}{\partial n_1}\right)_{T,P} dn_1 = 0. \quad (28.9)$$

Suppose that when the dn_1 moles of solid (solute) are transferred from the solid phase to the solution, as described above, it is necessary to transfer dn_2 moles of the component 2 (solvent) from the vapor to the solution in order to maintain the equilibrium. By proceeding in a manner precisely similar to that used in deriving equation (28.9), it is found that

$$-(\bar{S}_2 - S_2'')dT + (\bar{V}_2 - V_2'')dP + \left(\frac{\partial \mu_2}{\partial n_2}\right)_{T,P} dn_2 = 0, \quad (28.10)$$

where S_2'' and V_2'' refer to the molar entropy and volume of component 2 in the (pure) vapor phase.

If equation (28.9) is multiplied by n_1, the number of moles of constituent 1 in the solution, and (28.10) is multiplied by n_2, the number of moles of the substance 2, and the results added, it is possible to eliminate the last term in each equation. The reason is that

$$n_1 \left(\frac{\partial \mu_1}{\partial n_1}\right)_{T,P} dn_1 + n_2 \left(\frac{\partial \mu_2}{\partial n_2}\right)_{T,P} dn_2$$

* As in § 26a, the subscript N is used to represent constant composition.

is equivalent to $n_1 d\mu_1 + n_2 d\mu_2$ for the solution at constant temperature and pressure, and by the Gibbs-Duhem equation (26.15) this is zero. Hence, it follows that

$$-(n_1\bar{S}_1 - n_1 S_1' + n_2\bar{S}_2 - n_2 S_2'')dT + (n_1\bar{V}_1 - n_1 V_1' + n_2\bar{V}_2 - n_2 V_2'')dP = 0. \quad (28.11)$$

The expression in the first parentheses, which may be rearranged to take the form $(n_1\bar{S}_1 + n_2\bar{S}_2) - (n_1 S_1' + n_2 S_2'')$, is equal to the increase of entropy accompanying the formation of a saturated solution from n_1 moles of solid 1 and n_2 moles of vapor 2, since $n_1\bar{S}_1 + n_2\bar{S}_2$ represents the entropy of the solution [cf. equation (26.6)], $n_1 S_1'$ is that of n_1 moles of the pure solid 1, and $n_2 S_2''$ that of the n_2 moles of pure vapor 2. This quantity may be represented by ΔS, and the quantity in the second parentheses may similarly be replaced by ΔV, the corresponding increase of volume. It follows, therefore, that equation (28.11) may be written as

$$-\Delta S dT + \Delta V dP = 0,$$

$$\frac{dP}{dT} = \frac{\Delta S}{\Delta V},$$

and since ΔS may be replaced by $\Delta H/T$, this becomes

$$\frac{dP}{dT} = \frac{\Delta H}{T\Delta V}, \quad (28.12)$$

which is identical in form with equation (27.4).

It is thus seen that an equation analogous to that of Clapeyron has been derived for a univariant system of two components, and a similar result could be secured for any number of components. As obtained above, equation (28.12) gives the variation with temperature of the vapor pressure of a saturated solution; ΔH is the change of heat content accompanying the formation of the given solution from the solid solute and the vapor of the solvent, and ΔV is the corresponding volume change. The same equation, when inverted, will give the influence of pressure on the eutectic temperature at which two pure solids 1 and 2 are in equilibrium with saturated solution. In this case ΔH and ΔV are the heat content and volume changes, respectively, for the formation of the equilibrium solution from the appropriate amounts of the two constituents in the solid state.

28d. Properties of the Surface Phase.—In the thermodynamic treatment of systems in equilibrium it has been postulated, up to the present, that the only force acting upon the system is that due to the external pressure. The work done is then only work of expansion, represented by PdV. It is possible, however, that other forces may have to be taken into consideration. For example, thermodynamic systems are invariably subject to gravitational and surface forces, but in most instances their influence is so small as to be negligible. Electrical and magnetic forces may also be operative in special circumstances. Of particular physicochemical interest are surface forces, the effect of which becomes apparent when the quantity of matter contained in the surface is relatively large in comparison with that of the system as a whole.

Consider a heterogeneous system, of one or more components, consisting of two phases; there will then be a surface of separation between the phases. The transition from one phase to the other, across the bounding surface, is probably a gradual one, so that the surface of separation is not to be regarded as sharp, but as a region of more or less definite thickness. For the present purpose, the exact thickness of the region is immaterial, provided it contains all the parts of the system which are within the influence of the surface forces. Thus the two bulk phases, e.g., two liquids or a liquid and vapor, may be thought of as being separated by a **surface phase**. It may be remarked that the so-called surface phase is not a true phase in the usual physical sense, but the description is convenient.

In order to derive the thermodynamic properties of the surface phase, consider a *hypothetical* state involving an exact geometrical surface placed between the two bulk phases. A suitable position for this imaginary surface will be proposed later. The **surface excess**, which may be positive or negative, of any constituent of the system is then defined as *the excess of that constituent in the actual surface phase over the amount there would have been if both bulk phases remained homogeneous right up to the geometrical surface.* Thus if n_i is the total number of moles of the constituent i in the system, c'_i and c''_i are the numbers of moles per unit volume, i.e., the concentrations, in the interior of the two bulk phases, and V' and V'' are the respective volumes of these phases right up to the imaginary geometrical surface, then the number of moles n_i^s of surface excess of that constituent is given by

$$n_i = c'_i V' + c''_i V'' + n_i^s. \tag{28.13}$$

The actual values n_1^s, n_2^s, ..., for the surface excesses of the various components of the system depend on the location postulated for the hypothetical surface. For the general treatment which follows, its exact position need not be specified, but subsequently this will be defined in a particularly convenient manner.

The surface free energy F^s may be expressed by a relationship similar to equation (28.13), viz.,

$$F = F' + F'' + F^s, \tag{28.14}$$

where F is the total free energy of the system, consisting of the two bulk phases and the surface phase; F' and F'' are the free energies of the bulk phases calculated on the assumption that they both remain homogeneous right up to the hypothetical, geometrical surface. The free energy F^s may thus be regarded as the contribution to the system made by the "surface excess" amounts of the various constituents or, in other words, of the surface phase. Other thermodynamic properties of the surface phase are defined in a manner exactly analogous to equation (28.14), but the surface chemical potential (partial molar free energy) μ^s, and the surface entropy S^s only will be employed here.

28e. Equilibrium of Surface Phase.—Consider a small change in the system described above. The free energy change dF is then equal to

$dF' + dF'' + dF^s$. The first two terms, which refer to the homogeneous bulk phases, are given by equation (26.13) as

$$dF' = -S'dT + V'dP' + \mu_1'dn_1' + \mu_2'dn_2' + \cdots \qquad (28.15)$$

and

$$dF'' = -S''dT + V''dP'' + \mu_1''dn_1'' + \mu_2''dn_2'' + \cdots. \qquad (28.16)$$

In evaluating dF^s for the surface phase, it is necessary to take into account the free energy change accompanying a change in the surface area. The work required to increase the area of the surface by an infinitesimal amount ds, at constant temperature, pressure and composition, is equal to γds, where γ is the quantity usually referred to as the **surface tension**. The latter is a measure of the reversible work which must be done for unit increase in the surface area under the given conditions. Since the surface work does not involve work of expansion against the external pressure, it may be identified with the net work and hence with the free energy change (§ 25b). The expression for dF^s for a small change in the system should thus include the term γds. On the other hand, since the surface contribution to the volume may be ignored, the quantity corresponding to VdP may be omitted. It follows, therefore, that

$$dF^s = -S^s dT + \gamma ds + \mu_1^s dn_1^s + \mu_2^s dn_2^s + \cdots, \qquad (28.17)$$

where $\mu_1^s, \mu_2^s, \ldots$, are the **surface chemical potentials** of the various constituents of the system. It is consequently seen from equations (28.15), (28.16) and (28.17), that dF, obtained by summing dF', dF'' and dF^s, is

$$dF = -SdT + V'dP' + V''dP'' + \gamma ds \\ + \Sigma \mu_i' dn_i' + \Sigma \mu_i'' dn_i'' + \Sigma \mu_i^s dn_i^s, \qquad (28.18)$$

where S is the total entropy of the system, i.e., $S' + S'' + S^s$.

For a system in which the only force acting is that due to the external pressure, the condition of equilibrium is given by equation (25.39) as $dF_{T,P} = 0$. When surface forces are significant, however, it is not difficult to show that this result must be modified by stipulating constant surface area, in addition to constant temperature and pressure, so that for equilibrium

$$dF_{T,P,s} = 0. \qquad (28.19)$$

Upon applying this conclusion to equation (28.18), it is seen that

$$\Sigma \mu_i' dn_i' + \Sigma \mu_i'' dn_i'' + \Sigma \mu_i^s dn_i^s = 0 \qquad (28.20)$$

at equilibrium. If there is no restriction concerning the passage of matter between the two bulk phases and the surface phase, the variations dn_i', dn_i'' and dn_i^s are independent, provided the sum for each component is equal to zero, since no matter passes in or out of the system as a whole. It follows, therefore, from equation (28.20) that for each constituent

$$\mu_i' = \mu_i'' = \mu_i^s. \qquad (28.21)$$

This result is an extension of equation (28.3); *the surface chemical potential of any constituent of a system is thus equal to its chemical potential in the bulk phases at equilibrium.*

28f. The Gibbs Adsorption Equation.—The surface free energy F^s is the sum of the contributions of the various constituents, i.e., $n_1^s \mu_1^s + n_2^s \mu_2^s + \cdots$, [cf. equation (26.6)], and of a quantity depending upon the area of the surface. The latter is equal to γs, where s is the total surface area; hence,

$$F^s = \gamma s + n_1^s \mu_1^s + n_2^s \mu_2^s + \cdots, \qquad (28.22)$$

where n_1^s, n_2^s, ..., are as defined by equation (28.13). Since F^s is a definite property of the surface, depending only upon the thermodynamic state of the system, dF^s is a complete differential, and hence differentiation of equation (28.22) gives

$$\begin{aligned} dF^s &= \gamma ds + s d\gamma + n_1^s d\mu_1^s + n_2^s d\mu_2^s + \cdots + \mu_1^s dn_1^s + \mu_2^s dn_2^s + \cdots \\ &= \gamma ds + s d\gamma + \sum n_i^s d\mu_i^s + \sum \mu_i^s dn_i^s. \end{aligned} \qquad (28.23)$$

Upon comparison with equation (28.17), it is seen that

$$S^s dT + s d\gamma + \sum n_i^s d\mu_i^s = 0,$$

and at constant temperature this becomes [cf. equation (26.15)]

$$s d\gamma + \sum n_i^s d\mu_i^s = 0, \qquad (28.24)$$

thus providing a relationship between the change in surface tension and the corresponding changes in the surface chemical potentials. Dividing through by s, the surface area, the result is

$$d\gamma + \frac{n_1^s}{s} d\mu_1^s + \frac{n_2^s}{s} d\mu_2^s + \cdots = 0,$$

or, replacing n_1^s/s by Γ_1, n_2^s/s by Γ_2, and so on,

$$d\gamma + \Gamma_1 d\mu_1^s + \Gamma_2 d\mu_2^s + \cdots = 0, \qquad (28.25)$$

where Γ_1, Γ_2, ..., are the excess surface "concentrations" of the various constituents of the system; these "concentrations" are really the excess amounts per *unit area* of surface. For a system of two components, e.g., a solution of a single solute, equation (28.25) becomes

$$d\gamma + \Gamma_1 d\mu_1^s + \Gamma_2 d\mu_2^s = 0. \qquad (28.26)$$

at constant temperature.

If the system is in equilibrium, at constant temperature, pressure and surface area, the surface chemical potential of any constituent must always be equal to its chemical potential in the bulk phase, i.e., the solution, by equation (28.21). It is thus permissible to write equation (28.26) in the form

$$d\gamma + \Gamma_1 d\mu_1 + \Gamma_2 d\mu_2 = 0, \qquad (28.27)$$

where μ_1 and μ_2 refer to the chemical potentials in the solution. It will be recalled (§ 28d) that n_1^s and n_2^s, and hence Γ_1 and Γ_2, depend upon the arbitrary position chosen for the geometrical surface. In connection with the study of dilute solutions it is convenient to choose the surface so as to make Γ_1 zero; that is to say, the surface excess of constituent 1, the solvent, is made equal to zero. In these circumstances equation (28.27) becomes

$$d\gamma + \Gamma_2 d\mu_2 = 0$$

$$\Gamma_2 = -\left(\frac{\partial \gamma}{\partial \mu_2}\right)_T. \tag{28.28}$$

It will be seen later [equation (31.2)] that the chemical potential μ_2 of any constituent of a solution, e.g., the solute, may be represented by $\mu^0 + RT \ln a_2$, where μ^0 is a constant for the substance at constant temperature, and a_2 is called the "activity" of the solute, the activity being an idealized concentration for free energy changes. Upon making the substitution for μ_2, equation (28.28) becomes

$$\begin{aligned}\Gamma_2 &= -\frac{1}{RT}\left(\frac{\partial \gamma}{\partial \ln a_2}\right)_T \\ &= -\frac{a_2}{RT}\left(\frac{\partial \gamma}{\partial a_2}\right)_T. \end{aligned} \tag{28.29}$$

For dilute solutions, a_2 may be replaced by c, the concentration of the solute, so that

$$\Gamma_2 \approx -\frac{c}{RT}\left(\frac{\partial \gamma}{\partial c}\right)_T. \tag{28.30}$$

The equations (28.28), (28.29) and (28.30) are forms of the **Gibbs adsorption equation**, first derived by J. Willard Gibbs (1878); it relates the surface excess of the solute to the variation of the surface tension of the solution with the concentration (or activity). If an increase in the concentration of the solute causes the surface tension of the solution to decrease, i.e., $(\partial \gamma / \partial c)_T$, is negative, Γ_2 will be positive, by equation (28.30), so that there is an actual excess of solute in the surface; in other words, **adsorption** of the solute occurs under these conditions. If $(\partial \gamma / \partial c)_T$ is positive, Γ_2 is negative and there is a deficiency of the solute in the surface; this phenomenon is referred to as **negative adsorption**.

28g. Vapor Pressure and Solubility of Small Particles.—Another effect of surface forces relates to the change in certain physical properties, e.g., vapor pressure and solubility, resulting from the difference in size of the particles of a solid or drops of a liquid. Consider, on the one hand, a spherical drop or particle of a pure substance, of radius r, in equilibrium with vapor at a pressure p. On the other hand, consider a flat surface of the same substance, the vapor pressure p_0 differing from that of the small particles. The free energy change dF for the transfer of dn moles of substance from the flat surface to the spheres is equivalent to the transfer of this quantity from pressure p_0 to pressure p, at constant temperature. If the vapor behaves as an ideal gas, then by equation (25.25)

$$dF = dnRT \ln \frac{p}{p_0}. \tag{28.31}$$

The increase of free energy for the process under consideration is to be attributed to the fact that the addition of material to the small drops causes an appreciable increase in the surface area, whereas for the flat surface the accompanying decrease is negligible. The increase of free energy for a change ds in the surface area is equal to γds, where γ is the surface tension; hence it is possible to write equation (28.31) as

$$dnRT \ln \frac{p}{p_0} = \gamma ds. \tag{28.32}$$

If V is the *molar* volume of the substance under consideration, and the spherical drop of radius r contains n moles, then

$$nV = \tfrac{4}{3}\pi r^3,$$

and upon differentiation it is found that

$$dn = \frac{4\pi r^2}{V} dr. \tag{28.33}$$

The surface area s of the drop is equal to $4\pi r^2$, and hence

$$ds = 8\pi r dr. \tag{28.34}$$

Consequently, by combining equations (28.32), (28.33) and (28.34), it is seen that

$$RT \ln \frac{p}{p_0} = \frac{2\gamma V}{r}. \tag{28.35}$$

The vapor pressure of the spherical drops or particles is thus greater than that of the flat surface, the proportion increasing as the radius of the particles decreases. The higher vapor pressure of small drops or particles accounts for their tendency to disappear by "distillation" on to larger particles. Large drops or particles thus tend to grow at the expense of smaller ones. If the vapor does not behave ideally, equation (28.35) is not exact; as will be evident from the next chapter, the correct form is obtained by using the "fugacity" in place of the vapor pressure.

Problem: The vapor pressure of a large (flat) body of water is 23.76 mm. at 25° C. Calculate the vapor pressure of drops of 10^{-5} cm. radius. The surface tension of water may be taken as 72.0 dynes cm.$^{-1}$ and its molar volume is 18.0 cc. mole^{-1}.

If γ is in dynes cm.$^{-1}$, V in cc. mole^{-1} and r in cm., the right-hand side of equation (28.35) would be in ergs mole^{-1}; hence, it is convenient to express R as 8.314×10^7 ergs deg.$^{-1}$ mole^{-1}. Consequently, at 298.2° K,

$$8.314 \times 10^7 \times 2.303 \times 298.2 \log \frac{p}{23.76} = \frac{2 \times 72.0 \times 18.0}{10^{-5}}$$

$$p = 24.01 \text{ mm.}$$

For particles in equilibrium with a saturated solution, the free energy of transfer from a flat surface, also in contact with its saturated solution, can be expressed in terms of the concentrations of the solutions, at least for dilute solutions (cf. Chapter XV). In this case, equation (28.35) takes the form

$$RT \ln \frac{c}{c_0} = \frac{2\gamma V}{r}, \qquad (28.36)$$

where c and c_0 represent the concentrations of saturated solutions in contact with small particles and a flat surface, e.g., large crystals, respectively. In this case γ is the tension at the interface between the solid (or liquid) solute and the solution. It is evident that fine particles can have an appreciably larger solubility than large crystals of the same substance. Strictly speaking, the activity of the solute should be used instead of the concentration in equation (28.36), but this refinement may be ignored for the present.

EXERCISES

1. The vapor pressures of carbon tetrachloride at several temperatures are as follows:

$t°$	25°	35°	45°	55° C
p	113.8	174.4	258.9	373.6 mm.

Plot $\log p$ against $1/T$ and from the slope evaluate the mean heat of vaporization of carbon tetrachloride in the given range.

2. From the results of the preceding exercise determine (approximately) the constant C in equation (27.16), and see how close the result is to that expected from Trouton's law. Using the values of ΔH_v and C just obtained estimate the normal boiling point of carbon tetrachloride. (The actual value is 76.8° C.)

3. At its normal boiling point (77.15° C) the orthobaric densities of ethyl acetate are 0.828 (liquid) and 0.00323 (vapor) g. cc.$^{-1}$. The rate of change of vapor pressure with temperature in the vicinity of the boiling point is 23.0 mm. deg.$^{-1}$. Calculate the heat of vaporization by (i) the Clapeyron equation, (ii) the Clapeyron-Clausius equation.

4. Below its boiling point the variation of the vapor pressure of benzene with temperature is given by

$$\log p(\text{mm.}) = 7.2621 - \frac{1402.46}{T} - \frac{51387.5}{T^2}$$

[Mathews, *J. Am. Chem. Soc.*, **48**, 562 (1926)] from which the boiling point is found to be 80.20° C. The specific volume of benzene vapor at its boiling point at 1 atm. is 356 cc. g.$^{-1}$ and that of the liquid is 1.2 cc. g.$^{-1}$. Calculate the heat of vaporization of benzene at this temperature and estimate the boiling point at 77.0 cm. pressure.

5. Use the equation for $\log p$ as a function of temperature in the preceding exercise to derive an expression for the variation of the heat of vaporization of benzene with temperature. (The vapor may be supposed to behave ideally.)

6. Show that if T_A and T_B are the temperatures at which two liquids have the same vapor pressure, then by the Ramsay-Young rule, $\log T_A = \log T_B + \text{const.}$; the plot of $\log T_A$ against $\log T_B$ should thus be linear.

The vapor pressures of mesitylene at various temperatures are as follows:

$t°$	60°	80°	100°	120° C
p	87.35	150.8	247.25	381.1 mm.

By means of these data and those given for carbon tetrachloride in Exercise 1, determine how closely the Ramsay-Young rule is obeyed. (Plot the vapor pressure as a function of temperature in each case, and determine the temperatures at which the two liquids have the vapor pressures 120, 240 and 360 mm., and then plot log T_A against log T_B.)

7. At 110° C, dp/dT for water is 36.14 mm. deg.$^{-1}$; the orthobaric specific volumes are 1209 (vapor) and 1.05 (liquid) cc. g.$^{-1}$. Calculate the heat of vaporization of water in cal. g.$^{-1}$ at 110° C.

8. The mean heat capacity of water vapor in the range from 100° to 120° C is 0.479 cal. g.$^{-1}$, and for liquid water it is 1.009 cal. g.$^{-1}$. Taking the heat of vaporization of water as 539 cal. g.$^{-1}$ at 100° C, determine the approximate value at 110°C, and compare the result with that obtained in the preceding exercise.

9. The melting point of benzene is found to increase from 5.50° to 5.78° C when the external pressure is increased by 100 atm. The heat of fusion of benzene is 30.48 cal. g.$^{-1}$. What is the change of volume per gram accompanying the fusion of benzene?

10. A liquid (mercury) normally boils at 357° C and its heat of vaporization is 68 cal. g.$^{-1}$. It is required to distil the liquid at 100° C; estimate the approximate pressure that would be used.

11. The variation of the vapor pressure of solid iodine is given by

$$\log p(\text{atm.}) = -\frac{3512.83}{T} - 2.013 \log T + 13.374.$$

[Giauque, *J. Am. Chem. Soc.*, **53**, 507 (1931)]. The heat of sublimation at 25° C is 58.6 cal. g.$^{-1}$, and the specific volume of the solid is 0.22 cc. g^{-1}. Estimate the molar volume of the vapor at its equilibrium pressure at 25° C, and compare with the ideal gas value.

12. The heat of vaporization of chlorobenzene at its boiling point (132.0° C) is 73.4 cal. g.$^{-1}$. Estimate the (approximate) pressure in cm. of mercury under which the liquid will boil at 130° C. Recalculate the result, taking $V_g - V_l$ as 277.5 cc. g.$^{-1}$ at the normal boiling point.

13. The true vapor pressure of ethyl acetate at 35° C is 59.0 mm. and its density is 0.788 g. cc.$^{-1}$. Determine the change of vapor pressure resulting from the introduction of an inert gas at 2 atm. pressure.

14. A hydrocarbon (*n*-heptane) is known to have a vapor pressure of 92 mm. at 40° C. Estimate its normal boiling point. (The experimental value is 98.5° C.)

15. The normal boiling point of *n*-hexane is 69.0° C. Estimate its vapor pressure at 30° C. (The experimental value is 185 mm.)

16. Dühring's rule has been found to apply to unsaturated solutions of a given solute, pure water being the reference liquid. An unsaturated solution of calcium chloride (30 g. per 100 g. solution) has a vapor pressure of 240 mm. at 75.7° C; pure water has this vapor pressure at 70.6° C. At what temperature will the given solution boil at 1 atm. pressure?

17. Ramsay and Young (1886) found that in the vicinity of the boiling point the quantity $T(dp/dT)$ is approximately constant for many liquids. Show that this is a consequence of the Ramsay-Young rule, and that it leads to the relationship $\Delta T = cT\Delta P$, where ΔT is the increase of boiling point for an increase ΔP in the external pressure; for liquids obeying Trouton's rule, c should be approximately constant and equal to 1.2×10^{-4} if ΔP is in mm. (Craft's rule). Estimate

the boiling point of benzene at 77.0 cm. and compare the result with that obtained in Exercise 4.

18. The vapor pressures of (a) solid, (b) liquid, hydrogen cyanide are given by [Perry and Porter, J. Am. Chem. Soc., **48**, 299 (1926)]:

(a) $\log p(\text{mm.}) = 9.33902 - 1864.8/T$ (from 243.7° to 258° K)
(b) $\log p(\text{mm.}) = 7.74460 - 1453.06/T$ (from 265° to 300.4° K).

Calculate (i) the heat of sublimation, (ii) the heat of vaporization, (iii) the heat of fusion, (iv) the triple point temperature and pressure, (v) the normal boiling point. (Note that the latent heats are approximately constant in the given temperature ranges.)

19. Calculate the difference in slope, in mm. deg.$^{-1}$, between the vapor pressure curves of solid and liquid hydrogen cyanide at the triple point, using the data in the preceding exercise.

20. Show that if the vapor behaves ideally,

$$\ln \frac{p_s}{p_l} = \frac{\Delta H_f}{R} \left(\frac{1}{T_m} - \frac{1}{T} \right),$$

where p_s and p_l are the vapor pressures of solid and supercooled liquid, respectively, at the same temperature T; ΔH_f is the (mean) molar heat of fusion and T_m is the melting point of the solid. Use the data in Exercise 10, Chapter X to calculate the mean heat of fusion of ice in the range from 0° to $-$ 10° C.

21. Give the complete derivation of equation (28.10).

22. Give in full the derivation of an expression for the variation of a binary eutectic temperature, i.e., for a two-component system, with pressure.

23. Prove the condition of equilibrium given by equation (28.19) which is used when surface forces must be taken into consideration.

24. At appreciable concentrations, the variation of the surface tension γ with concentration c of aqueous solutions of the lower fatty acids is given by $\gamma = A + B \log c$, where A is a constant for each acid and B is approximately the same for all the acids. Show that the extent of adsorption of a fatty acid at the surface of its aqueous solution is then roughly independent of the concentration of the solution and of the nature of the acid. Suggest a physical interpretation of this result.

CHAPTER XII

FUGACITY AND ACTIVITY

29. Fugacity of a Single Gas

29a. Definition of Fugacity.—By utilizing the free energy function, G. N. Lewis (1901) introduced the concept of "fugacity," which has proved of great value for representing the actual behavior of real gases, as distinct from the postulated behavior of ideal gases. It has been applied especially, as will be seen in § 32c, in the study of chemical equilibria involving gases at high pressures. The fugacity is chiefly employed in connection with gas mixtures, but the introductory treatment will be restricted to pure gases; at a later stage (§ 30b) it will be extended to systems consisting of more than one component.[1]

According to equation (25.24), for an infinitesimal, reversible stage of an isothermal change involving work of expansion only,

$$dF = VdP. \tag{29.1}$$

If the system consists of 1 mole of an ideal gas, V may be replaced by RT/P, so that

$$dF = RT\frac{dP}{P} = RTd\ln P, \tag{29.2}$$

where P is the pressure of the gas. For a gas which does not behave ideally, equation (29.2) will not hold, but a function f, known as the **fugacity**, may be (partially) defined in such a manner that the relationship

$$dF = RTd\ln f \tag{29.3}$$

is always satisfied, irrespective of whether the gas is ideal or not. Integration of (29.3) at constant temperature gives

$$F = RT\ln f + C, \tag{29.4}$$

where F is the molar free energy of the gas and f is its fugacity; the integration constant C is dependent upon the temperature and the nature of the gas.

Actually, equation (29.4) defines the *ratio* of the fugacities at two different pressures, i.e., the *relative* fugacity, at a given temperature. This may be seen by considering the definite integral of equation (29.3), viz.,

$$F_2 - F_1 = RT\ln\frac{f_2}{f_1}, \tag{29.5}$$

[1] Lewis, *Proc. Am. Acad. Arts Sci.*, **37**, 49 (1901); *Z. phys. Chem.*, **38**, 205 (1901); G. N. Lewis and M. Randall, "Thermodynamics and the Free Energy of Chemical Substances," 1923, Chap. XVII.

where F_1 and F_2 are the molar free energies of the gas in two states, i.e., pressures, at the same temperature, and f_1 and f_2 are the corresponding fugacities. The experimentally determinable quantity is the free energy difference $F_2 - F_1$ (or ΔF), and this is seen by equation (29.5) to give the ratio of the fugacities f_2/f_1. In order to express the fugacity in any state, it is necessary therefore to assign it a specific value in a particular **reference state**.

For an *ideal gas* the difference of molar free energy in two states at the same temperature is given by equation (25.25), which may be written as

$$F_2 - F_1 = RT \ln \frac{P_2}{P_1},$$

and comparison of this result with equation (29.5) shows that for an ideal gas the fugacity is proportional to the pressure. It is convenient to take the proportionality constant as unity, so that for an ideal gas $f/P = 1$, and the fugacity is always equal to the pressure. For a real gas, the fugacity and pressure are, in general, not proportional to one another, and f/P is not constant. As the pressure of the gas is decreased, however, the behavior approaches that for an ideal gas, and so the *gas at very low pressure* is chosen as the reference state and it is postulated that *the ratio f/P of the fugacity to the pressure then approaches unity;* thus,

$$\lim_{P \to 0} \frac{f}{P} = 1$$

or

$$\frac{f}{P} \to 1 \quad \text{as} \quad P \to 0.$$

It will be seen shortly that this postulate, which makes *the fugacity of a real gas equal to its pressure at very low pressure*, permits the evaluation of actual fugacities at various pressures. It may be mentioned that since gas pressures are usually expressed in atm., fugacities are recorded in the same units.

29b. Determination of Fugacity: Graphical Method.—For an ideal gas the fugacity is equal to the pressure at all pressures, but for a real gas this is only the case at very low pressures when it behaves ideally. To determine the fugacity of a gas at any pressure where it deviates from ideal behavior, the following procedure has been used. By combining equations (29.1) and (29.3), both of which apply to any gas, it follows that at constant temperature

$$RT d \ln f = V dP \tag{29.6}$$

or

$$\left(\frac{\partial \ln f}{\partial P}\right)_T = \frac{V}{RT}, \tag{29.7}$$

where V is the actual molar volume of the gas at the temperature T and pressure P. For an ideal gas the volume of 1 mole is RT/P, and for a real gas the quantity α, which is a function of the temperature and pressure, may be defined by

$$\alpha = \frac{RT}{P} - V. \tag{29.8}$$

Hence from equation (29.6),

$$RT d\ln f = RT \frac{dP}{P} - \alpha dP,$$

$$d\ln f = d\ln P - \frac{\alpha}{RT} dP,$$

$$d\ln \frac{f}{P} = - \frac{\alpha}{RT} dP.$$

If this result is integrated between a low, virtually zero, pressure and a given pressure P, at constant temperature, the result is

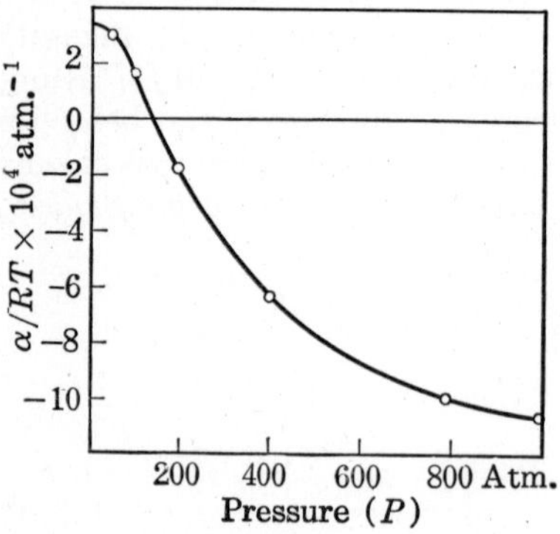

$$\ln \frac{f}{P} = - \frac{1}{RT} \int_0^P \alpha dP.$$

or

$$\ln f = \ln P - \frac{1}{RT} \int_0^P \alpha dP, \tag{29.9}$$

since, as postulated above, f/P becomes equal to unity, and hence $\ln(f/P)$ becomes zero, at zero pressure.[2] To calculate the fugacity, therefore, it is necessary to plot α, derived from experimentally determined molar volumes of the gas at various pressures, against the pressure; the area under the curve between the pressures of zero and P gives the value of the integral in equation (29.9).

Fig. 17. Determination of fugacity of nitrogen gas

Problem: Utilize the following data to calculate the fugacities of nitrogen gas at the various pressures at 0° C.

P	50	100	200	400	800	1,000 atm.
PV/RT	0.9846	0.9846	1.0365	1.2557	1.7959	2.0641

It can be readily seen from equation (29.8), which defines α, that

$$\frac{\alpha}{RT} = \left(1 - \frac{PV}{RT}\right) \frac{1}{P}.$$

Since the PV/RT values for various P's are given above, it is possible to derive

---

[2] Tunell, *J. Phys. Chem.*, **35**, 2885 (1931), has suggested that it would be preferable to define the fugacity by means of equation (29.9); the conditions $f/P \to 1$ and $f \to 0$ as $P \to 0$ then follow automatically.

the corresponding α/RT values; these may be plotted against P (Fig. 17),* and hence the integral of α/RT between zero and any pressure P may be evaluated graphically. By equation (29.9) this is equal to $-\ln(f/P)$, and hence f/P and f can be determined; the results are given below.

P	α/RT	Integral	f/P	f
50 atm.	3.08×10^{-4} atm.$^{-1}$	0.0206	0.979	48.95 atm.
100	1.54	0.0320	0.967	96.7
200	-1.82	0.0288	0.971	194.2
400	-6.39	-0.0596	1.061	424.4
800	-9.95	-0.3980	1.489	1191
1,000	-10.64	-0.6060	1.834	1834

29c. Determination of Fugacity from Equation of State.—If f is the fugacity of a gas at pressure P, and f^* is the value at a low pressure P^*, then integration of equation (29.6) or (29.7) gives

$$\ln \frac{f}{f^*} = \frac{1}{RT}\int_{P^*}^{P} V dP. \qquad (29.10)$$

The variable of the integrand in equation (29.10) is now changed by integrating by parts; thus,

$$\int_{P^*}^{P} V dP = PV\Big]_{V^*}^{V} - \int_{V^*}^{V} P dV$$

$$= PV - P^*V^* - \int_{V^*}^{V} P dV,$$

where V^* is the molar volume corresponding to the low pressure P^*. Since the gas then behaves almost ideally, it is permissible to replace P^*V^* by RT, and upon substituting the result in equation (29.10), it is seen that

$$\ln \frac{f}{f^*} = \frac{1}{RT}\left(PV - RT - \int_{V^*}^{V} P dV\right).$$

Utilizing the postulate that f/P approaches unity at low pressure, i.e., f^*/P^* is virtually unity, it follows that $\ln(f/f^*)$ may be replaced by $\ln(f/P^*)$, that is, by $\ln f - \ln P^*$; hence,

$$\ln f = \ln P^* + \frac{1}{RT}\left(PV - RT - \int_{V^*}^{V} P dV\right). \qquad (29.11)$$

By means of an equation of state, it is possible to express P as a function of V, at constant temperature, and hence the integral in (29.11) can be evaluated analytically.

* The shape of the curve at low pressures has been adjusted to the fact that α then tends to an approximately constant value (cf. § 29d).

The procedure may be illustrated by reference to a *van der Waals gas*, for which the equation of state is

$$P = \frac{RT}{V-b} - \frac{a}{V^2}, \qquad (29.12)$$

where a and b may be regarded as constants, independent of the pressure, that have been derived from experimental P-V data at the given temperature. By equation (29.12),

$$PdV = \left(\frac{RT}{V-b} - \frac{a}{V^2}\right) dV,$$

and hence

$$\int_{V^*}^{V} PdV = \int_{V^*}^{V} \frac{RT}{V-b} dV - \int_{V^*}^{V} \frac{a}{V^2} dV$$

$$= RT \ln \frac{V-b}{V^*-b} + \frac{a}{V} - \frac{a}{V^*}.$$

Since V^* is very large, $V^* - b$ may be replaced by V^*, which is equal to RT/P^*, and a/V^* can be neglected; thus,

$$\int_{V^*}^{V} PdV = RT \ln \frac{V-b}{V^*} + \frac{a}{V}$$

$$= RT \ln \frac{V-b}{RT} + RT \ln P^* + \frac{a}{V}. \qquad (29.13)$$

It can be readily shown from the van der Waals equation (29.12) that

$$PV - RT = \frac{RTb}{V-b} - \frac{a}{V},$$

and combination of this result with equations (29.11) and (29.13) gives

$$\ln f = \ln \frac{RT}{V-b} + \frac{b}{V-b} - \frac{2a}{RTV}. \qquad (29.14)$$

Consequently, the fugacity of a van der Waals gas at any pressure can be calculated from the volume at that pressure, at the specified constant temperature, provided the van der Waals constants for the given gas are known. The values of a and b to be used here are those which have been derived from actual P-V measurements at the required temperature. Because of the incomplete quantitative nature of the van der Waals equation (cf. Chapter II), these will differ from one temperature to another; in any event they will not be identical with the tabulated a and b values, for the latter are usually based on the critical data (§ 5d). It is only when other information is lacking that these may be used to obtain an approximate indication of the fugacity.

The fugacities of oxygen at a number of pressures at 0° C have been calculated by G. N. Lewis and M. Randall[3] using equation (29.14); a was taken as 1.009 liter^2 atm. mole^{-2} and b as 2.64×10^{-2} liter mole^{-1}.* The

[3] Lewis and Randall, ref. 1, p. 196.

* The conventional values of a and b for oxygen, derived from critical data, are 1.32 and 3.12×10^{-2}, respectively (see Table I).

results are recorded in Table XX; the figures given under the heading P_{id} are the pressures which an ideal gas would exert if it occupied the same volume as the actual gas at the given temperature.

TABLE XX. FUGACITY OF OXYGEN AT 0° C

P	f	P_{id}	f/P	P/P_{id}
50 atm.	48.0	52.0	0.960	0.961
100	92.5	108	0.925	0.929
200	174	220	0.87	0.91
400	338	381	0.85	1.05
600	540	465	0.90	1.29

Even though a and b are derived from actual P-V-T data, the fugacities obtained from equation (29.14) may not be too reliable over a range of pressures because of the approximate nature of the van der Waals equation. By using a more exact equation of state, such as the Beattie-Bridgeman equation, better values for the fugacities can be obtained, but since this equation involves five empirical constants, in addition to R, the treatment is somewhat more complicated than that given above.[4]

29d. Approximate Calculation of Fugacity.—It is an experimental fact, with which the van der Waals equation is in agreement (cf. Chapter II, Exercise 7), that at not too high pressures the value of PV for any gas is a linear function of its pressure at constant temperature; thus,

$$PV = RT - AP,$$

where A may be taken as constant at a given temperature. From this equation it is seen that α, defined by equation (29.8), is given by

$$\alpha = \frac{RT}{P} - V = A,$$

so that α is (approximately) constant over a range of pressures, provided they are not too high. Utilizing this result, equation (29.9) becomes

$$\ln f = \ln P - \frac{\alpha P}{RT}$$

or

$$\ln \frac{f}{P} = - \frac{\alpha P}{RT}. \qquad (29.15)$$

At moderate pressures f/P is not very different from unity (cf. Table XX), and so it is possible to make use of the fact that $\ln x$ is approximately equal to $x - 1$ when x approaches unity; hence, equation (29.15) becomes

$$\frac{f}{P} - 1 = - \frac{\alpha P}{RT}.$$

[4] Maron and Turnbull, *Ind. Eng. Chem.*, **33**, 69, 246 (1941); see also, Brown, *ibid.*, **33**, 1536 (1941); Maron and Turnbull, *J. Am. Chem. Soc.*, **64**, 44 (1942).

Upon introducing the definition of α, it is seen that

$$f = \frac{P^2 V}{RT}, \qquad (29.16)$$

so that the approximate fugacity of the gas can be determined directly from its pressure and molar volume. This result may be put into another form of interest by introducing $P_{\text{id.}}$, referred to above; this is equal to RT/V, and hence,

$$\frac{f}{P} = \frac{P}{P_{\text{id.}}}. \qquad (29.17)$$

It can be seen from Table XX that this relationship is approximately true for oxygen at 0° C up to pressures of about 100 atm., and so equation (29.16) may be utilized to give reasonable values of the fugacity at moderate pressures. For gases which depart from ideal behavior to a greater extent than does oxygen, however, equation (29.17), and hence (29.16), does not hold up to such high pressures, except at higher temperatures.

It may be mentioned here that although the fugacities in Table XX are always less than the corresponding pressures, this is not always the case (cf. problem in § 29b). For hydrogen and helium, for example, even at moderate pressures, at ordinary temperatures, PV is greater than RT, and hence it is evident from equation (29.16) that the fugacity will then be greater than the pressure. The results obtained for hydrogen at 25° C are recorded in Table XXI.[5]

TABLE XXI. FUGACITY OF HYDROGEN AT 25° C

Pressure	25	50	100	200	500	1,000 atm.
Fugacity	25.4	51.5	106.1	225.8	685	1,899 atm.

29e. Generalized Method for Determining Fugacities.—The definition of α given by equation (29.8) may be written in an alternative form, viz.,

$$\alpha = \frac{RT}{P} - V = \frac{RT}{P}\left(1 - \frac{PV}{RT}\right).$$

The factor PV/RT is the compressibility factor κ, considered in earlier sections (§§ 5i, 20e), so that

$$\alpha = \frac{RT}{P}(1 - \kappa).$$

If this is inserted in equation (29.9), it follows that

$$\ln f = \ln P + \int_0^P \frac{\kappa - 1}{P} dP, \qquad (29.18)$$

[5] Deming and Shupe, *Phys. Rev.*, **40**, 848 (1932); see also, *idem., ibid.*, **37**, 638 (1931); **38**, 2245 (1931); **56**, 108 (1939).

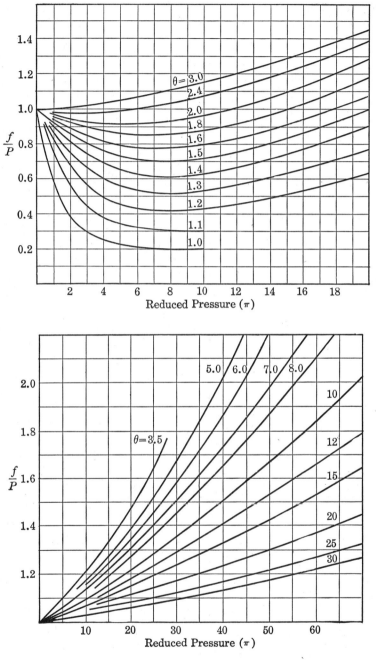

Fig. 18. Generalized fugacity curves

or, replacing the P terms in the integral by the corresponding reduced pressure π,

$$\ln \frac{f}{P} = \int_0^\pi \frac{\kappa - 1}{\pi} d\pi, \qquad (29.19)$$

π being equal to P/P_c, where P_c is the critical pressure of the gas.

It was seen in § 5i that at a given reduced temperature and pressure the compressibility factors κ of all gases are approximately equal; hence equation (29.19) is a generalized expression which may be plotted graphically on a chart so as to give the value of $\ln (f/P)$, or of f/P itself, for any gas in terms of the reduced temperature and pressure. In constructing the chart the values of κ are derived from Fig. 4, and the integral in equation (29.19) is evaluated graphically.[6] An example of such a chart is given in Fig. 18 in which f/P is plotted as ordinate with the reduced pressure π as the abscissa and the reduced temperature θ as the parameter. For reasons which will be understood later (§ 30b), the ratio of the fugacity of the gas to its pressure, i.e., f/P, is called the **activity coefficient** of the gas at the given pressure.

Problem: Utilize Fig. 18 to estimate the fugacity of nitrogen gas at 0° C and pressures of 50, 100, 200 and 400 atm.

Since T_c for nitrogen is 126° K, it follows that θ is 273/126 = 2.17; P_c is 33.5 atm., and the respective values of π are given below. By utilizing these results, the following data for f/P are obtained from Fig. 18, and hence the f's can be calculated.

P	π	f/P	f
50 atm.	1.5	0.98	49 atm.
100	3.0	0.97	97
200	6.0	0.98	196
400	12	1.07	428

(The results are seen to be in very good agreement with those obtained by the accurate procedure based on P-V data in the problem in § 29b.)

29f. Variation of Fugacity with Temperature and Pressure.—For two states of a gas in which the molar free energies are F and F^*, and the corresponding fugacities are f and f^*, equation (29.5) becomes

$$F - F^* = RT \ln \frac{f}{f^*}, \qquad (29.20)$$

where F^* and f^* refer to a very low pressure when the gas behaves almost ideally. Upon dividing through equation (29.20) by T, and rearranging,

[6] Newton, *Ind. Eng. Chem.*, **27**, 302 (1935); see also, Selheimer, Souders, Smith and Brown, *ibid.*, **24**, 515 (1932); Lewis and Luke, *ibid.*, **25**, 725 (1933); Lewis, *ibid.*, **28**, 257 (1936); for a nomograph applicable below the critical point, see Thomson, *ibid.*, **35**, 895 (1943).

the result is

$$R \ln \frac{f}{f^*} = \frac{F}{T} - \frac{F^*}{T}. \tag{29.21}$$

By equation (25.28),

$$\left[\frac{\partial (F/T)}{\partial T}\right]_P = -\frac{H}{T^2},$$

and so it is readily found by differentiation of equation (29.21) with respect to temperature, at constant pressure, that

$$\left(\frac{\partial \ln f}{\partial T}\right)_P = \frac{H^* - H}{RT^2}, \tag{29.22}$$

since f^* is equal to the gas pressure at very low pressures and hence is independent of temperature. In this expression H is the molar heat content of the gas at the given pressure P, and H^* is the value at a very low, i.e., zero, pressure; hence, $H^* - H$ is the increase of heat content accompanying the expansion of the gas to zero pressure from the pressure P, at constant temperature. This is, in fact, the quantity discussed in § 20e, so that the results in Fig. 14 may be utilized to calculate the change in the fugacity of a gas with temperature, at constant pressure. In a sense, Fig. 18 already includes the variation of fugacity with temperature, but more reliable values may be obtained from Fig. 14 if the fugacity is known accurately at one temperature. However, because of the variation of $H^* - H$ itself with temperature, the direct use of equation (29.22), in conjunction with Fig. 14, is not simple.

The temperature dependence of fugacity may be related to the Joule-Thomson coefficient $\mu_{\text{J.T.}}$ of the gas. Since $(\partial H/\partial P)_T$ is equal to $-\mu_{\text{J.T.}} C_P$ [equation (11.5)],

$$H^* - H = -\int_0^P \left(\frac{\partial H}{\partial P}\right)_T dP = \int_0^P \mu_{\text{J.T.}} C_P dP, \tag{29.23}$$

the low pressure, for which H^* applies, being taken as zero, and combination of this result with equation (29.22) gives

$$\left(\frac{\partial \ln f}{\partial T}\right)_P = \frac{1}{RT^2} \int_0^P \mu_{\text{J.T.}} C_P dP. \tag{29.24}$$

As a first approximation, C_P may be treated as independent of the pressure, and if $\mu_{\text{J.T.}}$ is expressed as a function of the pressure, it is possible to carry out the integration in equation (29.24); alternatively, the integral may be evaluated graphically. It is thus possible to determine the variation of the fugacity with temperature.

The influence of pressure on the fugacity of a gas at constant temperature is given by equation (29.7) which has been used in connection with the determination of fugacities. It is consequently unnecessary to discuss this expression further.

29g. Fugacity of Solids and Liquids. Every solid or liquid may be regarded as having a definite vapor pressure at a given temperature; although this vapor pressure may be extremely small for so-called nonvolatile substances, there is, nevertheless, always a definite pressure at which a solid or liquid is in equilibrium with its vapor, at constant temperature. Since the system is in equilibrium, the molar free energy of the liquid (or solid) must be the same as that of the vapor (§ 27a). It follows, therefore, from equation (29.4), that the fugacity of the liquid (or solid) will be equal to that of the vapor with which it is in equilibrium provided, of course, that the reference state is taken to be the same in each case. If the pressure of the vapor is not too high, its fugacity will be equal to the vapor pressure; hence *the fugacity of a liquid (or solid) is measured, approximately, by its vapor pressure.* This rule is frequently employed, although it is not exact. If the vapor pressure is high enough for departure from ideal behavior to be considerable, the fugacity may be derived by means of equation (29.16).

Problem: The vapor pressure of liquid chlorine is 3.66 atm. at 0° C, and the molar volume of the vapor under these conditions is 6.01 liters mole^{-1}. Calculate the fugacity of liquid chlorine at 0° C.

The fugacity of the liquid is equal to that of the vapor at 0° C and 3.66 atm., with V equal to 6.01 liters mole^{-1}; hence, by equation (29.16), with T equal to 273° K and R as 0.0820 liter-atm. deg.$^{-1}$ mole^{-1}, it follows that

$$f = \frac{(3.66)^2 \times 6.01}{0.082 \times 273} = 3.59 \text{ atm.}$$

The fugacity of liquid chlorine is thus 3.59 atm. at 0° C, which is somewhat less than its vapor pressure.

The variation of the fugacity of a solid or liquid with pressure or temperature is expressed by equations similar to those applicable to gases. Since equations (29.1) and (29.3), the latter partially defining the fugacity, will hold for solid, liquid or gaseous substances, the result obtained by combining them, viz., equation (29.7), gives the effect of pressure on the fugacity at constant temperature in each case. It should be noted, however, that V refers to the molar volume of *the particular phase under consideration*, since this is its significance in equation (29.1).

By following the procedure given in § 29f, with f representing the fugacity of pure liquid or solid, an equation exactly analogous to (29.22) is obtained for the variation of the fugacity with temperature at constant pressure. As before, H^* is the molar heat content of the gas, i.e., vapor, at low pressure, but H is now the molar heat content of the *pure liquid or solid* at the pressure P. The difference $H^* - H$ has been called the **ideal heat of vaporization**, for it is the heat absorbed, per mole, when a very small quantity of liquid or solid vaporizes into a vacuum. The pressure of the vapor is not the equilibrium value, but rather an extremely small pressure where it behaves as an ideal gas.

30. Mixtures of Gases

30a. Mixture of Ideal Gases.—In the thermodynamic study of mixtures of gases it is necessary to utilize the partial molar free energy, i.e., the chemical potential, in place of the free energy. The treatment may be introduced by considering, first, a system consisting of a mixture of ideal gases. According to equation (26.28), the variation with total pressure P of the chemical potential μ_i of any constituent i at a partial pressure p_i in a mixture of ideal gases, is given by

$$\left(\frac{\partial \mu_i}{\partial P}\right)_{T,N} = \frac{RT}{p_i}. \tag{30.1}$$

If the change of pressure ∂P is due entirely to a change in the constituent i, the pressure of all the others being constant, it may be replaced by ∂p_i, so that equation (30.1), at constant temperature, may be written as

$$d\mu_i = RT \frac{dp_i}{p_i} = RT d \ln p_i. \tag{30.2}$$

Upon integration this gives

$$\mu_i = \mu^* + RT \ln p_i, \tag{30.3}$$

where μ^* is the integration constant, the value of which depends on the nature of the gas, and *also on the temperature*. It will be seen from equation (30.3) that μ^* is the chemical potential of the gas i, at the given temperature, when its partial pressure p_i is unity. According to equation (30.3), the chemical potential of any constituent of a mixture of ideal gases is determined by its partial pressure in the mixture.

The partial pressure p_i of an ideal gas is related to its concentration c_i in the mixture by $p_i = RTc_i$ (§ 5b), and if this is introduced into equation (30.3) the result is

$$\begin{aligned}\mu_i &= (\mu^* + RT \ln RT) + RT \ln c_i \\ &= \mu_c^* + RT \ln c_i,\end{aligned} \tag{30.4}$$

where μ_c^* for the given gas is also dependent on the temperature.

Another possibility is to express p_i as $N_i P$, where N_i is the mole fraction of the gas i in the mixture and P is the total pressure [equation (5.8)]. Insertion of $N_i P$ for p_i in equation (30.3) then gives

$$\begin{aligned}\mu_i &= (\mu^* + RT \ln P) + RT \ln N_i \\ &= \mu_N^* + RT \ln N_i,\end{aligned} \tag{30.5}$$

where the quantity μ_N^* now depends on *both temperature and the total pressure*.

30b. Mixtures of Real Gases.—For a mixture of real gases, the foregoing results are no longer applicable, but the introduction of the fugacity concept rectifies the situation. By comparison of equation (29.3) with (30.2), it is possible to write for a constituent i of a mixture of nonideal gases, at constant temperature,

$$d\mu_i = RT d \ln f_i, \tag{30.6}$$

where f_i is the fugacity of the gas i in the given mixture. Upon integration, the result, analogous to equation (29.4) for a single gas, is

$$\mu_i = \mu^* + RT \ln f_i, \tag{30.7}$$

where the integration constant μ^* depends on the nature of the gas and the temperature of the system. The equation (30.7) partially defines the fugacity in a mixture, and in order to complete the definition it is necessary to specify the reference state. Comparison of equations (30.3) and (30.7) shows that, as for a single gas (§ 29a), the fugacity of an *ideal gas* in a mixture is proportional to its partial pressure, and, as before, the proportionality constant is chosen as unity, so that f_i/p_i is equal to unity, i.e., the fugacity and partial pressure are taken to be identical, at all pressures. For a real gas, the reference state is taken as *the system at a very low total pressure*, and it is postulated that *f_i/p_i approaches unity as the total pressure of the mixture approaches zero*. On the basis of this postulate, the constant μ^* in equation (30.7) becomes equivalent to that in (30.3), and it may be regarded either as the chemical potential of a real gas when its fugacity in the mixture is unity [equation (30.7)], or as the chemical potential of the same gas if it behaved ideally at unit (1 atm.) partial pressure [equation (30.3)], at the same temperature.

For certain purposes, it is useful to employ a different procedure to express the chemical potential of a real gas in a mixture; thus, equation (30.3), for an ideal gas, is modified so as to take the form

$$\mu_i = \mu^0 + RT \ln a_i, \tag{30.8}$$

where a_i is called the **activity** of the given gas in the mixture, and μ^0 is an arbitrary constant. This equation may be taken as partially defining the activity of any constituent of a mixture, the complete definition requiring specification of the condition which determines μ^0.

Since μ^* and μ^0 are both constants for a given substance at a specified temperature, it is evident from equations (30.7) and (30.8) that *the activity is proportional to the fugacity*, the actual value depending on the arbitrary choice of the proportionality constant. This conclusion is of general applicability, as will be seen later.

According to equation (30.8), μ^0 is the chemical potential of the gas i when its activity a_i is unity, but as the value of μ^0 can be chosen arbitrarily, the condition of unit activity is also arbitrary. The *state of unit activity* is called the **standard state**, and there is some freedom in its choice (§ 37a); however, in practice there are certain states that are more convenient than others in this connection. For a gaseous system it is advantageous to choose *the standard state of unit activity as that in which the fugacity of the gas is unity*, at the given temperature. In other words, the standard state is that of the gas behaving ideally at unit (1 atm.) pressure, i.e., the gas at unit fugacity. This particular choice makes μ^0 in equation (30.8) identical with μ^* in (30.7), for the latter gives the chemical potential of the gas under consideration in the state of unit fugacity. It is then possible to write equation (30.8) in the form

$$\mu_i = \mu^* + RT \ln a_{p(i)}, \tag{30.9}$$

where $a_{p(i)}$ is the activity of the constituent i on the basis of the proposed standard state.

Since the actual chemical potential μ_i must have a definite value, irrespective of any choice of standard state, it follows that equations (30.7) and (30.9) must be identical. As already seen, the constants are made equal by the particular standard state chosen, and so *the activity $a_{p(i)}$ of the given gas in the mixture must be equal to its fugacity f_i in that system*. It should be clearly understood, however, that this coincidence of fugacity and activity is a consequence of the particular standard state defined above. It makes the arbitrary proportionality constant, which relates the activity to the fugacity, equal to unity. The two quantities are then expressed in terms of the same units, e.g., atm.

In an ideal gas mixture, $a_{p(i)}$ should be equal to the partial pressure p_i, since the latter is then identical with the fugacity; the ratio $a_{p(i)}/p_i$, or the equivalent quantity f_i/p_i, may be taken as a measure of the approach to ideality. This dimensionless ratio, which tends to unity as the gas approaches ideal behavior, i.e., at very low total pressure, is called the **activity coefficient**, and is represented by the symbol γ_p. Thus, omitting the subscript i to avoid the multiplicity of symbols, equation (30.9), for any constituent of a mixture of gases, may be written as

$$\mu = \mu^* + RT \ln \gamma_p p, \tag{30.10}$$

where p is the partial pressure of the given constituent. For the present purpose the partial pressure may be defined, as indicated in § 5i, by equation (5.8), i.e., $p_i = N_i P$, where N_i is the mole fraction of the gas and P is the total pressure of the mixture.

The standard state chosen in the foregoing is a hypothetical one, viz,. an actual gas behaving ideally at 1 atm. pressure, which may be difficult to comprehend. Its use may be avoided, however, by means of a **reference state** which leads to exactly the same results. The reference state chosen is identical with that employed in connection with fugacity, namely, the gas at very low total pressure of the mixture. It is then postulated that *the ratio of the activity of any gas to its partial pressure becomes unity in the reference state*, i.e., $a_{p(i)}/p_i$ approaches unity as the total pressure becomes very small. It will be evident that this postulate makes the activity of a gas in a mixture identical with its fugacity, just as does the standard state proposed above.

In some cases the standard state is chosen as the ideal gas at unit molar concentration, and the chemical potential is expressed as

$$\mu_i = \mu_c^0 + RT \ln a_{c(i)}, \tag{30.11}$$

where μ_c^0 is the chemical potential the gas i would have if it behaved ideally at unit concentration. This is equivalent to postulating that *the ratio $a_{c(i)}/c_i$ of the activity to the molar concentration becomes unity in the reference state of low total pressure.*† With this choice of standard state, μ_c^* of equation (30.4) is identical with μ_c^0 of equation (30.11), and hence

$$\mu_c^0 = \mu_c^* = \mu^* + RT \ln RT,$$

so that equation (30.11) becomes

$$\mu_i = \mu^* + RT \ln RT + RT \ln a_{c(i)}. \tag{30.12}$$

Comparison of this result with equation (30.9) shows that

$$RT \ln RT + RT \ln a_{c(i)} = RT \ln a_{p(i)} = RT \ln f_i,$$

and so, omitting the subscripts i for simplicity,

$$a_c = \frac{a_p}{RT} = \frac{f}{RT}. \tag{30.13}$$

It is seen that the proportionality constant which relates the activity to the fugacity of a given gas is now equal to $1/RT$.

The activity coefficient γ_c in the present case is defined by a_c/c; this also approaches unity as the gas approximates to ideal behavior, i.e., at very low total pressure, but it is not equal to γ_p. By equation (30.13)

$$\gamma_c = \frac{a_c}{c} = \frac{f}{RTc} \tag{30.14}$$

and, as seen above,

$$\gamma_p = \frac{a_p}{p} = \frac{f}{p}, \tag{30.15}$$

but p for a nonideal gas is not equal to RTc, and so the two activity coefficients will not be identical, although they will approach one another, and unity, as the gas pressure is diminished.

30c. Variation of Fugacity with Pressure.—The expressions for the variation with pressure and temperature of the fugacity of a gas in a mixture are similar to those previously obtained for a pure gas [cf. equations (29.7) and (29.22)]. Since μ^* of a gas depends only on the temperature, differentiation of equation (30.7) with respect to the total pressure, at constant temperature and composition of the gas mixture,‡ gives

$$\left(\frac{\partial \mu_i}{\partial P}\right)_{T,N} = RT \left(\frac{\partial \ln f_i}{\partial P}\right)_{T,N}. \tag{30.16}$$

---

† It will be observed that whereas the standard state is that in which the defined activity is unity, the corresponding reference state is that in which the defined *activity coefficient* is unity.

‡ As in § 26a, the subscript N is used to imply constant composition.

Hence, by equation (26.26), i.e., $(\partial \mu_i/\partial P)_{T,N} = \bar{V}_i$, where $\bar{V}_i$ is the partial molar volume of the gas i in the given mixture, it follows that

$$\left(\frac{\partial \ln f_i}{\partial P}\right)_{T,N} = \frac{\bar{V}_i}{RT}. \tag{30.17}$$

When the standard state is chosen so as to make the activity of a gas equal to its fugacity, as explained above, *this expression also gives the variation of the activity with the total pressure.*

To determine the dependence of the *activity coefficient* γ_p of a gas in a mixture on the pressure, equation (30.15) is written in the logarithmic form,

$$\ln \gamma_p = \ln f_i - \ln p_i$$
$$= \ln f_i - \ln P - \ln N_i,$$

the defined partial pressure p_i being replaced by the product of the mole fraction N_i and the total pressure P. Upon differentiating with respect to P it is seen that, at constant temperature and composition,

$$\left(\frac{\partial \ln \gamma_p}{\partial P}\right)_{T,N} = \left(\frac{\partial \ln f_i}{\partial P}\right)_{T,N} - \left(\frac{\partial \ln P}{\partial P}\right)_{T,N}, \tag{30.18}$$

the mole fraction N_i being independent of the pressure. Upon introducing equation (30.17), it follows from (30.18) that

$$\left(\frac{\partial \ln \gamma_p}{\partial P}\right)_{T,N} = \frac{\bar{V}_i}{RT} - \frac{1}{P} = \frac{\bar{V}_i - V_i^*}{RT}, \tag{30.19}$$

where V_i^*, equal to RT/P, is the molar volume of the pure gas i at the total pressure P of the mixture on the supposition that it behaves ideally.

For an *ideal gas* the activity coefficient γ_p is unity at all pressures, and hence, by equation (30.19), $\bar{V}_i - V_i^*$ is always zero, so that $\bar{V}_i$ is equal to V_i^*. Further, V_i^* is now identical with the actual molar volume V_i of the gas, since the latter is supposed to be ideal. It follows, therefore, that the partial molar volume $\bar{V}_i$ of a gas in a mixture of ideal gases is equal to its molar volume V_i in the pure state, at the same temperature and (total) pressure.

The total volume V of a mixture of gases, at constant temperature, pressure and composition, is given by equation (26.6) as

$$V = n_1\bar{V}_1 + n_2\bar{V}_2 + \cdots + n_i\bar{V}_i + \cdots,$$

where $n_1, n_2, \ldots, n_i, \ldots$, are the numbers of moles of the various gases whose partial molar volumes in the given mixture are $\bar{V}_1, \bar{V}_2, \ldots, \bar{V}_i, \ldots$, respectively. As seen above, if all the gases are ideal, the partial molar volume of each is equal to its molar volume at the same temperature and pressure, so that

$$V = n_1 V_1 + n_2 V_2 + \cdots + n_i V_i + \cdots,$$

where $V_1, V_2, \ldots, V_i, \ldots$, are the molar volumes of the constituent gases in the pure state. The right-hand side of this expression is, incidentally, also equal to the sum of the volumes of the individual gases before mixing.

It is seen, therefore, that there would be no volume change upon mixing a number of ideal gases. For real gases, $\bar{V}_i$ is, in general, not equal to V_i, and hence the total volume of the mixture will usually differ from the sum of volumes of the separate gases, and mixing will be accompanied by a change of volume (cf., however, § 5k).

30d. Variation of Fugacity with Temperature.—If equation (30.7) is divided through by T and rearranged, the result is

$$R \ln f_i = \frac{\mu_i}{T} - \frac{\mu^*}{T}.$$

Making use of equation (26.25), it is found upon differentiation with respect to temperature, at constant pressure and composition, that

$$\left(\frac{\partial \ln f_i}{\partial T}\right)_{P,N} = \frac{\bar{H}_i^* - \bar{H}_i}{RT^2}, \qquad (30.20)$$

where $\bar{H}_i$ is the partial molar heat content of the constituent i in the mixture, at the given temperature and pressure. As stated in § 30b, μ^* is equivalent to the chemical potential of the gas if it behaved ideally at a pressure of 1 atm.; hence $\bar{H}_i^*$ refers to the partial molar heat content under these conditions. It will be shown below that for an ideal gas in a mixture the partial molar heat content is equal to the molar heat content of the pure gas at the same temperature. It is consequently permissible to replace $\bar{H}_i^*$ in equation (30.20) by H_i^*, where the latter is the molar heat content of the pure gas i at very low pressures when it behaves ideally, i.e., in the reference state; hence the equation for the variation of fugacity with temperature may be written

$$\left(\frac{\partial \ln f_i}{\partial T}\right)_{P,N} = \frac{H_i^* - \bar{H}_i}{RT^2}. \qquad (30.21)$$

This equation also gives the dependence of the activity a_p on the temperature, since this activity is equal to the fugacity. Further, since a_p is equal to $\gamma_p p$, where γ_p is the activity coefficient, *the same equation represents the variation of the activity coefficient with temperature,* p being constant.

Since equation (30.20), like (30.21), also represents the variation with temperature of the activity coefficient of a gas in a mixture, it follows that for an *ideal gas* $\bar{H}_i^* - \bar{H}_i$ is constant, and consequently $\bar{H}_i$ is equal to $\bar{H}_i^*$ at all pressures. By utilizing exactly similar arguments to those employed in § 30c when considering the volume of a mixture of ideal gases, it can be shown that the partial molar heat content $\bar{H}_i$ of any gas in a mixture of ideal gases is equal to its molar heat content H_i in the pure state at the same temperature. It is unnecessary to specify the pressure because the molar heat content of an ideal gas is independent of the pressure (§ 9e). Since $\bar{H}_i$ for an ideal gas is identical with H_i, it follows that the total heat content of a mixture of ideal gases is equal to the sum of the heat contents of the individual gases, and there is consequently no heat change upon mixing, at constant temperature.

30e. Determination of Fugacity in Gas Mixtures: The Lewis-Randall Rule.—

The principle involved in the determination of the fugacity of a gas in a mixture is analogous to that developed in § 29b for a pure gas. According to equation (30.17), at constant temperature and composition,

$$RT d \ln f_i = \bar{V}_i dP,$$

and if a quantity α_i, which is a function of the temperature, total pressure and composition of the mixture, is defined for the gas i by

$$\alpha_i = \frac{RT}{P} - \bar{V}_i, \qquad (30.22)$$

where P is the total pressure, it is readily found that

$$d \ln f_i = d \ln P - \frac{\alpha_i}{RT} dP.$$

Upon integration, at constant temperature and composition, between the limits of the very low total pressure P^* and the appreciable pressure P, it is seen that

$$\ln \frac{f_i}{f_i^*} = \ln \frac{P}{P^*} - \frac{1}{RT} \int_{P^*}^{P} \alpha_i dP. \qquad (30.23)$$

The partial pressure p_i^* of the gas is equal to $N_i P^*$, where N_i is the mole fraction of the gas in the given mixture; hence, equation (30.23) becomes

$$\ln \frac{f_i}{f_i^*} = \ln N_i + \ln \frac{P}{p_i^*} - \frac{1}{RT} \int_{P^*}^{P} \alpha_i dP$$

or

$$\ln f_i = \ln N_i + \ln P + \ln \frac{f_i^*}{p_i^*} - \frac{1}{RT} \int_{P^*}^{P} \alpha_i dP.$$

If the low pressure P^* may be taken as virtually zero, f_i^*/p_i^* is then unity, as postulated in § 30b, and hence $\ln (f_i^*/p_i^*)$ is zero, so that

$$\ln f_i = \ln N_i + \ln P - \frac{1}{RT} \int_{0}^{P} \alpha_i dP. \qquad (30.24)$$

If P-V data are available for the gas mixture, it is possible to determine the partial molar volume $\bar{V}_i$ of any constituent (see Chapter XVIII), and hence α_i at various total pressures may be calculated from equation (30.22). The integral in equation (30.24) can thus be evaluated graphically, and hence the fugacity f_i of the gas whose mole fraction is N_i in the given mixture, at the total pressure P, can be determined.[7]

There is an interesting modification of equation (30.24) possible in the special case in which there is no volume change when the gases are mixed at constant temperature, at all pressures. In this event, the partial molar volume $\bar{V}_i$ of each gas in the mixture must be equal to its molar volume.

---

[7] Gibson and Sosnick, *J. Am. Chem. Soc.*, **49**, 2172 (1927); see also, Gillespie, *ibid.*, **47**, 305, 3106 (1925); **48**, 28 (1926); Lurie and Gillespie, *ibid.*, **49**, 1146 (1927); Gillespie, *Chem. Rev.*, **18**, 359 (1936).

The last two terms in equation (30.24) are then equivalent to the fugacity f_i' of the *pure gas i at the total pressure P of the mixture* [cf. equation (29.9)]. It follows, therefore, that

$$\ln f_i \approx \ln \text{N}_i + \ln f_i'$$

or

$$f_i \approx \text{N}_i f_i'. \tag{30.25}$$

The fugacity of a gas in a mixture would then be equal to the product of its mole fraction in the mixture and its fugacity in the pure state at the total pressure of the mixture. This rule was proposed from general considerations by G. N. Lewis and M. Randall (1923). It follows from the arguments presented above that it can apply only to a mixture of gases which is formed without a volume change, sometimes called an "ideal mixture," not necessarily consisting of ideal gases. Although the formation of gas mixtures is often accompanied by a volume change, the rule represented by equation (30.25) is frequently used, on account of its simplicity, to give an approximate indication of the fugacity of a gas in a mixture (cf. § 32c). Actual determinations of fugacities have shown that the Lewis-Randall rule holds with a fair degree of accuracy at pressures up to about 100 atm. for a number of common gases.[8]

31. Liquid Mixtures

31a. Fugacity in Liquid Mixtures.—The fugacity of a constituent of a liquid mixture may be expressed in terms of the fugacity of the vapor of that constituent in equilibrium with the mixture, just as was seen to be the case for a pure liquid or solid (§ 29g). Since the chemical potential of a given substance must be the same in all phases at equilibrium, at constant temperature, it follows that the fugacities must be equal, provided the same reference state, namely the vapor at very low pressures, is used in each case. If the vapor pressures are not too high, the vapor may be regarded as behaving ideally, and so the fugacity of any constituent of a mixture is approximately equal to its partial vapor pressure in equilibrium with the mixture; this approximation is frequently employed, especially for vapor pressures of the order of 1 atm. or less.

The equations derived in §§ 30c, 30d thus also give the variation with pressure and temperature of the fugacity of a constituent of a liquid (or solid) solution. In equation (30.17), $\bar{V}_i$ is now the partial molar volume of the particular constituent *in the solution*, and in (30.21), $\bar{H}_i$ is the corresponding partial molar heat content. The numerator $H_i^* - \bar{H}_i$ thus represents the change in heat content, per mole, when the constituent is vaporized from the solution into a vacuum (cf. § 29g), and so it is the "ideal" heat of vaporization of the constituent i from the given solution, at the specified temperature and total pressure.

[8] G. N. Lewis and M. Randall, ref. 1, pp. 225–227; see also, papers mentioned in ref. 7, and Merz and Whittaker, *J. Am. Chem. Soc.*, **50**, 1522 (1928); Krichevsky, *ibid.*, **59**, 2733 (1937).

31b. Activities and Activity Coefficients in Liquid Solutions.

—Since the chemical potential of any constituent of a liquid solution must be equal to that of the vapor in equilibrium with it, it follows from equation (30.7) that

$$\mu_i = \mu^* + RT \ln f_i \tag{31.1}$$

is the general expression for the chemical potential of any constituent of a solution. Alternatively, this quantity may be stated in terms of the activity by means of the equation

$$\mu_i = \mu^0 + RT \ln a_i, \tag{31.2}$$

where a_i is the activity in the given solution, and μ^0 is the chemical potential when the activity of i is unity. Comparison of equations (31.1) and (31.2) shows, in agreement with the statement made in § 30b, that the activity of the constituent i of the given solution is proportional to its fugacity in that solution. The actual value of the activity, and of the proportionality constant relating it to the fugacity, depends upon the choice of the standard state of unit activity.

For dilute solutions the best choice of standard state for the solute is different from that adopted here (see § 37b, III), but for the solvent in dilute solutions and for all constituents of solutions of completely miscible, or almost completely miscible, liquids, the standard state of unit activity is usually selected as that of *the pure liquid at the same temperature and 1 atm. pressure*. The activity scale is chosen, therefore, so as to make the activity of any constituent of a mixture equal to unity for that substance in the pure liquid state at 1 atm. pressure. According to the postulated standard state, the value of μ^0 for a given liquid depends only on the temperature and is *independent of the pressure*.

The standard state described above is equivalent to choosing the pure liquid form of any constituent at 1 atm. as the reference state in which the activity is equal to its mole fraction, i.e., unity. Thus, at 1 atm. pressure, the ratio of the activity a_i to the mole fraction N_i, i.e., a_i/N_i, tends to unity as the pure liquid state is approached. Except for an ideal solution (cf. § 34a) at atmospheric pressure, the ratio a_i/N_i, which defines the activity coefficient γ_N of i in the given solution, varies with the composition and may differ from unity. In general, the activity a_i may be stated as the product of the activity coefficient and the mole fraction of the particular constituent, i.e., $\gamma_N N_i$.

The chemical potential of a pure liquid at a given temperature and 1 atm. pressure, i.e., μ^0, can be expressed by means of equation (31.1) as

$$\mu^0 = \mu^* + RT \ln f_i^0, \tag{31.3}$$

where f_i^0 is the fugacity of the pure liquid in its standard state at the specified temperature. If this result is inserted into equation (31.2), it is seen that

$$\mu_i = \mu^* + RT \ln f_i^0 + RT \ln a_i, \tag{31.4}$$

and comparison with equation (31.1) yields

$$a_i = \frac{f_i}{f_i^0}, \qquad (31.5)$$

which may be taken as an alternative definition of the activity. It is seen that the proportionality constant relating the activity to the fugacity of a constituent of a liquid mixture is $1/f_i^0$, where f_i^0 is numerically equal to the fugacity of the pure liquid at 1 atm. pressure, at the temperature of the mixture.

31c. Variation of Activity in Liquid Mixtures with Temperature and Pressure.—If equation (31.2) is divided through by T, it is found upon rearrangement that

$$R \ln a_i = \frac{\mu_i}{T} - \frac{\mu^0}{T},$$

and hence, upon differentiation with respect to temperature, at constant pressure and composition, utilizing equation (26.25), the result is

$$\left(\frac{\partial \ln a_i}{\partial T}\right)_{P,N} = \frac{H_i^0 - \bar{H}_i}{RT^2} \qquad (31.6)^*$$

where $\bar{H}_i$ is the partial molar heat content of the constituent i in the given solution, and H_i^0, written in place of $\bar{H}_i^0$ to which it is equivalent, is its molar heat content in the pure state at 1 atm. pressure, since μ^0 is the chemical potential of the pure liquid i at this pressure. It will be seen later that *for an ideal solution* the partial molar heat content is equal to the molar heat content (cf. also § 30d), so that $\bar{H}_i$ and H_i^0 are identical. In these circumstances, the activity at constant composition and pressure, is independent of the temperature.

It should be noted that equation (31.6) also gives *the variation of the activity coefficient* γ_N *with temperature*. This follows from the definition of γ_N as a_i/N_i; since N_i, the mole fraction of the given constituent, is constant, the variation of the activity coefficient with temperature will be exactly the same as that of the activity at constant composition and pressure.

Upon differentiating equation (31.2) with respect to pressure, at constant temperature and composition, and utilizing equation (26.6), i.e., $(\partial \mu_i/\partial P)_{T,N} = \bar{V}_i$, it follows that

$$\left(\frac{\partial \ln a_i}{\partial P}\right)_{T,N} = \frac{1}{RT}\left(\frac{\partial \mu_i}{\partial P}\right)_{T,N} = \frac{\bar{V}_i}{RT}, \qquad (31.7)\dagger$$

where $\bar{V}_i$ is the partial molar volume of the constituent i in the solution. It should be noted that since a pressure of 1 atm. was specified in the definition of the standard state, μ^0 is independent of pressure, as indicated earlier, and hence $(\partial \mu^0/\partial P)_T$ is zero. In accordance with the remarks made above,

---

* The same result can be obtained directly from equations (30.21) and (31.5).
† This result can also be obtained from equations (30.17) and (31.5).

equation (31.7) also gives *the variation of the activity coefficient* γ_N *with pressure.*

Further reference to activities and activity coefficients in connection with the properties of various solutions will be made in later sections. Sufficient, however, has been given here for the present purpose, which is the application of the concept of activity to the study of chemical equilibrium to be taken up in the next chapter.

EXERCISES

1. Utilize the following data for 0° C to calculate the fugacity of carbon monoxide at 50, 100, 400 and 1,000 atm. by graphical integration based on equation (29.9):

P	25	50	100	200	400	800	1000 atm.
PV/RT	0.9890	0.9792	0.9741	1.0196	1.2482	1.8057	2.0819

[Bartlett, *J. Am. Chem. Soc.*, **52**, 1374 (1932)]. Compare the results with those given by the approximate rule applicable at low pressures.

2. Using the PV/RT values for 50 atm. and 400 atm. given in Exercise 1, calculate the van der Waals a and b for carbon monoxide. From these determine the fugacities at the various pressures and compare the results with those obtained in the preceding exercise.

3. Show that at moderate and low pressures (and moderate temperatures) f/P of a pure gas is approximately equal to its compressibility factor (κ). Verify this for π equal to 2 and values of θ of 1.20 or more by means of Figs. 4 and 18.

4. The compressibility of a gas may be represented by

$$PV/RT = A + BP + CP^2 + DP^3,$$

where A, B, C, D are functions of the temperature; hence, derive an expression for the fugacity as a function of the pressure at a given temperature. For nitrogen at 0° C, A is 1.000, B is -5.314×10^{-4}, C is 4.276×10^{-6} and D is -3.292×10^{-9} with P in atm. up to 400 atm. [Bartlett, *J. Am. Chem. Soc.*, **49**, 687 (1927)]. Evaluate the fugacity of the gas at 300 atm. pressure.

5. At 25° C, the vapor pressure of water is 23.76 mm. and the specific volume of the vapor under these conditions is 43,400 cc. g.$^{-1}$. Calculate the free energy change for the transfer of 1 mole of water from liquid at 25° C to vapor at unit fugacity. What error is involved in treating the vapor as an ideal gas (cf. Exercise 9, Chapter X)?

6. By utilizing the form of the van der Waals equation applicable at low pressures, show that the function α, defined by equation (29.8), is usually not zero at zero pressure.

7. Use the generalized fugacity diagram (Fig. 18) to determine the fugacity of (i) hydrogen at 25° C and 200 atm., (ii) oxygen at 0° C and 200 atm., (iii) carbon dioxide at 100° C and 250 atm.

8. Utilize the relationship between the free energy change and the fugacities in the final and initial states to show how the generalized fugacity diagram can be used to determine the free energy change for an isothermal expansion or compression of any gas. Calculate the free energy change in cal. for the compression of nitrogen gas from 1 to 200 atm. at 25° C.

9. Calculate the fugacity of liquid water at 100° C, at which temperature and 1 atm. pressure the specific volume of the vapor is 1675 cc. g.$^{-1}$, relative to the usual reference state of the gas at low pressure. What is the free energy change corresponding to the transfer of 1 mole of water vapor from 1 atm. at 100° C to unit fugacity at the same temperature? Is the value significant in magnitude?

10. The following data were obtained for the molar volumes V_1 and V_2 in ml. for pure hydrogen and nitrogen, respectively, at various pressures P and 0° C. On the other hand, $\bar{V}_1$ and $\bar{V}_2$ are the partial molar volumes in a mixture containing 0.6 mole hydrogen to 0.4 mole nitrogen, where P is now the total pressure.

P	V_1	V_2	$\bar{V}_1$	$\bar{V}_2$
50 atm.	464.1	441.1 ml.	466.4	447.5 ml.
100	239.4	220.6	241.3	226.7
200	127.8	116.4	129.1	120.3
300	90.5	85.0	91.1	86.9
400	72.0	70.5	72.5	71.8

Determine by the graphical method the fugacity of nitrogen in the mixture at the various total pressures, and compare the results with those obtained for the pure gases; hence, test the Lewis-Randall rule for the fugacity of a gas in a mixture [cf. Merz and Whittaker, *J. Am. Chem. Soc.*, **50**, 1522 (1928)].

11. Use the Lewis-Randall rule and Fig. 18 to calculate the fugacities of carbon monoxide, oxygen and carbon dioxide in a mixture containing 23, 34 and 43 mole %, respectively, of these gases at 400° C and a total pressure of 250 atm.

12. Give the alternative derivations of equations (31.6) and (31.7).

13. Combine equation (29.11) with the Berthelot equation to derive an expression for the fugacity of a Berthelot gas. Calculate the fugacities of carbon monoxide at various pressures at 0° C, and compare the results with those obtained in Exercises 1 and 2.

14. By considering the influence of temperature and pressure on the fugacities, which must remain the same in both phases, derive a form of the Clapeyron equation (27.4) for the equilibrium between two phases of a single substance.

15. The following values of PV/RT were obtained for nitrogen at high pressures at 0° C:

P	1500	2000	2500	3000 atm.
PV/RT	2.720	3.327	3.920	4.947

Extend the results of the problem in § 29b and Fig. 17 to determine the fugacity of nitrogen gas at 3000 atm. pressure. (Note the high value of f/P.)

16. Show that in its standard state the chemical potential μ^0 of a gas is independent of the pressure.

CHAPTER XIII

FREE ENERGY AND CHEMICAL REACTIONS

32. The Equilibrium Constant

32a. Chemical Equilibrium.—When a particular chemical reaction can proceed simultaneously in both directions, a state of equilibrium will be reached when the system appears stationary at a given temperature and pressure. As is well known, it is highly probable that the apparently stationary state is one of dynamic equilibrium in which the reactions are proceeding in opposite directions at the same rate. When the state of equilibrium is attained, a condition which may take a considerable time if the reactions are relatively slow, the system still contains certain amounts of the reactants, as well as of the products. Many chemical reactions, on the other hand, appear to proceed to virtual completion, so that undetectable quantities of the reactants remain when the reaction is complete. However, even in such cases it is probable that the reverse reaction takes place, although to a very small extent, and a state of equilibrium is attained. By the use of the free energy concept, thermodynamics has been employed to throw light on the problems of chemical equilibrium. It is possible, for example, to define the conditions of equilibrium, and to show how they may vary with temperature and pressure. Further, by the use of heat content and entropy data, the actual state of chemical equilibrium can be determined.

32b. The Equilibrium Constant.—Consider a closed system in which the perfectly general reaction represented by

$$a\text{A} + b\text{B} + \cdots \rightleftharpoons l\text{L} + m\text{M} + \cdots$$

has been allowed to reach a state of equilibrium at a given temperature and external (total) pressure. Suppose an infinitesimal change is made to occur in this system, from left to right; thus dn_A moles of A, dn_B moles of B, etc., are consumed, while dn_L moles of L, dn_M moles of M, etc., are formed. The free energy change accompanying this process is given by [cf. equation (26.14)]

$$dF_{T,P} = (\mu_\text{L} dn_\text{L} + \mu_\text{M} dn_\text{M} + \cdots) - (\mu_\text{A} dn_\text{A} + \mu_\text{B} dn_\text{B} + \cdots), \quad (32.1)$$

where the μ's are the chemical potentials (partial molar free energies) of the indicated species. Since the closed system is in equilibrium, this free energy change $dF_{T,P}$ must be equal to zero [equation (25.39)], so that

$$(\mu_\text{L} dn_\text{L} + \mu_\text{M} dn_\text{M} + \cdots) - (\mu_\text{A} dn_\text{A} + \mu_\text{B} dn_\text{B} + \cdots) = 0. \quad (32.2)$$

The various quantities dn_A, dn_B, ..., dn_L, dn_M, ..., taking part in the chemical reaction indicated above must be in the respective proportions of $a, b, ..., l, m, ...$, so that equation (32.2) for the equilibrium condition may be written as

$$(l\mu_L + m\mu_M + \cdots) - (a\mu_A + b\mu_B + \cdots) = 0. \tag{32.3}$$

By equation (26.29) the left-hand side of (32.3) is equivalent to ΔF, the free energy change accompanying the complete reaction under equilibrium conditions; hence,

$$\Delta F = 0 \tag{32.4}$$

at equilibrium [cf. equation (25.40)]. The equations (32.3) and (32.4) represent *the fundamental condition for chemical (and physical) equilibria of all types.*

The chemical potential μ of any constituent of a mixture, gaseous or liquid, may be represented by an equation of the form [cf. equations (30.8) and (31.2)]

$$\mu = \mu^0 + RT \ln a, \tag{32.5}$$

where μ^0 is the chemical potential of the given substance in the chosen standard state of unit activity, and a is its activity in the mixture under consideration. If the values of the chemical potential given by equation (32.5) are introduced into (32.3), the equilibrium condition becomes

$$l(\mu_L^0 + RT \ln a_L) + m(\mu_M^0 + RT \ln a_M) + \cdots \\ - a(\mu_A^0 + RT \ln a_A) - b(\mu_B^0 + RT \ln a_B) + \cdots = 0,$$

so that

$$RT \ln \frac{a_L^l \times a_M^m \times \cdots}{a_A^a \times a_B^b \times \cdots} = (a\mu_A^0 + b\mu_B^0 + \cdots) - (l\mu_L^0 + m\mu_M^0 + \cdots)$$
$$= -\Delta F^0, \tag{32.6}$$

using equation (26.29). At constant temperature the right-hand side of equation (32.6) is constant;* since RT is also constant, it follows that

$$\frac{a_L^l \times a_M^m \times \cdots}{a_A^a \times a_B^b \times \cdots} = \text{constant, i.e., } K, \tag{32.7}$$

where the constant K is called the **equilibrium constant** of the reaction. The expression is sometimes known as the **law of equilibrium**, for it provides a simple relationship between the activities of the reactants and products when equilibrium is attained in a chemical reaction at a given temperature. In the form given in equation (32.7) the law holds for any equilibrium irrespective of whether it involves one or more phases; the results are applicable, however, to the whole (closed) system, and not to each phase separately. The special forms to be taken by the expression for the equilibrium constant in a number of reactions of different types will now be considered.

* Since standard states are commonly defined in terms of unit fugacity for gases and 1 atm. pressure for liquids, solids and solutions, the μ^0 values, and hence ΔF^0, are independent of pressure (cf. § 31b and Chapter XII, Exercise 16).

32c. Equilibrium in Homogeneous Gaseous Systems.

If the reaction is one involving gases only, it will be homogeneous, taking place in a single phase. If the standard state of unit activity is taken as the state of unit fugacity, i.e., ideal gas at 1 atm. pressure, the activity of each reacting substance may then be replaced by its fugacity (cf. § 30b) in the equilibrium system, so that the equation (32.7) for the equilibrium constant becomes

$$K_f = \frac{f_L^l \times f_M^m \times \cdots}{f_A^a \times f_B^b \times \cdots}, \tag{32.8}$$

where the f's are the respective fugacities. It will be noted that the equilibrium constant is now represented by K_f, the subscript f being used to indicate that the activities are expressed by the fugacities.

Since the ratio of the fugacity f_i to the partial pressure p_i of any gas in a mixture is equal to the activity coefficient γ_i (§ 30b), it is possible to replace each fugacity factor by the product $p \times \gamma$. If this substitution is made, equation (32.8) becomes

$$K_f = \frac{p_L^l \times p_M^m \times \cdots}{p_A^a \times p_B^b \times \cdots} \cdot \frac{\gamma_L^l \times \gamma_M^m \times \cdots}{\gamma_A^a \times \gamma_B^b \times \cdots}. \tag{32.9}$$

The data necessary for the evaluation of the activity coefficients (or fugacities) of the individual gases in a mixture (§ 30e) are not usually available, and an alternative treatment for allowing for departure from ideal behavior is frequently adopted.

According to the approximate rule referred to in § 30e, the fugacity f_i of any gas in a mixture, in which its mole fraction is N_i, is given by equation (30.25), viz.,

$$f_i \approx N_i f'_i, \tag{32.10}$$

where f'_i is the fugacity of the pure gas i at the total pressure of the mixture. Making this substitution in equation (32.8), it is possible to write

$$K'_f = \frac{N_L^l \times N_M^m \times \cdots}{N_A^a \times N_B^b \times \cdots} \cdot \frac{f'^l_L \times f'^m_M \times \cdots}{f'^a_A \times f'^b_B \times \cdots}, \tag{32.11}$$

the symbol K'_f being used, in place of K_f, to show that the expression in equation (32.11) is based on an approximation [equation (32.10)] and hence K'_f cannot be exactly constant. If each of the f' terms is divided by P, the total pressure, the result, i.e., f'_i/P, is the corresponding activity coefficient γ'_i of the particular gas *in the pure state when its pressure is equal to P*; hence equation (32.11) may be written as

$$K'_f = \frac{N_L^l \times N_M^m \times \cdots}{N_A^a \times N_B^b \times \cdots} \cdot \frac{\gamma'^l_L \times \gamma'^m_M \times \cdots}{\gamma'^a_A \times \gamma'^b_B \times \cdots} \cdot P^{\Delta n}, \tag{32.12}$$

where Δn, the change in the number of molecules in the gas reaction, is equal to $(l + m + \cdots) - (a + b + \cdots)$. The first and third factors in equation (32.12) can be obtained by experiment, and so these may be combined to give K'_p; thus,

$$K'_p = \frac{N_L^l \times N_M^m \times \cdots}{N_A^a \times N_B^b \times \cdots} \cdot P^{\Delta n}, \qquad (32.13)$$

and hence, equation (32.12) may be written in the form

$$K'_f = K'_p \times J_{\gamma'}, \qquad (32.14)$$

where the function $J_{\gamma'}$ * of the activity coefficients is defined by

$$J_{\gamma'} = \frac{\gamma'^l_L \times \gamma'^m_M \times \cdots}{\gamma'^a_A \times \gamma'^b_B \times \cdots}. \qquad (32.15)$$

It may be noted that if the gases taking part in the reaction were ideal, the activity coefficient factor in equation (32.9) would be unity; further, the partial pressure p_i of each gas would be equal to $N_i P$, by equation (5.8). In this event, the result would be identical with the expression in (32.13), so that the quantity defined by K'_p would represent the true equilibrium constant. When the departure of the reacting gases from ideal behavior is not large, e.g., when the total pressures are of the order of 1 atm. or less, a very satisfactory approximation to the equilibrium constant K_f can be obtained from equation (32.13). At high pressures, however, the values of K'_p deviate considerably from constancy, as will be seen below, but a great improvement is possible by the introduction of the activity coefficient factor, as in equation (32.14).

Although it is not commonly used in the study of gaseous equilibria, mention may be made of the form taken by the equilibrium constant when the activities a_c are expressed on the basis of the standard state of unit activity as equal to that of an ideal gas at unit molar concentration (§ 30b); thus, equation (32.7) may be written

$$K_c = \frac{(a_c)_L^l \times (a_c)_M^m \times \cdots}{(a_c)_A^a \times (a_c)_B^b \times \cdots}. \qquad (32.16)$$

According to equation (30.13), a_c is equal to f/RT, where f is the fugacity, so that

$$K_c = \frac{f_L^l \times f_M^m \times \cdots}{f_A^a \times f_B^b \times \cdots} \cdot (RT)^{-\Delta n}. \qquad (32.17)$$

The first term on the right-hand side is obviously equal to K_f, as defined by equation (32.8), and hence

$$K_f = K_c \times (RT)^{\Delta n}. \qquad (32.18)$$

For a gas reaction in which there is no change in the number of molecules, Δn is zero, and K_f and K_c are identical; the equilibrium constant is then independent of the chosen standard state.

An alternative form of equation (32.16) is obtained by replacing each activity factor by the product of the concentration and the appropriate activity coefficient,

---

* The symbol J will be used for a function having the same form as the equilibrium constant, the nature of the variable being indicated by the subscript.

i.e., $a_c = c \times \gamma_c$; then

$$K_c = \frac{c_L^l \times c_M^m \times \cdots}{c_A^a \times c_B^b \times \cdots} \cdot \frac{(\gamma_c)_L^l \times (\gamma_c)_M^m \times \cdots}{(\gamma_c)_A^a \times (\gamma_c)_B^b \times \cdots}. \tag{32.19}$$

The concentration function may be represented by K'_c and the activity coefficient function by J_{γ_c}, so that

$$K_c = K'_c \times J_{\gamma_c}. \tag{32.20}$$

This equation is exact, but for many approximate purposes J_{γ_c} is taken as unity, and then the law of equilibrium is expressed in the form

$$K'_c = \frac{c_L^l \times c_M^m \times \cdots}{c_A^a \times c_B^b \times \cdots}, \tag{32.21}$$

where K'_c is not precisely constant.

32d. The Ammonia Equilibrium.

A homogeneous gaseous equilibrium which has been studied in some detail, over a considerable range of pressure and temperature, is the reaction

$$\tfrac{1}{2}N_2 + \tfrac{3}{2}H_2 \rightleftharpoons NH_3$$

employed in the Haber process for the production of ammonia. For the equilibrium,

$$K_f = \frac{f_{NH_3}}{f_{N_2}^{1/2} \times f_{H_2}^{3/2}}, \tag{32.22}$$

and a proper application of this equation requires a knowledge of the fugacity of each gas in the particular equilibrium mixture. Since the data are not available, it is more convenient for practical purposes to employ the somewhat less exact equation (32.12) or (32.14); in the present case this takes the form

$$K'_f = K'_p \times \frac{\gamma'_{NH_3}}{\gamma'^{1/2}_{N_2} \times \gamma'^{3/2}_{H_2}}, \tag{32.23}$$

where the experimental quantity K'_p is given by

$$K'_p = \frac{N_{NH_3}}{N_{N_2}^{1/2} \times N_{H_2}^{3/2}} \cdot \frac{1}{P}. \tag{32.24}$$

In the investigation of the equilibrium, mixtures containing nitrogen and hydrogen in the molar ratio of 1 to 3 were allowed to come to equilibrium at various temperatures and pressures; the proportion of ammonia in the resulting mixture was then determined by analysis. From this the mole fractions N_{NH_3}, N_{N_2} and N_{H_2} at equilibrium could be readily calculated, and hence K'_p could be evaluated by equation (32.24). A selection of the results obtained in this manner by A. T. Larson and R. L. Dodge (1923–24) is given in Table XXII for a temperature of 450° C.[1]

[1] Larson and Dodge, *J. Am. Chem. Soc.*, **45**, 2918 (1923); Larson, *ibid.*, **46**, 367 (1924).

TABLE XXII. THE $N_2 + \tfrac{3}{2}H_2 \rightleftharpoons NH_3$ EQUILIBRIUM AT 450° C

Pressure	K_p'	$J_{\gamma'}$	K_f'
10 atm.	6.59×10^{-3}	0.988	6.51×10^{-3}
30	6.76	0.969	6.55
50	6.90	0.953	6.57
100	7.25	0.905	6.56
300	8.84	0.750	6.63
600	12.94	0.573	7.42
1000	23.28	0.443	10.32

An examination of the values of K_p' shows that they are fairly constant up to a pressure of about 50 atm.; below this pressure the deviations of the gases from ideal behavior at 450° C are evidently not very large. At higher pressures, however, the deviations are considerable, and the value of K_p' is seen to increase rapidly with increasing pressure. By using the method described in § 29e, based on Fig. 18 which gives the activity coefficient f/P at any (reduced) pressure directly, the function $J_{\gamma'}$ has been calculated for various values of the total pressure. The results, together with the corresponding values of K_f', that is, of $K_p' \times J_{\gamma'}$, are recorded in Table XXII. It is seen that K_f' is remarkably constant up to 300 atm., and fairly constant up to 600 atm. total pressure. At 1000 atm. there is appreciable deviation from constancy, and this is undoubtedly due to the failure at high pressures of the fugacity rule for mixtures, as given by equation (32.10).[2]

32e. Homogeneous Reactions in Liquid Solution.—For a reaction taking place in a homogeneous liquid system, the standard state of unit activity for each substance is taken as the pure liquid at the temperature of the reaction and 1 atm. pressure, as described in § 31b. The activity of each species may then be set equal to the product of its mole fraction and the appropriate activity coefficient in the equilibrium mixture, i.e., $a = N\gamma_N$, so that equation (32.7) gives the expression for the equilibrium constant as

$$K_N = \frac{N_L^l \times N_M^m \times \cdots}{N_A^a \times N_B^b \times \cdots} \cdot \frac{(\gamma_N)_L^l \times (\gamma_N)_M^m \times \cdots}{(\gamma_N)_A^a \times (\gamma_N)_B^b \times \cdots}. \quad (32.25)$$

The approximately constant equilibrium function

$$K_N' = \frac{N_L^l \times N_M^m \times \cdots}{N_A^a \times N_B^b \times \cdots} \quad (32.26)$$

is often employed, but this should be multiplied by the activity coefficient function J_{γ_N}, represented by the second factor in equation (32.25), to give the true equilibrium constant. Unfortunately, no liquid phase reaction has been studied in sufficient detail to enable equation (32.25) to be verified completely, but the equilibrium mole fraction function K_N' has been determined for some esterification processes. The approximate constancy ob-

---

[2] Newton, *Ind. Eng. Chem.*, **27**, 302 (1935); Newton and Dodge, *ibid.*, **27**, 577 (1935); see also, Gillespie, *Chem. Rev.*, **18**, 359 (1936); R. N. Pease, "Equilibrium and Kinetics of Gas Reactions," 1942, pp. 13 et seq.

tained in these cases is mainly due to the partial cancellation of the activity coefficients on both sides of the reaction, so that J_{γ_N} is not very different from unity.

32f. Homogeneous Reactions in Dilute Solution.—Most equilibrium reactions that have been studied in dilute solution involve electrolytes, and for these substances a special standard state is used (§ 37b, IV). However, it is possible to derive a modification of equation (32.25) which is applicable to all reactions in sufficiently dilute solution. If a solution contains n_A moles of A, n_B moles of B, ..., n_L moles of L, n_M moles of M, etc., dissolved in n_0 moles of an inert solvent, the mole fraction of any constituent, A for instance, is given by

$$N_A = \frac{n_A}{n_0 + n_A + n_B + \cdots + n_L + n_M + \cdots} = \frac{n_A}{\sum n}. \quad (32.27)$$

The concentration (molarity) c_A, in moles per liter, of the same constituent is obtained by dividing n_A by the volume of the solution in liters; the latter can be derived from the total mass of the constituents of the solution and its density ρ. Thus, if M_0 is the molecular weight of the solvent, and $M_A, M_B, \ldots, M_L, M_M$, etc., are the molecular weights of the substances taking part in the reaction, the mass of the solution is given by

$$\text{Mass of Solution} = n_0 M_0 + n_A M_A + n_B M_B + \cdots$$
$$+ n_L M_L + n_M M_M + \cdots = \sum nM. \quad (32.28)$$

The volume of the solution in liters is then obtained upon dividing the mass in grams by ρ to obtain the volume in ml., and then by 1000 to convert into liters; thus,

$$\text{Volume of solution} = \frac{\sum nM}{1000\rho} \text{ liters}, \quad (32.29)$$

and consequently, the concentration of A is given by

$$c_A = \frac{n_A}{\text{Vol. of solution}} = \frac{1000 \rho n_A}{\sum nM}. \quad (32.30)$$

If this expression is compared with equation (32.27) it is seen that

$$N_A = \frac{c_A}{1000\rho} \cdot \frac{\sum nM}{\sum n}, \quad (32.31)$$

a result which is of completely general applicability for the relationship between the mole fraction of a given constituent of a solution and its molarity, i.e., its concentration in moles per liter.

If the solution is dilute, the number of moles of solvent is greatly in excess of the total number of moles of the reacting substances, so that

$$\sum nM \approx n_0 M_0 \quad \text{and} \quad \sum n \approx n_0,$$

and hence equation (32.31) reduces to

$$N_A \approx \frac{M_0}{1000\rho} c_A. \qquad (32.32)$$

For a dilute solution, also, ρ is not greatly different from the density ρ_0 of the pure solvent, and since this and its molecular weight M_0 are constant, it follows from equation (32.32) that the *concentration (molarity) of any solute in a dilute solution is approximately proportional to its mole fraction.* It is possible, therefore, to modify equation (32.25) in the following manner. In the first place, if the solution is dilute, and particularly if it contains no electrolytes, the activity coefficient factor may be taken as unity, so that equation (32.25) reduces to (32.26). If the mole fractions are replaced by the corresponding molarities, utilizing equation (32.32), it is found that

$$K'_N = \frac{c_L^l \times c_M^m \times \cdots}{c_A^a \times c_B^b \times \cdots} \cdot \left(\frac{M_0}{1000\rho_0}\right)^{\Delta n}, \qquad (32.33)$$

where Δn, as before, is equal to $(l + m + \cdots) - (a + b + \cdots)$. Since the final factor in the expression is constant, it follows that

$$K'_c = \frac{c_L^l \times c_M^m \times \cdots}{c_A^a \times c_B^b \times \cdots}, \qquad (32.34)$$

where K'_c is approximately constant at equilibrium, provided the solution is dilute.

The use of equation (32.34) may be justified by means of the results obtained by J. T. Cundall (1891) for the dissociation of nitrogen tetroxide in dilute chloroform solution. The equilibrium is represented by

$$N_2O_4 \rightleftharpoons 2NO_2,$$

and the concentration equilibrium function K'_c is given by

$$K'_c = \frac{c_{NO_2}^2}{c_{N_2O_4}}.$$

The data over a range of concentrations, for a temperature of 8.2° C, are given in Table XXIII.[3] The relative constancy of the values of K'_c in the

TABLE XXIII

$c_{N_2O_4}$	c_{NO_2}	K'_c
0.129 mole liter^{-1}	1.17×10^{-3} mole liter^{-1}	1.07×10^{-5}
0.227	1.61	1.14
0.324	1.85	1.05
0.405	2.13	1.13
0.778	2.84	1.04

last column shows that the solutions are sufficiently dilute for the activity coefficient factor to be close to unity.

[3] Cundall, *J. Chem. Soc.*, **59**, 1076 (1891); **67**, 794 (1895).

Another form of representation of the equilibrium in dilute solutions is based on the expression of concentrations in terms of **molality**, that is, *the number of moles of solute to 1000 grams of solvent*. The mass of the solvent corresponding to n_A moles of solute A is $n_0 M_0$; hence the molality m_A is given by

$$m_A = \frac{1000 n_A}{n_0 M_0}, \tag{32.35}$$

and consequently,

$$n_A = \frac{n_0 M_0}{1000} m_A. \tag{32.36}$$

If this result is introduced into equation (32.27), the mole fraction is

$$N_A = \frac{m_A}{1000} \cdot \frac{n_0 M_0}{\sum n}. \tag{32.37}$$

If the solution is dilute, $\sum n$ may be taken as approximately equal to n_0, as seen above; equation (32.37) then reduces to

$$N_A \approx \frac{M_0}{1000} m_A, \tag{32.38}$$

so that *the molality of any solute in a dilute solution is approximately proportional to its mole fraction in that solution*. It is thus possible to derive from equation (32.25) a molality equilibrium function, viz.,

$$K'_m = \frac{m_L^l \times m_M^m \times \cdots}{m_A^a \times m_B^b \times \cdots}. \tag{32.39}$$

32g. Chemical Equilibria in Heterogeneous Systems.—The most common heterogeneous chemical equilibria involve gases and solids, or liquid solutions and solids. In all cases the general equation (32.7) is applicable, but the situation is considerably simplified by the universally adopted convention that *the activity of every pure solid or pure liquid is to be taken as unity at atmospheric pressure*. This is, of course, in accord with the usual choice of the standard state of unit activity of a liquid as the pure liquid at the same temperature and 1 atm. pressure; an analogous standard state, viz., pure solid at 1 atm. pressure, is chosen for substances present in the solid form. It should be understood that if the solid or liquid involved in a heterogeneous system is not present in the pure state, but as solid or liquid solution, its activity is no longer unity; as a first approximation, it may then be taken as equal to the mole fraction.

The result of taking the activity of a pure solid or pure liquid as unity is that the corresponding factor may be omitted from the expression for the equilibrium constant. Strictly speaking, this is only true if the total pressure on the system is 1 atm., in accordance with the defined standard states. At other pressures, the activities of solid and liquid phases are constant, but their values are not unity. If the corresponding factors are then omitted from the expression for the equilibrium constant, the result is a quantity which varies with pressure, as well as with temperature. This quantity is not the true equilibrium constant, and it should not be used in equations such as (33.6), etc., to be considered shortly, to determine standard free energy changes. Since the effect of pressure on the activity of a liquid or solid is usually small [cf. equation (31.7)], and there is frequently a partial cancellation of terms for reactants and products, the constants obtained at pressures other than atmospheric, on the assumption that the activities of pure solids and liquids are unity, do not differ appreciably from the true equilibrium constants. In practice, therefore, the variation of the constant with pressure is frequently neglected.

A simple illustration of a heterogeneous reaction, in which, incidentally, the effect of pressure is likely to be small, is

$$\text{Fe}(s) + \text{H}_2\text{O}(g) = \text{FeO}(s) + \text{H}_2(g),$$

the Fe and FeO being present as pure solids.[4] Consequently, strictly at a total pressure of 1 atm., and with a good degree of approximation at all moderate pressures,

$$K = \frac{a_{\text{H}_2}}{a_{\text{H}_2\text{O}}},$$

and hence,

$$K_f = \frac{f_{\text{H}_2}}{f_{\text{H}_2\text{O}}} \approx \frac{p_{\text{H}_2}}{p_{\text{H}_2\text{O}}} \approx \frac{N_{\text{H}_2}}{N_{\text{H}_2\text{O}}}.$$

Numerous instances of reactions of this type are to be found in the literature of physical chemistry, and need not be considered here. It is of interest to mention, however, that vaporization (cf. § 33g), freezing, solubility, and other heterogeneous physical equilibria can be treated in the same manner, although other methods are often preferable (see Chapter XIV).

33. Free Energy Change in Chemical Reactions

33a. The Reaction Isotherm.—Consider, once again, the general reaction

$$a\text{A} + b\text{B} + \cdots \rightleftharpoons l\text{L} + m\text{M} + \cdots.$$

The free energy of any mixture of a moles of A, b moles of B, etc., is given in terms of the chemical potentials, i.e., the partial molar free energies, by means of equation (26.6), as

$$F(\text{reactants}) = a\mu_\text{A} + b\mu_\text{B} + \cdots,$$

and the free energy of a mixture of l moles of L, m moles of M, etc., is

$$F(\text{products}) = l\mu_\text{L} + m\mu_\text{M} + \cdots,$$

at constant temperature, pressure and composition, in each case. These expressions are applicable to the systems of reactants and products *at any*

[4] For review, see Austin and Day, *Ind. Eng. Chem.*, **33**, 23 (1941).

arbitrary concentrations, not necessarily the equilibrium values. The free energy increase ΔF accompanying the reaction depicted above, at constant temperature and pressure, is thus given by

$$\Delta F_{T,P} = F(\text{products}) - F(\text{reactants})$$
$$= (l\mu_L + m\mu_M + \cdots) - (a\mu_A + b\mu_B + \cdots). \quad (33.1)$$

Utilizing the general equation (32.5), i.e., $\mu = \mu^0 + RT \ln a$, for the chemical potential of any substance, μ^0 being the value in the standard state, it follows from equations (33.1) and (26.29) that

$$\Delta F_{T,P} = \Delta F_T^0 + RT \ln \frac{a_L^l \times a_M^m \times \cdots}{a_A^a \times a_B^b \times \cdots}, \quad (33.2)*$$

where *the activities are any arbitrary values*.

It is evident from a comparison of equation (32.6) and (32.7) that

$$\Delta F_T^0 = -RT \ln K. \quad (33.3)$$

The same result can be derived from equation (33.2), for if the arbitrary activities were the equilibrium values, the last term would be identical with $RT \ln K$, while $\Delta F_{T,P}$ would then be zero, thus leading to the expression in equation (33.3). If this is substituted in (33.2), it is found that, in general,

$$\Delta F_{T,P} = -RT \ln K + RT \ln \frac{a_L^l \times a_M^m \times \cdots}{a_A^a \times a_B^b \times \cdots} \quad (33.4)$$

$$= -RT \ln K + RT \ln J_a, \quad (33.5)$$

where, as indicated in the footnote in § 32c, J_a is a function of the same form as the equilibrium constant involving, in this case, the arbitrary values of the activities of reactants and products. The important equations (33.4) and (33.5) are forms of the **reaction isotherm** first derived by J. H. van't Hoff (1886). *It gives the increase of free energy accompanying the transfer of reactants at any arbitrary concentrations (activities) to products at arbitrary concentrations (activities).* As seen above, if the arbitrary concentrations are chosen so as to correspond to the equilibrium state, the two terms on the right-hand side become identical, and hence $\Delta F_{T,P}$ is then zero.

33b. Standard Free Energy of Reaction.—The standard free energy increase ΔF_T^0, often referred to as the standard free energy, of the reaction at a specified temperature T, is given by equation (33.3) which, because of its importance, will be repeated here; thus,

$$\Delta F_T^0 = -RT \ln K. \quad (33.6)$$

It gives the increase of free energy when the reactants, all in their standard states of unit activity, are converted into the products, in their standard states. The relationship (33.6) is extremely useful, as will be seen shortly,

---

* As stated earlier, ΔF^0 is independent of pressure, if the usual definitions of standard states are employed; hence, the symbol ΔF_T^0 is sufficient.

for it provides a convenient method for tabulating the equilibrium constants of reactions, especially for processes which appear to go to virtual completion.

Substitution of the value for ΔF_T^0, given by equation (33.6), into (33.5), leads to

$$\Delta F_{T,P} = \Delta F_T^0 + RT \ln J_a, \qquad (33.7)$$

and the reaction isotherm is frequently encountered in this form. In general, the subscripts T and P are omitted, it being understood that constant temperature and pressure are implied.

As seen earlier, the actual values of the activities and of the equilibrium constant depend on the choice of the standard state; this must also be the case, therefore, for the standard free energy changes, as defined above. *It is consequently necessary to specify clearly the standard state employed in every case.* For gas reactions, the standard state is usually that of unit fugacity, i.e., ideal gas at 1 atm. pressure; the form of equation (33.7) is then

$$\Delta F = \Delta F_f^0 + RT \ln J_f, \qquad (33.8)$$

where

$$\Delta F_f^0 = -RT \ln K_f. \qquad (33.9)$$

If the gases behave ideally, or approximately ideally, the fugacities may be replaced by the corresponding partial pressures, so that equation (33.8) becomes

$$\Delta F \approx \Delta F_p^0 + RT \ln J_p, \qquad (33.10)$$

in which form the reaction isotherm is sometimes employed for gas reactions.

If the standard state is chosen as the ideal gas at unit concentration, the corresponding equation is

$$\Delta F = \Delta F_c^0 + RT \ln J_c \qquad (33.11)$$

where

$$\Delta F_c^0 = -RT \ln K_c, \qquad (33.12)$$

the definition of K_c being given by equation (32.19).

For reactions in solution, the reaction isotherm may take the form

$$\Delta F = \Delta F_N^0 + RT \ln J_N, \qquad (33.13)$$

where

$$\Delta F_N^0 = -RT \ln K_N, \qquad (33.14)$$

the equilibrium constant K_N being defined by equation (32.25). In connection with these expressions, the standard state of each substance is chosen so that the activity is equal to the mole fraction in the reference state.

33c. The Direction of Chemical Change.—From the chemical point of view the great significance of the reaction isotherm lies in the fact that it provides a means of determining whether a particular reaction is possible or not under a given set of conditions. It was seen in § 25f that for a process, at constant pressure and temperature, to be spontaneous it must be accompanied by a decrease of free energy; that is to say, for a spontaneous reaction, $\Delta F_{T,P}$ must be negative. If the value of ΔF under a given set of conditions is positive, the reaction cannot possibly take place under those conditions, although it may do so if the conditions are altered.

An examination of the general equation (33.5) shows that the sign of ΔF depends on the relative values of the quanities K and J_a; if the reaction is to be spontaneous, the latter must be less than the former, for then ΔF will be negative. If the conditions are such that J_a is greater than K, the reaction will not be possible under those conditions, but there are two ways in which the situation can be changed. First, the arbitrary activities, i.e., pressures or concentrations, of the substances concerned in the reaction may be changed so as to decrease the value of J_a below that of K; this means that the activities of the products must be decreased or those of the reactants increased, or both. Second, the temperature may be changed in such a manner as to increase K; in this way, without altering the arbitrary activity function J_a, the latter may become less than K, so that ΔF is negative and the reaction is possible. The foregoing arguments are applicable to processes of all types; for heterogeneous reactions, the activity of each pure solid is taken as unity, as mentioned earlier. This is equivalent to its omission from $\ln J_a$, since the logarithm of unity is zero.

An analogous significance to that just described can be given to the standard free energy ΔF^0 of a reaction. If ΔF^0 is negative, the reaction with all the reactants and the products in their respective chosen standard states can take place spontaneously. On the other hand, if ΔF^0 is positive, the reaction cannot occur under those conditions.

It can now be seen that thermodynamics provides a simple solution of the problem, concerning the conditions determining the direction of chemical change, which had puzzled investigators for many years until J. H. van't Hoff (1883) indicated the importance of the free energy change. Incidentally, it may be mentioned that the same idea was undoubtedly in the minds of J. W. Gibbs (1876) and of H. von Helmholtz (1882), although it was not stated explicitly.

Problem: It is required to pass carbon monoxide at 10 atm. and water vapor at 5 atm. pressure into a reaction chamber at 700° C and to withdraw carbon dioxide and hydrogen at partial pressures of 1.5 atm. Is this possible theoretically? The equilibrium constant for the reaction is known to be 0.71.

The reaction is

$$CO(g) + H_2O(g) = CO_2(g) + H_2(g),$$

and since there is no change in the number of molecules, the equilibrium constant is independent of the standard state. However, since the arbitrary concentrations (activities) of the reactants and products are stated in terms of pressures, it will be assumed that the equilibrium constant is K_f or K_p. For the present purpose, it is sufficient to assume ideal behavior, so that the free energy change is given by equation (33.10); this takes the form

$$\Delta F = -RT \ln K_p + RT \ln \frac{p_{CO_2} \times p_{H_2}}{p_{CO} \times p_{H_2O}},$$

where the p's are the arbitrary values given in the problem. All that it is required to know is the relative values of K_p and $J_p{}^*$, but it is instructive to calculate the actual free energy change. This is usually expressed in calories, so that R is taken as 1.987 cal. deg.$^{-1}$ mole^{-1}; the temperature T is 700° C or 973° K, and K_p is 0.71 as given above. Hence, converting the logarithms,

$$\Delta F = -4.576 \times 973 \log 0.71 + 4.576 \times 973 \log \frac{1.5 \times 1.5}{10 \times 5}$$
$$= -4.576 \times 973 \, (\log 0.71 - \log 0.045)$$
$$= -5{,}340 \text{ cal.}$$

Since ΔF is negative, the process is theoretically possible.

The reaction isotherm is of particular value in the study of chemical processes, but it has also been used in connection with physical changes, to which it is equally applicable. The general criterion of a decrease of free energy for a process to be spontaneous is applicable, of course, to all processes, physical or chemical.

Although any reaction accompanied by a free energy decrease is theoretically possible, it is important to note that this is no indication that the process will occur with a measurable speed. In a series of analogous reactions the rates at which the processes occur are often roughly in the order of the free energy decreases, but in general, for different reactions, there is no connection between the magnitude of the decrease of free energy and the rate of the reaction. For example, at ordinary temperature and pressure the free energy change for the combination of hydrogen and oxygen has a very large negative value, yet the reaction is so slow that no detectable amount of water would be formed in years. The passage of an electric spark or the presence of a suitable catalyst, however, facilitates in this case the occurrence of a reaction which the free energy change shows to be theoretically possible.

33d. Variation of Equilibrium Constant with Pressure.—The extent to which the equilibrium constant of a reaction is affected by pressure can be determined by utilizing equation (33.6), in the form

$$\ln K = -\frac{\Delta F^0}{RT}. \qquad (33.15)$$

Differentiation with respect to the total pressure P, at constant temperature, gives

$$\left(\frac{\partial \ln K}{\partial P}\right)_T = -\frac{1}{RT}\left[\frac{\partial (\Delta F^0)}{\partial P}\right]_T = 0.$$

Since standard states have been defined so as to be independent of the pressure of the system, the standard free energy change ΔF^0, and consequently the equilibrium constant K, will not vary with the external pressure.

---

* The latter is $1.5 \times 1.5/10 \times 5$, i.e., 0.045, and since this is less than K_p, i.e., 0.71, the reaction is obviously possible.

It should be noted that although K, and hence K_f, is independent of the pressure, the equilibrium functions K'_f and K'_p for a homogeneous gas reaction should vary with the pressure, as is actually found to be the case (see Table XXII). This is, of course, due to the dependence of the activity coefficients on the pressure at constant temperature, as indicated in § 32d. Since a change of pressure is usually accompanied by a change in the composition of the system it is not possible to use equation (30.19) to determine the effect of pressure on the activity coefficient factor. The calculation can be made by thermodynamic methods, along lines which will be described in Chapter XIV in connection with physical equilibria, but the results have no direct practical application without introducing a number of approximations.

Similar considerations apply to the influence of pressure on other equilibrium functions, such as K'_N, etc., for reactions in solution. In general, however, the changes due to increase of pressure can be neglected at all moderate pressures.

It is opportune to recall here the remarks made in § 32g in connection with heterogeneous reactions. If the activities of pure solids and liquids taking part in the process are taken as unity, then the true equilibrium constant is obtained only if the total pressure of the system is 1 atm. Unless allowance is made for the effect of pressure on the activities, the constants obtained at other pressures vary with pressure in a manner dependent on the volume change of the solid and liquid phases involved in the reaction [cf. equation (31.7)].

33e. Effect of Pressure on Position of Equilibrium.—Although the true equilibrium constant of a gas reaction is not affected by the pressure, the actual *position of equilibrium* will be altered if the reaction is one involving a change in the number of molecules. The subject is dealt with adequately in physical chemistry texts, in connection with the Le Chatelier principle of mobile equilibrium. As is well known, increase of pressure will favor the reaction which is accompanied by a decrease in the number of molecules. Such a change is necessary in order that the equilibrium constant may remain unaltered. However, there is another aspect of this matter that merits attention here. Because of the effect of pressure on K'_p, there will be an additional change in the equilibrium composition. The increase of K'_p with pressure in the ammonia equilibrium, for example, means that there is a larger proportion of ammonia present when equilibrium is attained at high pressures than would have been the case if the system had consisted of ideal gases. This factor, therefore, operates to the advantage of the yield of ammonia in the Haber process. The effect of pressure on K'_p due to departure from ideal behavior will alter the position of equilibrium even when there is no change in the number of molecules.

Problem: Estimate qualitatively the effect of a pressure of 500 atm. on the gaseous equilibrium $CO + H_2O \rightleftharpoons CO_2 + H_2$ at 600° C.

If the four gases behaved ideally, pressure would have no effect on the position of equilibrium, since there is no change in the number of molecules accompanying the homogeneous reaction. However, because of deviations from ideal behavior, a shift in the composition is possible when the pressure is changed. The equilibrium constant may be written as

$$K_f = \frac{p_{CO_2} \times p_{H_2}}{p_{CO} \times p_{H_2O}} \cdot \frac{\gamma_{CO_2} \times \gamma_{H_2}}{\gamma_{CO} \times \gamma_{H_2O}},$$

and it is desired to find the value of the activity coefficient factor J_γ at 600° C, i.e., 873° K, and 500 atm. At 1 atm. pressure, the value may be taken as virtually unity. At the higher pressure use may be made of the fugacity chart in Fig. 18, and for this purpose the reduced temperature (θ) and pressure (π) of each gas are required.

For CO_2: T_c is 304° K, $\theta = 873/304 = 2.87$
            P_c is 72.9 atm., $\pi = 500/72.9 = 6.85$ } $\gamma_{CO_2} = 1.09$.

For H_2: T_c is 33.2° K, $\theta = 873/(33.2 + 8) = 21.2$
            P_c is 12.8 atm., $\pi = 500/(12.8 + 8) = 24.0$ } $\gamma_{H_2} = 1.10$.

For CO: T_c is 134° K, $\theta = 873/134 = 6.51$
            P_c is 34.6 atm., $\pi = 500/34.6 = 14.4$ } $\gamma_{CO} = 1.23$.

For H_2O: T_c is 647° K, $\theta = 873/647 = 1.35$
             P_c is 218 atm., $\pi = 500/218 = 2.29$ } $\gamma_{H_2O} = 0.77$.

The values of γ given at the right are those based on the approximation that the activity coefficient of a gaseous constituent of a mixture is equal to the value for the pure gas at the total pressure of the mixture. In this case,

$$J_\gamma = \frac{1.09 \times 1.10}{1.23 \times 0.77} = 1.27.$$

Since J_γ has increased when the pressure is increased to 500 atm., it follows that the product $p_{CO_2} \times p_{H_2}$ must decrease, or $p_{CO} \times p_{H_2O}$ increase, in order to retain K_f constant. Thus, increase of pressure to 500 atm. causes the equilibrium to shift to the left, because of deviations from ideal behavior.

33f. Variation of Equilibrium Constant with Temperature.—The effect of temperature on the equilibrium constant may be determined by differentiation of equation (33.15) with respect to temperature. Since K is independent of the external pressure, the constant pressure condition and the accompanying partial differential notation can be omitted, so that

$$\frac{d \ln K}{dT} = -\frac{1}{R} \cdot \frac{d(\Delta F^0/T)}{dT}$$

and hence, by equation (25.33), it follows that

$$\frac{d \ln K}{dT} = \frac{\Delta H^0}{RT^2}, \qquad (33.16)$$

where

$$\Delta H^0 = (lH_L^0 + mH_M^0 + \cdots) - (aH_A^0 + bH_B^0 + \cdots)$$

or, in general,

$$\Delta H^0 = \sum nH^0(\text{final state}) - \sum nH^0(\text{initial state}).$$

The equation (33.16) is the rigorous form of a relationship originally derived by J. H. van't Hoff; it is consequently sometimes referred to as the **van't Hoff equation**. It is to be understood, of course, that the particular standard states chosen for the activities of the reactants and products in the expression for K, i.e., in equation (32.7), must also apply to ΔH^0, as defined above.

For a *homogeneous gas reaction* the equilibrium constant is K_f and equation (33.16) takes the form

$$\frac{d \ln K_f}{dT} = \frac{\Delta H^0}{RT^2}, \tag{33.17}$$

where ΔH^0 is the heat of reaction adjusted to a low value of the total pressure, when the gases behave ideally. It was seen in § 20e that at moderate pressures the heat content of a real gas does not differ appreciably from the ideal gas value, and for most purposes it is possible to replace ΔH^0 in equation (33.17) by ΔH, the ordinary heat of reaction in the vicinity of atmospheric pressure.

If the standard state is that of an ideal gas at unit molar concentration, the equilibrium constant K_c is related to K_f by equation (32.18), viz.,

$$K_f = K_c \times (RT)^{\Delta n},$$

so that, taking logarithms,

$$\ln K_f = \ln K_c + \Delta n \ln R + \Delta n \ln T,$$

and hence, upon differentiation with respect to temperature, and utilizing equation (33.16),

$$\frac{d \ln K_f}{dT} = \frac{d \ln K_c}{dT} + \frac{\Delta n}{T} = \frac{\Delta H^0}{RT^2}.$$

It follows, therefore, that

$$\frac{d \ln K_c}{dT} = \frac{\Delta H^0 - \Delta n(RT)}{RT^2}. \tag{33.18}$$

Since ΔH^0 is the heat of reaction at constant pressure, adjusted to ideal behavior, it follows that $\Delta H^0 = \Delta E^0 + P\Delta V = \Delta E^0 + \Delta n(RT)$, so that equation (33.18) becomes

$$\frac{d \ln K_c}{dT} = \frac{\Delta E^0}{RT^2}. \tag{33.19}$$

The variation of the functions K_f' and K_p' with temperature is not quite the same as that given above for K_f, because of the change in the activity coefficient factor with temperature. Since the composition of the system does not remain constant when the temperature is changed, equation (30.21) cannot be employed, but a method similar to that described in §§ 34f, 34i could be used. However, without introducing approximations, the results are of no practical value. Provided the pressures are not too high, it is therefore the common practice to employ equation (33.17) in the approximate form

$$\frac{d \ln K'_p}{dT} \approx \frac{\Delta H}{RT^2}, \tag{33.20}$$

where ΔH is taken as the ordinary heat of reaction at constant pressure, usually 1 atm. The variation of K'_p with pressure is ignored.

For a *homogeneous reaction in a liquid system* the equilibrium constant is conveniently expressed in the form of equation (32.25), the standard state of each substance being the pure liquid at the equilibrium temperature and 1 atm. pressure. Differentiation with respect to temperature then gives [cf. equation (33.16)]

$$\frac{d \ln K_N}{dT} = \frac{\Delta H^0}{RT^2}, \tag{33.21}$$

where ΔH^0 is the heat of reaction starting with pure liquid reactants and ending with pure liquid products, at 1 atm. pressure and temperature T. It will be noted that, as before, the constant pressure condition has been omitted from equation (33.21), since the equilibrium constant on the basis of the chosen standard states is independent of the pressure. For many purposes it is sufficiently accurate to replace K_N by the equilibrium function K'_N involving mole fractions; at the same time ΔH, the ordinary heat of reaction is substituted for ΔH^0, giving the approximate result

$$\frac{d \ln K'_N}{dT} \approx \frac{\Delta H}{RT^2}. \tag{33.22}$$

For reactions in dilute solution, it is convenient to use the concentration equilibrium function K'_c, defined by equation (32.34). Upon taking logarithms, it follows from equations (32.33) and (32.34) that

$$\ln K'_N = \ln K'_c + \Delta n \ln \left(\frac{M_0}{1000\rho_0} \right). \tag{33.23}$$

Provided the temperature range is not too great, the variation of the density ρ_0 of the solvent will not be large, and the last term in equation (33.23) may be taken as approximately independent of temperature. The difference between $d \ln K'_N/dT$ and $d \ln K'_c/dT$ may thus be neglected, and it is possible to write

$$\frac{d \ln K'_c}{dT} \approx \frac{\Delta H}{RT^2} \tag{33.24}$$

It is important that this result, for a reaction in dilute solution, should not be confused with equation (33.19), which gives the variation of K_c for a homogeneous gaseous equilibrium; the latter involves ΔE^0, and not ΔH.

33g. Heterogeneous Reactions: Influence of Temperature.

In the foregoing discussion various forms of the van't Hoff equation, some exact and others approximate, have been derived for equilibria of different types. It is necessary to bear in mind, however, that in the form of equation (33.16) the relationship is applicable precisely to any equilibrium, chemical or physical.

33g. Heterogeneous Reactions: Influence of Temperature.—The equations derived above for homogeneous reactions apply equally to heterogeneous equilibria. The various ΔH values refer to the complete reaction, including the solid phases, although, as mentioned in § 32g, the activity is unity for every pure solid taking part in the reaction; the corresponding factor may thus be omitted from the expression for the equilibrium constant. Because they present certain features of interest, brief reference will be made here to some heterogeneous equilibria, and others will be considered later.

In reactions involving one or more solids and one gas, as, for example,

$$CaCO_3(s) = CaO(s) + CO_2(g)$$
$$\tfrac{1}{2}CuSO_4 \cdot 5H_2O(s) = \tfrac{1}{2}CuSO_4 \cdot 3H_2O(s) + H_2O(g),$$

the true equilibrium constant, e.g., K_f, is equal to the fugacity of the gas in equilibrium with the solid *at a total pressure of 1 atm.* (cf. § 32g). The variation of $\ln f$ with the temperature is then given by equation (33.16), with ΔH^0 equal to the heat of reaction adjusted to the value for the gas at low pressure. In studying these equilibria, the experimentally determined quantity is the partial pressure p, which is equivalent to K'_p, and as a general rule no attempt is made to keep the total pressure at 1 atm., or even constant, although the dissociation pressure should vary to some extent with the total pressure. The dependence of the pressure p on temperature is thus usually expressed by the approximate equation (33.20) in the form

$$\frac{d \ln p}{dT} = \frac{\Delta H}{RT^2}, \qquad (33.25)$$

where ΔH is the ordinary heat of reaction at the temperature T.

Heterogeneous physical equilibria, e.g., between a pure solid and its vapor or a pure liquid and its vapor, can be treated in a manner similar to that just described. If the total pressure of the system is 1 atm., the fugacity of the vapor is here also equivalent to the equilibrium constant. The variation of $\ln f$ with temperature is again given by equation (33.16), where ΔH^0 is now the *ideal* molar heat of vaporization of the liquid (or of sublimation of the solid) at the temperature T and a pressure of 1 atm. If the total pressure is not 1 atm., but is maintained constant at some other value, the dependence of the fugacity on the temperature can be expressed by equation (29.22), since the solid or liquid is in the pure state; thus,

$$\left(\frac{\partial \ln f}{\partial T}\right)_P = \frac{\Delta H^*_v}{RT^2}, \qquad (33.26)$$

where, according to the remarks in § 29g, the numerator on the right-hand side is equal to the ideal molar heat of vaporization (or sublimation) at temperature

T and pressure P. It should be noted that equation (33.26) is the exact form of the Clausius-Clapeyron equation (27.12). If the vapor is assumed to be ideal, so that the fugacity may be replaced by the vapor pressure, and the total pressure is taken as equal to the equilibrium pressure, the two equations become identical. In this simplification the assumption is made that the activity of the liquid or solid does not vary with pressure; this is exactly equivalent to the approximation used in deriving the Clausius-Clapeyron equation, that the volume of the liquid or solid is negligible.

33h. Integration of the van't Hoff Equation.—If the change of heat content ΔH^0 were independent of temperature, general integration of the van't Hoff equation would give [cf. equation (27.15)]

$$\ln K = -\frac{\Delta H^0}{RT} + \text{constant},$$

or

$$\log K = -\frac{\Delta H^0}{4.576 T} + C, \tag{33.27}$$

so that the plot of $\log K$ against $1/T$ should be a straight line of slope $-\Delta H^0/4.576$ with ΔH^0 in calories. Integration between the temperature limits of T_1 and T_2, the corresponding values of the equilibrium constant being K_1 and K_2, gives

$$\ln \frac{K_2}{K_1} = \frac{\Delta H^0}{R}\left(\frac{T_2 - T_1}{T_1 T_2}\right)$$

or, converting the logarithms and expressing R in cal. deg.$^{-1}$ mole^{-1},

$$\log \frac{K_2}{K_1} = \frac{\Delta H^0}{4.576}\left(\frac{T_2 - T_1}{T_1 T_2}\right). \tag{33.28}$$

This equation, or (33.27), may be used to evaluate the heat of reaction if the equilibrium constant is known at two temperatures; alternatively, if ΔH^0 is available, the equilibrium constant at any temperature can be determined provided it is known at one particular temperature.

The results obtained in this manner are approximate only, for heats of reaction are known to vary with the temperature, as seen in §§ 12j, 12k.[5] The standard change of heat content can be expressed as a function of temperature in accordance with equation (12.16), so that

$$\Delta H^0 = \Delta H_0^0 + \Delta \alpha T + \tfrac{1}{2}\Delta \beta T^2 + \tfrac{1}{3}\Delta \gamma T^3 + \cdots, \tag{33.29}$$

where $\Delta \alpha$, $\Delta \beta$, $\Delta \gamma$, etc., are derived from the heat capacities of the reactants and products and their variation with temperature. If this expression is

[5] For a more exact significance of ΔH^0 in equation (33.28), even when the heat content varies with temperature, see Douglas and Crockford, *J. Am. Chem. Soc.*, **57**, 97 (1935); Walde, *J. Phys. Chem.*, **45**, 431 (1939).

substituted into (33.16), the result is

$$\frac{d \ln K}{dT} = \frac{\Delta H_0^0}{RT^2} + \frac{\Delta \alpha}{RT} + \frac{\Delta \beta}{2R} + \frac{\Delta \gamma}{3R} T + \cdots,$$

and upon integration it is found that

$$\ln K = -\frac{\Delta H_0^0}{RT} + \frac{\Delta \alpha}{R} \ln T + \frac{\Delta \beta}{2R} T + \frac{\Delta \gamma}{6R} T^2 + \cdots + I', \quad (33.30)$$

where K is either K_f or K_N, for reactions in the gas or liquid phase, respectively, and I' is the integration constant.

If ΔH^0 and $\Delta \alpha$, $\Delta \beta$, etc., are available from thermal measurements, it is possible to derive ΔH_0^0 by utilizing the procedure described in § 12k; if K is known at any one temperature, it is possible to evaluate the integration constant I', and the variation of $\ln K$ (or $\log K$) with temperature can then be expressed in the form of equation (33.30). The accuracy of the resulting expression is limited largely by the thermal data, for these are often not known with great certainty. Care should be taken to ensure that the standard states used in connection with the heat of reaction ΔH^0 are also those employed for the equilibrium constant. Actually, the standard states chosen in §§ 30b, 31b correspond with those almost invariably employed in both equilibrium studies and heat of reaction measurements.

Problem: Taking the equilibrium constant (K_f) for the $\frac{1}{2}N_2 + \frac{3}{2}H_2 \rightleftharpoons NH_3$ equilibrium to be 0.00655 at 450° C (Table XXII), and utilizing the heat of reaction and heat capacity data in § 12k, derive a general expression for the variation of the equilibrium constant with temperature. Determine the value of K_f at 327° C.

According to the result obtained in the problem in § 12k, which is supposed to refer to the standard states, it follows that

$$\Delta H^0 = -9.13 - 7.46 \times 10^{-3}T + 3.69 \times 10^{-6}T^2 - 0.47 \times 10^{-9}T^3 \text{ kcal.}$$
$$= -9{,}130 - 7.46T + 3.69 \times 10^{-3}T^2 - 0.47 \times 10^{-6}T^3 \text{ cal.,}$$

and hence,

$$\Delta H_0^0 = -9{,}130; \quad \Delta \alpha = -7.46; \quad \tfrac{1}{2}\Delta \beta = 3.69 \times 10^{-3}; \quad \tfrac{1}{3}\Delta \gamma = -0.47 \times 10^{-6} \text{ cal.}$$

Upon insertion of these values into equation (33.30), it is seen that

$$\ln K_f = \frac{9{,}130}{RT} - \frac{7.46}{R} \ln T + \frac{3.69 \times 10^{-3}}{R} T - \frac{0.47 \times 10^{-6}}{2R} T^2 + I'.$$

At 450° C, i.e., 723° K, the value of K_f is 0.00655, and utilizing these data it is found that I' is 12.07, so that, taking R as 1.987 cal. deg.$^{-1}$ mole^{-1}, the expression for K_f becomes

$$\ln K_f = 2.303 \log K_f = \frac{4{,}600}{T} - 8.64 \log T$$
$$+ 1.86 \times 10^{-3}T - 0.12 \times 10^{-6}T^2 + 12.07.$$

At 327° C, i.e., 600° K, ln K_f is thus found to be -3.19, and hence K_f is 0.041. (The experimental value is about 0.040.)

It can be seen that if $\Delta\alpha$, $\Delta\beta$, etc., are known, the right-hand side of equation (33.30) still contains two unknowns, viz., ΔH_0^0 and I'. If the equilibrium constant K has been determined at two or more temperatures, therefore, equation (33.30) can be solved for the unknowns, and hence ln K can be expressed as a function of the temperature. If the value of K is known at several temperatures, a simple graphical procedure is possible which permits the use of all the available data. By writing equation (33.30) in the form

$$-R \ln K + \Delta\alpha \ln T + \tfrac{1}{2}\Delta\beta T + \tfrac{1}{6}\Delta\gamma T^2 + \cdots = \frac{\Delta H_0^0}{T} - I'R, \quad (33.31)$$

it is seen that the plot of the left-hand side against $1/T$ should yield a straight line of slope equal to ΔH_0^0. With ΔH_0^0 known, the integration constant I' can be readily evaluated.[6]

Attention may be called to the fact that the procedure just described provides a useful method for determining heats of reaction from equilibrium constant data. Since ΔH_0^0 has been derived above, and $\Delta\alpha$, $\Delta\beta$, etc., are available, it is possible to express ΔH^0 as a function of the temperature by means of equation (33.29). The standard heat of reaction can then be determined at any temperature in the range in which $\Delta\alpha$, $\Delta\beta$, etc., are applicable.

Problem: Utilize the following values of K_f for the ammonia equilibrium to derive an expression for ΔH^0 as a function of the temperature. Determine the standard heat of reaction, i.e., heat of formation of ammonia, at 25° C.

Temp.	350°	400°	450°	500° C
K_f	2.62×10^{-2}	1.27×10^{-2}	6.55×10^{-3}	3.78×10^{-3}

Inserting the values for $\Delta\alpha$, $\Delta\beta$ and $\Delta\gamma$ given in the preceding problem, equation (33.31) becomes

$$-4.576 \log K - 7.46 \ln T + 3.69 \times 10^{-3} T - \tfrac{1}{2} \times 0.47 \times 10^{-6} T^2 = \frac{\Delta H_0^0}{T} - I'R.$$

The values of the left-hand side (L. H. S.), obtained from the equilibrium constants given above, and the corresponding $1/T$'s are then found to be as follows:

L. H. S.	-38.56	-37.53	-36.58	-35.82
$1/T$	1.605×10^{-3}	1.486×10^{-3}	1.383×10^{-3}	1.294×10^{-3}

These results are plotted in Fig. 19; the points are seen to fall almost exactly on a straight line, the slope of which is $-8,810$ cal. This is consequently the value of ΔH_0^0, so that by equation (33.29),

$$\Delta H^0 = -8,810 - 7.46T + 3.69 \times 10^{-3} T^2 - 0.47 \times 10^{-6} T^3 \text{ cal.}$$

[6] Randall, et al., J. Chem. Ed., 8, 1062 (1931); Ind. Eng. Chem., 23, 388 (1931).

At 25° C, i.e., 298° K, it is found that

$$\Delta H^0 = -\,8{,}810 - 2{,}223 + 328 - 12 = -\,10{,}717 \text{ cal.}$$

(It will be noted that ΔH_0^0 obtained here differs by about 320 cal. from that adopted in the preceding problem. The discrepancy may be due partly to an error in the thermal ΔH^0 value upon which ΔH_0^0 in the previous problem is based, and partly to uncertainties in $\Delta\alpha$, $\Delta\beta$, etc., which give the variation of heat capacities with temperature. It is also possible that the K_f values are not exact.)

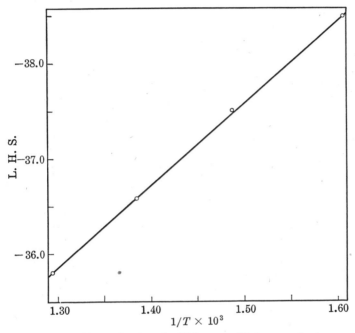

Fig. 19. Determination of ΔH_0^0 from equilibrium constants

33i. Variation of Standard Free Energy with Temperature.—If equation (33.30) is multiplied through by RT, the result is $RT \ln K$; by equation (33.6) this is equal to $-\Delta F^0$, the standard free energy of the reaction under consideration. It follows, therefore, that

$$\Delta F^0 = \Delta H_0^0 - \Delta\alpha T \ln T - \tfrac{1}{2}\Delta\beta T^2 - \tfrac{1}{6}\Delta\gamma T^3 - \cdots + IT, \quad (33.32)$$

where the constant I is equal to $-I'R$. This highly important equation gives the variation of the standard free energy of a reaction with temperature, and since $-\Delta F^0$ is equal to $RT \ln K$ it is also, in effect, an expression for the equilibrium constant as a function of temperature. The value of ΔF^0 given by equation (33.32) applies to the reaction as written for the determination of the equilibrium constant. For example, if the ammonia equilibrium is written as

$$\tfrac{1}{2}N_2 + \tfrac{3}{2}H_2 \rightleftharpoons NH_3$$

the equilibrium constant K_f is $f_{NH_3}/f_{N_2}^{\frac{1}{2}} \times f_{H_2}^{\frac{3}{2}}$, as stated earlier, and ΔF^0 is the standard free energy increase for the formation of 1 mole of ammonia. If the reaction were written

$$N_2 + 3H_2 \rightleftharpoons 2NH_3,$$

the equilibrium constant would be $f_{NH_3}^2/f_{N_2} \times f_{H_2}^3$, which would be the square of the preceding value. Since ΔF^0 is equal to $-RT \ln K$, the free energy change would consequently be doubled, as it should be for the formation of two moles of ammonia instead of one.

In order to utilize equation (33.32) it is necessary to know ΔH_0^0, $\Delta \alpha$, etc., and the integration constant I. It will be assumed, as before, that heat capacity data are available, so that $\Delta \alpha$, $\Delta \beta$, etc., may be regarded as known. The procedure for the derivation of ΔH_0^0 and I, the latter being equal to $-I'R$, is then similar to that described in § 33h in connection with equation (33.30); this is based either on a single value of the equilibrium constant together with ΔH^0 at any temperature or, alternatively, on the equilibrium constants at two or more temperatures. If the constants in equation (33.30) have previously been evaluated it is, of course, a simple matter to derive the corresponding equation for ΔF^0.

Problem: Derive an expression for the standard free energy of formation of 1 mole of ammonia as a function of temperature, and evaluate ΔF^0 at 25° C.

By utilizing the results obtained in the first problem in § 33h, with ΔH_0^0 equal to $-9,130$ cal. and I' equal to 12.07, it is readily seen that

$$\Delta F^0 = -9,130 + 7.46T \ln T - 3.69 \times 10^{-3}T^2 + 0.235 \times 10^{-6}T^3 - 12.07RT.$$

This gives ΔF^0 in cal. per mole of ammonia. Setting R equal to 1.987, and taking T as 25° C, i.e., 298° K, it is found that

$$\Delta F_{298}^0 = -9,130 + 12,660 - 328 + 6 - 7,145 = -3,936 \text{ cal.}$$

It may be noted that if a slightly different value of ΔH_0^0 were employed, e.g., $-8,810$ cal. as derived in the second problem in § 33h, I' and hence I would be also changed, with the result that ΔF_{298}^0 would not be greatly affected.

Attention may be drawn to the fact that the constant I in equation (33.32) is related to the entropy change ΔS accompanying the reaction, and might be evaluated if ΔS were known. This may be seen, in a general manner, by comparing equation (33.32) with (25.8) for the standard states, i.e., $\Delta F^0 = \Delta H^0 - T\Delta S^0$. However, it is more convenient to use entropy values in another manner for calculating free energy changes, as will be seen shortly.

33j. Simultaneous Equilibria: Addition of Free Energies.—In some systems there may be two or more equilibria which are established simultaneously. In the "water gas" reaction

(1) $CO_2 + H_2 = CO + H_2O,$

for example, at high temperatures, there will be a partial dissociation of carbon dioxide, viz.,

(2) $\quad CO_2 = CO + \tfrac{1}{2}O_2$

and of water vapor, viz.,

(3) $\quad H_2O = H_2 + \tfrac{1}{2}O_2$,

so that three simultaneous equilibria, at least, will exist. For reaction (1) the equilibrium constant K_1 is given by

$$K_1 = \frac{a_{CO} \times a_{H_2O}}{a_{CO_2} \times a_{H_2}},$$

whereas, for reactions (2) and (3), the constants are

$$K_2 = \frac{a_{CO} \times a_{O_2}^{\frac{1}{2}}}{a_{CO_2}} \quad \text{and} \quad K_3 = \frac{a_{H_2} \times a_{O_2}^{\frac{1}{2}}}{a_{H_2O}}.$$

Since these equilibria occur in the same system, the equilibrium activity of any substance involved in K_1, K_2 and K_3 will be the same; hence, it follows, immediately, that

$$K_1 = \frac{K_2}{K_3}. \tag{33.33}$$

It is thus possible to calculate any one of these equilibrium constants if the other two are known. This fact has proved useful in the determination of equilibria which cannot be readily studied directly by experiment.

By taking logarithms of equation (33.33) and multiplying each side by $-RT$, it is seen that

$$-RT \ln K_1 = -RT \ln K_2 + RT \ln K_3,$$

and consequently, by equation (33.6),

$$\Delta F_1^0 = \Delta F_2^0 - \Delta F_3^0.$$

An examination of the chemical equations for the three equilibria given above shows that, as written, reaction (1) is equivalent to reaction (2) minus reaction (3). Consequently, if standard free energy equations are written out in a manner similar to that used for thermochemical (heat of reaction) equations in § 12d, etc., they can be added and subtracted in an analogous manner.

Problem: The standard free energy ΔF_f^0 for reaction (2) is 61.44 kcal. and that for reaction (3) is 54.65 kcal. at 25° C. Calculate K_f for the "water gas" reaction at this temperature.

The standard free energy is equal to $\Delta F_2^0 - \Delta F_3^0$, i.e., $61.44 - 54.65 = 6.79$ kcal., or 6,790 cal. Since $\Delta F_f^0 = -RT \ln K_f$, it follows at 25° C, i.e., 298.2° K,

$$6,790 = -4.576 \times 298.2 \log K_f,$$

with R, like ΔF^0, expressed in calorie units; hence,

$$\log K_f = -4.98,$$
$$K_f = 1.05 \times 10^{-5} \text{ at } 25° \text{ C.}$$

33k. Standard Free Energies of Formation.—Because of their additivity and their relationship to equilibrium constants, and for other reasons, it is useful to tabulate standard free energies, like heats of reaction; the methods available for the determination of free energy changes for chemical reactions will be reviewed in the next section. The results are conveniently recorded in the form of standard free energies of formation of compounds, and these are derived from the free energies of reactions in a manner similar to that used to obtain standard heats of formation from the heats of reaction in § 12f. There is no method known for the experimental determination of absolute free energies, but this is of no consequence, since the solution of thermodynamic problems requires a knowledge of free energy *changes*. For the purpose of evaluating free energies of compounds, *the convention is adopted of taking the free energies of all elements in their standard states to be zero at all temperatures*. The standard states of solid and liquid elements are their respective pure stable forms at 1 atm.; for gaseous elements the standard states are the ideal gases at 1 atm. pressure, i.e., at unit fugacity. On the basis of these conventions, the standard free energy of a compound is equal to its standard free energy of formation, that is, to the increase of free energy accompanying the formation of 1 mole of the compound from its elements, all the substances being in their respective standard states. For example, for the reaction between hydrogen and oxygen at 25° C,

$$H_2(g) + \tfrac{1}{2}O_2(g) = H_2O(g), \qquad \Delta F^0_{298} = -54.65 \text{ kcal.},$$

the standard states being the gases at unit fugacity, so that the standard free energy of formation of water vapor is -54.65 kcal. mole^{-1} at 25° C. This means that the free energy of 1 mole of water vapor in its standard state is less by 54.65 kcal. than the sum of the free energies of 1 mole of hydrogen gas and $\tfrac{1}{2}$ mole of oxygen gas in their respective standard states at 25° C; hence,

$$\Delta F^0_{298} = F^0(H_2O, g) - [F^0(H_2, g) + \tfrac{1}{2}F^0(O_2, g)] = -54.65 \text{ kcal.}$$

If the standard free energies of the elements are set equal to zero, by convention, the standard free energy of 1 mole of water vapor becomes equal to the standard free energy of formation from its elements.

The standard free energies of formation of a number of compounds at 25° C are given in Table XXIV.[7] They can be utilized, like heats of formation (standard heat contents) in Chapter V, to calculate the standard free energies of many reactions, for several of which the values could not be obtained in any other way. This is especially the case for reactions, par-

---

[7] Data mainly from W. M. Latimer, "The Oxidation States of the Elements, etc.," 1938; see also International Critical Tables, Vol. V; G. S. Parks and H. M. Huffman, "The Free Energies of Some Organic Compounds," 1932; Kelley, *U. S. Bur. Mines Bull.*, **384** (1935); 406, 407 (1937); Pitzer, *Chem. Rev.*, **27**, 39 (1940); Parks, *ibid.*, **27**, 75 (1940); Thacker, Folkins and Miller, *Ind. Eng. Chem.*, **33**, 584 (1941); Wagman, Kilpatrick, Taylor Pitzer and Rossini, *J. Res. Nat. Bur. Stand.*, **34**, 143 (1945); Prosen, Pitzer and Rossini, *ibid.*, **34**, 255, 403 (1945); Wagman, Kilpatrick, Pitzer and Rossini, *ibid.*, **35**, 467 (1945); Kilpatrick, Prosen, Pitzer and Rossini, *ibid.*, **36**, 559 (1946).

ticularly those involving organic compounds, which proceed to virtual completion, and for which equilibrium constants could not be determined experimentally.

It is sometimes required to know the free energy of formation of a substance in one physical state when that in another is given. For example, the free energy of formation of liquid water is recorded in Table XXIV, since

TABLE XXIV. STANDARD FREE ENERGIES OF FORMATION
AT 25° C IN KCAL. PER MOLE *

Inorganic Compounds

Substance	ΔF^0	Substance	ΔF^0	Substance	ΔF^0
$HF(g)$	− 31.8	$NO(g)$	20.66	$Sb_2O_3(s)$	− 149.0
$HCl(g)$	− 22.74	$H_2S(g)$	− 7.87	$As_2O_3(s)$	− 137.7
$HBr(g)$	− 12.54	$SO_2(g)$	− 71.7	$BaCO_3(s)$	− 271.6
$CO(g)$	− 32.81	$NH_3(g)$	− 3.94	$Bi_2O_3(s)$	− 116.6
$CO_2(g)$	− 94.26	$H_2O(l)$	− 56.70	$CaCO_3(s)$	− 207.4

Organic Compounds

Substance	ΔF^0	Substance	ΔF^0
Methane(g)	− 12.14	Methyl alcohol(l)	− 40.0
Ethane(g)	− 7.86	Ethyl alcohol(l)	− 40.2
Propane(g)	− 5.61	Acetaldehyde(l)	− 31.9
Ethylene(g)	16.34	Acetic acid(l)	− 94.5
Acetylene(g)	50.7	Benzene(l)	29.06

* For further data, see Table 5 at end of book.

this is the standard state at ordinary temperatures, but for certain purposes the standard free energy of water vapor, i.e., at unit fugacity, is required. For this purpose, it is necessary to determine the standard free energy change of the process $H_2O(l) = H_2O(g)$. Since the molar free energy of the liquid water is equal to the molar free energy of the vapor with which it is in equilibrium (§ 27a), the problem is simply to calculate the difference in free energy of 1 mole of water vapor at its equilibrium (vapor) pressure, at the given temperature, and in the standard state of unit fugacity, i.e., ideal gas at 1 atm. For this purpose, use may be made of equation (29.5), viz.,

$$\Delta F = F_2 - F_1 = RT \ln \frac{f_2}{f_1}, \qquad (33.34)$$

where f_2 is unit fugacity, and f_1 is that corresponding to the vapor pressure. For most purposes, except near the critical point, the ratio of the fugacities may be replaced by the ratio of the vapor pressures, that is, ideal behavior is assumed (cf. Chapter XII, Exercise 5); equation (33.34) then becomes

$$\Delta F = RT \ln \frac{p_2}{p_1}, \qquad (33.35)$$

which, incidentally, is identical with the expression in § 25d for the free energy change accompanying the expansion of an ideal gas. The vapor pressure of water at 25° C is 23.76 mm., and since equation (33.35) involves a ratio of pressures, the units are immaterial. Thus p_2 may be set equal to 760 mm. and p_1 to 23.76 mm., so that at 298.2° K, with R in cal. deg.$^{-1}$ mole^{-1},

$$\Delta F = 1.987 \times 298.2 \times 2.303 \log \frac{760.0}{23.76} = 2,054 \text{ cal.}$$

The standard free energy of formation of liquid water is $-$ 56.70 kcal., and hence for water vapor, ΔF^0 is $-$ 56.70 $+$ 2.05, i.e., $-$ 54.65 kcal., at 25° C.

Problem: The standard free energy change for the reaction $CO(g) + 2H_2(g) = CH_3OH(g)$ is $-$ 5.88 kcal. at 25° C. Calculate the standard free energy of formation of liquid methanol at 25° C, the vapor pressure being 122 mm. at that temperature.

For the process $CH_3OH(g) = CH_3OH(l)$, the standard free energy change at 298.2° K, assuming ideal behavior of the vapor, is given by

$$\Delta F = RT \ln \frac{p_2}{p_1} = 4.576 \times 298.2 \log \frac{122}{760}$$
$$= -1,084 \text{ cal.,}$$

since p_2, the vapor pressure of the final state, i.e., liquid, is 122 mm., and p_1, for the initial state, i.e., vapor at 1 atm., is 760 mm. Hence, for the reaction

$$CO(g) + 2H_2(g) = CH_3OH(l), \quad \Delta F^0 = -5.88 - 1.08 = -6.96 \text{ kcal.}$$
$$(-33.0) \quad (0) \quad (x)$$

Writing the free energy of formation of each species below its formula, it is seen that for liquid methanol at 25°, ΔF^0_{298} is given by

$$x - (-33.0 + 0) = -6.96,$$
$$x = -40.0 \text{ kcal.}$$

331. Determination of Standard Free Energies.—Three main procedures have been used for the evaluation of standard free energies of reactions. The first is based on the experimental determination of equilibrium constants, and their combination in the manner indicated above. By the procedure described in § 33i, an expression can then be obtained for ΔF^0 as a function of temperature, so that the value at any particular temperature can be evaluated.

The second method, which has been largely used for reactions involving ionized substances, is based on the measurement of the E.M.F.'s of certain galvanic cells. It was seen in § 25b that the decrease of free energy accompanying any process at constant temperature and pressure, i.e., $- \Delta F_{T,P}$, is equal to the maximum (reversible) work, other than work of expansion, obtainable from that process. If the particular reaction can be performed in such a manner that, apart from work of expansion, all the work is electrical in nature, then the free energy change is equal in magnitude to the electrical work. Many processes can be carried out reversibly in a suitable galvanic cell, and the (maximum) E.M.F. of the cell may be used to derive the accompanying free energy change.

The subject will be considered more fully in Chapter XIX, but for the present it will be sufficient to accept the fact that cells can be devised in which particular chemical and physical changes take place. In order that the E.M.F. may be employed for free energy calculations, it is necessary that the cell should function reversibly; the tests of reversibility are as follows. When the cell is connected to an external source of E.M.F., which is adjusted so as exactly to balance the E.M.F. of the cell, i.e., so that no current flows, there should be no chemical change in the cell. If the external E.M.F. is decreased by an infinitesimally small amount, current will flow from the cell and a chemical change, proportional in extent to the quantity of electricity passing, should take place. On the other hand, if the external E.M.F. is increased by a small amount, the current should pass in the opposite direction and the cell reaction should be exactly reversed. To determine the maximum, or reversible, E.M.F. of such a cell, it is necessary that the cell should be operated reversibly, that is, the current drawn should be infinitesimally small so that the system is always virtually in equilibrium. Fortunately, the potentiometer method, almost universally employed for the accurate determination of E.M.F.'s, involves the exact balancing of the E.M.F. of the cell by an external E.M.F. of known value, so that the flow of current is zero or infinitesimal. The external E.M.F. applied from the potentiometer thus gives an accurate measure of the reversible (maximum) E.M.F. of the cell. It may be mentioned that if large currents flow through the cell when its E.M.F. is measured, concentration gradients arise within the cell, because of the slowness of diffusion, etc., and the E.M.F. observed is less than the reversible value.

If the E.M.F. of a voltaic cell is E int. volts, and the process taking place within it is accompanied by the passage of N faradays of electricity, i.e., NF coulombs, where F represents 96,500 int. coulombs, the work done by the cell is NFE int. volt-coulombs, or int. joules (cf. § 3b). If the cell is a reversible one, as described above, and E is its reversible, i.e., maximum, E.M.F., at a given temperature and pressure, usually atmospheric, it follows from the arguments presented earlier that

$$-\Delta F = NFE, \qquad (33.36)$$

where ΔF is the increase in the free energy accompanying the process taking place in the cell. The value of ΔF refers, of course, to a definite temperature and pressure, and instead of writing $\Delta F_{T,P}$ this is usually understood. The number N of faradays required for the cell reaction can be readily obtained by inspection of the chemical equation. In general, this number of faradays is equal to the number of equivalents of electricity, or number of unit charges, transferred in the cell reaction (cf. § 45b).

At the present moment, the chief interest is the evaluation of the standard free energy ΔF^0 of a process. It will be seen in Chapter XIX that this can be calculated from ΔF and the activities of the substances present in the cell, by utilizing a form of the reaction isotherm [equation (33.5)]. It can be stated, however, that if the substances involved in the cell reaction are all present in their respective standard states, the resulting E.M.F. will be the standard value E^0. The standard free energy change is then given by the appropriate form of equation (33.36), viz.,

$$-\Delta F^0 = NFE^0. \tag{33.37}$$

The measurement of E.M.F.'s, of reversible cells thus provides a simple, and accurate, method for evaluating standard free energy changes. Various applications of this procedure will be given in later chapters.

The third method commonly employed for obtaining standard free energies depends on the use of entropy values; this will be discussed in some detail in the following section, as it provides a highly important practical application of the results obtained in the earlier study of entropy.

33m. Standard Free Energies and Entropy Changes.—It is apparent from the general equation

$$\Delta F^0 = \Delta H^0 - T\Delta S^0, \tag{33.38}$$

that ΔF^0 for any reaction could be calculated if ΔH^0 and ΔS^0 were known. In many cases where direct determinations of free energy changes are not possible, it is a relatively simple matter to obtain the heat of reaction ΔH^0 with reactants and products in their standard states. This is true, for instance, for many reactions involving organic compounds. If the corresponding entropy change ΔS^0 were known, the problem of obtaining the standard free energy change would be solved.

If the entropies in their standard states are known for all the substances concerned in a reaction, it is possible to evaluate the entropy change ΔS^0 for the process at the same temperature. It was seen in Chapter IX that the entropies of solids, liquids and gases can be obtained from heat capacity measurements, utilizing the third law of thermodynamics. For gases, the entropies can frequently be derived from the respective partition functions, and these can be converted into the values for the solid and liquid states by means of the entropies accompanying the phase changes as obtained from the corresponding latent heats. Another procedure which has been found useful is to determine the entropy of an element or compound from a knowledge of the standard entropy change of the reaction and the entropies of all but one of the substances taking part in the reaction. The entropy change for the whole reaction can, of course, be evaluated if ΔH^0 and ΔF^0 are known, by means of equation (33.38).

If the reaction can be made to take place in a reversible galvanic cell, a direct method for the determination of the standard entropy change is possible. For a substance in its standard state, equation (25.20) takes the form

$$\frac{dF^0}{dT} = -S^0,$$

the constant pressure condition being omitted as unnecessary; hence, for a process involving reactants and products in their respective standard states,

$$\Delta S^0 = -\frac{d(\Delta F^0)}{dT}. \qquad (33.39)$$

If the reaction can take place in a reversible cell, whose E.M.F. is E^0, it follows, by combining equations (33.37) and (33.39), that

$$\Delta S^0 = NF\frac{dE^0}{dT}. \qquad (33.40)$$

The standard entropy change of the reaction can thus be determined from the temperature coefficient of the E.M.F. of the cell.

Problem: For the cell $Ag(s)\,|\,AgCl(s)\,KCl$ soln. $Hg_2Cl_2(s)\,|\,Hg(l)$, in which the reaction is $Ag(s) + \frac{1}{2}Hg_2Cl_2(s) = AgCl(s) + Hg(l)$, the temperature coefficient of the E.M.F. is 3.38×10^{-4} volt deg.$^{-1}$ at 25° C. The entropies at this temperature of $Ag(s)$, $AgCl(s)$ and $Hg(l)$ are known from heat capacity measurements to be 10.2, 23.0 and 18.5 E.U. mole^{-1} (or g. atom^{-1}), respectively. Calculate ΔS^0 for the reaction, and the standard molar entropy of $Hg_2Cl_2(s)$ at 25° C.

It may be noted that in this particular cell all the substances taking part in the reaction are in their respective standard states, thus simplifying the problem. In this case dE^0/dT is equal to dE/dT, which is 3.38×10^{-4} volt deg.$^{-1}$. Since silver is univalent, the cell process would require the passage of 1 faraday, i.e., N is 1 equiv. Since the entropy is required in cal. deg.$^{-1}$ mole^{-1}, it is convenient to express F as 23,070 cal. volt^{-1} g. equiv.$^{-1}$ (Table 1, Appendix), so that

$$\Delta S^0 = 1 \times 23{,}070 \times 3.38 \times 10^{-4}$$
$$= 7.80 \text{ cal. deg.}^{-1}, \text{ i.e., E.U.}$$

For the given reaction,

$$\Delta S^0 = [S^0(AgCl) + S^0(Hg)] - [S^0(Ag) + \tfrac{1}{2}S^0(Hg_2Cl_2)],$$
$$7.80 = (23.0 + 18.5) - [10.2 + \tfrac{1}{2}S^0(Hg_2Cl_2)],$$
$$S^0(Hg_2Cl_2) = 47.0 \text{ E.U. mole}^{-1}.$$

By definition, the entropy change in a reaction is equal to the heat change, when the process is carried out reversibly, divided by the absolute temperature. If the amount of heat liberated in a reversible cell when it operates could be measured, the entropy change for the reaction could be determined. Because of experimental difficulties this method does not appear to have been used.

33n. Application of Free Energy and Entropy Data.—With the standard entropies for many elements and compounds known (see Tables XV and XIX), a highly valuable means is available for the determination of free energy changes, and hence of equilibrium constants, for numerous reactions. For many of the processes no direct experimental methods for obtaining these quantities have yet been developed. An important application of the data is to decide whether a particular reaction is possible, or not, at a certain temperature before attempting to carry it out experimentally.

Problem: Determine the standard free energy change of the reaction $CH_4(g) + 2O_2(g) = CO_2(g) + 2H_2O(l)$ at 25° C, using the following standard entropies and heats of formation:

	$CH_4(g)$	$O_2(g)$	$CO_2(g)$	$H_2O(l)$
S^0	44.5	49.00	51.06	16.75 cal. deg.$^{-1}$ mole^{-1}
ΔH^0	-17.9	0	-94.0	-68.3 kcal. mole^{-1}.

For the reaction as written above,

$$\Delta S^0 = [51.06 + (2 \times 16.75)] - [44.5 + (2 \times 49.00)] = -57.9 \text{ E.U.}$$
$$\Delta H^0 = [-94.0 + (-2 \times 68.3)] - [-17.9 + (2 \times 0)] = -212.7 \text{ kcal.}$$

At 25° C,
$$T\Delta S^0 = -298 \times 57.9 = -17{,}250 \text{ cal.} = -17.3 \text{ kcal.}$$

Hence,
$$\Delta F^0 = \Delta H^0 - T\Delta S^0 = -212.7 + 17.3$$
$$= -195.4 \text{ kcal. at } 25° \text{ C.}$$

Problem: A mixture of hydrogen and carbon monoxide in the molecular proportion of 2 to 1 is allowed to come to equilibrium, in the presence of a catalyst, at 600° K and 250 atm. Disregarding side reactions, determine the extent of conversion into methanol according to the reaction $2H_2 + CO = CH_3OH(g)$, using heat capacity, heat content and entropy data, together with the generalized fugacity chart.[8]

At 25° C, ΔH^0 of formation of $CH_3OH(g)$ is -48.1 kcal., that of CO is -26.4 kcal., and that of hydrogen is zero by convention. Hence, for the methanol reaction, ΔH^0 is $-48.1 - (-26.4 + 0) = -21.7$ kcal., or $-21{,}700$ cal.
The values of C_P in cal. deg.$^{-1}$ mole^{-1} are as follows:

$$C_P(CH_3OH) = 4.394 + 24.274 \times 10^{-3}T - 68.55 \times 10^{-7}T^2$$
$$2 \times C_P(H_2) = 13.894 - 0.399 \times 10^{-3}T + 9.62 \times 10^{-7}T^2$$
$$C_P(CO) = 6.342 + 1.836 \times 10^{-3}T + 2.80 \times 10^{-7}T^2,$$

so that
$$\Delta C_P = -15.842 + 22.837 \times 10^{-3}T - 80.97 \times 10^{-7}T^2.$$

Hence,
$$\Delta\alpha = -15.84, \quad \Delta\beta = 22.84 \times 10^{-3}, \quad \Delta\gamma = -80.97 \times 10^{-7}.$$

Inserting these results, together with ΔH^0 at 298.2° K, into equation (33.29),

$$-21{,}700 = \Delta H_0^0 - 15.84 \times 298.2 + 11.42 \times 10^{-3} \times (298.2)^2$$
$$- 26.99 \times 10^{-7} \times (298.2)^3$$
$$= \Delta H_0^0 - 4{,}723 + 1{,}015 - 72,$$
$$\Delta H_0^0 = -17{,}920 \text{ cal.}$$

The standard entropies are 56.63 for $CH_3OH(g)$, $2 \times 31.21 = 62.42$ for $2H_2$, and 47.30 E.U. for CO, so that for the reaction ΔS^0 is -53.09 cal. deg.$^{-1}$. Hence, at 298.2° K,
$$\Delta F^0 = \Delta H^0 - T\Delta S^0$$
$$= -21{,}700 + 298.2 \times 53.09 = -5{,}870 \text{ cal.}$$

[8] Cf. Ewell, *Ind. Eng. Chem.*, **32**, 147 (1940).

Using this value of ΔF^0, together with ΔH_0^0, $\Delta\alpha$, $\Delta\beta$ and $\Delta\gamma$, it is possible to determine the integration constant I in equation (33.32); thus, with T equal to 298.2° K,

$$-5{,}870 = -17{,}920 + 15.84 \times 298.2 \times \ln 298.2 - 11.42 \times 10^{-3} \times (298.2)^2$$
$$+ 13.5 \times 10^{-7} \times (298.2)^3 + 298.2 I$$
$$= -17{,}920 + 26{,}920 - 1{,}015 + 36 + 298.2 I,$$
$$I = -46.59.$$

At 600° K, the value of ΔF^0 is given by

$$\Delta F^0 = -17{,}920 + 15.84 \times 600 \times \ln 600 - 11.42 \times 10^{-3} \times (600)^2$$
$$+ 13.5 \times 10^{-7} \times (600)^3 - 46.59 \times 600$$
$$= -17{,}920 + 60{,}820 - 4{,}111 + 292 - 27{,}950$$
$$= 11{,}130 \text{ cal.}$$

Since $\Delta F^0 = -RT \ln K_f$, with fugacities in atm. in this case, it follows that K_f at 600° K is equal to 8.9×10^{-5}.

The data for the determination of the activity coefficient function $J_{\gamma'}$ at 600° K at 250 atm. are as follows:

	T_c	P_c	θ	π	γ'
CH_3OH	513.2° K	98.7 atm.	1.17	2.53	0.58
H_2	33.2°(+ 8°)	12.8(+ 8)	14.56	12.02	1.10
CO	134.4°	34.6	4.65	7.23	1.13

$$J_{\gamma'} = \frac{0.58}{(1.10)^2 \times 1.13} = 0.42.$$

If this may be taken as equal to the true activity function, then as a good approximation $K_f = J_{\gamma'} \times K_p'$; hence, K_p' is $8.9 \times 10^{-5}/0.42 = 2.1 \times 10^{-4}$.

If the system initially contains 2 moles of H_2 and 1 mole of CO, and x is the extent of conversion at equilibrium, the numbers of moles at equilibrium are $2 - 2x(H_2)$, $1 - x(CO)$ and $x(CH_3OH)$. Hence, the respective mole fractions are $(2 - 2x)/(3 - 2x)$, $(1 - x)/(3 - 2x)$ and $x/(3 - 2x)$, and so for the reaction $2H_2 + CO \rightleftharpoons CH_3OH(g)$, for which Δn is -2,

$$K_p' = \frac{N_{CH_3OH}}{N_{H_2}^2 \times N_{CO}} P^{-2} = \frac{x(3 - 2x)^2}{(1 - x)(2 - 2x)^2 P^2}.$$

Since K_p' is 2.1×10^{-4} and P is 250 atm., K_f and K_p' being based on the fugacity of 1 atm. as standard state, solution of the equation by successive approximation gives x equal to 0.67. There is consequently 67 per cent conversion of hydrogen and carbon monoxide into methanol under the given conditions.

Attention may be called to the omission of the entropies and heat contents of ions from the foregoing discussion, so that the data for the calculation of the free energies of ionic reactions are not yet available. However, the subject will be taken up in Chapter XIX, when the omission will be rectified.

33o. Confirmation of Third Law of Thermodynamics.—It was stated in § 23d that confirmation of the third law is to be found in the agreement with experiment obtained by using thermal entropies in conjunction with other data. Numerous examples of such agreement are to be found in the literature of thermodynamics, but two simple cases will be considered here. In the problem in § 33m, the entropy of mercurous chloride was derived from E.M.F. measurements and from the third law entropies of mercury, silver and silver chloride. The fact that an almost identical result, viz., 47 cal. deg.$^{-1}$ mole^{-1}, has been obtained by direct heat capacity measurements on solid mercurous chloride, assuming the entropy to be zero at the absolute zero, may be regarded as confirmation of this (third law) postulate. Very satisfactory agreement has been found in other instances of a similar type.

Another procedure for testing the third law of thermodynamics is to combine heat content with entropy data for a given reaction, and so to determine the free energy change, the value of which is known from direct measurement. The standard free energy change for the formation of silver oxide, i.e., for the reaction $2Ag(s) + \frac{1}{2}O_2(g) = Ag_2O(s)$, can be derived from the dissociation pressure (cf. Exercise 11); this is found to be -2.59 kcal. at 25° C. The third law entropies of silver oxide, silver and oxygen are 29.1, 10.2 and 49.0 cal. deg.$^{-1}$ mole^{-1}, respectively, at 25° C, and hence for the reaction in which 1 mole of silver oxide is formed, ΔS^0 is $29.1 - (2 \times 10.2 + \frac{1}{2} \times 49.0)$, i.e., -15.8 cal. or -0.0158 kcal. deg.$^{-1}$. The standard heat of formation of the oxide is known to be -7.30 kcal. mole^{-1}, and hence the free energy of formation, given by $\Delta H^0 - T\Delta S^0$, should be $-7.30 - (298.2 \times -0.0158) = -2.59$ kcal. mole^{-1}, in complete agreement with the direct experimental value. Analogous results have been obtained in many cases involving both inorganic and organic compounds.

33p. Free Energy Functions.—In connection with homogeneous gas reactions in particular, an alternative method of presenting what amounts essentially to entropy data is to utilize certain free energy functions based on the use of partition functions. It was seen in § 25g [equation (25.42)], that

$$F - E_0 = - RT \ln \frac{Q}{N}, \tag{33.41}$$

where F is the molar free energy and Q is the partition function of the given species; E_0 is the molar energy content when all the molecules are in their lowest energy state, i.e., at 0° K; N is the Avogadro number. The partition functions invariably refer to the ideal gaseous state, and if the pressure of the gas is taken as 1 atm., the values apply to the usual standard state. It follows, therefore, that equation (33.41) may be written as

$$F^0 - E_0^0 = - RT \ln \frac{Q^0}{N},$$

so that

$$-\frac{F^0 - E_0^0}{T} = R \ln \frac{Q^0}{N}. \tag{33.42}$$

By definition, $H = E + PV$, and for 1 mole of an ideal gas, this may be written as $H = E + RT$, or $H^0 = E^0 + RT$ if the substance is in its standard state, i.e., 1 atm. pressure. It follows, therefore, that at 0° K, when RT is zero,

$$H_0^0 = E_0^0.$$

It is consequently possible to substitute H_0^0 for E_0^0 in equation (33.42), giving

$$-\frac{F^0 - H_0^0}{T} = R \ln \frac{Q^0}{N}. \qquad (33.43)$$

The quantity $-(F^0 - E_0^0)/T$ or $-(F^0 - H_0^0)/T$, that is, the left-hand side of equation (33.42) or (33.43), is known as the **free energy function** of the substance; its value for any gaseous substance, at a given temperature, can be readily derived from the partition function for that temperature, at 1 atm. pressure. The data for a number of substances, for temperatures up to 1500° K, have been determined in this manner and tabulated (Table XXV).[9]

TABLE XXV. FREE ENERGY FUNCTIONS IN CAL. DEG.$^{-1}$ MOLE^{-1}
$-(F^0 - H_0^0)/T$

Substance	298.16°	400°	600°	800°	1000°	1500° K
H_2	24.423	26.422	29.203	31.186	32.738	35.590
O_2	42.061	44.112	46.968	49.044	50.697	53.808
N_2	38.817	40.861	43.688	45.711	47.306	50.284
Graphite	0.517	0.824	1.477	2.138	2.771	4.181
CO	40.350	42.393	45.222	47.254	48.860	51.884
CO_2	43.555	45.828	49.238	51.895	54.109	58.461
$H_2O(g)$	37.172	39.508	43.688	45.711	47.306	50.622
Cl_2	45.951	48.148	51.298	53.614	55.453	58.876
HCl	37.734	39.771	42.588	44.597	46.171	49.096
NO	42.985	45.141	48.100	50.314	51.878	54.979
H_2S	41.174	43.53	46.83	49.27	51.24	55.06
CH_4	36.46	38.86	42.39	45.21	47.65	52.84
C_2H_6	45.25	48.20	53.06	57.28	61.12	69.49
C_2H_4	44.05	46.7	50.8	54.4	57.5	64.2
C_2H_2	40.01	42.49	46.38	49.50	52.14	57.43
NH_3	37.989	40.380	43.826	46.450	48.634	53.033
SO_2	50.95	53.49	57.21	60.05	62.39	66.91
$SO_3(g)$	51.94	54.81	59.31	62.98	66.12	—

It will be observed that Table XXV contains values for the free energy function of graphite; this has not been obtained from equation (33.43), which is applicable to gases only. The method for the calculation of $-(F^0 - H_0^0)/T$ for solids is based on the use of heat capacities. For a pure solid, since S_0 is zero, by the third law of thermodynamics, equation (23.1) becomes

$$S^0 = \int_0^T \frac{C_P}{T} dT,$$

[9] For main sources of data, see references in Wilson, *Chem. Rev.*, **27**, 17 (1940); also, Wagman, *et al.*, ref. 7, Kilpatrick, *et al.*, ref. 7, Pitzer, ref. 7.

the symbol S^0 being employed since the pure solid, at 1 atm. pressure, represents the standard state. Further, since C_P is equal to $(\partial H/\partial T)_P$, it follows upon integration that

$$H^0 - H_0^0 = \int_0^T C_P dT. \qquad (33.44)$$

Upon combining these two expressions with the relationship $F^0 = H^0 - TS^0$, for the standard state, it follows immediately that

$$-\frac{F^0 - H_0^0}{T} = \int_0^T \frac{C_P}{T} dT - \frac{1}{T}\int_0^T C_P dT.$$

Hence, if C_P of the solid is known as a function of the temperature, $-(F^0 - H_0^0)/T$ can be obtained by analytical or graphical integration.

33q. Calculation of Standard Free Energies.

The free energy functions can be utilized to calculate equilibrium constants or the standard free energy changes of gas reactions in the following manner. The change in the free energy function accompanying a particular reaction is $-\Delta(F^0 - H_0^0)/T$, and this may be written as

$$-\Delta\left(\frac{F^0 - H_0^0}{T}\right) = -\frac{\Delta F^0}{T} + \frac{\Delta H_0^0}{T},$$

so that

$$\Delta F^0 = T\Delta\left(\frac{F^0 - H_0^0}{T}\right) + \Delta H_0^0. \qquad (33.45)$$

The first term on the right-hand side can be derived directly from the table of free energy functions, and hence for the determination of the standard free energy change ΔF^0 it is necessary to know ΔH_0^0 for the given reaction. Several methods are available for arriving at the proper values of ΔH_0^0, but these all require a knowledge of heat content changes, that is, of heats of reaction. From this fact it will be evident that the foregoing method of obtaining ΔF_0^0 is really equivalent to the use of the entropy change.

As seen earlier (§ 12k), ΔH_0^0 for a reaction may be evaluated from thermal measurements, including heat capacities at several temperatures. However, instead of using experimental heat capacity data to derive ΔH_0^0 from ΔH^0 values, the results may be obtained indirectly from partition functions (cf. § 16c). The energy content of an ideal gas is independent of the pressure, at a given temperature; hence, $E - E_0$ in equation (16.8) may be replaced by $E^0 - E_0^0$, so that

$$E^0 - E_0^0 = RT^2\left(\frac{\partial \ln Q}{\partial T}\right)_V, \qquad (33.46)$$

and since $H^0 = E^0 + RT$ for 1 mole, and E_0^0 and H_0^0 are identical, it follows that

$$H^0 - H_0^0 = RT^2\left(\frac{\partial \ln Q}{\partial T}\right)_V + RT. \qquad (33.47)$$

Values of $H^0 - H_0^0$, known as the **relative heat content**, for a number of gases at various temperatures, calculated from partition functions by means of equation (33.47), have been recorded (Table XXVI).[10] For pure solids,

TABLE XXVI. RELATIVE HEAT CONTENTS IN KCAL. PER MOLE *
$H^0 - H_0^0$

Substance	298.16°	400°	600°	800°	1000°	1500° K
H_2	2.024	2.731	4.129	5.537	6.966	10.694
O_2	2.070	2.792	4.279	5.854	7.497	11.776
N_2	2.072	2.782	4.198	5.669	7.203	11.254
Graphite	0.252	0.503	1.198	2.082	3.075	5.814
CO	2.073	2.784	4.210	5.700	7.257	11.359
CO_2	2.238	3.195	5.322	7.689	10.222	17.004
$H_2O(g)$	2.365	3.190	4.873	6.666	8.580	13.876
Cl_2	2.11	—	—	—	—	—
HCl	2.00	—	—	—	—	—
NO	2.206	2.923	4.385	5.834	7.584	11.701
H_2S	2.386	3.22	4.99	6.92	9.01	14.79
CH_4	2.397	3.323	5.549	8.321	11.560	21.130
C_2H_6	2.865	4.27	8.03	12.78	18.37	34.56
C_2H_4	2.59	—	—	—	—	—
C_2H_2	2.41	—	—	—	—	—
NH_3	2.407	3.2992	5.2656	7.5720	10.123	17.460
SO_2	2.53	3.53	5.75	8.17	10.70	17.29
$SO_3(g)$	2.77	4.10	7.22	10.78	14.63	—

* The entropy at any temperature may be derived from the free energy function and the relative heat content by utilizing the relationship $S^0 = -[(F^0 - H_0^0)/T] + [(H^0 - H_0^0)/T]$.

$H^0 - H_0^0$ at any temperature may be derived from heat capacity measurements by means of equation (33.44).

If ΔH^0 for the reaction is known at any temperature (Chapter V), it is then a simple matter to obtain ΔH_0^0 by means of the relationship

$$\Delta H_0^0 = \Delta H^0 - \Delta (H^0 - H_0^0), \qquad (33.48)$$

the last term being evaluated from Table XXVI.

A third method for obtaining ΔH^0 has been employed in certain instances by utilizing heats of dissociation derived from spectroscopic measurements. These results refer to the substances in their lowest energy state, and hence correspond to ΔH_0; the corrections required to convert the results to ΔH_0^0, i.e., to the standard states, are usually small, and can be calculated if necessary.

Problem: For the reaction $H_2(g) + \tfrac{1}{2}O_2(g) = H_2O(g)$, the value of ΔH^0 at 25° C is -57.80 kcal. Utilize Tables XXV and XXVI to determine the standard free energy change of the reaction.

From Tables XXV and XXVI the following data are obtained at 25° C, i.e., 298.2° K:

$$\begin{array}{lccc} & H_2(g) & \tfrac{1}{2}O_2(g) & H_2O(g) \\ -(F^0 - H_0^0)/T & 24.423 & 21.031 & 37.172 \\ H^0 - H_0^0 & 2.024 & 1.035 & 2.365 \end{array}$$

$-\Delta[(F^0 - H_0^0)/T] = -8.282$ cal. deg.$^{-1}$
$\Delta(H^0 - H_0^0) = -0.694$ kcal.

[10] See ref. 9.

Since ΔH^0 is -57.80 kcal., ΔH_0^0 is $-57.80 - (-0.69) = -57.11$ kcal. By equation (33.45), $\Delta F^0 = -(-298.2 \times 8.282 \times 10^{-3}) + (-57.11) = -54.64$ kcal.

33r. Equilibrium Constant and Partition Functions.—There is another approach to the problem of evaluating free energy changes of gas reactions from partition functions which, although formulated somewhat differently, is fundamentally the same as that described above. Since, by equation (33.43),

$$F^0 = -RT \ln \frac{Q^0}{N} + H_0^0,$$

it follows that for any homogeneous gas reaction,

$$\Delta F^0 = -RT \ln J_{Q^0/N} + \Delta H_0^0,$$

using a notation similar to that in § 32c (see footnote, p. 276). Since ΔF^0 is equal to $-RT \ln K_f$, in this case, it is seen that

$$\ln K_f = \ln J_{Q^0/N} - \frac{\Delta H_0^0}{RT}. \tag{33.49}$$

For the general reaction

$$a\text{A} + b\text{B} + \cdots = l\text{L} + m\text{M} + \cdots,$$

equation (33.49) becomes

$$\ln K_f = \ln \frac{(Q^0/N)_\text{L}^l \times (Q^0/N)_\text{M}^m \times \cdots}{(Q^0/N)_\text{A}^a \times (Q^0/N)_\text{B}^b \times \cdots} - \frac{\Delta H_0^0}{RT},$$

and hence,

$$K_f = \frac{(Q^0/N)_\text{L}^l \times (Q^0/N)_\text{M}^m \times \cdots}{(Q^0/N)_\text{A}^a \times (Q^0/N)_\text{B}^b \times \cdots} e^{-\Delta H_0^0/RT}. \tag{33.50}$$

The method of calculating K_f based on equation (33.50) thus also requires a knowledge of ΔH_0^0, which can be obtained by the methods described in § 33q. If this and the partition functions of the reactants and products of the process are known, the equilibrium constants can be readily calculated.

For many purposes it is not necessary to know the actual partition functions, for they may be derived with sufficient accuracy by means of the equations given in Chapter VI. The calculations may be illustrated by reference to the simple case of the dissociation of a gaseous molecule into two atoms, viz.,

$$\text{A}_2 \rightleftharpoons 2\text{A},$$

for which

$$K_f = \frac{(Q^0/N)_1^2}{(Q^0/N)_2} e^{-\Delta H_0^0/RT}, \tag{33.51}$$

where the subscripts 1 and 2 refer to the atoms and molecules, respectively.

The partition function of the atomic species consists of the electronic and translational contributions only, but for the diatomic molecule A_2 the partition function involves the electronic, translational and rotational factors, and also the contribution of one vibrational mode. The translational partition function is given by equation (16.16) as

$$Q_t = \frac{(2\pi m k T)^{3/2}}{h^3} V, \tag{33.52}$$

where V is the molar volume. For the present purpose the partition function is required for each substance in its standard state. In order to obtain K_f, the standard state is that of the ideal gas at 1 atm. pressure. It is, therefore, convenient to write Q_t in a slightly different form, by replacing V by RT/P, where P is the pressure of the gas; hence, equation (33.52) becomes

$$Q_t = \frac{(2\pi mkT)^{3/2}}{h^3} \cdot \frac{RT}{P}. \qquad (33.53)$$

As seen in § 16b, the partition functions are based on the postulate of ideal behavior of gases; hence if P in equation (33.53) is set equal to 1 atm. the result will be the translational contribution to the partition function of the gas in the standard state, i.e.,

$$Q_t^0 = \frac{(2\pi mkT)^{3/2}}{h^3} RT, \qquad (33.54)$$

with R in cc.-atm. deg.$^{-1}$ mole^{-1}. This result is applicable to both atoms and molecules, using the appropriate value of the mass m of each unit particle, i.e., the atom A or the molecule A_2, respectively.

The rotational and vibrational factors in the partition function of the diatomic molecule A_2 are given by equations (16.24) and (16.30), respectively. These are independent of the pressure (or volume) and require no adjustment or correction to the standard state.

Utilizing the foregoing facts, and representing the masses of the atoms and molecules by m_1 and m_2, respectively, equation (33.51) becomes

$$K_f = \frac{\left[Q_{e1} \dfrac{(2\pi m_1 kT)^{3/2}}{h^3} \cdot \dfrac{RT}{N}\right]^2}{Q_{e2} \left[\dfrac{(2\pi m_2 kT)^{3/2}}{h^3} \cdot \dfrac{RT}{N}\right] \dfrac{8\pi^2 I kT}{\sigma h^2} (1 - e^{-hc\omega/kT})^{-1}} e^{-\Delta H_0^0/RT},$$

where Q_{e1} and Q_{e2} are the respective electronic factors; I is the moment of inertia of the molecule A_2, σ is its symmetry number, i.e., 2, and ω is the vibration frequency in cm.$^{-1}$. In addition to ΔH_0^0, which is derived in the manner already described, the masses m_1 and m_2, obtained from the atomic and molecular weights, respectively, and the universal constants π, k, h, R, N and c, the other information required for the evaluation of the equilibrium constant K_f can be obtained from spectroscopic data. In some cases, e.g., dissociation of hydrogen, oxygen, nitrogen and the halogens, the value of ΔH_0^0 can also be derived from spectroscopic studies, and so the equilibrium constant can be calculated without any chemical or thermal measurements.

Problem: Calculate the equilibrium constant K_f for the dissociation of molecular iodine into atoms at 1000° C. The electronic state of the molecules may be taken as a singlet, and all the atoms may be regarded as being in the state for which the resultant quantum number j is $\frac{3}{2}$. The value of ΔH_0^0 is known from spectroscopic measurements to be 35,480 cal. The moment of inertia of molecular iodine is 743×10^{-40} g. cm.^{2} and the vibration frequency is 214 cm.$^{-1}$.

For atomic iodine, Q_{e1} is $2j + 1$, i.e., 4.00; m_1 is $127/N$, where N is the Avogadro number. For molecular iodine, Q_{e2} is 1, m_2 is $254/N$, and the symmetry number is 2. Inserting these values in the equation given above, together with

h, c and k in c.g.s. units and R in the brackets in cc.-atm. deg.$^{-1}$ mole^{-1}, it is found that at 1000° C, i.e., 1273° K,

$$K_f \text{ (for } I_2 = 2I) = 0.184.$$

The example given above provides a simple illustration of the use of moments of inertia and vibration frequencies to calculate equilibrium constants. The method can, of course, be extended to reactions involving more complex substances. For polyatomic, nonlinear molecules the rotational contributions to the partition functions would be given by equation (16.34), and there would be an appropriate term of the form of (16.30) for each vibrational mode.[11]

33s. Equilibrium Constants of Metathetic Reactions.—When a reaction involves a metathesis, or double decomposition, a very simple, if somewhat approximate, expression can be derived for the equilibrium constant, particularly if diatomic molecules only take part in the process, as in the case, for example, of the gas reaction

$$\underset{1}{AB} + \underset{2}{CD} = \underset{3}{AC} + \underset{4}{BD}.$$

For such a reaction, in which there is no change in the number of molecules, the equilibrium constant is independent of the standard state (§ 32c), so that it is possible to write equation (33.50) as

$$K = \frac{(Q^0/N)_3 \times (Q^0/N)_4}{(Q^0/N)_1 \times (Q^0/N)_2} e^{-\Delta H_0^0/RT} = \frac{Q_3 \times Q_4}{Q_1 \times Q_2} e^{-\Delta H_0^0/RT}, \qquad (33.55)$$

where, for simplicity, the symbols Q_1, Q_2, Q_3 and Q_4 are used to represent the partition functions, in their standard states, of the diatomic molecules AB, CD, AC and BD, respectively. For the diatomic molecules the electronic factors may be taken as unity; if they are not individually unity, the cancellation of the terms in equation (33.55) will lead to a value not very different from unity. The expression for the translational partition function [equation (33.54)] may be simplified in form; thus,

$$Q_t^0 = \frac{(2\pi mkT)^{3/2}}{h^3} RT = aM^{3/2},$$

where M is the molecular weight of the gas, and a is the same for all gases at definite temperature, independent of the nature of the substance. In a similar manner, the equation for the rotational partition function for a diatomic molecule may be written as

$$Q_r = \frac{8\pi^2 IkT}{\sigma h^2} = b\frac{I}{\sigma},$$

where b is also a constant, at a given temperature; I is the moment of inertia of the molecule. The four vibrational factors, i.e., $(1 - e^{-hc\omega/kT})^{-1}$, are not very different from unity, and, in any event, because of cancellation of terms, the complete contribution is virtually unity, and so they may be omitted. Making the appropriate substitutions, equation (33.55) becomes

$$K = \left(\frac{M_3 M_4}{M_1 M_2}\right)^{3/2} \frac{I_3 I_4}{I_1 I_2} \cdot \frac{\sigma_1 \sigma_2}{\sigma_3 \sigma_4} e^{-\Delta H_0^0/RT}$$

[11] For extensive reviews, see Zeise, *Z. Elek.*, **39**, 758, 895 (1933); **40**, 662, 885 (1934); **48**, 425, 476, 693 (1942).

or

$$\ln K = -\frac{\Delta H_0^0}{RT} + \tfrac{3}{2} \ln \frac{M_3 M_4}{M_1 M_2} + \ln \frac{I_3 I_4}{I_1 I_2} + \ln \frac{\sigma_1 \sigma_2}{\sigma_3 \sigma_4}. \quad (33.56)$$

The value of the equilibrium constant may thus be derived from ΔH_0^0 for the reaction, and the molecular weights, moments of inertia, and the symmetry numbers of the substances taking part. Equations of the type of (33.56) have been employed particularly in the study of isotopic exchange reactions, where the error due to the cancellation of the vibrational partition functions is very small, especially if the temperatures are not too high.[12]

Problem: Determine the equilibrium constant of the homogeneous gaseous isotopic exchange reaction between hydrogen and deuterium iodide, viz.,

$$H_2 + 2DI = 2HI + D_2$$

at 25° C. From spectroscopic measurements, ΔH_0^0 is found to be 83 cal., and the moments of inertia are as follows: $H_2(0.459 \times 10^{-40})$, $DI(8.55 \times 10^{-40})$, $HI(4.31 \times 10^{-40})$, $D_2(0.920 \times 10^{-40}$ g. cm.$^2)$.

The symmetry numbers for H_2 and D_2 are each 2, and for HI and DI they are both 1; consequently the symmetry numbers cancel, so that the corresponding factor is unity, and its logarithm in equation (33.56) is zero. The molecular weights are 2 for H_2, 129 for DI, 128 for HI and 4 for D_2, so that by equation (33.56), after converting the logarithms, at 298° K,

$$\log K = -\frac{\Delta H_0^0}{2.303 RT} + \tfrac{3}{2} \log \frac{M_{HI}^2 \times M_{D_2}}{M_{H_2} \times M_{DI}^2} + \log \frac{I_{HI}^2 \times I_{D_2}}{I_{H_2} \times I_{DI}^2}$$

$$= -\frac{83.0}{4.576 \times 298} + \tfrac{3}{2} \log \frac{(128)^2 \times 4}{2 \times (129)^2} + \log \frac{(4.31)^2 \times 0.920}{0.459 \times (8.55)^2}$$

$$= 0.0876,$$

$$K = 1.22.$$

EXERCISES *

1. At 250° C and 170 atm. pressure the value of K_p' for the methanol equilibrium, i.e., $CO + 2H_2 = CH_3OH(g)$, was found to be 2.1×10^{-2}. Apply the corrections for deviation from ideal behavior, and hence calculate K_f at 250° C.

2. The following equilibrium constants K_f, with fugacities in atm., were obtained at 700° C, the Fe, FeO and C being present as solids:

$$Fe + H_2O(g) = FeO + H_2 \qquad K_f = 2.35$$
$$Fe + CO_2 = FeO + CO \qquad 1.52$$
$$C + H_2O(g) = CO + H_2 \qquad 1.55.$$

Utilizing the free energy concept, and disregarding departure from ideal behavior, answer the following questions: (i) Will a mixture consisting of equimolecular proportions of water vapor and hydrogen at 1 atm. total pressure be capable of re-

---

[12] Cf. Urey and Rittenberg, *J. Chem. Phys.*, 1, 137 (1933); Rittenberg, Bleakney and Urey, *ibid.*, 2, 48 (1934); Urey and Greiff, *J. Am. Chem. Soc.*, 57, 321 (1935); Libby, *J. Chem. Phys.*, 11, 101 (1943).

* For heat content, heat capacity, entropy and free energy data, see tables in text or at end of book.

ducing FeO at 700° C? What is likely to be the effect of increasing the total pressure? (ii) Will a gas mixture consisting of 12 mole % carbon monoxide, 9 mole % carbon dioxide and 71 mole % of an inert gas (e.g., nitrogen) at a total pressure of 1 atm. tend to deposit carbon at 700° C? Above or below what pressure is this likely to occur? [Cf. Austin and Day, *Ind. Eng. Chem.*, **33**, 23 (1941).].

3. The values of K_p' for the ammonia equilibrium at 500° C and various pressures were found (Larson and Dodge, ref. 1) to be as follows:

P	50	100	300	600 atm.
K_p'	3.88×10^{-3}	4.02×10^{-3}	3.98×10^{-3}	6.51×10^{-3}

Calculate the fugacity correction (i) by using the generalized fugacity diagram, (ii) by means of the van der Waals constants (cf. § 29c); hence, determine the true equilibrium constant at 500° C. (Cf. Pease, ref. 2.)

4. Show that in the reaction $\tfrac{1}{2}N_2 + \tfrac{3}{2}H_2 = NH_3$, assuming ideal behavior of the gases, the maximum (equilibrium) conversion of nitrogen and hydrogen into ammonia, at any definite temperature and pressure, is obtained when the reacting gases are in the proportion of 1 to 3. (Suppose the *equilibrium* system consists of N_2 and H_2 molecules in the ratio of 1 to r; let x be the *mole fraction* of NH_3 formed at equilibrium. Hence, set up an expression for K_p in terms of mole fractions and total pressure, and find the condition which makes dx/dr equal to zero, i.e., for maximum conversion to NH_3.)

5. The following values were obtained for the equilibrium constant of the homogeneous gas reaction $H_2 + CO_2 = CO + H_2O$:

$t°$	600°	700°	800°	900° C
K	0.39	0.64	0.95	1.30

Use a graphical method, based on heat capacities, etc., to derive ΔH_0^0 for this reaction, and hence develop expressions for ΔF^0 and ΔH^0 as functions of the temperature. What are the standard free energy and entropy changes for the reaction at 25° C?

6. Check the results derived for 25° C in the preceding exercise by using tabulated free energy and entropy data. Also check the value for $\Delta(H^0 - H_0^0)$, i.e., $\Delta H^0 - \Delta H_0^0$, obtained.

7. For the reaction $C(s) + H_2O(g) = CO + H_2$, the equilibrium constant K_f is 2.78×10^{-2} at 800° K. Calculate the free energy change in cal. accompanying the reaction of steam at 100 atm. with carbon to form carbon monoxide and hydrogen each at 50 atm, (i) assuming ideal behavior of the gases, (ii) allowing for departure from ideal behavior. What is the standard free energy change of the reaction at 800° K?

8. Use ΔF^0 values at 25° C, together with heat capacity data, to derive an expression for the variation with temperature of the equilibrium constant of the reaction $SO_2 + \tfrac{1}{2}O_2 = SO_3(g)$. What would be the value of K_f at 450° C?

9. A mixture of sulfur dioxide and oxygen, in the molecular proportion of 2 to 1, is allowed to come to equilibrium at 450° C at (i) 1 atm. (ii) 200 atm., total pressure; estimate the composition of the final mixture in each case, allowing for deviations from ideal behavior. (Use the value of K_f obtained in the preceding exercise.)

10. Utilizing heat content, heat capacity and entropy data, together with the generalized fugacity diagram, make a thermodynamic study of the reaction

$C_2H_4(g) + H_2O(g) = C_2H_5OH(g)$. Determine the extent of conversion of an equimolar mixture of ethylene and water vapor into ethanol at 450° C and 200 atm. [cf. Aston, *Ind. Eng. Chem.*, **34**, 514 (1942)].

11. The dissociation pressures of silver oxide at several temperatures were found by Benton and Drake [*J. Am. Chem. Soc.*, **54**, 2186 (1932)] to be as follows:

$t°$	173°	178°	183.1°	188.2° C
p	422	509	605	717 mm.

In this temperature range C_P for Ag is given by $5.60 + 1.5 \times 10^{-3}T$, for Ag_2O by $13.87 + 8.9 \times 10^{-3}T$, and for O_2 by $6.50 + 1.0 \times 10^{-3}T$. Determine ΔH^0, ΔF^0 and ΔS^0 for the reaction $Ag_2O = 2Ag + \tfrac{1}{2}O_2$ at 25° C.

12. The E.M.F. of the cell $H_2(g) \mid KOH\ aq \mid HgO(s), Hg(l)$, in which the reaction for the passage of two faradays is $H_2(g) + HgO(s) = H_2O(l) + Hg(l)$, is 0.9264 volt at 25° C. Use this result together with the known free energy of formation of mercuric oxide to calculate the standard free energy of formation of liquid water at 25° C. (The fugacity of the hydrogen gas may be taken as 1 atm.)

13. The temperature coefficient of the standard E.M.F. of the cell $Pb \mid PbCl_2$ sat. soln. $\mid AgCl(s)$, Ag is -1.86×10^{-4} volt deg.$^{-1}$; the cell reaction for the passage of two faradays is $Pb + 2AgCl(s) = PbCl_2(s) + 2Ag$. Determine the standard entropy change for the reaction at 25° C. Calculate ΔS^0 by using the third law entropies in Table XV, and state whether the results tend to confirm the third law of thermodynamics.

14. Using heat content data, determine ΔH^0 at 25° C for the reaction considered in the preceding exercise; hence calculate ΔF^0 for the reaction at 25° C. What is the E.M.F. of the cell at this temperature? (The experimental value is 0.4900 volt.)

15. The reaction $Ag(s) + \tfrac{1}{2}Hg_2Cl_2(s) = AgCl(s) + Hg(l)$ takes place in a reversible cell whose E.M.F. is 0.0455 volt at 25° C and 1 atm. pressure. Calculate the free energy change of the reaction. By using equation (25.21), estimate (i) the free energy change of the cell reaction, (ii) the E.M.F. of the cell, at 1000 atm. and 25° C. (The following densities, which may be taken as the mean values over the pressure range, may be used: Ag, 10.50; AgCl, 5.56; Hg, 13.55; $HgCl_2$ 7.15 cc. g.$^{-1}$.)

16. Utilize the free energy function and relative heat content tables to determine the standard free energy change of the reaction $CH_4 + CO_2 = 2CO + 2H_2$ at 1000° K. (Use heat content data from Table V to obtain ΔH^0 at 25° C.)

17. Suppose nothing were known of the $\tfrac{1}{2}N_2 + \tfrac{3}{2}H_2 = NH_3$ reaction except the heat of reaction at 25° C. Using the moments of inertia and vibration frequencies of the molecules given in Chapter VI, calculate the free energy functions and relative heat contents, by equations (33.43) and (33.47), and utilize them to determine the equilibrium constant of the reaction at 500° C (cf. Exercise 3). [See Stephenson and McMahon, *J. Am. Chem. Soc.*, **61**, 437 (1939).]

18. By making use of the expression for the variation of C_P for graphite with temperature, calculate its free energy function and heat content, relative to the value at 0° K, at 1000° K.

19. Calculate the equilibrium degree of dissociation of molecular hydrogen into atoms at 3000° K and 1 atm. pressure. The j value for atomic hydrogen in the ground state is $\tfrac{1}{2}$ and no higher electronic states need be considered; molecular hydrogen exists in a singlet state. The value of ΔH_0^0 for the reaction $H_2 = 2H$ is known from spectroscopic measurements to be 102.8 kcal.

20. Determine the equilibrium constant of the isotopic exchange reaction $H_2 + D_2 = 2HD$ at 25° C, given that ΔH_0^0 is 157 cal. (The experimental value is 3.28.)

21. Estimate the extent of dissociation of water vapor into hydrogen atoms and hydroxyl radicals, by calculating the equilibrium constant of the reaction $H_2O = H + OH$, at 2000° K, using the method of partition functions. The electronic statistical weights in the normal state are 1 for H_2O, 2 for H and 4 for OH; higher energy states may be neglected. (Utilize the vibration frequencies and moments of inertia in Chapter VI, and take ΔH_0^0 as equal to the bond energy of the O—H bond which is broken in the reaction.)

22. In the reaction $Fe(s) + H_2O(g) = FeO(s) + H_2(g)$, the density of the iron is greater than that of the oxide. How does the constant f_{H_2}/f_{H_2O} vary with pressure?

23. Determine the standard free energy change for the process $CH_3OH(g) = CH_3OH(l)$ at 25° C on the assumption that the vapor behaves as a Berthelot gas. The vapor pressure of liquid CH_3OH is 122 mm. at 25° C. (Cf. problem in § 33k, where the vapor is treated as an ideal gas.)

CHAPTER XIV

THE PROPERTIES OF SOLUTIONS

34. Ideal Solutions

34a. Properties of Ideal Solutions.—An **ideal liquid solution** is defined as *one which obeys an idealized form of Raoult's law over the whole range of composition at all temperatures and pressures*. According to the original form of this law, the partial vapor pressure p_i of any constituent of a liquid solution is equal to the product of its mole fraction N_i and its vapor pressure $p_i^{\square}$ in the pure state, i.e.,

$$p_i = N_i p_i^{\square}.$$

For thermodynamic purposes, however, it is preferable to use the idealized form

$$f_i = N_i f_i^{\square}, \tag{34.1}$$

where f_i is the fugacity of the constituent i, either in the vapor or in the solution in equilibrium with it, since they are identical (§ 31a), and $f_i^{\square}$ is the fugacity of the pure liquid at the same temperature and pressure.† It will be seen below, at least for a solution of two components, that if equation (34.1) applies to one constituent over the whole range of composition it must also apply to the other. It follows, therefore, that *in an ideal solution the fugacity of each component is proportional to the mole fraction in the given solution at all concentrations;* the proportionality constant is equal to the fugacity of that component in the pure state at the temperature and pressure of the system.[1]

The use of the foregoing definition of an ideal solution implies certain properties of such a solution. The variation of the fugacity $f_i^{\square}$ of a pure liquid i with temperature, at constant pressure and composition, is given by equation (29.22), viz.,

$$\left(\frac{\partial \ln f_i^{\square}}{\partial T}\right)_P = \frac{H_i^* - H_i^{\square}}{RT^2}, \tag{34.2}$$

where $H_i^{\square}$ is the molar heat content of the pure liquid at pressure P, and H_i^* is that of the ideal vapor, i.e., at low pressure, so that $H_i^* - H_i^{\square}$ is the

---

† It is the common practice to represent the fugacity of a pure liquid by the symbol f^0; this is misleading, since the fugacity in the standard state is thus implied. To avoid the possibility of confusion the symbol $f^{\square}$ is used here for the pure liquid, and f^0 is retained for the standard state, i.e., pure liquid *at 1 atm. pressure*. At this pressure $f^{\square}$ and f^0 become identical, but in general the quantity $f^{\square}$ is dependent on the pressure, whereas f^0 is not.

[1] For a review of the properties of ideal solutions, see J. H. Hildebrand, "Solubility of Non-electrolytes," 2nd ed., 1936, Chapter II.

ideal heat of vaporization (§ 29g). For the same substance when present in solution, equation (30.21) is applicable (cf. § 31a); thus,

$$\left(\frac{\partial \ln f_i}{\partial T}\right)_{P,N} = \frac{H_i^* - \bar{H}_i}{RT^2}, \tag{34.3}$$

where $\bar{H}_i$ is here the partial molar heat content of i in the liquid solution. Upon subtracting equation (34.2) from (34.3), the result is

$$\left[\frac{\partial \ln (f_i/f_i^\square)}{\partial T}\right]_{P,N} = \frac{H_i^\square - \bar{H}_i}{RT^2}. \tag{34.4}$$

This expression will hold for any solution, but in the special case of an ideal solution, $f_i/f_i^\square$ is equal to N_i, the mole fraction of that constituent, by equation (34.1). Since the mole fraction, for a mixture of definite composition, is independent of the temperature, it follows that for an ideal solution $H_i^\square - \bar{H}_i$ must be zero, i.e., $H_i^\square$ and $\bar{H}_i$ are identical. Consequently, the partial molar heat content of any constituent of an ideal solution ($\bar{H}_i$) is equal to the molar heat content of that substance in the pure liquid state ($H_i^\square$). As a result, equation (26.6), which for heat contents takes the form

$$H = n_1\bar{H}_1 + n_2\bar{H}_2 + \cdots,$$

becomes

$$H = n_1 H_1^\square + n_2 H_2^\square + \cdots.$$

The total heat content H of the mixture is thus equal to the sum of individual heat contents of the pure liquid constituents; hence, there is no heat change upon mixing the components of an ideal solution. In this respect, therefore, an ideal liquid solution resembles an ideal gas mixture (§ 30d).

By differentiating $\ln f_i$ and $\ln f_i^\square$ with respect to pressure, at constant temperature and composition, utilizing equations (29.7) and (30.17), and adopting a procedure similar to that given above, it can be readily shown that

$$\left[\frac{\partial \ln (f_i/f_i^\square)}{\partial P}\right]_{T,N} = \frac{\bar{V}_i - V_i^\square}{RT}. \tag{34.5}$$

As before, $f_i/f_i^\square$ is constant for an ideal solution, so that the partial molar volume $\bar{V}_i$ of each constituent must always be equal to the molar volume $V_i^\square$ of the substance in the pure liquid state, at the same temperature and pressure. It follows then that there is no change of volume when the liquid components of an ideal solution are mixed, so that in this respect, also, an ideal liquid solution resembles an ideal gas mixture (§ 30c).

34b. The Duhem-Margules Equation.—For a system consisting of a liquid solution of two components in equilibrium with their vapors, at *constant temperature and pressure*, the condition for an infinitesimal change of composition is given by the Gibbs-Duhem equation (26.15) in the form

$$n_1 d\mu_1 + n_2 d\mu_2 = 0, \tag{34.6}$$

where n_1 and n_2 are the numbers of moles of the constituents 1 and 2, respectively, present in the solution, and μ_1 and μ_2 are their partial molar free energies or chemical potentials. If equation (34.6) is divided by $n_1 + n_2$, the result is

$$\frac{n_1}{n_1 + n_2} d\mu_1 + \frac{n_2}{n_1 + n_2} d\mu_1 = 0,$$

$$N_1 d\mu_1 + N_2 d\mu_2 = 0, \qquad (34.7)$$

where N_1 and N_2 are the mole fractions of the respective components.

The chemical potential of any constituent of a solution depends on the temperature, (total) pressure and composition of the solution; if the temperature and pressure are maintained constant, however, the chemical potential is determined by the composition only. It is then possible, therefore, to write, for an infinitesimal change of composition,

$$d\mu_i = \left(\frac{\partial \mu_i}{\partial N_i}\right)_{T,P} dN_i,$$

and upon substitution in equation (34.7) it follows that

$$N_1 \left(\frac{\partial \mu_1}{\partial N_1}\right)_{T,P} dN_1 + N_2 \left(\frac{\partial \mu_2}{\partial N_2}\right)_{T,P} dN_2 = 0, \qquad (34.8)$$

$$\left(\frac{\partial \mu_1}{\partial \ln N_1}\right)_{T,P} dN_1 + \left(\frac{\partial \mu_2}{\partial \ln N_2}\right)_{T,P} dN_2 = 0. \qquad (34.9)$$

Since the sum of the two mole fractions is equal to unity, i.e., $N_1 + N_2 = 1$, it is seen that

$$dN_1 + dN_2 = 0 \quad \text{or} \quad dN_1 = -dN_2,$$

so that equation (34.9) can be written

$$\left(\frac{\partial \mu_1}{\partial \ln N_1}\right)_{T,P} - \left(\frac{\partial \mu_2}{\partial \ln N_2}\right)_{T,P} = 0, \qquad (34.10)$$

giving a useful form of the Gibbs-Duhem equation.

According to equation (31.1), the chemical potential of any constituent of a liquid mixture is represented by

$$\mu_i = \mu^* + RT \ln f_i,$$

where f_i represents the fugacity of the given constituent in the liquid or in the vapor phase with which it is in equilibrium; μ^* is a constant for the substance at constant temperature. Upon differentiation at constant temperature, it is seen that

$$d\mu_i = RT d\ln f_i \qquad (34.11)$$

[cf. equation (30.6)], and insertion of this result into (34.10) gives

$$\left(\frac{\partial \ln f_1}{\partial \ln N_1}\right)_{T,P} = \left(\frac{\partial \ln f_2}{\partial \ln N_2}\right)_{T,P}. \qquad (34.12)$$

This is the precise form of what is known as the **Duhem-Margules equation**, derived independently, and in various ways, by J. W. Gibbs (1876), P. Duhem (1886), M. Margules (1895), and R. A. Lehfeldt (1895). It is frequently encountered and employed in a less exact form which is based on the approximation that the vapor behaves as an ideal gas. In this event, the fugacity of each component in the vapor may be replaced by its respective partial (vapor) pressure, so that equation (34.12) becomes

$$\left(\frac{\partial \ln p_1}{\partial \ln N_1}\right)_{T,P} = \left(\frac{\partial \ln p_2}{\partial \ln N_2}\right)_{T,P}, \qquad (34.13)$$

where p_1 and p_2 are the partial vapor pressures of the two constituents in equilibrium with the liquid containing the mole fractions N_1 and N_2, respectively, of these constituents.

It is important to note that equation (34.12) is applicable to *any liquid solution of two constituents*, irrespective of whether the solution (or vapor) is ideal or not. In the derivation of this equation no assumption or postulate was made concerning the properties of the solution; the results are based only on thermodynamic considerations, and hence they are of completely general applicability. The form given in equation (34.13) is also independent of the ideality or otherwise of the solution, but it involves the supposition that the vapor in equilibrium with it behaves ideally.

34c. Application of Raoult's Law to Both Constituents of an Ideal Solution.—By means of the Duhem-Margules equation, it can be shown that if the Raoult equation (34.1) is applicable to one constituent of an ideal binary solution, over the whole range of composition, it must also apply to the other constituent. Suppose the law holds for the constituent 1, so that

$$f_1 = N_1 f_1^\square.$$

Upon taking logarithms and differentiating with respect to N_1, at constant temperature and pressure, it is found that

$$d \ln f_1 = d \ln N_1$$

or

$$\left(\frac{\partial \ln f_1}{\partial \ln N_1}\right)_{T,P} = 1, \qquad (34.14)$$

since $f_1^\square$ is constant for a given temperature and external pressure. If this result, which is applicable at all concentrations, is compared with equation (34.12), it is seen that

$$\left(\frac{\partial \ln f_2}{\partial \ln N_2}\right)_{T,P} = 1, \qquad (34.15)$$

also at all concentrations. Upon integration, bearing in mind that f_2 becomes equal to $f_2^\square$ when N_2 is unity, it follows that

$$f_2 = N_2 f_2^\square.$$

It is clear, therefore, that *if Raoult's law is applicable to one of the constituents of a liquid mixture, at all compositions, it must be equally applicable to the other constituent.*

Attention may be called here to the restriction in connection with Raoult's law that the total (external) pressure, under which the system is in equilibrium, should remain constant as the composition is changed, at a given temperature. This has been implied in the foregoing treatment. At moderate pressures, however, the fugacity (or vapor pressure) is virtually independent of the external pressure (§ 27m); hence, in many of the practical applications of Raoult's law given below, the restriction of constant total pressure is not emphasized, although for strict accuracy it should be understood.[2]

34d. Vapor Pressure Curves.—Although precise thermodynamic treatment always requires the use of the fugacity in the equations given above, it is the vapor pressure which is the corresponding property of practical interest. Provided the pressure is not too high, the vapor will not depart greatly from ideal behavior, and it is permissible to use Raoult's law in its original form

$$p_i = \mathrm{N}_i p_i^\square, \qquad (34.16)$$

as given at the commencement of this chapter. If equation (34.16) is applicable over the whole range of concentrations for one constituent, it must hold for the other, so that

$$p_1 = \mathrm{N}_1 p_1^\square \quad \text{and} \quad p_2 = \mathrm{N}_2 p_2^\square,$$

for all mixtures of the two constituents 1 and 2. The plot of the partial pressure of each constituent against its mole fraction in the liquid phase, at constant temperature, should be a straight line passing

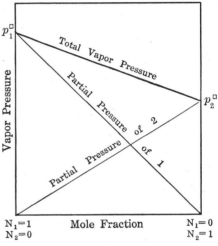

Fig. 20. Vapor pressure curves for ideal solution

through the origin, as shown by the thin lines in Fig. 20. The total vapor pressure P is the sum of the two partial pressures, i.e., $P = p_1 + p_2$; hence, since N_2 is equal to $1 - \mathrm{N}_1$,

$$\begin{aligned} P &= \mathrm{N}_1 p_1^\square + \mathrm{N}_2 p_2^\square \\ &= \mathrm{N}_1 p_1^\square + (1 - \mathrm{N}_1) p_2^\square = p_2^\square + \mathrm{N}_1 (p_1^\square - p_2^\square), \end{aligned} \qquad (34.17)$$

so that the total vapor pressure also varies in a linear manner with the mole fraction of either component in the liquid phase.

[2] Cf. Krichevsky and Kasarnowsky, *J. Am. Chem. Soc.*, **57**, 2168 (1935); Bury, *Trans. Farad. Soc.*, **36**, 795 (1940).

34e. Composition of Liquid and Vapor in Equilibrium.—The composition of the vapor phase in equilibrium with any liquid solution is readily derived from the fact that the number of moles (or the mole fraction) of each constituent in the vapor must be proportional to its partial pressure, assuming the vapor to behave as an ideal gas. If N_1' and N_2' represent the mole fractions of the two components in the vapor, then

$$\frac{N_1'}{N_2'} = \frac{p_1}{p_2} = \frac{N_1 p_1^\square}{N_2 p_2^\square}, \qquad (34.18)$$

provided Raoult's law is applicable.

Problem: Mixtures of benzene and toluene behave almost ideally; at 30° C, the vapor pressure of pure benzene is 118.2 mm. and that of pure toluene is 36.7 mm. Determine the partial pressures and weight composition of the vapor in equilibrium with a liquid mixture consisting of equal weights of the two constituents.

Let w be the weight of each constituent in the mixture; if the molecular weights are M_1 and M_2, respectively, the mole fractions N_1 and N_2 are

$$N_1 = \frac{w/M_1}{(w/M_1) + (w/M_2)} = \frac{M_2}{M_1 + M_2}, \quad \text{and} \quad N_2 = \frac{M_1}{M_1 + M_2}.$$

The molecular weight of benzene (M_1) is 78.0, and of toluene (M_2) it is 92.0, so that

$$N_1 = \frac{92.0}{78.0 + 92.0} = 0.541, \quad \text{and} \quad N_2 = 0.459.$$

The partial vapor pressures, assuming Raoult's law and ideal behavior of the vapors, are then $p_1 = 0.541 \times 118.2 = 64.0$ mm., and $p_2 = 0.459 \times 36.7 = 16.8$ mm.

The ratio of the numbers of moles (or mole fractions) of the two constituents in the vapor is thus 64.0/16.8; the proportions by weight are then obtained by multiplying by the respective molecular weights, viz.,

$$\frac{\text{wt. of benzene in vapor}}{\text{wt. of toluene in vapor}} = \frac{64.0 \times 78}{16.8 \times 92} = 3.23.$$

Consequently, although the liquid contains equal weights of benzene and toluene, the vapor, at 30° C, should contain 3.23 parts by weight of the former to 1 part of the latter. (For alternative method of obtaining the weight composition, see Exercise 23.)

By replacing N_2' by $1 - N_1'$, and N_2 by $1 - N_1$ in equation (34.18), and then combining it with equation (34.17) so as to eliminate N_1, it follows that

$$P = \frac{p_1^\square p_2^\square}{p_1^\square - N_1'(p_1^\square - p_2^\square)}. \qquad (34.19)$$

From this result it is possible to calculate the total vapor pressure P for any value N_1', the mole fraction of the constituent 1 in the vapor. It is evident from equation (34.19) that *the total vapor pressure is not a linear function of the mole fraction composition of the vapor*, although it is such a function of the composition of the liquid phase, in the case of an ideal solution. By utilizing either equation (34.17) or (34.19) it is possible to plot the variation of the total pressure with the mole fraction of either constituent of the vapor. The result is shown in Fig. 21, in which the upper (linear) and lower (nonlinear) curves represent the total vapor pressure as a function of the mole fraction composition of the liquid (N_1) and of the vapor (N_1'),

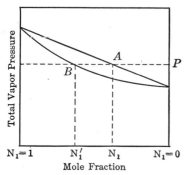

FIG. 21. Composition of liquid and vapor for ideal solution

respectively. At any value of the total pressure P, the point A consequently gives the composition of the liquid which is in equilibrium with a vapor of composition represented by the point B.

It can be seen from equation (34.18) that for an ideal liquid system, the vapor will always be relatively richer than the liquid in the more volatile constituent, i.e., the one with the higher vapor pressure. For example, if, as in Fig. 21, the component 1 is the more volatile, $p_1^\square$ will be greater than $p_2^\square$, and hence N_1'/N_2' will exceed N_1/N_2, and the vapor will contain relatively more of the component 1 than does the liquid with which it is in equilibrium. This fact is fundamental to the separation of liquids by fractional distillation. The limitations arising in connection with nonideal systems will be considered in §§ 35b, 35c.

34f. Influence of Temperature.—When the temperature of a liquid solution is changed the composition of the vapor in equilibrium with it will be affected accordingly. The relation between the mole fraction of a given constituent in the liquid and vapor phases can be readily determined by making use of the fact the fugacity must be the same in each phase if the system is to remain in equilibrium. Let f_i' and f_i represent the fugacities of the constituent i in the vapor and liquid phases, respectively, and suppose the temperature is increased by dT, at constant pressure, the system remaining in equilibrium. The change in f_i', i.e., df_i' or $d \ln f_i'$, must then be equal to the change in f_i, i.e., df_i or $d \ln f_i$. Since the composition of each phase changes at the same time, it follows that

$$\left(\frac{\partial \ln f_i'}{\partial T}\right)_{P, N'} dT + \left(\frac{\partial \ln f_i'}{\partial N_i'}\right)_{T, P} dN_i'$$
$$= \left(\frac{\partial \ln f_i}{\partial T}\right)_{P, N} dT + \left(\frac{\partial \ln f_i}{\partial N_i}\right)_{T, P} dN_i. \quad (34.20)$$

The variation of $\ln f_i'$, for the vapor phase, with temperature is given by equation (30.21), and that of $\ln f_i$, for the liquid phase, by an equation of the same form (cf. § 31a), so that equation (34.20) becomes

$$\left(\frac{\partial \ln f_i'}{\partial N_i'}\right)_{T, P} dN_i' - \left(\frac{\partial \ln f_i}{\partial N_i}\right)_{T, P} dN_i = \frac{\bar{H}_i' - \bar{H}_i}{RT^2} dT, \quad (34.21)$$

where $\bar{H}'_i$ is the partial molar heat content of the constituent i in the vapor phase and $\bar{H}_i$ is that in the liquid phase.

The result just obtained is applicable to any liquid-vapor equilibrium, irrespective of the behavior of the gas phase or the solution; if, however, these are assumed to be ideal, equation (34.21) can be greatly simplified. For an ideal gas mixture, the fugacity f'_i, or partial pressure, of any constituent is proportional to its mole fraction N'_i, at constant temperature and total pressure (§ 5b); it can be readily seen, therefore, that

$$\left(\frac{\partial \ln f'_i}{\partial \text{N}'_i}\right)_{T,P} = \frac{1}{\text{N}'_i}.$$

Similarly, it follows from equation (34.1) that for the liquid solution

$$\left(\frac{\partial \ln f_i}{\partial \text{N}_i}\right)_{T,P} = \frac{1}{\text{N}_i}.$$

These results can be inserted into equation (34.21), and at the same time the partial molar heat content terms $\bar{H}_i$ and $\bar{H}'_i$ may be replaced by the corresponding molar heat contents H_i and H'_i, respectively, for the pure substances, since ideal behavior has been postulated for both phases (cf. §§ 30d, 34a); it is then seen that

$$\left[\frac{\partial \ln (\text{N}'_i/\text{N}_i)}{\partial T}\right]_P = \frac{H'_i - H_i}{RT^2} = \frac{\Delta H_v}{RT^2}, \qquad (34.22)$$

the numerator on the right-hand side being equal to ΔH_v, the molar heat of vaporization of the constituent i at the temperature T and pressure P. If the ratio of the mole fraction in the vapor phase to that in the liquid, i.e., N'_i/N_i, is represented by k_i, it follows that for any constituent of an ideal system

$$\left(\frac{\partial \ln k_i}{\partial T}\right)_P = \frac{\Delta H_v}{RT^2}. \qquad (34.23)$$

It was seen in § 34e that if component 1 is the more volatile, the ratio N'_1/N'_2 will exceed N_1/N_2. For purposes of fractional distillation it is desirable to make this difference as large as possible; in other words, the quantity $\text{N}'_1\text{N}_2/\text{N}_1\text{N}'_2$, which is equivalent to k_1/k_2, should be large. By equation (34.23)

$$\left(\frac{\partial \ln (k_1/k_2)}{\partial T}\right)_P = \frac{\Delta H_1 - \Delta H_2}{RT^2}, \qquad (34.24)$$

and hence it is of interest to find the conditions which will tend to increase the value of k_1/k_2. According to Trouton's law (§ 27j), $\Delta H_v/T_b$ is approximately constant for all substances; hence, the lower the boiling point T_b, that is to say, the more volatile the liquid, the smaller the value of ΔH_v, the heat of vaporization. Since constituent 1 is the more volatile, ΔH_1 is less than ΔH_2, and the right-hand side of equation (34.24) will usually be negative. The value of k_1/k_2 will thus increase with decreasing temperature, and hence better fractionation can be achieved by distilling at a low temperature. For practical purposes, this means that distillation should be carried out under reduced pressure, for in this way the liquid can be made to boil at a lower temperature than normal.

34g. The Solubility of Gases.—A solution of a gaseous substance in a liquid may be treated from the thermodynamic standpoint as a liquid mixture, and if it behaves ideally Raoult's law will be applicable. If the gas is designated as constituent 2, then $f_2 = N_2 f_2^\square$, where f_2 is the fugacity in the solution, and $f_2^\square$ is the fugacity of the pure substance in the *liquid state* at the same temperature and external pressure. If the gas is assumed to be ideal, the fugacities may be replaced by the respective pressures, so that

$$p_2 = N_2 p_2^\square, \tag{34.25}$$

which may be interpreted as giving the saturation solubility, i.e., N_2 mole fraction, of an ideal gas at a partial pressure p_2 and a specified temperature; $p_2^\square$ is then the vapor pressure of the liquefied gas at the same temperature and total pressure. It is thus possible to calculate the solubility of an ideal gas in an ideal solution at any temperature below the critical point, provided the vapor pressure of the liquid form at the same temperature is known. In this connection, the very small effect of the total external pressure on the vapor pressure (cf. § 27m) is neglected.[3]

Problem: The vapor pressure of liquid ethane at 25° C is 42 atm. Calculate the ideal solubility of ethane in a liquid solvent at the same temperature at a pressure of 1 atm., assuming the gas to behave ideally.

Utilizing equation (34.25), $p_2^\square$ is 42 atm., and p_2 is specified as 1 atm.; hence,

$$N_2 = \tfrac{1}{42} = 0.024 \text{ mole fraction.}$$

The ideal solubility of ethane at 25° C and 1 atm. pressure in any solvent should be 0.024 mole fraction.

The solubilities of ethane in hexane and heptane are not very different from the ideal value calculated from equation (34.25). In solvents which are not chemically similar to the solute, however, appreciable departure from ideal behavior would probably occur, and the solubilities would differ from the ideal value calculated above.

In order to extend the method of calculating ideal solubilities of gases to temperatures above the critical point, when the liquid cannot exist and direct determination of the vapor pressure $p_2^\square$ is not possible, it is necessary to estimate a hypothetical vapor pressure by suitable extrapolation. This is best done with the aid of the integrated form of the Clausius-Clapeyron equation. If the vapor pressures at any two temperatures are known, the hypothetical value at a temperature above the critical point may be evaluated on the assumption that the heat of vaporization remains constant.

Problem: The vapor pressure of liquid methane is 25.7 atm. at − 100° C and 11.84 atm. at − 120° C. Estimate the ideal solubility of methane in any solvent at 25° C and 1 atm. pressure.

[3] Hildebrand, *op. cit.*, ref. 1, p. 30.

At $-100°$ C, i.e., $173°$ K, p is 25.7 atm., and at $-120°$ C, i.e., $153°$ K, p is 11.84 atm.; hence by the integrated Clausius-Clapeyron equation (27.14),

$$\log \frac{25.7}{11.84} = \frac{\Delta H_v}{4.576} \left(\frac{173 - 153}{153 \times 173} \right),$$

$$\Delta H_v = 2{,}037 \text{ cal. mole}^{-1}.$$

Utilizing this value of the heat of vaporization, which is assumed to remain constant, the hypothetical vapor pressure p at $25°$ C, i.e., $298°$ K, is derived from the analogous expression

$$\log \frac{p}{25.7} = \frac{2{,}037}{4.576} \left(\frac{298 - 173}{173 \times 298} \right),$$

$$p = 308 \text{ atm.}$$

This gives $p_2^\square$ to be used in equation (34.25); hence, if the gas pressure p_2 is 1 atm., the ideal solubility of methane at $25°$ C and 1 atm., is $N_2 = 1/308$, i.e., 0.00325, mole fraction, in any solvent. (The observed solubility in hexane is 0.0031 and in xylene it is 0.0026.)

According to equation (34.25), the *ideal solubility* of an *ideal gas* under a pressure of 1 atm. should be equal to $1/p_2^\square$, where $p_2^\square$ is the vapor pressure of the liquefied gas at the experimental temperature and pressure. It follows, therefore, that the smaller this vapor pressure the greater will be the solubility of the gas at a given temperature. Gases which are difficult to liquefy, that is to say, those having low boiling points, will have high vapor pressures; such gases will consequently have low solubilities. It is thus to be concluded that if solutions of gases in liquids behaved ideally, the easily liquefiable gases would be the more soluble in a given solvent at a particular temperature. Even though most solutions of gases are not ideal, this general conclusion is in agreement with experimental observation.

34h. Influence of Temperature on Gas Solubility.—The influence of temperature on the solubility of a gas is given by equation (34.22) in the form

$$\left[\frac{\partial \ln (N_2'/N_2)}{\partial T} \right]_P = \frac{\Delta H_v}{RT^2}, \tag{34.26}$$

provided both gas and liquid solutions are ideal. In the special case in which the partial pressure of the gas, and hence its mole fraction N_2', remains constant,* equation (34.26) becomes

$$\left(\frac{\partial \ln N_2}{\partial T} \right)_{P, p_2} = - \frac{\Delta H_v}{RT^2}, \tag{34.27}$$

where ΔH_v is the molar heat of vaporization of the liquefied gas at the given temperature T and total pressure P.

It should be noted that $-\Delta H_v$ may also be identified with the heat of solution of the gas. The latter can be determined, in principle, by liquefying

---

* This condition would be approximated at total constant pressure if the vapor pressure of the solvent were negligible in comparison with the pressure of the gas.

the gas, at the experimental temperature and pressure, and adding the resulting liquid to the liquid solvent so as to make the saturated solution. If the solution is ideal, there is no change of heat content in the second stage of this process (§ 34a); hence the heat of solution is equal to the heat change in the first stage, and this is equal in magnitude, but opposite in sign, to the heat of vaporization of the liquid constituent 2 under the specified conditions. A further consequence of the foregoing argument is that the heat of solution of a given (ideal) gas forming an ideal solution is independent of the nature of the solvent.

Since it is unlikely that the latent heat of vaporization of the liquefied gas can be determined at normal temperatures and pressures, it is more convenient for practical purposes to regard $-\Delta H_v$ as the heat of solution per mole of gas. If this may be taken as independent of temperature, equation (34.27) can be integrated so as to yield an expression for the variation of the solubility of the gas with temperature. On the other hand, if the solubilities at two or more temperatures are known, the heat of solution can be calculated. It may be noted that since the heat of vaporization ΔH_v is always positive, the ideal solubility of a gas, at a given pressure, should decrease with increasing temperature. The relatively few cases in which this rule fails, e.g., certain aqueous solutions of hydrogen halides, are to be attributed to marked deviations from ideal behavior (cf. § 35a).

34i. Solid-Liquid Equilibria.—When one of the components of a binary solution separates as a solid phase, the system may be considered from two points of view. If the solid phase is that of the solute, i.e., the constituent present in smaller proportion, the composition of the solution may be regarded as giving the **saturation solubility** of that constituent at the temperature and pressure of the system. On the other hand, if the solvent, i.e., the constituent present in excess, separates as a solid, the temperature is said to be the **freezing point** of the solution at the particular composition and pressure. If the two components of the system are such that it is not desirable to distinguish between solvent and solute, the latter point of view is usually adopted. However, as far as thermodynamics is concerned the two types of system are fundamentally the same, since each involves an equilibrium between a binary liquid solution and a solid phase of one of the constituents. The subject will, therefore, be first treated in a general manner and the results will later be applied to special cases.

In the preceding paragraph, it has been tacitly assumed that a pure solid separates from the liquid solution in each case. Although this is frequently true, there are many instances in which the solid phase is a solid solution containing both constituents of the system. The most general case will be examined first, and the simplifications which are possible when the solid consists of a single substance will be introduced later. When the temperature of a system consisting of a binary liquid solution in equilibrium with a solid solution is changed, the compositions of both phases will be altered if the system is to remain in equilibrium. However, in spite of the change in composition and temperature, the fugacity of a given constituent must always have the same value in both phases. By following a procedure similar to that used in connection with the analogous problem in § 34f, it is found that [cf. equation (34.21)]

$$\left(\frac{\partial \ln f'_i}{\partial \mathrm{N}'_i}\right)_{T,P} d\mathrm{N}'_i - \left(\frac{\partial \ln f_i}{\partial \mathrm{N}_i}\right)_{T,P} d\mathrm{N}_i = \frac{\bar{H}'_i - \bar{H}_i}{RT^2} dT, \quad (34.28)$$

where $\bar{H}_i$ is the partial molar heat content of the constituent i in the liquid phase and $\bar{H}'_i$ is that in the solid phase.* As before, this result can be simplified by postulating ideal behavior for both liquid and solid solutions; that is to say, it is supposed that at constant temperature and pressure, the fugacity of any constituent in each phase is proportional to its mole fraction in that phase. In these circumstances,

$$\left(\frac{\partial \ln f'_i}{\partial \mathrm{N}'_i}\right)_{T,P} = \frac{1}{\mathrm{N}'_i} \quad \text{and} \quad \left(\frac{\partial \ln f_i}{\partial \mathrm{N}_i}\right)_{T,P} = \frac{1}{\mathrm{N}_i},$$

and, at the same time, H_i and $\bar{H}'_i$ become equal to the molar heat contents of pure liquid and solid i, respectively; hence, equation (34.28) reduces to

$$\left[\frac{\partial \ln (\mathrm{N}'_i/\mathrm{N}_i)}{\partial T}\right]_P = \frac{H'_i - H_i}{RT^2} = -\frac{\Delta H_f}{RT^2}, \quad (34.29)$$

where ΔH_f is the molar heat of fusion of i at the given temperature and pressure.

In the special case in which the solid which separates from the liquid solution is a pure substance, N'_i is unity at all temperatures and pressures; equation (34.29) then becomes

$$\left(\frac{\partial \ln \mathrm{N}_i}{\partial T}\right)_P = \frac{\Delta H_f}{RT^2}. \quad (34.30)$$

This equation gives the variation of the mole fraction solubility with temperature or of the freezing point with composition, at constant pressure, according as i is the solute or solvent, respectively. The former aspect of this equation will be considered below, and the latter will be deferred to a later section.

34j. Influence of Temperature on Solubility.—If component 2 is taken to be the solute, in accordance with the usual convention, equation (34.30), in the form

$$\left(\frac{\partial \ln \mathrm{N}_2}{\partial T}\right)_P = \frac{\Delta H_f}{RT^2}, \quad (34.31)$$

gives the variation with temperature of the concentration of a saturated solution, i.e., the variation of solubility with temperature; ΔH_f is the molar heat of fusion of the pure solid solute at the given temperature and pressure.

* In the present chapter, a prime is used to designate the phase (vapor or solid) which is in equilibrium with the liquid solution.

IDEAL SOLUTIONS

It will be recalled that this result is based on two assumptions: first, that the solution behaves ideally, and second, that the solute which separates from the solution is always pure solute and not a solid solution.

If the heat of fusion is independent of temperature, equation (34.31) can be integrated at constant pressure, e.g., 1 atm., to yield

$$\ln N_2 = -\frac{\Delta H_f}{RT} + C, \qquad (34.32)$$

where C is an integration constant. This expression gives the solubility N_2, in mole fractions, as a function of the temperature. If the solution behaves ideally over the whole composition range, it is possible to derive the value of C in equation (34.32). Thus, when the mole fraction N_2 of the solute in the liquid phase is unity, the pure solid is in equilibrium with liquid of the same composition; the temperature is then the melting point T_m. Since $\ln N_2$ is now zero, it follows from equation (34.32) that C is equal to $\Delta H_f/RT_m$. Upon inserting this result, it is seen that

$$\ln N_2 = \frac{\Delta H_f}{R}\left(\frac{1}{T_m} - \frac{1}{T}\right)$$

or

$$\log N_2 = \frac{\Delta H_f}{4.576}\left(\frac{T - T_m}{TT_m}\right), \qquad (34.33)$$

where ΔH_f is the molar heat of fusion in calories.

According to equation (34.32) or (34.33) the plot of the logarithm of the solubility of a solid, expressed in mole fractions, i.e., $\log N_2$, against the reciprocal of the absolute temperature, i.e., $1/T$, should be a straight line terminating at the melting point; the slope of this line should be equal to $-\Delta H_f/4.576$. Experimental observations have shown that in many cases the plot of $\log N_2$ against $1/T$ is, in fact, linear over a considerable temperature range. It is only in a few cases, however, that the slope has the theoretical value; this is not surprising, for the saturated solutions would have to be ideal for equation (34.33) to hold exactly. When the solid and solute are similar in nature the solutions behave ideally, and solubilities in fair agreement with experiment can be calculated from equation (34.33), utilizing only the melting point and heat of fusion of the solid solute.[4]

Problem: The melting point of naphthalene is 80.2° C, and its molar heat of fusion at this temperature is 4,540 cal. mole^{-1}. Determine its ideal solubility at 20° C.

Assuming ΔH_f to remain constant at 4,540 cal. mole^{-1}, equation (34.33) may be employed with T equal to $20 + 273 = 293°$ K, and T_m as $80 + 273 = 353°$ K;

[4] Schroeder, *Z. phys. Chem.*, **11**, 449 (1893); see also, Washburn and Read, *Proc. Nat. Acad. Sci.*, **1**, 191 (1915); Johnston, Andrews, *et al.*, *J. Phys. Chem.*, **29**, 882 *et seq.* (1925); Ward, *ibid.*, **39**, 2402 (1935); Morris and Cook, *J. Am. Chem. Soc.*, **57**, 2402 (1935); Hildebrand, *op. cit.*, ref. 1, Chapter X; Germann and Germann, *Ind. Eng. Chem.*, **36**, 93 (1944).

hence,

$$\log N_2 = \frac{4{,}540}{4.576}\left(\frac{293 - 353}{293 \times 353}\right) = -0.5755,$$

$$N_2 = 0.266.$$

The ideal solubility of naphthalene in any liquid solvent at 20° C should thus be 0.266 mole fraction; the experimental values in solvents which are chemically similar to the solute, and which might be expected to form approximately ideal solutions, viz., benzene and toluene, are 0.241 and 0.224 mole fraction, respectively. In hydroxylic solvents, which form nonideal solutions, the values are quite different, for example, 0.018 in methanol and 0.0456 in acetic acid.

Two general deductions concerning the solubility of solids may be made from equation (34.33). First, if the heats of fusion of two substances are not very different, the one with the larger value of T_m, i.e., with the higher melting point, should have the smaller mole fraction solubility in any solvent. Second, if two solids have similar melting points, the one with the lower heat of fusion should be more soluble than the other; this is because the melting point T_m must be greater than the experimental temperature T, and hence the quantity in the parentheses is always negative.

For an ideal solution the heat of solution is equal to the heat of fusion of the solid at the given temperature and pressure, and consequently for a given solute the former should be independent of the nature of the solvent. That this should be the case can be readily shown by means of a procedure exactly analogous to that employed in § 34h in connection with the heat of solution of gases.

35. Nonideal Solutions

35a. Deviations from Ideal Behavior.—Theoretical considerations show that if a mixture of two liquids is to behave ideally, the two types of molecules must be similar. The environment of any molecule in the solution, and hence the force acting upon it, is then not appreciably different from that existing in the pure liquid. It is to be anticipated, therefore, that under these conditions the partial vapor pressure (or fugacity) of each constituent, which is a measure of its tendency to escape from the solution, will be directly proportional to the number of molecules of the constituent in the liquid phase. This is equivalent to stating that a mixture of two liquids consisting of similar molecules would be expected to obey Raoult's law. Such is actually the case, for the relatively few liquid systems which are known to behave ideally, or to approximate to ideal behavior, consist of similar molecules, e.g., ethylene bromide and propylene bromide, n-hexane and n-heptane, n-butyl chloride and bromide, ethyl bromide and iodide, and benzene and toluene.

If the constituents of a mixture differ appreciably in nature, deviations from ideal behavior are to be expected and are, in fact, observed. These deviations are most frequently "positive" in nature, so that the actual partial vapor pressure (or fugacity) of each constituent is greater than it should be if Raoult's law were obeyed. It can be readily seen from the Duhem-Margules equation that *if one constituent of a mixture exhibits positive deviations from ideal behavior, the other constituent must do likewise.* Thus, if f_1 is greater than, instead of being equal to, $\text{N}_1 f_1^\square$ [cf. equation (34.1)], indicating positive deviations for constituent 1, it follows from the argument in § 34c that

$$\left(\frac{\partial \ln f_1}{\partial \ln \text{N}_1}\right)_{T,P} = \left(\frac{\partial \ln f_2}{\partial \ln \text{N}_2}\right)_{T,P} > 1,$$

so that f_2 must be greater than $\text{N}_2 f_2^\square$, implying positive deviations from ideal behavior for constituent 2. Incidentally, it will be evident that if the deviation from Raoult's law is "negative" for one component, that is to say, if its partial pressure (or fugacity) is less than that required by Raoult's law, the same must be true for the other component. Negative deviations from ideal behavior are only observed in systems in which the different molecules have a very strong attraction for one another.

For a nonideal solution exhibiting positive deviations $f_i/f_i^\square$ for each component is greater than its mole fraction N_i. It is an experimental fact that as the temperature is increased most liquid solutions tend toward ideal behavior. This means that for a system of given composition for which the deviations from Raoult's law are positive, the ratio $f_i/f_i^\square$ usually decreases with increasing temperature. According to equation (34.4), therefore, which holds for solution of all types, the numerator $H_i^\square - \bar{H}_i$ must be negative; thus $\bar{H}_i$ is greater than $H_i^\square$. The total heat content of the solution, consisting of n_1 moles of constituent 1 and n_2 moles of 2, is equal to $n_1 \bar{H}_1 + n_2 \bar{H}_2$, while the sum of the heat contents of the separate substances is $n_1 H_1^\square + n_2 H_2^\square$. Since $\bar{H}_1$ is greater than $H_1^\square$ and $\bar{H}_2$ is greater than $H_2^\square$, it is evident that the heat content of the solution must be larger than that of the constituents before mixing. In other words, in this particular case, heat must be absorbed when the pure liquids are mixed. The general conclusion to be drawn, therefore, is that *upon mixing two liquids which yield a system exhibiting positive deviations from Raoult's law there is an absorption of heat.*

For a system which manifests negative deviations from ideal behavior the ratio $f_i/f_i^\square$ increases as the temperature is raised; consequently $H_i^\square - \bar{H}_i$ for each constituent is positive. By using arguments analogous to those presented above it is a simple matter to show that *the formation of a solution exhibiting negative deviations from Raoult's law is accompanied by an evolution of heat.*[5]

35b. Vapor Pressure Curves for Nonideal Systems.—The general nature of the vapor pressure curves showing positive and negative deviations are depicted in Fig. 22, A and B, respectively; these results refer to a constant temperature. At any given composition, the slopes of the two partial vapor pressure curves are related by the Duhem-Margules equation. Thus, if the

---

[5] For a review of the behavior of nonideal solutions, see Hildebrand, *op. cit.*, ref. 1, Chapter III.

vapors are supposed to behave ideally, which is possible even if the solution is not ideal, the Duhem-Margules equation (34.13) may be written in the form

$$\frac{N_1}{p_1} \cdot \frac{dp_1}{dN_1} = \frac{N_2}{p_2} \cdot \frac{dp_2}{dN_2}, \qquad (35.1)$$

constant temperature and pressure being understood. The quantities N_i/p_i and dp_i/dN_i for each constituent, corresponding to any composition of the solution, can be readily derived if the partial vapor pressure curves of the

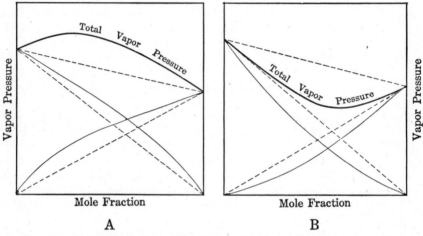

FIG. 22. Nonideal solutions: (A) positive deviations, (B) negative deviations

two constituents have been determined. For any given solution the value of the product of N_i/p_i and dp_i/dN_i should then be the same for both constituents. This is frequently used as a test of the reliability of experimental partial vapor pressure data for nonideal mixtures. Where it appears necessary, however, allowance should be made for the deviation of the vapor from ideal behavior.[6]

In both Figs. 22, A and B the uppermost curve gives the total vapor pressure as a function of the composition of the liquid. The corresponding curve as a function of the vapor composition will lie below it in each case, so that the vapor contains more of the constituent the addition of which causes an increase in the total vapor pressure. If an expression for the partial vapor pressure in terms of the mole-fraction composition of the liquid is available (cf. § 35d), an analogous expression for the vapor composition can be derived by utilizing the relationship based on the postulated ideal behavior of the vapor, i.e., that N_1'/N_2' is equal to p_1/p_2 (§ 34e), where N_1' and N_2' refer to the respective mole fractions in the vapor phase.

[6] Zawidski, *Z. phys. Chem.*, **35**, 129 (1900); Beatty and Calingaert, *Ind. Eng. Chem.*, **26**, 504, 905 (1934); Carlson and Colburn, *ibid.*, **34**, 581 (1942); Scatchard and Raymond, *J. Am. Chem. Soc.*, **60**, 1278 (1938).

If the vapor pressures of the two pure constituents are close together, then any appreciable positive deviation from Raoult's law will lead to a maximum in the total vapor pressure curve; similarly, a negative deviation will, in the same circumstances, be associated with a minimum in the curve. In any event, even if the vapor pressures of the pure constituents are appreciably different, marked positive or negative deviations can lead to a maximum or a minimum, respectively, in the total vapor pressure curve. Such maxima and minima are the cause of the formation of the familiar **constant boiling mixtures** or **azeotropic mixtures**. A liquid mixture having the composition represented by a maximum or a minimum will distil without change of composition, for the proportions of the two constituents are then the same in the liquid and vapor phases. That this must be the case will be shown in the next section.

35c. Liquid and Vapor Compositions.—Some general rules concerning the relative compositions of liquid and vapor in equilibrium, which are applicable to systems of all types, may be derived from the Duhem-Margules equation, using the form of (35.1). Since the increase in the mole fraction of one component of a binary mixture must be equal to the decrease for the other component, $d\text{N}_1$ is equal to $-d\text{N}_2$, as seen in § 34b; hence equation (35.1) may be written as

$$\frac{\text{N}_1}{p_1} \cdot \frac{dp_1}{d\text{N}_1} + \frac{\text{N}_2}{p_2} \cdot \frac{dp_2}{d\text{N}_1} = 0. \tag{35.2}$$

The total vapor pressure P is equal to $p_1 + p_2$, and differentiation with respect to N_1 gives

$$\frac{dP}{d\text{N}_1} = \frac{dp_1}{d\text{N}_1} + \frac{dp_2}{d\text{N}_1}, \tag{35.3}$$

and thus, by equation (35.2),

$$\frac{dP}{d\text{N}_1} = \frac{dp_2}{d\text{N}_1}\left(1 - \frac{\text{N}_2 p_1}{\text{N}_1 p_2}\right). \tag{35.4}$$

Suppose it is required to find the condition that makes $dP/d\text{N}_1$ positive, that is to say, the total vapor pressure is to increase as the mole fraction of constituent 1 in the system is increased; it is necessary to determine under what circumstances the right-hand side of equation (35.4) is positive. The factor $dp_2/d\text{N}_1$, equal to $-dp_2/d\text{N}_2$, must always be negative, since the partial pressure of any component never decreases as its mole fraction in the liquid phase is increased. Consequently, the condition for $dP/d\text{N}_1$ to be positive, is that

$$\frac{\text{N}_2 p_1}{\text{N}_1 p_2} > 1 \quad \text{or} \quad \frac{p_1}{p_2} > \frac{\text{N}_1}{\text{N}_2}.$$

If the vapors behave as ideal gases, p_1/p_2 is equal to N_1'/N_2', by equation (34.18), and hence it is necessary that

$$\frac{\text{N}_1'}{\text{N}_2'} > \frac{\text{N}_1}{\text{N}_2}. \tag{35.5}$$

In other words, if $dP/d\text{N}_1$ is to be positive, the vapor must be relatively richer in the constituent 1 than is the liquid with which it is in equilibrium. *The vapor thus contains relatively more of the substance the addition of which to the liquid mixture results in an increase of the total vapor pressure.* The general applicability of this conclusion has been already seen in § 34e for ideal solutions, and in § 35b for those exhibiting marked nonideality (Figs. 21 and 22).

For a maximum or minimum in the total vapor pressure curve, $dP/d\text{N}_1$ must be zero; hence, by equation (35.4), either $dp_2/d\text{N}_1$ must be zero, or $\text{N}_2 p_1$ must equal $\text{N}_1 p_2$. The former condition is unlikely, since it would mean that the partial vapor pressure would remain constant in spite of a change of composition, and so for a maximum or minimum in the total vapor pressure curve $\text{N}_2 p_1 = \text{N}_1 p_2$ or $\text{N}_1/\text{N}_2 = p_1/p_2$. If the gases behave ideally, p_1/p_2 is equal to N_1'/N_2', and the vapor will have the same composition as the liquid in equilibrium with it, as stated above.

35d. General Equations for Liquid Mixtures.—By combining equation (31.2) for the activity of any component of a liquid solution in terms of its chemical potential, viz., $\mu_i = \mu_i^0 + RT \ln a_i$, and equation (31.5) which defines the activity as the relative fugacity, i.e., $a_i = f_i/f_i^0$, the result is

$$\mu_i = \mu_i^0 + RT \ln (f_i/f_i^0),$$

and hence

$$f_i = f_i^0 e^{(\mu_i - \mu_i^0)/RT}.$$

Utilizing the fact that x is equal to $e^{\ln x}$, this equation may be written as

$$f_i = \text{N}_i f_i^0 e^{[(\mu_i - \mu_i^0)/RT] - \ln \text{N}_i}, \tag{35.6}$$

which is an expression of general applicability to each constituent of any mixture, ideal or nonideal.

An equation, somewhat similar to (35.6), was suggested by M. Margules (1895) to express the variation of vapor pressure with composition of liquid mixtures in general; replacing the vapor pressure by the fugacity, this can be written as

$$f_1 = \text{N}_1 f_1^{\square} e^{\frac{1}{2}\beta_1 \text{N}_2^2 + \frac{1}{3}\gamma_1 \text{N}_2^3 + \cdots} \tag{35.7}$$

and

$$f_2 = \text{N}_2 f_2^{\square} e^{\frac{1}{2}\beta_2 \text{N}_1^2 + \frac{1}{3}\gamma_2 \text{N}_1^3 + \cdots}, \tag{35.8}$$

where $f_1^{\square}$ and $f_2^{\square}$ are, as usual, the fugacities of the pure liquids. It should be understood that these equations have no real theoretical basis; they were proposed because they appeared to represent the experimental data in a satisfactory manner. The constants β_1, β_2, γ_1 and γ_2, etc., which are not independent, can be derived from the slopes of the fugacity—actually partial vapor pressure—curves at $\text{N}_i = 0$ or $\text{N}_i = 1$. For an ideal solution, these constants would all be zero, so that equations (35.7) and (35.8) would then become identical with (34.1). For nonideal solutions, however, the sign and magnitude of β_1, γ_1, etc., depend on the nature, i.e., positive or negative, and extent of departure from ideal behavior.

In order to describe the procedure for evaluating the constants without making the treatment too complicated, it will be supposed, as is frequently the case, that all terms beyond the first in the exponents of equation (35.7) and (35.8) can be neglected. Upon taking logarithms of these expressions and differentiating with respect to $\ln N_1$ and $\ln N_2$, respectively, at constant temperature and external pressure, it is seen that

$$\left(\frac{\partial \ln f_1}{\partial \ln N_1}\right)_{T,P} = 1 - \beta_1 N_1 N_2 + \cdots$$

and

$$\left(\frac{\partial \ln f_2}{\partial \ln N_2}\right)_{T,P} = 1 - \beta_2 N_1 N_2 + \cdots.$$

According to equation (34.12) these results must be identical, and hence, to the approximation that γ_1, γ_2, etc., are negligible, β_1 is equal to β_2. The equations for the variation of fugacity with the mole fraction thus become

$$f_1 = N_1 f_1^\square e^{\frac{1}{2}\beta N_2^2} \quad \text{and} \quad f_2 = N_2 f_2^\square e^{\frac{1}{2}\beta N_1^2}, \tag{35.9}$$

where β is the same in both cases. Under most ordinary conditions, when the vapor pressures are not too high, the fugacities may be replaced by the respective partial vapor pressures. The equations (35.7), (35.8) and (35.9) then give the variation of the vapor pressure with composition; they are frequently used in this form.[7]

By inserting the value of p_1 or p_2 for a solution of known composition, it is possible to determine β from equation (35.9), since $p_1^\square$ or $p_2^\square$ may be regarded as available. In most cases, an alternative method is used. Replacing the fugacities in the equations (35.9) by the respective partial pressures, and differentiating each with respect to N_1, addition of the results gives dP/dN_1, in accordance with equation (35.3); thus,

$$\frac{dP}{dN_1} = (p_1^\square e^{\frac{1}{2}\beta N_2^2} - p_2^\square e^{\frac{1}{2}\beta N_1^2})(1 - \beta N_1 N_2).$$

Two special cases may now be considered, viz., $N_1 = 0$ and $N_1 = 1$; the results are

$$\frac{dP}{dN_1} = p_1^\square e^{\frac{1}{2}\beta} - p_2^\square, \quad \text{when} \quad N_1 = 0 \quad \text{and} \quad N_2 = 1,$$

and

$$\frac{dP}{dN_1} = p_1^\square - p_2^\square e^{\frac{1}{2}\beta}, \quad \text{when} \quad N_1 = 1 \quad \text{and} \quad N_2 = 0.$$

It is thus possible to derive β directly from the slope, at either end, of the plot of the total vapor pressure against the mole fraction.

Another form of the general empirical equations for the variation of the fugacity with composition, which is said to be more convenient for certain purposes, was proposed by J. J. van Laar (1910); thus,

$$f_1 = N_1 f_1^\square e^{\alpha_1 N_2^2/(\beta_1 N_1 + N_2)^2} \quad \text{and} \quad f_2 = N_2 f_2^\square e^{\alpha_2 N_1^2/(N_1 + \beta_2 N_2)^2}. \tag{35.10}$$

The relationships between the constants α_1, β_1 and α_2, β_2, which are different from those in the Margules equations (35.7) and (35.8), can be derived by means of the Duhem-Margules equation in a manner similar to that described above. It is then found that

$$\alpha_1 \beta_2 = \alpha_2 \quad \text{and} \quad \alpha_2 \beta_1 = \alpha_1.$$

[7] Cf. Zawidski, ref. 6; Porter, *Trans. Farad. Soc.*, **16**, 336 (1920); see also, Lewis and Murphree, *J. Am. Chem. Soc.*, **46**, 1 (1924); Levy, *Ind. Eng. Chem.*, **33**, 928 (1941).

so that there are only two independent constants, the values of which can be obtained from vapor pressure curves.[8]

The Margules and van Laar equations, with vapor pressures in place of fugacities, are useful for the analytical expression of experimental data, particularly for extrapolation and interpolation purposes. It is of interest to note that when the vapor pressure (fugacity)-composition curve of one component of a binary mixture is known, the relationships among the constants α, β, etc., automatically determine the nature of the curve for the other component. This is, of course, a consequence of the Duhem-Margules equation. Both the Margules and van Laar formulae are capable of representing positive and negative deviations from ideal behavior. In the former case, i.e., positive deviations, the exponential factor is greater than unity, so that the exponent is positive; in the latter case, i.e., negative deviations, this factor is less than unity, the exponent being negative.

35e. Partially Miscible Liquids.—The systems considered so far have consisted of completely miscible liquids forming a single layer. In the event of large positive deviations from ideal behavior, however, a separation into two layers, that is to say, partial miscibility, becomes possible. This can be seen, for example, by means of the equations (35.9), in which, for a system

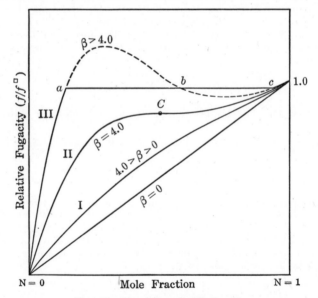

Fig. 23. Partially miscible liquids

exhibiting positive deviations from Raoult's law, β has a positive value. As this quantity, which is a constant for a given pair of liquids at a definite temperature, is increased, e.g., by lowering the temperature or by changing the components, it can be readily shown that the variation of fugacity (or

[8] See Carlson and Colburn, ref. 6.

partial vapor pressure) with the composition of each substance may be represented by curves of the types I, II and III in Fig. 23.[9] In order to make the results comparable, the relative fugacity $f/f^\square$, rather than f, is plotted against the mole fraction of the corresponding constituent in each case. When β is equal to 4.0 (curve II) the curve exhibits a horizontal inflection, and if β exceeds this value the fugacity curve has a sigmoid character, as seen in curve III. This curve implies that there are three different solutions, a, b and c, having the same fugacity, which should be capable of coexisting in equilibrium at the given temperature and external pressure. Actually, the situation is similar to that arising in connection with the van der Waals equation (cf. § 5d); the point b cannot be realized, and the fugacity, or partial vapor pressure, curve, as obtained experimentally, is flat between a and c. There are then two liquid layers, having different compositions, which are in equilibrium with each other. In between the compositions represented by the points a and c the two liquids are partially miscible at the given temperature. Large positive deviations from ideal behavior can thus lead to incomplete solubility of one liquid in another. The point of horizontal inflection C on curve II corresponds to the **critical solution** (or **consolute**) **temperature** for the given system; below this temperature the liquids are not completely miscible in all proportions.*

In Fig. 23 there are shown the relative fugacity (or partial vapor pressure) curves for one constituent of the system; quite similar curves will be obtained for the other. The points of horizontal inflection C will, of course, be identical, as regards temperature and composition of the liquid. Similarly, when the conditions are such that two liquid layers are formed, the points a, b and c, for a given temperature, will occur at the same compositions on the curves for the two constituents of the system.

It is evident from Fig. 23 that when two layers are present, the partial vapor pressure of a given constituent will have a constant value, at a particular temperature, which is the same in both layers (points a and c). The total vapor pressure, equal to the sum of the separate partial pressures, will likewise be constant, at a given temperature, irrespective of the over-all composition of the system, provided the two layers are present. The same conclusion can, of course, be reached by means of the phase rule.[10]

36. Dilute Solutions

36a. Henry's Law.—It has been found experimentally that *as the mole fraction of a given constituent of a solution approaches unity, the fugacity of that constituent approximates to the value for an ideal system* [equation (34.1)],

---

[9] Hildebrand, *op. cit.*, ref. 1, p. 52; see also, Porter, ref. 7; Butler, *et al.*, *J. Chem. Soc.*, 674 (1933).

* For certain pairs of liquids β decreases as the temperature is lowered, within a particular temperature range; such systems exhibit a *lower* consolute temperature, *above* which the liquids are not completely miscible.

[10] See, for example, S. Glasstone, "Textbook of Physical Chemistry", 2nd ed., 1946, Chapter X.

even though the behavior at lower mole fractions departs markedly from the ideal.* This fact is indicated in Fig. 22, A and B, by the actual partial pressure or fugacity curve becoming asymptotic with the corresponding ideal curve as the mole fraction approaches unity. Employing the terms solvent and solute to indicate the constituents present in greater and smaller amounts, respectively, in any solution, the observation referred to above may be stated in the form: *In a dilute solution the behavior of the solvent approaches that required by Raoult's law*, even though it may depart markedly from ideal behavior in more concentrated solutions.

It is necessary now to consider the behavior of the solute, i.e., the constituent present in small amount, in the dilute solution; two methods of approach to the problem are of interest. Taking component 1 as the solvent and 2 as the solute, in accordance with the usual convention, it is seen that in the dilute solution, as N_1 approaches unity and N_2 approaches zero, the empirical equations (35.9) become

$$f_1 = N_1 f_1^\square \quad \text{as} \quad N_1 \to 1 \quad \text{and} \quad N_2 \to 0 \quad (36.1)$$

$$f_2 = N_2 f_2^\square e^{\frac{1}{2}\beta} \quad \text{as} \quad N_1 \to 1 \quad \text{and} \quad N_2 \to 0. \quad (36.2)$$

According to (36.1), the behavior of the solvent tends toward that required by Raoult's law, as stated above. Further, since β is constant, equation (36.2), for the solute, is equivalent to $f_2 = N_2 f_2^\square k'$, where k' is equal to $e^{\frac{1}{2}\beta}$ and hence is also constant. It is obvious that in dilute solution the solute cannot obey Raoult's law unless k' is unity, that is, unless β is very small or zero. In other words, although the solvent in a dilute solution satisfies Raoult's law, nevertheless, the solute does not do so unless the system as a whole, i.e., over the whole range of composition, exhibits little or no departure from ideal behavior. This conclusion is in harmony with the results depicted in Fig. 22, A and B. However, although the solute in the dilute solution does not necessarily obey Raoult's law, it does conform to the simple expression $f_2 = N_2 f_2^\square k'$, which may be written as

$$f_2 = N_2 k \quad \text{as} \quad N_1 \to 1 \quad \text{and} \quad N_2 \to 0, \quad (36.3)$$

where k, equal to $f_2^\square k'$, is a constant; this constant becomes identical with $f_2^\square$ when k' is unity, that is, when the system is ideal over the whole range of concentrations. The result given in equation (36.3) is an idealized expression of **Henry's law**, viz., *the fugacity of a solute in dilute solution is proportional to its mole fraction*. A **dilute solution** may thus be defined as *one in which the solvent obeys Raoult's law and the solute satisfies Henry's law*.

From the remarks in § 34g it can be seen that equation (36.3) may be interpreted as implying that for a saturated solution of a gas the fugacity of the solute is proportional to its mole fraction in the solution. If the gas behaves ideally, and the solution is dilute, it follows that the molar concentration of the

* For a solution containing a solute which dissociates, it is necessary to treat the products of dissociation as distinct molecules when evaluating the mole fractions of solvent and solute.

saturated solution, i.e., the solubility, is proportional to the pressure of the gas, at constant temperature. This is equivalent to the original form of Henry's law (W. Henry, 1803), which dealt only with gas solubility, but the statement of the law given above is a generalized form of wider applicability.

A derivation of Henry's law [equation (36.3)] is possible by the use of the Duhem-Margules equation. It is readily seen that if $f_1 = N_1 f_1^0$, as is the case for the solvent in a dilute solution, then for the solute, i.e., component 2,

$$\left(\frac{\partial \ln f_2}{\partial \ln N_2}\right)_{T,P} = 1 \quad \text{as} \quad N_1 \to 1 \quad \text{and} \quad N_2 \to 0, \quad (36.4)$$

by equation (34.15). In § 34c this result, like Raoult's law, was taken as applicable over the whole range of composition, but if the solvent behaves ideally only when its mole fraction approaches unity, the integration of equation (36.4) can be carried *only over the same limited range*. By general integration, this gives

$$\ln f_2 = \ln N_2 + \text{constant},$$

or

$$f_2 = N_2 k \quad \text{as} \quad N_1 \to 1 \quad \text{and} \quad N_2 \to 0,$$

which is identical with equation (36.3). It follows, therefore, that *in any solution for which Raoult's law is applicable to the solvent, Henry's law must hold for the solute, over the same concentration range*. Incidentally, it can be readily shown that the reverse must also be true; *if Henry's law is applicable to the solute, then the solvent must obey Raoult's law over the same range of composition.** These are characteristic properties of a dilute solution.

The foregoing conclusions are depicted graphically in Fig. 24, which shows the partial vapor pressure (or fugacity) curves of three types. Curve I is for an ideal system obeying Raoult's law over the whole concentration range; for such solutions k in equation (36.3) is equal to f_2^0, and Henry's law and Raoult's law are identical. For a system exhibiting positive deviations, curve II may be taken as typical; in the dilute range,

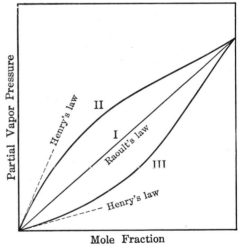

FIG. 24. Raoult's law and Henry's law

---

* The restriction of constant external pressure should apply to Henry's law, just as to Raoult's law (§ 34c). At moderate pressures, however, the influence of the total pressure may be neglected.

i.e., at the left-hand side of the diagram, there is marked departure from Raoult's law, but the curve has an approximately linear slope, indicating that it satisfies Henry's law [equation (36.3)]. The same is true for curve III which represents the fugacity of a substance exhibiting negative deviations from Raoult's law. In the region in which Henry's law is obeyed by one constituent of a solution, i.e., the solute, the fugacity (or vapor pressure) curve of the other constituent, i.e., the solvent, which is not shown in the figure, would approach that for an ideal system obeying Raoult's law.

36b. The Freezing Points of Dilute Solutions.—Since in any dilute solution the solvent behaves ideally, in the respect that it obeys Raoult's law, it follows that equation (34.30), in the form

$$\left(\frac{\partial \ln \text{N}_1}{\partial T}\right)_P = \frac{\Delta H_f}{RT^2}, \tag{36.5}$$

should apply *at the freezing point* of a dilute solution in equilibrium with *pure solid solvent*. In this expression N_1 is mole fraction of the solvent in the solution which is in equilibrium with solid solvent at the temperature T and pressure P; ΔH_f is the heat of fusion of the solid solvent under the same conditions. It is seen, therefore, that equation (36.5) gives the variation with composition of the freezing point of a dilute solution, *irrespective of whether it behaves ideally or not at higher concentrations*.

If the heat of fusion is taken to be independent of temperature, an approximation which can reasonably be made for dilute solutions, integration of equation (36.5) gives

$$\ln \text{N}_1 = -\frac{\Delta H_f}{RT} + C, \tag{36.6}$$

where C is the integration constant. When the liquid phase consists of pure solvent, N_1 is unity and the freezing point is then T_0, which is that of the solvent; hence it is seen from equation (36.6) that the integration constant is equal to $\Delta H_f/RT_0$. If this result is inserted into equation (36.6) it becomes

$$\ln \text{N}_1 = \frac{\Delta H_f}{R}\left(\frac{1}{T_0} - \frac{1}{T}\right) = \frac{\Delta H_f}{R}\left(\frac{T - T_0}{TT_0}\right), \tag{36.7}$$

where T is the freezing point of the solution containing N_1 mole fraction of solvent. Since N_1 must be less than unity, $\ln \text{N}_1$ is negative, and since ΔH_f is positive, T must be less than T_0; in other words, *the freezing point of the solution is always less than that of the pure solvent*, provided the solid phase is the pure solvent and not a solid solution, as postulated above. Since N_1 decreases with increasing concentration of the solution, it is clear from equation (36.7) that the freezing point must decrease at the same time.

Since equation (36.7) may be expected to apply to any dilute solution, further simplifications, which are applicable to such solutions, can be made. For a dilute solution the difference between T and T_0 is small, so that TT_0 may be replaced by T_0^2, and if $T_0 - T$, called the **lowering of the freezing**

point of the solution, is represented by θ, equation (36.7) becomes

$$\ln \text{N}_1 = -\frac{\Delta H_f}{RT_0^2}\theta. \qquad (36.8)$$

If the solution is sufficiently dilute, N_1 is only slightly different from unity, so that $\ln \text{N}_1$ is approximately equal to $\text{N}_1 - 1$. Further, since $1 - \text{N}_1$ is equal to N_2, the mole fraction of the solute, it follows that $\ln \text{N}_1$ in equation (36.8) may be replaced by $-\text{N}_2$, giving

$$\text{N}_2 = \frac{\Delta H_f}{RT_0^2}\theta,$$

or

$$\theta = \frac{RT_0^2}{\Delta H_f}\text{N}_2. \qquad (36.9)$$

In a *dilute solution*, therefore, *the depression of the freezing point is proportional to the mole fraction of the solute*.

36c. Determination of Molecular Weights.—Because of the relationship between the mole fraction and the molecular weight of the solute, it is possible to use the freezing point lowering of a solution, in conjunction with equation (36.9), to determine the molecular weight of the solute. Since the solution must be dilute for this equation to be applicable, a further simplification is possible. The mole fraction N_2 is equal to $n_2/(n_1 + n_2)$, where n_1 and n_2 are the numbers of moles of solvent and solute, respectively, in the solution; since the solution is dilute n_2 may be neglected in comparison with n_1, and hence N_2 may be taken as equal to n_2/n_1. If w_1 and w_2 are the respective weights of the solvent and solute in the solution, and M_1 and M_2 are their molecular weights, then

$$\text{N}_2 \approx \frac{n_2}{n_1} = \frac{w_2 M_1}{w_1 M_2}.$$

Upon substituting this result into equation (36.9), it is found that

$$\theta \approx \frac{RT_0^2}{\Delta H_f} \cdot \frac{w_2 M_1}{w_1 M_2}. \qquad (36.10)$$

This equation may be written in the form

$$\theta \approx \frac{RT_0^2 M_1}{1000 \Delta H_f} \cdot \frac{1000 w_2}{w_1 M_2}$$

$$\approx \lambda m, \qquad (36.11)$$

where λ, which is equal to $RT_0^2 M_1/1000\Delta H_f$, is a constant for the given solvent. The factor $1000 w_2/w_1 M_2$ is seen to be equivalent to the number of moles of solute, i.e., m, dissolved in 1000 g. of solvent; it is hence equal to the molality m of the solution, as indicated in equation (36.11). It will be apparent that if this equation were applicable at such a concentration, λ would be the lowering of the freezing point for a molal solution, i.e., when $m = 1$; hence λ is referred to as the **molal freezing point depression constant**. Actually a molal solution would be too concentrated for equation (36.11) to be applicable, and so λ may be defined as equal to θ/m as the solution approaches infinite dilution; thus,

$$\lambda = \lim_{m \to 0} \frac{\theta}{m} = \frac{RT_0^2 M_1}{1000 \Delta H_f}, \qquad (36.12)$$

where ΔH_f is the heat of fusion per mole of the solid solvent, whose *formula* weight is M_1, into pure liquid at its freezing point T_0. The value of λ for any solvent can thus be determined in two ways; either from the known freezing point T_0 and the heat of fusion or by determining θ/m at various molalities, from actual freezing-point measurements, and extrapolating the results to infinite dilution.

It will be observed that equation (36.11) is used as the basis of the familiar procedure for the determination of molecular weights by the freezing point method. It is obviously strictly applicable only to very dilute solutions; at appreciable concentrations the approximations made in its derivation are no longer justifiable. These are as follows: first, that ΔH_f is independent of temperature; second, that $T_0 T$ is equal to T_0^2; third, that $\ln N_1$ may be replaced by $- N_2$; and fourth, that N_2 may be set equal to n_2/n_1. The two latter approximations are avoided in equation (36.8), and this gives somewhat better results than does (36.11) in solutions of moderate concentration, provided the solvent still obeys Raoult's law.

Problem: The freezing point of benzene is 5.4° C and its latent heat of fusion is 30.2 cal. g.$^{-1}$. A solution containing 6.054 g. of triphenylmethane in 1000 g. of benzene has a freezing point which is 0.1263° below that of the pure solvent. Calculate the molecular weight of the solute.

Since both R and ΔH_f can be expressed in terms of calories, the molal depression constant λ for benzene is given by

$$\lambda = \frac{RT_0^2 M_1}{1000 \Delta H_f} = \frac{1.987 \times (278.6)^2}{1000 \times 30.2}$$
$$= 5.11,$$

noting that $\Delta H_f / M_1$ is equal to the heat of fusion per gram. By equation (36.11) the molality m of the solution is equal to θ/λ, so that

$$m = 0.1263/5.11 = 0.0247.$$

Since the solution contains 6.054 g. of solute to 1000 g. of solvent, 6.054 must represent $0.0247 M_2$, where M_2 is the molecular weight of the solute; hence,

$$M_2 = 6.054/0.0247 = 245.$$

36d. Separation of Solid Solutions.—If upon cooling a liquid solution the solid phase consists of a solid solution, instead of pure solvent, it is necessary to make use of equation (34.29). Assuming the solutions to be dilute enough for the fugacity of the solvent, i.e., the constituent present in excess, to be proportional to its mole fraction in both solid and liquid phases, this equation becomes

$$\left[\frac{\partial \ln (\text{N}_1'/\text{N}_1)}{\partial T}\right]_P = -\frac{\Delta H_f}{RT^2}, \quad (36.13)$$

where ΔH_f is the molar heat of fusion of the solvent. If the latter remains constant in the small temperature range between T, the freezing point of the solution, and T_0, that of the pure solvent, at the same pressure, integration of equation (36.13) yields

$$\ln \frac{\text{N}_1}{\text{N}_1'} = \frac{\Delta H_f}{R}\left(\frac{T - T_0}{TT_0}\right) \quad (36.14)$$

at constant pressure. For dilute liquid and solid solutions, $\ln \text{N}_1$ may be replaced by $-\text{N}_2$, $\ln \text{N}_1'$ by $-\text{N}_2'$, and TT_0 by T_0^2; $T_0 - T$ is equal to θ, and equation (36.14) then reduces to

$$\theta = \frac{RT^2}{\Delta H_f}(\text{N}_2 - \text{N}_2'), \quad (36.15)$$

which may be compared with equation (36.9), for which N_2', the mole fraction of solute in the solid phase, is zero.

The equation (36.15) gives the change in freezing point with composition (N_2) of the dilute solution, when the solid phase is a solid solution of known composition (N_2'). It should be noted that if N_2' is less than N_2, the value of θ is positive, and the solution has a lower freezing point than the pure solvent. However, it frequently happens that N_2' is greater than N_2; that is to say, there is a larger proportion of solute in the solid phase than in the liquid. In these circumstances θ is negative, and the freezing point of the solution is above that of the pure solvent.

Another form of equation (36.15) may be obtained by writing k for the distribution ratio N_2'/N_2 of the solute between solid and liquid phases; the result is

$$\theta = \frac{RT_0^2}{\Delta H_f}\text{N}_2(1 - k).$$

By treating this expression in the same manner as equation (36.9), it is found that

$$\theta = \lambda m(1 - k), \quad (36.16)$$

so that it is possible to determine the distribution of the solute between the two phases from freezing point measurements.[11]

36e. The Boiling Points of Solutions.—Just as a solution has a lower freezing point than the pure solvent, provided no solute separates in the solid phase, so it has a higher boiling point, provided the solute is nonvolatile and hence is not present in the vapor phase. The problem of the variation of the boiling point of a solution with composition, at constant pressure, may be treated by means of equations already derived, provided the solution is dilute enough for the solvent to behave ideally. If the solute is nonvolatile, as postulated above, the vapor consists entirely of solvent molecules, and hence N_1' is unity. In this event, equation (34.22), which applies to a

[11] van't Hoff, *Z. phys. Chem.*, **5**, 322 (1890).

liquid-vapor equilibrium such as would exist at the boiling point of a solution, can be written as

$$\left(\frac{\partial \ln \text{N}_1}{\partial T}\right)_P = -\frac{\Delta H_v}{RT^2}, \qquad (36.17)$$

where ΔH_v is the molar heat of vaporization of the solvent at the specified temperature and pressure. By assuming this quantity to remain constant in a small temperature range, integration of equation (36.17) gives

$$\ln \text{N}_1 = -\frac{\Delta H_v}{R}\left(\frac{T - T_0}{TT_0}\right), \qquad (36.18)$$

where T is the boiling point of the solution and T_0 is that of the pure solvent at the same pressure. If the system does not contain an inert gas this constant pressure is that of the atmosphere, for the temperature is then the normal boiling point. Because N_1 is always less than unity, $\ln \text{N}_1$ is negative, and hence $T - T_0$ must be positive; consequently, *the boiling point of the solution is greater than that of the solvent. The presence of a nonvolatile solute thus raises the boiling point.*

If the rise of boiling point $T - T_0$ is represented by θ, it follows that for a dilute solution [cf. equation (36.8)],

$$\ln \text{N}_1 = -\frac{\Delta H_v}{RT_0^2}\theta, \qquad (36.19)$$

and hence for a very dilute solution,

$$\theta = \frac{RT_0^2}{\Delta H_v}\text{N}_2,$$

$$\approx \lambda m,$$

by analogy with equations (36.9) and (36.11). The **molal boiling point elevation constant** λ is equal to $RT_0^2 M_1/1000\Delta H_v$, where ΔH_v is the molar heat of vaporization at the boiling point T_0 of the pure solvent, of formula weight M_1. The results given above may be used for determining the molecular weight of a solute from the rise of boiling point of the solution.

If the solute is volatile, so that the vapor in equilibrium with the solution contains both constituents, it is necessary to make use of equation (34.22). Provided the solution is dilute enough for the solvent to behave ideally in the liquid phase, and the vapor behaves as an ideal gas, this equation becomes

$$\ln \frac{\text{N}_1}{\text{N}_1'} = -\frac{\Delta H_v}{R}\left(\frac{T - T_0}{TT_0}\right), \qquad (36.20)$$

where N_1' is the mole fraction of solvent molecules in the vapor phase. By treating this result in the same manner as the analogous equation (36.18), and writing θ for $T - T_0$, the elevation of the boiling point, it is readily found that for a dilute solution

$$\theta = \frac{RT_0^2}{\Delta H_v}(N_2 - N_2')$$

$$= \frac{RT_0^2}{\Delta H_v} N_2(1 - k),$$

where k, equal to N_2'/N_2, is the distribution ratio of the volatile solute between the vapor and the solution. As before, for a very dilute solution, this can be converted into a form analogous to equation (36.16), viz.,

$$\theta = \lambda m(1 - k). \tag{36.21}$$

It can be seen from this expression that if k is small, that is to say, if the solute is not appreciably volatile, $1 - k$ is positive, and the presence of the solute results in a rise of boiling point On the other hand, if k is larger than unity, as might be the case for a volatile solute, θ would be negative, so that the solution has a lower boiling point than the solvent. This is, of course, of common occurrence in the distillation of mixtures of two volatile liquids.

36f. Temperature and Solubility in Dilute Solutions.—For a saturated solution of a gas in a nonvolatile solvent, or of a pure solid in a liquid solvent, the equations (34.21) and (34.28) reduce to

$$\left(\frac{\partial \ln f_2}{\partial N_2}\right)_{T,P} dN_2 = \frac{\bar{H}_2 - H_2'}{RT^2} dT \tag{36.22}$$

where H_2' is the molar heat content of the pure gas or solid solute, and $\bar{H}_2$ is its partial molar heat content in the saturated solution, at the given temperature and pressure. If the solution is dilute, equation (36.4) will be applicable, and then (36.22) becomes

$$\left(\frac{\partial \ln N_2}{\partial T}\right)_P = \frac{\bar{H}_2 - H_2'}{RT^2}. \tag{36.23}$$

In this expression, which gives the influence of temperature on solubility in dilute solution, the saturation solubility N_2 in mole fractions, may be replaced by the molality m or the molarity c, since these quantities are proportional to one another in dilute solution.

It should be noted that since the solute does not necessarily obey Raoult's law, although it satisfies Henry's law, it is not permissible to replace $\bar{H}_2$ by the molar heat content of the pure liquid, as was done in §§ 34f, 34i. The numerator on the right-hand side of equation (36.23) thus cannot be identified with a latent heat. It will be seen in § 44b that it is actually equivalent to the differential heat of solution (cf. § 12l) of the solute in the saturated solution under the given conditions. For a dilute solution this is usually not appreciably different from the integral heat of solution. It is thus possible to determine the heat of solution of a sparingly soluble gas or solid by measuring the solubility at two or more temperatures and then utilizing the integrated form of equation (36.23). As indicated at the outset, the assumption is made that one of the phases consists of either pure solid or pure gas. The same result would apply to a sparingly soluble liquid solute, provided the solvent did not dissolve in it to any appreciable extent.

EXERCISES

1. If the vapors behave ideally, the equations (35.9) take the form

$$p_1 = N_1 p_1^\square e^{\frac{1}{2}\beta N_2^2} \quad \text{and} \quad p_2 = N_2 p_2^\square e^{\frac{1}{2}\beta N_1^2}.$$

Hence, derive expressions for the total vapor pressure of a liquid mixture in terms of the mole fraction composition of (i) the liquid, (ii) the vapor.

2. Taking $p_1^\square$ as 100 mm., $p_2^\square$ as 200 mm. and β as 2.0, determine the total vapor pressure at mole fractions 0.2, 0.4, 0.6 and 0.8 by using the results of the preceding exercise. Plot the values against the mole fraction composition of (i) the liquid, (ii) the vapor.

3. At 70° C, the vapor pressures of carbon tetrachloride and benzene are 617.43 and 551.03 mm., respectively; at 50° C, the values are 312.04 and 271.34 mm., respectively [Scatchard, Mochel and Wood, *J. Am. Chem. Soc.*, **62**, 712 (1940)]. Assuming ideal behavior, since the actual deviations are small, plot the curve giving the variation of the (mole fraction) composition of the vapor with that of the liquid at each temperature. Would the separation of the two components by fractional distillation be more efficient at high or at low temperature?

4. Give in full the derivation of the statement in the text that heat is evolved upon mixing two liquids which form a system exhibiting negative deviations from ideal behavior.

5. The total vapor pressures of mixtures of chlorobenzene (1) and 1-nitropropane (2) at 120° C are as follows [Lacher, Buck and Parry, *J. Am. Chem. Soc.*, **63**, 2422 (1941)]:

N_1	0	0.096	0.282	0.454	0.507	0.675	0.765	0.844	1.00
P	545	565.2	590.2	597.2	597.1	596.6	589.4	579.9	544 mm.

Plot the results and determine the value of β in the simple form of the Margules equation. Estimate the partial pressure of each constituent in equilibrium with an equimolar liquid mixture.

6. Show that in the two-constant Margules equations (35.7) and (35.8) $\beta_2 = \beta_1 + \gamma_1$ and $\gamma_1 = -\gamma_2$.

7. Prove that the constants in the van Laar equation (35.10) are related by $\alpha_1 \beta_2 = \alpha_2$ and $\alpha_2 \beta_1 = \alpha_1$.

8. Derive expressions for $\log \gamma$, where γ is the activity coefficient a/N, on the basis of (i) the simple Margules equation (35.9), (ii) the van Laar equation (35.10). Show that for a liquid mixture exhibiting positive deviations from Raoult's law, the activity coefficient of each constituent, on the basis of the usual standard state, must be greater than unity, whereas for negative deviations it must be less than unity.

9. A mixture of 1 mole of ethyl bromide and 2 moles of ethyl iodide, which behaves ideally, is evaporated into a closed space at 40° C, when the vapor pressures of the pure components are 802 and 252 mm., respectively. Calculate the total pressure at the beginning and end of the process. (Note that at the end of the vaporization the composition of the vapor will be identical with that of the initial liquid.)

10. The vapor pressures of n-hexane and n-heptane at several temperatures are as follows:

	50°	60°	70°	80°	90°	100° C
n-hexane	401	566	787	1062	1407	1836 mm.
n-heptane	141	209	302	427	589	795

Assuming ideal behavior, draw the curve representing the boiling points of the mixtures at 1 atm. pressure as a function of the mole fraction composition, e.g., 0.2, 0.4, 0.6 and 0.8, of (i) the liquid, (ii) the vapor. [Plot the total vapor pressure-composition curves for various constant temperatures, and hence obtain the necessary data for 760 mm. The calculations may also be made at another pressure, e.g., 560 mm. and the change noted (cf. § 34f).]

11. The vapor pressure of liquid ethylene is 40.6 atm. at 0° C and 24.8 atm. at $-20°$ C. Estimate the ideal solubility of the gas in a liquid at 25° C and 1 atm. pressure. How many grams of ethylene should dissolve in 1,000 g. of benzene under these conditions, if the gas and solution behave ideally?

12. The absorption coefficient (of solubility) α of a gas is the volume of gas reduced to 0° C and 1 atm. pressure which will be dissolved by unit volume of solvent at the experimental temperature under a partial pressure of the gas of 1 atm. Show that for a dilute solution, N_2 in equation (36.23) may be replaced by α. The absorption coefficient of nitrogen is 0.01685 at 15° and 0.01256 at 35° C. Determine the mean differential heat of solution per mole of nitrogen in the saturated solution in this temperature range.

13. The pressures of sulfur dioxide in equilibrium with a solution containing 0.51 g. SO_2 in 100 g. water are 49.0 mm. at 35.6° C, 57.0 mm. at 41.0° C, 70.0 mm. at 47.0° C; for a solution containing 1.09 g. SO_2 in 100 g. water, the pressures are 70.5 mm. at 26.8° C, 96.0 mm. at 33.6° C, 124.5 mm. at 39.4° C and 147.0 mm. at 44.2° C [Beuschlein and Simenson, *J. Am. Chem. Soc.*, **62**, 612 (1940)]. Show that in dilute solutions Henry's law takes the form $m = kp$, where m is the molality of the solution and p the partial pressure of the solute; hence, use the foregoing data to test the validity of the law for the sulfur dioxide solutions at 35° and 45°C. Calculate the mean differential heat of solution of the gas in the range from 26.8° to 47.0° C at a gas pressure of 70.0 mm., and consider how you would expect this to differ from the heat of condensation, i.e., $-\Delta H_v$.

14. The molar heat of fusion ΔH_f of naphthalene at the absolute temperature T is given by $\Delta H_f = \Delta H_0 + 4.8T$, where ΔH_0 is a constant [Ward, *J. Phys. Chem.*, **40**, 761 (1934)]; at the melting point (80.2° C), ΔH_f is 4,540 cal. Determine the ideal solubility of naphthalene at 25° C, allowing for the variation of the heat of fusion with temperature.

15. The heat of fusion of the hydrocarbon anthracene ($C_{14}H_{10}$) at its melting point (216.6° C) is 38.7 cal. g.$^{-1}$. Its mole fraction solubility at 20° C is 0.0081 in benzene and 0.0009 in ethanol. Account for the results.

16. The freezing point depression of a 3.360 molal solution of urea in water is 5.490° [Chadwell and Politi, *J. Am. Chem. Soc.*, **60**, 1291 (1938)]. The heat of fusion of ice is 79.80 cal. g.$^{-1}$. Test the applicability of equations (36.7), (36.8), (36.9) and (36.11) to the results, and hence draw conclusions as to the reliability of the various approximations.

17. The melting points of benzene and naphthalene are 5.4° and 80.2° C, and the heats of fusion are then 2,360 and 4,540 cal. mole^{-1}. Assuming the system to be ideal and the heats of fusion to remain constant, determine the temperature and composition of the eutectic system, i.e., when the liquid mixture is in equi-

librium with both solids. [Use equation (34.33); a graphical procedure will be found convenient for obtaining the required solution.]

18. The melting point of p-dichlorobenzene is 53.2° C and that of naphthalene is 80.2° C; the eutectic temperature is 30.2° C when the mole fraction of the naphthalene in the liquid phase is 0.394. Calculate the molar heats of fusion of the two components of the system, assuming ideal behavior.

19. Show that if a solution is dilute and obeys Henry's law, and the vapor behaves ideally, the molal solubility of the solid (m_s) and of the supercooled liquid (m_l) solute at the same temperature T should be given by

$$\ln \frac{m_s}{m_l} = \frac{\Delta H_f}{R}\left(\frac{1}{T_m} - \frac{1}{T}\right),$$

where T_m is the melting point of the solid and ΔH_f is its heat of fusion (cf. Exercise 20, Chapter XI). The heat of fusion of solid nitrobenzene at its melting point (5.7° C) is 2,770 cal. mole^{-1}. The solubility of the solid in water at 0° C is 12.6×10^{-3} mole per 1000 g. water; calculate the solubility of the supercooled liquid at the same temperature. [The experimental value is 13.5×10^{-3} mole per 1000 g. water (Saylor, Stuckey and Gross, *J. Am. Chem. Soc.*, **60**, 373 (1938)).]

20. The solubilities of stannic iodide in carbon disulfide at several temperatures are as follows [Dorfman and Hildebrand, *J. Am. Chem. Soc.*, **49**, 729 (1927)]:

Temperature	10.0°	25.0°	40.0° C
Solubility	49.01	58.53	67.56 g. per 100 g. solution

Estimate the melting point and heat of fusion of stannic iodide. (The experimental values are 143.5° C and approximately 3,800 cal. mole^{-1}.)

21. A solution of 9.14 g. of iodine in 1000 g. benzene has a freezing point 0.129° below that of pure benzene (5.400° C); the solid which separates is a solid solution of iodine in benzene. The heat of fusion of benzene is 30.2 cal. g.$^{-1}$. Calculate the ratio in which the iodine distributes itself between liquid and solid phases in the vicinity of 5° C. Compare the result with that obtained by analytical determination as follows: the solid phase contains 0.317 g. and the liquid 0.945 g. iodine per 100 g. benzene [Beckmann and Stock, *Z. phys. Chem.*, **17**, 120 (1895)].

22. Starting with an equation of the form of (34.20), give all the stages in the derivation of equation (34.29).

23. Show that the weight composition of the vapor, treated as an ideal gas, in equilibrium with an ideal solution is given by

$$\frac{w_1'}{w_2'} = \frac{p_1^\square w_1}{p_2^\square w_2},$$

where w_1 and w_2 are the weights of the constituents in the liquid phase, and w_1' and w_2' are those in the vapor; hence, solve the problem in § 34e.

24. Show that for an ideal system of two liquids, the free energy of mixing, i.e., the difference between the free energy per mole of mixture and the sum of the free energies of the separate constituents, is given by $\Delta F = N_1 RT \ln N_1 + N_2 RT \ln N_2$. Hence, use the result concerning the heat of mixing in § 34a to show that the entropy of mixing two liquids forming an ideal system is $\Delta S = -R(N_1 \ln N_1 + N_2 \ln N_2)$ (cf. Exercise 1, Chapter IX).

25. Show that the variation with pressure of the solubility of a pure solid forming an ideal solution in a liquid is given by

$$\left(\frac{\partial \ln \text{N}_2}{\partial P}\right)_T = -\frac{\Delta V_f}{RT},$$

where ΔV_f is the volume change accompanying the fusion of 1 mole of solid solute at T and P. What conclusion may be drawn concerning the sign and magnitude of the effect of pressure on solubility?

26. Show that although the partial molar heat content of the constituent of an ideal solution is independent of the composition (§ 34a), this is not the case for the partial molar free energy and entropy. Derive expressions for $(\partial \mu_i/\partial \text{N}_i)_{T,P}$ and $(\partial \bar{S}_i/\partial \text{N}_i)_{T,P}$ for an ideal solution.

CHAPTER XV

ACTIVITIES AND ACTIVITY COEFFICIENTS

37. Standard States

37a. Choice of Standard State.—A knowledge of activities and activity coefficients is of both theoretical and practical interest in the study of solutions; the present chapter will, therefore, be devoted to a consideration of the methods by which these quantities may be determined. As seen in § 31b, the activity of a substance in any state is a relative property, the actual value being relative to a chosen standard state. The nature of the standard state is of no thermodynamic significance, as may be seen in the following manner. Let a'_α and a''_α represent the activities of a given substance in two different solutions, the values being in terms of a specified standard state; let a'_β and a''_β be the corresponding values for another standard state. Since these activities are relative to the activity in the chosen standard states, it follows that a''_α/a'_α must be equal to a''_β/a'_β. If μ' is the chemical potential (partial molar free energy) of the substance in the first solution and μ'' is that in the other solution, the free energy increase accompanying the transfer of 1 mole of the substance from the first solution to the second, under such conditions that the compositions remain virtually unchanged, is given by [cf. equation (26.29)]

$$\Delta F = \mu'' - \mu'.$$

By equation (31.2), the chemical potentials are

$$\mu' = \mu^0_\alpha + RT \ln a'_\alpha \quad \text{and} \quad \mu'' = \mu^0_\alpha + RT \ln a''_\alpha,$$

in terms of the one standard state, and

$$\mu' = \mu^0_\beta + RT \ln a'_\beta \quad \text{and} \quad \mu'' = \mu^0_\beta + RT \ln a''_\beta,$$

in terms of the alternative standard state. It follows, then, from the foregoing equations that

$$\Delta F = RT \ln \frac{a''_\alpha}{a'_\alpha}$$

in one case, and

$$\Delta F = RT \ln \frac{a''_\beta}{a'_\beta}$$

in the other case. As seen above, however, a''_α/a'_α is equal to a''_β/a'_β, and so the free energy change for a given process, which is the thermodynamically important quantity, expressed in terms of the activities, will have the same value, as indeed it must, irrespective of the choice of the standard state.

In view of the foregoing arguments, it is evident that the standard state chosen in any particular instance should be the most convenient one in the given circumstances. It is quite permissible, in fact, to choose different standard states for a given substance in two (or more) phases in equilibrium; the effect of the choice is merely to alter the value of the equilibrium constant, without affecting its constancy. It may be pointed out that the activities of a given constituent of two phases in equilibrium are equal *when referred to the same standard state*. However, when different standard states are chosen for the two phases, the activities are not equal, although the chemical potentials must be identical (cf. § 28a). Thus, for a constituent in phase I,

$$\mu_\mathrm{I} = \mu_\alpha^0 + RT \ln a_\alpha,$$

while for this substance in the phase II, in equilibrium with phase I,

$$\mu_\mathrm{II} = \mu_\beta^0 + RT \ln a_\beta.$$

The chemical potentials in the two phases in equilibrium must be the same, i.e., $\mu_\mathrm{I} = \mu_\mathrm{II}$, and hence, for the given substance

$$\mu_\alpha^0 + RT \ln a_\alpha = \mu_\beta^0 + RT \ln a_\beta.$$

If the chosen standard states are identical, μ_α^0 and μ_β^0 are the same, since these are the chemical potentials in the standard states; consequently a_α and a_β, the activities in the two phases, will be equal. On the other hand, if the standard states selected for the phases I and II are not the same, a_α and a_β will be different, the difference depending on the corresponding values of μ_α^0 and μ_β^0.

37b. Convenient Standard States: I. Gases.—For gases and vapors the most convenient standard state, which is almost invariably employed in thermodynamic studies, is the one referred to in an earlier section (§ 30b). *The standard state of unit activity is usually chosen as that of unit fugacity*, so that for a gas the activity is identical with the fugacity. As seen earlier, this is equivalent to choosing the gaseous system at very low total pressure as the reference state and postulating that the ratio of the activity of any constituent to its partial pressure, i.e., the activity coefficient a/p, then approaches unity. The determination of the fugacities of gases has already been discussed in some detail (Chapter XII), and since the activity of a gas is almost invariably defined so as to be identical with the fugacity, it is not necessary to consider the subject further.

II. Solid or Liquid Solvent.—For a liquid or solid which acts as the solvent of a solution, that is to say, it is the constituent present in excess, the most convenient standard state is also the one used in earlier sections. *The activity of the pure liquid (or pure solid) solvent, at atmospheric pressure, is taken as unity at each temperature.* As stated in § 31b, the corresponding reference state is the pure liquid (or solid) at 1 atm. pressure, the activity coefficient, defined by a/N, being then equal to unity. With increasing dilution of a solution the mole fraction N_2 of the solute tends to zero and that of the solvent, i.e., N_1, tends to unity. It follows, therefore, that at 1 atm. pressure a_1/N_1 approaches unity as the solution becomes more dilute; thus,

$$\frac{a_1}{N_1} \to 1 \quad \text{as} \quad N_1 \to 1 \quad \text{and} \quad N_2 \to 0,$$

on the basis of the chosen standard state.

By equation (31.5), the activity of the solvent is equivalent to f_1/f_1^0 where f_1 is the fugacity in a given solution and f_1^0 is numerically equal to that in the standard state, i.e., pure liquid at 1 atm. pressure at the given temperature. Hence, it is seen from equation (34.1) that for an ideal solution the activity of the solvent should always be equal to its mole fraction, provided the total pressure is 1 atm. In other words, in these circumstances the activity coefficient a_1/N_1 should be unity at all concentrations. For a nonideal solution, therefore, the deviation of a_1/N_1 from unity at 1 atm. pressure may be taken as a measure of the departure from ideal (Raoult law) behavior. Since the activities of liquids are not greatly affected by pressure, this conclusion may be accepted as generally applicable, provided the pressure is not too high.

III. Solutes: A. Pure Liquid as Standard State.—Several different standard states are used for solutes according to the circumstances. If the solution is of the type in which two liquids are completely, or almost completely, miscible, so that there is no essential difference between solvent and solute, the standard state of the solute is chosen as the pure liquid at atmospheric pressure, just as for the solvent. The defined activity coefficient a_2/N_2 then approaches unity as N_2 tends to unity, and its deviation from this value, at 1 atm. pressure, gives an indication of the deviation from ideal behavior.

In many instances, especially when the solute has a limited solubility in the solvent, as is often the case for solid solutes, particularly electrolytes or where dilute solutions are under consideration, it is more convenient to choose an entirely different standard state. Three such states are used: one is based on compositions expressed in mole fractions, but the others are more frequently employed, because the compositions of dilute and moderately dilute solutions are usually stated in terms of molality, i.e., moles per 1000 g. solvent, or in terms of molarity, i.e., moles per liter of solution.[1]

III. B. Infinitely Dilute Solution as Reference State: Composition in Mole Fraction.—In the first case, *the reference state is chosen as the infinitely dilute solution*, with the postulate that at the given temperature *the ratio of the activity of the solute to its mole fraction, i.e., a_2/N_2, then becomes unity at atmospheric pressure*. The activity is consequently defined so that at the temperature of the solution and 1 atm. pressure

$$\frac{a_2}{N_2} \to 1 \quad \text{as} \quad N_2 \to 0. \tag{37.1}$$

[1] G. N. Lewis and M. Randall, "Thermodynamics and the Free Energy of Chemical Substances," 1923, pp. 256 et seq.; see also, Adams, *Chem. Rev.*, **19**, 1 (1936); Goranson, *J. Chem. Phys.*, **5**, 107 (1937).

The standard state is here a purely hypothetical one, just as is the case with gases (§ 30b); it might be regarded as the state in which the mole fraction of the solute is unity, but certain thermodynamic properties, e.g., partial molar heat content and heat capacity, are those of the solute in the reference state, i.e., infinite dilution (cf. § 37d). If the solution behaved ideally over the whole range of composition, the activity would always be equal to the mole fraction, even when $N_2 = 1$, i.e., for the pure solute (cf. Fig. 24, I). In this event, the proposed standard state would represent the pure liquid solute at 1 atm. pressure. For nonideal solutions, however, the standard state has no reality, and so it is preferable to define it in terms of a reference state.

The significance of the foregoing definition of the activity of a solute may be seen by making use of the fact (§ 31b) that the activity is proportional to the fugacity, the value of the proportionality constant depending on the chosen reference or standard state. Representing this constant by $1/k$, it is seen that the activity a_2 of the solute and its fugacity f_2 may be related by

$$a_2 = \frac{f_2}{k} \quad \text{or} \quad f_2 = a_2 k. \tag{37.2}$$

In dilute solution, when N_2 approaches zero, a_2/N_2 approaches unity, according to the postulated standard state; hence equation (37.2) can be written as

$$f_2 = N_2 k \quad \text{as} \quad N_2 \to 0.$$

This result is identical with Henry's law [equation (36.3)], and so on the basis of the present standard state, a_2/N_2 will be unity, or a_2 will equal N_2, for all solutions obeying Henry's law. The activity coefficient γ_N, equal to a_2/N_2, of the solute in any solution may thus be used to indicate its obedience of Henry's law. It is for this reason that the activity coefficient defined in this manner has been called the **rational activity coefficient**.

If the solution were ideal over the whole range of composition, k in equation (37.2) would, of course, be equal to the fugacity of the pure liquid solute at 1 atm. pressure, and Henry's law and Raoult's law would be identical (§ 36a). However, although the behavior of a *solute* in solution may deviate considerably from Raoult's law, it almost invariably satisfies Henry's law at high dilutions. Consequently, for the study of not too concentrated solutions, the standard state under consideration has some advantages over that in III. A.

III. C. Composition in Molality or Molarity.—Because the compositions of solutions, especially of electrolytes, in the study of which the activity concept has proved of the utmost significance, are expressed in terms of molality or molarity,* alternative standard states have been widely used. In the event that the composition of the solution is given by the molality of the solute, *the reference state is the infinitely dilute solution, the activity a_2*

---

* The use of the molality is to be preferred in thermodynamic studies, since it is independent of temperature and pressure; this is not the case for the molarity, because the volume of the solution is affected by a change of temperature or pressure.

of the solute being defined so that its ratio to the molality m, i.e., a_2/m, is then unity at atmospheric pressure, at the temperature of the system. The definition of the activity is consequently based on the postulate that

$$\frac{a_2}{m} \to 1 \quad \text{as} \quad m \to 0. \tag{37.3}$$

The standard state is here also a hypothetical one; it is equivalent to a 1 molal solution in which the solute has some of the partial molar properties, e.g., heat content and heat capacity, of the infinitely dilute solution. It has been referred to as the "hypothetical ideal 1 molal solution". At high dilutions the molality of a solution is directly proportional to its mole fraction (§ 32f), and hence dilute solutions in which the activity of the solute is equal to its molality also satisfy Henry's law. Under such conditions, the departure from unity of the activity coefficient γ_m, equal to a_2/m, like that of γ_N, is a measure of the deviation from Henry's law.

At infinite dilution the activity coefficients γ_N and γ_m are both unity, in accordance with equations (37.1) and (37.3), respectively, which define the standard states. In dilute solutions the values will also be approximately equal, but not necessarily unity because of failure to obey Henry's law. At appreciable concentrations, however, the molality of the solution is no longer proportional to the mole fraction of the solute, and even if the solution obeyed Henry's law γ_m would not be unity. In such solutions γ_N and γ_m will be appreciably different.

When the composition of the solution is given in terms of molarity, i.e., moles of solute per liter of solution, the standard state chosen is analogous to that proposed above. The activity is defined in such a manner that at the given temperature *the ratio of the activity a_2 of the solute to its molarity c, i.e., a_2/c, approaches unity in the infinitely dilute solution at 1 atm. pressure;* thus,

$$\frac{a_2}{c} \to 1 \quad \text{as} \quad c \to 0. \tag{37.4}$$

The standard state is again a hypothetical one; it corresponds to a solution containing 1 mole of solute per liter, but in which certain partial molar properties are those of the solute at infinite dilution.

As seen in §32f, the molarity of a very dilute solution is proportional to its mole fraction, and hence for such solutions the activity coefficient γ_c, defined by a_2/c, represents the compliance with Henry's law. As in the preceding case, γ_c and γ_N are both unity at infinite dilution, and the values are approximately equal in dilute solutions. With increasing concentration, however, γ_c and γ_N differ, and although the latter still indicates the adherence to Henry's law, the former, like γ_m, does not. Thus, at high dilutions γ_N, γ_m and γ_c approach one another, their values approximating to unity, but at appreciable concentrations the three coefficients differ; the actual relationships between them will be considered in the next section.

37c. Relationship between Activity Coefficients of Solute.—For a solution of molality m, the number of moles of solute is m whereas the number of moles of solvent is $1000/M_1$, where M_1 is its molecular weight. The mole fraction N of solute is then equal to $m/(m + 1000/M_1)$. On the other hand, for a solution of molarity c, the number of moles of solute is c, and the number of moles of solvent is $(1000\rho - cM_2)/M_1$, where ρ is the density of the solution and M_2 is the molecular weight of the solute. The mole fraction of solute in terms of molarity is thus $c/[c + (1000\rho - cM_2)/M_1]$. A slight rearrangement of these results then gives

$$N = \frac{mM_1}{mM_1 + 1000} = \frac{cM_1}{c(M_1 - M_2) + 1000\rho}. \qquad (37.5)$$

In very dilute solution, when the mole fraction of solute is N^*, the molality m^* and the molarity c^*, the density ρ_0 being that of the pure solvent, equation (37.5) reduces to

$$N^* = \frac{m^*M_1}{1000} = \frac{c^*M_1}{1000\rho_0}. \qquad (37.6)$$

The difference of the free energy of the solute in the solution represented by N, m and c, and that in the very dilute solution N^*, m^* and c^*, must have a definite value irrespective of the standard states (§ 37a). This is determined by the ratio of the activities in the two solutions, viz., a/a^*; the activity a may be represented by $N\gamma_N$, $m\gamma_m$ and $c\gamma_c$, depending on the particular standard state, but a^* is equal to N^*, m^* and c^*, respectively, since the solution is dilute enough for the activity coefficient to be unity in each case. It follows, therefore, that

$$\frac{a}{a^*} = \frac{N\gamma_N}{N^*} = \frac{m\gamma_m}{m^*} = \frac{c\gamma_c}{c^*},$$

and hence by utilizing equations (37.5) and (37.6) it is found that

$$\gamma_N = \gamma_m(1 + 0.001mM_1) = \gamma_c \frac{\rho + 0.001c(M_1 - M_2)}{\rho_0}, \qquad (37.7)$$

which gives the relationship among the three activity coefficients. It is evident from this result that in dilute solutions, e.g., when c and m are less than about 0.1, the values of γ_N, γ_m and γ_c are almost identical.

By treating γ_m and γ_c in an analogous manner, it is found that

$$\gamma_m = \gamma_c \frac{\rho - 0.001cM_2}{\rho_0}. \qquad (37.8)$$

This expression is more convenient than equation (37.7) when the conversion of γ_m to γ_c, or vice versa, is required, without the necessity of introducing γ_N.

37d. Partial Molar Heat Contents in Standard and Reference States.—
The general expression for the variation of the activity coefficient with temperature, at constant pressure and composition, may be derived by a

procedure similar to that used to obtain equation (31.6). Since the molality and mole fraction are independent of temperature, the temperature variation of activity coefficients expressed in terms of these quantities, i.e., γ_N or γ_m, is given by

$$\left(\frac{\partial \ln \gamma_i}{\partial T}\right)_{P,N} = \frac{\bar{H}_i^0 - \bar{H}_i}{RT^2}, \qquad (37.9)$$

where $\bar{H}_i$ is the partial molar heat content of the substance i in the given solution and $\bar{H}_i^0$ is the value in the standard state. It will be observed from the discussion in § 37b, that the standard state is always defined in such a manner as to make the activity coefficient equal to unity in the reference state, e.g., pure liquid for a solvent or infinite dilution for a solute, at 1 atm. pressure, at all temperatures. In the reference state, therefore, γ_i is independent of the temperature, so that the left-hand side of equation (37.9) is zero. Consequently, *the partial molar heat content of any constituent of a solution is the same in the reference state as in the standard state.* This is the justification for the remark in § 37b, III B, C, where the infinitely dilute solution was chosen as the reference state, that the partial molar heat content of the solute in the standard state is equal to that at infinite dilution. The same is true for the partial molar heat capacity at constant pressure, since this is the derivative of the heat content with respect to temperature. It should be understood that the foregoing conclusions apply, strictly, at 1 atm. pressure, for it is only then that the activity coefficient in the reference state is unity at all temperatures.

Since the volume of a solution changes with temperature, the molarity, i.e., moles per liter, must change at the same time. Hence, equation (37.9) does not give the variation of γ_c with temperature; however, from this equation, in which γ_i is γ_m or γ_N, and (37.7) or (37.8), noting that in the latter c is merely a number, it follows that

$$\left(\frac{\partial \ln \gamma_c}{\partial T}\right)_{P,N} = \frac{\bar{H}_i^0 - \bar{H}_i}{RT^2} + \left[\frac{\partial \ln (\rho_0/\rho)}{\partial T}\right]_{P,N}. \qquad (37.10)$$

At infinite dilution, γ_c is equal to unity at all temperatures, by definition, and at the same time ρ becomes identical with ρ_0, so that, in this case also, the standard partial molar heat content of the solute is equal to its value in the infinitely dilute solution at 1 atm. pressure.

In order to avoid the possibility of misunderstanding, it may be stated here that the partial molar free energy and partial molar entropy of the solute in the standard state are entirely different from the values in the infinitely dilute solution (see Exercise 24).

38. Determination of Activities

38a. Activity of Solvent from Vapor Pressure.—The activity of any constituent of a solution is given, in general, by the ratio of the fugacities, i.e., f_i/f_i^0, f_i being the value in the solution and f_i^0 is that in the standard state, viz., pure liquid at the same temperature and 1 atm. pressure. If the vapor pressure of the solvent is not too high, the fugacities may be replaced, without appreciable error, by the respective vapor pressures. At the same time, the effect of the external pressure on the activity may be neglected, so that it is possible to write

$$a_1 \approx \frac{p_1}{p_1^0} \approx \frac{p_1}{p_1^\square}, \qquad (38.1)$$

where p_1 is the partial vapor pressure of the solvent in equilibrium with the solution in which its activity is a_1, and $p_1^\square$ is the vapor pressure of the pure solvent at the same temperature and (approximately) the same pressure. The equation (38.1) thus provides a relatively simple method for determining the activity of the solvent in a solution; it has been applied to aqueous solutions, to solutions of organic liquids and to mixtures of molten metals.[2] The activity coefficient at any composition may be obtained by dividing the activity a_1 of the solvent by its mole fraction N_1.

Problem: An exactly 1 molal aqueous solution of mannitol has a vapor pressure of 17.222 mm. of mercury at 20° C; at the same temperature, the vapor pressure of pure water is 17.535 mm. Calculate the activity and activity coefficient of water in the given solution.

At the low pressures, the water vapor may be regarded as behaving ideally, so that

$$a_1 \approx \frac{p_1}{p_1^\square} = \frac{17.222}{17.535} = 0.9822.$$

A molal aqueous solution contains 1 mole of solute and $1000/18.016 = 55.51$ moles of solvent; the mole fraction of solvent is thus

$$N_1 = \frac{55.51}{1.00 + 55.51} = 0.9823.$$

The activity and mole fraction are so close that the activity coefficient is virtually unity. (It is of interest to note that the mole fraction N_2 of the solute in a 1 molal aqueous solution is only $1.0000 - 0.9823 = 0.0177$, and so it is not surprising to find that the solvent obeys Raoult's law.)

38b. Activity of Volatile Solute from Vapor Pressure.

If the solvent and solute are essentially miscible, the standard state of the latter may be chosen as the pure liquid (cf. § 37b, III A); the activity may then be determined by a procedure identical with that just described, based on equation (38.1) in the form $a_2 \approx p_2/p_2^\square$. For dilute solutions, however, it is more convenient to use the infinitely dilute solution as the reference state. By equation (37.2) the activity a_2 is equal to f_2/k, where f_2 is the fugacity of the solute in the given solution, and k is a constant. Replacing the

---

[2] See, for example, Hildebrand, et al., J. Am. Chem. Soc., **36**, 2020 (1914); **37**, 2452 (1915); **42**, 545 (1920); **49**, 3011 (1927); Hirst and Olson, ibid., **51**, 2398 (1929); Lacher and Hunt, ibid., **63**, 1754 (1941); Lacher, Buck and Parry, ibid., **63**, 2422 (1941).

fugacity f_2 by the partial vapor pressure p_2 of the solute, which can be measured, it follows that

$$a_2 \approx \frac{p_2}{k}. \tag{38.2}$$

The value of k may be derived by utilizing the fact that at high dilutions the activity of the solute, represented by a_2^*, is equal to the mole fraction N_2^*, so that if p_2^* is the corresponding vapor pressure, equation (38.2) takes the form

$$N_2^* \approx \frac{p_2^*}{k}.$$

By combining this result with equation (38.2) it is seen that at an appreciable concentration,

$$a_2 \approx p_2 \frac{N_2^*}{p_2^*}. \tag{38.3}$$

This expression may be put into a modified form by dividing each side by N_2, so that

$$\gamma_N = \frac{a_2}{N_2} \approx \frac{p_2}{N_2} \bigg/ \frac{p_2^*}{N_2^*}. \tag{38.4}$$

If p_2/N_2 for a number of solutions is determined and the values are extrapolated to $N_2 = 0$, the result is equal to p_2^*/N_2^*, which is required in equations (38.3) and (38.4) for the evaluation of the activity or activity coefficient, respectively.

Expressions exactly analogous to (38.3) and (38.4) are obtained for the activity and activity coefficient based on a_2/m or a_2/c becoming unity at infinite dilution. All that is necessary is to replace N_2 (and N_2^*) by m or c, respectively, according to the standard state chosen.

When it is not possible to make accurate measurements of the partial vapor pressure of the solute at low concentrations, such as would be necessary to obtain p_2^*/N_2^*, an alternative procedure can sometimes be adopted. Thus, if the activity at any composition of the solute is known from other measurements, such as those described below, the constant k in equation (38.2) can be evaluated by means of the vapor pressure of the solute in the same solution. The method has been utilized to determine the activities and activity coefficients of hydrochloric acid in aqueous solution.[3]

38c. Activity of Solvent from Freezing Points.—If a_s is the activity of pure solid solvent *in terms of the pure liquid solvent*, at the same temperature and 1 atm. pressure, as the standard state, it is possible to write, by equation (31.2),

$$\mu_s = \mu_l^0 + RT \ln a_s,$$

[3] Lewis and Randall, ref. 1, Chapter XXII; for application to electrolytes, see Chapter XXVI, also *idem.*, *J. Am. Chem. Soc.*, **43**, 1112 (1921); Randall and Young, *ibid.*, **50**, 989 (1928).

where μ_l^0 refers to the chemical potential of the liquid in the standard state, and μ_s is that of the solid at the temperature T. Upon dividing through this equation by T, so as to obtain

$$R \ln a_s = \frac{\mu_s}{T} - \frac{\mu_l^0}{T},$$

and then differentiating with respect to T, making use of equation (26.25), the result is

$$\left(\frac{\partial \ln a_s}{\partial T}\right)_P = \frac{H_l^0 - H_s}{RT^2} = \frac{\Delta H_f}{RT^2}, \qquad (38.5)$$

where the partial molar heat contents have been replaced by the corresponding molar values, since they refer to pure liquid and solid. If the constant pressure, applicable to a_s and H_s, is chosen as 1 atm., then the numerator on the right-hand side is equal to the molar heat of fusion at this pressure and the temperature T.

At the freezing point of any solution, assuming pure solid solvent to separate, the fugacity of the solid will be identical with that of the solvent in the solution, since the system is in equilibrium. In the preceding paragraph, the same standard state was chosen for the solid as is conventionally adopted for the solvent in a solution; consequently, the activity of the solvent at the freezing point must be the same as that of the solid phase in equilibrium with it (cf. § 37a). Hence, the variation of the activity a_1 of the solvent at the freezing point of solutions of varying concentration, at 1 atm. pressure, is given by equation (38.5) in the form

$$\frac{d \ln a_1}{dT} = \frac{\Delta H_f}{RT^2}, \qquad (38.6)$$

which may be compared with equation (34.30) for an ideal solution.

For the complete integration of equation (38.6), allowance must be made for the variation of ΔH_f with temperature, instead of regarding it as constant as when dealing with dilute solutions in § 36b. By the Kirchhoff equation, which is applicable in the form of (12.7) since ΔH_f refers to a constant (1 atm.) pressure process,

$$\left[\frac{\partial(\Delta H_f)}{\partial T}\right]_P = (C_P)_l - (C_P)_s = \Delta C_P, \qquad (38.7)$$

where $(C_P)_l$ and $(C_P)_s$ represent the molar heat capacities of the pure liquid and solid solute at 1 atm. pressure. Over a short range of temperature the heat capacities may be taken as constant, so that integration of equation (38.7) gives

$$\Delta H_f = L_0 + \Delta C_P(T - T_0), \qquad (38.8)$$

where L_0 is used, for simplicity, for the molar latent heat of fusion at T_0, the freezing point of the solvent. As in § 36b, the temperature difference $T_0 - T$ may be replaced by θ, the lowering of the freezing point, so that

equation (38.8) becomes
$$\Delta H_f = L_0 - \theta \Delta C_P, \qquad (38.9)$$
and substitution in (38.6) gives
$$d \ln a_1 = \frac{L_0 - \theta \Delta C_P}{RT^2} dT.$$

Since T is equal to $T_0 - \theta$, and dT to $-d\theta$, this equation is equivalent to
$$-d \ln a_1 = \frac{L_0 - \theta \Delta C_P}{R(T_0 - \theta)^2} d\theta. \qquad (38.10)$$

In order to simplify integration of equation (38.10), the device is used of expressing $1/(T_0 - \theta)^2$ in the form of a power series; thus,
$$\frac{1}{(T_0 - \theta)^2} = \frac{1}{T_0^2}\left(1 - \frac{\theta}{T_0}\right)^{-2}$$
$$= \frac{1}{T_0^2}\left(1 + \frac{2\theta}{T_0} + \frac{3\theta^2}{T_0^2} + \cdots\right).$$

Hence equation (38.10) becomes
$$-d \ln a_1 = \frac{1}{RT_0^2}\left(1 + \frac{2\theta}{T_0} + \cdots\right)(L_0 - \theta \Delta C_P) d\theta$$
$$= \frac{1}{RT_0^2}\left[L_0 + \left(\frac{2L_0}{T_0} - \Delta C_P\right)\theta + \cdots\right] d\theta, \qquad (38.11)$$

and this can be integrated between the limits of a_1 from unity, i.e., pure solvent,* to a_1, the corresponding values of the freezing point depression θ being zero and θ, respectively; thus,
$$-\ln a_1 = \frac{L_0 \theta}{RT_0^2} + \frac{\theta^2}{RT_0^2}\left(\frac{L_0}{T_0} - \frac{\Delta C_P}{2}\right) + \cdots. \qquad (38.12)$$

Since θ is usually small, the terms involving θ^3, θ^4, etc., can be neglected and the foregoing expression gives the activity of the solvent at the freezing point of the given solution.

By means of equation (38.12), it is possible to determine the activity of the solvent a_1 in any solution, from a knowledge of the depression of the freezing point, and certain other properties. The procedure may be illustrated by a consideration of aqueous solutions. With water as solvent, L_0 is the molar heat of fusion of ice at 0° C and 1 atm. pressure, viz., 1438 cal. mole^{-1}; further, $(C_P)_l$ for water may be taken with sufficient accuracy as 18 and $(C_P)_s$ for ice as 9 cal. deg.$^{-1}$ mole^{-1}, so that ΔC_P is 9 cal. deg.$^{-1}$ mole^{-1}. Inserting these values into equation (38.12), with T_0 equal to 273.16° K,

* It should be noted that it is at this point that the postulated standard state for the solvent is introduced.

it is found that

$$\ln a_1' = -9.702 \times 10^{-3}\theta - 5.2 \times 10^{-6}\theta^2.$$

The activity a_1 can thus be readily calculated from the freezing point depression at atmospheric pressure.

Problem: The freezing point depression of 0.1 molal aqueous KCl solution is 0.345° C. Determine the activity of the water in this solution at the freezing point, with reference to pure water as the standard state.

In this case,

$$\ln a_1 = 2.303 \log a_1 = -9.702 \times 10^{-3} \times 0.345 - 5.2 \times 10^{-6} \times (0.345)^2$$
$$= -3.3478 \times 10^{-3}$$
$$\log a_1 = -1.454 \times 10^{-3}$$
$$a_1 = 0.9966.$$

It is important to note that because of the low mole fraction of solute, the activity of the water is very close to unity in the 0.1 molal solution of electrolyte. It is consequently the common practice to take the activity of the solvent as unity in any dilute solution, e.g., about 0.1 molal or less.

The results obtained from equation (38.12) give the activity of the solvent at temperatures which vary with the concentration of the solution. It is desirable, therefore, to adjust the results so that the values for the different solutions refer to the same temperature. This can be done by means of equation (31.6), which for the present purpose becomes

$$\left(\frac{\partial \ln a_1}{\partial T}\right)_{P,N} = \frac{H^0_{1l} - \bar{H}_1}{RT^2}, \qquad (38.13)^*$$

where H^0_{1l} is the molar heat content of the solvent in its standard state, i.e., pure liquid at 1 atm., and $\bar{H}_1$ is its partial molar heat content in the solution. The quantity $\bar{H}_1 - H^0_{1l}$, known as the relative partial molar heat content of the solvent,† which will be described more fully in Chapter XVIII, is usually represented by the symbol $\bar{L}_1$; hence equation (38.13) may be written

$$\left(\frac{\partial \ln a_1}{\partial T}\right)_{P,N} = -\frac{\bar{L}_1}{RT^2}. \qquad (38.14)$$

If a_1' represents the activity of the solvent at a variable temperature T', and a_1'' is the value at some standard temperature T'', e.g., 25° C, integration of equation (38.14) gives

$$\ln \frac{a_1''}{a_1'} = -\frac{1}{R} \int_{T'}^{T''} \frac{\bar{L}_1}{T^2} dT. \qquad (38.15)$$

If the relative partial molar heat content $\bar{L}_1$ of the solvent (i.e., the differential heat of dilution) is small, as it is for dilute solutions, the activity coefficient of the

---

* It is important that the distinction between the similar equations (38.5) and (38.13) should be clearly understood. The former gives the variation with temperature of the activity of the solvent in a solution at its freezing point, which varies with the composition. The latter applies to the activity of the solvent in a solution of constant composition.

† It is also the differential heat of dilution of the given solution (cf. § 44b.)

solvent does not vary appreciably with temperature. For other solutions, however, it is necessary to know $\bar{L}_1$ and also its variation with temperature, in order to evaluate the integral in equation (38.15). The procedure has been mainly used in connection with the study of the activities of solutes, particularly electrolytes, and it will be described more fully in § 39c.[4]

38d. Activity of Solvent from Boiling Points.—The treatment for the determination of activities from boiling point measurements is exactly analogous to that given in the preceding section in connection with the freezing points of solutions. The activity a_g of the solvent in the vapor (gaseous) state can be expressed in terms of the pure liquid as the standard state; thus,

$$\mu_g = \mu_l^0 + RT \ln a_g,$$

where μ_l^0 is the chemical potential of the pure liquid at 1 atm. pressure at the given temperature. Upon dividing through by T, and differentiating with respect to temperature, at constant pressure, as in § 38c, it is found that

$$\left(\frac{\partial \ln a_g}{\partial T} \right)_P = \frac{H_l^0 - \bar{H}_g}{RT^2}. \tag{38.16}$$

If the vapor behaves ideally, or if the solute is nonvolatile, as is usually the case with electrolytes, the partial molar heat content $\bar{H}_g$ of the gaseous solvent may be replaced by its molar heat content H_g. Further, if the pressure is taken as 1 atm., $H_g - H_l^0$ is equal to the molar heat of vaporization ΔH_v of the solvent at its normal boiling point.

At the boiling point, the liquid and gaseous phases are in equilibrium, and hence the activity a_1 of the solvent in the former will be equal to that in the latter, as given above, since the same standard state is used in each case. It follows, therefore, that the variation of the activity of the solvent at the boiling point for solutions of differing concentrations, at 1 atm. pressure, is given by equation (38.16) as

$$\frac{d \ln a_1}{dT} = - \frac{\Delta H_v}{RT^2}. \tag{38.17}$$

This expression, except for the negative sign, is similar to equation (38.6), applicable at the freezing point, and by treating (38.17) in an analogous manner to that described in § 38c, it is possible to obtain equations of the same form as (38.11) and (38.12). The symbol θ now represents the rise of boiling point; its sign, being opposite to that of the lowering of the freezing point, compensates for the negative sign in equation (38.17).

Because boiling point determinations on solutions have been carried out over a considerable temperature range, by changing the external pressure, it is necessary to use a more precise expression than equation (38.9) to represent the variation of ΔH_v with temperature. For this purpose the empirical relationship

$$\Delta H_v = L_0 - b\theta - c\theta^2 - d\theta^3, \tag{38.18}$$

[4] Lewis and Randall, ref. 3.

where b, c and d are constants, has been proposed. Upon insertion of this expression in equation (38.17) and integrating, the result may be put in the form

$$-\ln a_1 = \frac{L_0 \theta}{RT_0^2} + \frac{\theta^2}{RT_0^2} b' + \frac{\theta^3}{RT_0^2} c' + \cdots, \qquad (38.19)$$

the constants b', c', etc., which are related to b, c, etc., being derived from actual heat of vaporization measurements at various temperatures. The activity of the solvent in the solution can then be calculated directly from the elevation of the boiling point. By making measurements at various pressures it is possible to obtain the values over a range of temperature, the effect of pressure on the activity being usually neglected. The results can then be used to calculate the relative partial molar heat content of the solvent, i.e., $\bar{L}_1$, by means of equation (38.14) or (38.15); further reference to this subject will be made in § 44g.[5]

38e. Activity from E.M.F. Measurements.—It was seen in § 33l that the E.M.F. of a reversible galvanic cell is related to the free energy change of the process taking place in the cell. The free energy change is dependent upon the activities of the substances involved in the cell reaction, and hence the measurement of E.M.F. provides a possibility for the evaluation of activities. The method has been largely used in the study of electrolytes, as will be seen later, but the general principle is of wide applicability. It will be employed here to determine the activity of one metal when dissolved in another metal, e.g., in a liquid alloy or amalgam.[6]

Consider a galvanic cell in which the two electrodes consist of homogeneous liquid amalgams of different proportions, mole fractions N_2 and N_2', of the same metal, e.g., thallium, dissolved in mercury; the electrolyte is an aqueous solution of a salt of the metal, i.e., thallium. The cell may be represented by

Tl amalgam (N_2') | Thallous salt solution | Tl amalgam (N_2).

The free energy change accompanying the transfer of 1 mole of thallium from the left-hand electrode to the right-hand electrode is equal to the difference in the partial molar free energies (chemical potentials) of the thallium in the two amalgams. Thus (cf. § 37a),

$$\Delta F = \mu_2 - \mu_2' = RT \ln \frac{a_2}{a_2'}, \qquad (38.20)$$

where a_2 and a_2' are the activities of the thallium in the two amalgams.

If the thallium molecule contains x atoms and z is the valence of the thallous ion in the cell solution, it requires the passage of xz faradays of electricity in order to dissolve 1 mole of thallium from one amalgam electrode,

[5] Saxton and Smith, *J. Am. Chem. Soc.*, **54**, 2625 (1932); Smith, *ibid.*, **61**, 500, 1123 (1939); **63**, 1351 (1941).
[6] Lewis and Randall, *J. Am. Chem. Soc.*, **43**, 233 (1921); see also, *idem.*, ref. 1, Chapter XXII.

by the reaction

$$\text{Tl}_x \text{ (in amalgam)} = x\text{Tl}^{z+} \text{ (in solution)} + xz\epsilon,$$

and to deposit it on the other by the reverse change. The increase of free energy accompanying the transfer process is therefore given by equation (33.36) as

$$\Delta F = -NFE = -xzFE, \qquad (38.21)$$

where xz is equivalent to N; F represents the faraday and E is the observed E.M.F. of the galvanic cell under consideration. In the special case where the metal is thallium, both x and z are unity, so that ΔF is equal to $-FE$. Comparing this result with equation (38.20), it follows that

$$E = -\frac{RT}{F} \ln \frac{a_2}{a_2'}. \qquad (38.22)$$

The E.M.F. of the cell is thus seen to be independent of the concentration of the thallous salt in the cell solution, and is determined by the ratio of the activities of the thallium in the two amalgams constituting the electrodes.

Although equation (38.22) gives the ratio of the activities in any two amalgams directly from a measurement of the E.M.F. of the appropriate cell, it is frequently desirable to express the actual activity of a metal in a given amalgam with reference to a particular standard state. Suppose the latter is chosen, in accordance with § 37b, III B, so that the activity coefficient a_2/N_2 is unity at infinite dilution; the following procedure may then be used. Equation (38.22) may be written as

$$\ln a_2 = -\frac{EF}{RT} + \ln a_2',$$

and subtracting $\ln \text{N}_2$ from both sides, this becomes

$$\ln \frac{a_2}{\text{N}_2} = -\frac{EF}{RT} - \ln \text{N}_2 + \ln a_2', \qquad (38.23)$$

or

$$\log \frac{a_2}{\text{N}_2} = \left(-\frac{EF}{2.303RT} - \log \text{N}_2 \right) + \log a_2'. \qquad (38.24)$$

A series of cells is now constructed in which the composition (N_2) of the amalgam constituting the right-hand electrode is varied, while that of the left-hand electrode (N_2'), usually the more dilute amalgam, is maintained constant. The E.M.F. (E) of each cell is measured, and the corresponding value of the term in parentheses on the right-hand side of equation (38.24) is plotted against the mole fraction N_2. Such a plot is shown in Fig. 25, the data being obtained from thallium amalgam cells at 20° C, with N_2' constant and equal to 0.00326 throughout. When N_2 is zero, i.e., at infinite dilution, the value of a_2/N_2 is unity, in accordance with the chosen standard state; since $\log (a_2/\text{N}_2)$ will then be zero, it is evident from equation (38.24) that the term in parentheses in this equation will be equal to $-\log a_2'$. In other words, the intercept on the vertical axis for $N_2 = 0$ in Fig. 25, which is 2.4689, is equal to $-\log a_2'$, so that a_2' is 0.003396. The value of $\log (a_2/N_2)$, i.e., the logarithm of the activity coefficient γ_N, in any amalgam can then

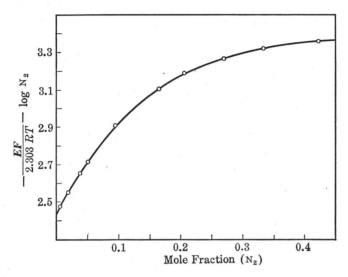

FIG. 25. Determination of activity of thallium in amalgams

be obtained [cf. equation (38.24)] by adding $\log a_2'$, i.e., -2.4689, to the term in parentheses, that is, the corresponding ordinate in Fig. 25. In this manner a_2/N_2, and hence a_2, for any amalgam within the experimental range can be determined. Some of the results obtained are quoted in Table

TABLE XXVII. ACTIVITY OF THALLIUM IN AMALGAMS AT 20° C

N_2	E	$-\dfrac{EF}{2.303RT} - \log N_2$	$\dfrac{a_2}{N_2}$	a_2
0	—	(2.4689)	1.000	0
0.00326	0 volt	2.4869	1.042	0.003396
0.01675	-0.04555	2.5592	1.231	0.02062
0.04856	-0.08170	2.7184	1.776	0.08624
0.0986	-0.11118	2.9177	2.811	0.2772
0.1680	-0.13552	3.1045	4.321	0.7259
0.2701	-0.15667	3.2610	6.196	1.674
0.4240	-0.17352	3.3558	7.707	3.268

XXVII.[7] The departure of the values of a_2/N_2 from unity indicate the deviations from Henry's law. It is evident that except in the most dilute solutions these are quite considerable.

[7] Data from Lewis and Randall, ref. 6.

Problem: The E.M.F. of the cell

Tl amalgam ($N_2' = 0.00326$) | Thallous solution | Tl amalgam ($N_2 = 0.0986$) is -0.11118 volt at 20° C. Calculate a_2/N_2 and a_2 for the thallium in the right-hand amalgam.

It will be seen in (§ 45d) that if the E.M.F. is in volts and F is in coulombs, R should then be in (int.) joules, and $F/2.3026R$ is equal to 1.984×10^{-4}; hence at 20° C, i.e., 293.16° K, equation (38.24) becomes

$$\log \frac{a_2}{N_2} = \left(-\frac{E}{1.984 \times 10^{-4} \times 293.16} - \log N_2 \right) + \log a_2'$$

$$= \left(-\frac{E}{0.05816} - \log N_2 \right) + \log a_2'.$$

As noted above, when N_2' is 0.00326, $\log a_2'$ is -2.4689; hence, when N_2 is 0.0986, so that $\log N_2$ is -1.0061 and E is -0.11118 volt,

$$\log \frac{a_2}{N_2} = \frac{0.11118}{0.05816} + 1.0061 - 2.4689$$

$$= 0.4488.$$

Hence, a_2/N_2 is 2.811, and a_2 is $2.811 \times 0.0986 = 0.2772$.

If it had been required to determine the deviations from Raoult's law it would have been necessary to choose pure liquid thallium at 20° C as the standard state of unit activity, instead of that employed in the preceding treatment. In this case the E.M.F. of the cell

Tl (pure liquid) | Thallous salt solution | Tl amalgam (N_2),

would be given by equation (38.22) as

$$E = -\frac{RT}{F} \ln a_2, \tag{38.25}$$

since the activity of the thallium in the left-hand electrode is now unity. The activity a_2 of the thallium in an amalgam, on the basis of the new standard state, could thus be calculated directly from the E.M.F. measurement. However, since thallium melts at 302° C, the cell depicted cannot be studied at 20° C, and an alternative device is used to obtain the activity in the amalgam with reference to that of the pure (supercooled) liquid thallium as unity. By plotting the values of a_2/N_2 given in Table XXVII against N_2, it is possible to extrapolate the results to N_2 equal to unity, that is, to pure liquid thallium. The value of a_2/N_2 is then found to be 8.3, and since N_2 is 1, the activity a_2 with reference to the standard based on the infinitely dilute solution, is 8.3. The new reference state, however, requires a_2 to be unity when N_2 is unity, and so the corresponding activities in the various amalgams can be obtained by dividing each of the results in the last column of Table XXVII by 8.3. This is a direct consequence of the conclusion reached in § 37a, that the *ratio* of the activities of a given component in two solutions is independent of the chosen standard state. The activity coefficients a_2/N_2 calculated in this manner show that in the more concentrated amalgams, with mole fraction of thallium exceeding about 0.4, the deviation from Raoult's law is of the order of a few per cent only.

Problem: Calculate the activity coefficient which indicates the departure from Raoult's law of thallium in the amalgam containing 0.424 mole fraction of thallium.

For this amalgam, it is seen that a_2 in Table XXVII is 3.268; the value in terms of pure liquid thallium as the standard state is then equal to 3.268/8.3. The activity coefficient is thus given by

$$\frac{a_2}{\text{N}_2} = \frac{3.268}{8.3 \times 0.424} = 0.929.$$

For a solution obeying Raoult's law, a_2/N_2 calculated in this manner is unity; the departure from ideal behavior is seen to be relatively small.

Although the method of obtaining the activity of a metal in a metallic solution directly from E.M.F. measurements is not usually possible at ordinary temperatures, it has been used considerably for studies at higher temperatures. For this purpose the electrolyte consists of a fused salt, instead of an aqueous solution. In general the E.M.F. of the cell

Metal A (liquid) | Fused salt of A | Solution of A in metal B (liquid)

is given by

$$E = -\frac{RT}{NF} \ln a_\text{A}, \qquad (38.26)$$

where a_A is the activity of the metal A in the right-hand electrode, with reference to pure liquid A as the standard state; N is equal to xz, where x is the number of atoms per molecule of A, and z is the valence. The activity a_A can thus be obtained directly from an E.M.F. measurement by means of equation (38.26). By varying the proportion of A in the right-hand electrode, determinations have been made over the whole range of composition from pure A to pure B.

The same principle as that just described has been used to evaluate the activity of a salt in a liquid (fused) mixture with another salt. Consider, for example, the galvanic cell

$$\text{Ag}(s) \,|\, \text{AgBr}(\text{N}_2) \text{ in fused LiBr} \,|\, \text{Br}_2(g),$$

in which the reaction is the combination of silver and bromine to form silver bromide at mole fraction N_2 (or activity a_2) dissolved in fused lithium bromide. The E.M.F. of this cell is given by (cf. § 45d)

$$E = E^0 - \frac{RT}{F} \ln a_2, \qquad (38.27)$$

where E^0 is the E.M.F. of the cell when the activity of the silver bromide is unity, i.e., when the electrolyte is pure fused silver bromide. It is thus possible to determine the activity a_2 of the latter salt in various fused mixtures, varying in composition from pure silver bromide ($\text{N}_2 = 1$) to pure lithium bromide ($\text{N}_2 = 0$).[8]

[8] Hildebrand, et al., *J. Am. Chem. Soc.*, **49**, 722 (1927); **51**, 462 (1929); **52**, 4641, 4650, 4655, (1930); Salstrom, *ibid.*, **54**, 2653, 4257 (1932); **58**, 1848 (1936); Strickler and Seltz, *ibid.*, **58**, 2084 (1936); Seltz, *Trans. Electrochem. Soc.*, **77**, 233 (1940).

Problem: At 500° C, the E.M.F. of the cell described above is 0.7865 volt when the electrolyte is pure AgBr; the E.M.F. is 0.8085 when the mole fraction is 0.5937. Calculate the activity coefficient in the latter case, the standard state being taken as pure liquid AgBr.

In equation (38.27), E is 0.8085 and E^0 is 0.7865; since T is 773° K, this equation becomes

$$0.8085 = 0.7865 - 1.984 \times 10^{-4} \times 773 \log a_2,$$
$$a_2 = 0.7188.$$

The activity coefficient is a_2/N_2, i.e., $0.7188/0.5937 = 1.211$, indicating positive deviation from Raoult's law.

In the present section it has been seen that E.M.F. measurements can be used to determine the activity of one of the components of a homogeneous liquid system. It will now be shown how the same results can be used to evaluate the activity or activity coefficient of the other constituent. The treatment is quite general, and can be employed whenever the activity of one constituent of a solution has been determined by any convenient method.

38f. Activity of One Component from that of the Other Component.—By combining the Gibbs-Duhem equation in the form $\text{N}_1 d\mu_1 + \text{N}_2 d\mu_2 = 0$ [equation (34.7)], with the expression for the chemical potential, viz., $\mu_i = \mu_i^0 + RT \ln a_i$ [equation (31.2)], it is seen that for a binary solution,

$$\text{N}_1 d \ln a_1 + \text{N}_2 d \ln a_2 = 0, \tag{38.28}$$

where N_1 and N_2 are the mole fractions of solvent and solute, respectively. This result is applicable at constant temperature and pressure irrespective of the standard states chosen for the two constituents, since μ_i^0 is constant, in any event, at a specified temperature. A slight rearrangement of equation (38.28) gives

$$d \ln a_1 = - \frac{\text{N}_2}{\text{N}_1} d \ln a_2, \tag{38.29}$$

and upon integrating, and converting the logarithms, this becomes

$$\log \frac{a_1}{a_1'} = - \int_{\text{N}_2'}^{\text{N}_2} \frac{\text{N}_2}{\text{N}_1} d \log a_2, \tag{38.30}$$

where a_1 and a_1' are the activities of the *solvent* in two solutions in which the mole fractions of the *solute* are N_2 and N_2', respectively. By plotting N_2/N_1 for any solution against the known value of $\log a_2$, the area under the curve between the limits of N_2' and N_2 represents the integral on the right-hand side of equation (38.30). In this way the ratio of the activities of the solvent, i.e., a_1/a_1', can be determined from the known activities a_2 of the solute.

Experience has shown that the procedure just described is not too satisfactory, especially when N_2 approaches zero; under the latter conditions a_2 for the solute, which is then equal to N_2, also approaches zero, and $\log a_2$ has a large negative value. An alternative method, which has been found to be more suitable, is the following. Since $N_1 + N_2 = 1$ for a binary mixture, $dN_1 + dN_2 = 0$, and hence,

$$N_1 \frac{dN_1}{N_1} + N_2 \frac{dN_2}{N_2} = 0,$$

that is,

$$N_1 d \ln N_1 + N_2 d \ln N_2 = 0.$$

If this is subtracted from equation (38.28) the result is

$$N_1 d \ln \frac{a_1}{N_1} + N_2 d \ln \frac{a_2}{N_2} = 0,$$

$$d \ln \frac{a_1}{N_1} = - \frac{N_2}{N_1} d \ln \frac{a_2}{N_2}. \tag{38.31}$$

Upon integrating and converting the logarithms, as before, it is found that

$$\log \frac{a_1}{N_1} - \log \frac{a_1'}{N_1'} = - \int_{N_2'}^{N_2} \frac{N_2}{N_1} d \log \frac{a_2}{N_2}. \tag{38.32}$$

By plotting $\log (a_2/N_2)$ against N_2/N_1, and determining the area under the curve between the limits of N_2' and N_2, the difference between the two corresponding values of $\log (a_1/N_1)$ can be obtained. Further, if the composition N_1' and N_2' represents infinite dilution, i.e., $N_1' = 1$ and $N_2' = 0$, then the activity a_1' of the solvent is unity, in accordance with the standard state usually chosen for a solvent;* a_1'/N_1' is then unity and $\log (a_1'/N_1')$ is zero, so that equation (38.32) becomes

$$\log \frac{a_1}{N_1} = - \int_0^{N_2} \frac{N_2}{N_1} d \log \frac{a_2}{N_2}. \tag{38.33}$$

The area under the curve from zero to any composition N_2 gives the value of $\log (a_1/N_1)$ for the solvent at that composition.

TABLE XXVIII. ACTIVITY OF MERCURY IN THALLIUM AMALGAMS AT $20°$ C

N_1	a_1/N_1	a_1
1.000	1.000	1.000
0.990	0.999	0.989
0.950	0.986	0.937
0.900	0.950	0.855
0.800	0.866	0.693
0.700	0.790	0.553
0.600	0.734	0.440
0.500	0.704	0.352

To illustrate the application of equation (38.33) to calculate the activity or activity coefficient of one constituent of a solution when that of the other is known, use may be made of the results for the thallium amalgams given in § 38e. The values of a_2/N_2 for thallium at various compositions are available (Table XXVII), and the data required for the evaluation of the

* The experimental pressure is assumed to be 1 atm.

integral are plotted in Fig. 26. The activities of the mercury (a_1) at a number of rounded compositions obtained by means of equation (38.33) are given in Table XXVIII.[9] The standard state is pure liquid mercury at 20° C and 1 atm. pressure; hence, the deviations of the results in the second column, i.e., a_1/N_1, from unity indicate the extent of the departure from Raoult's law, since the measurements were made at atmospheric pressure.

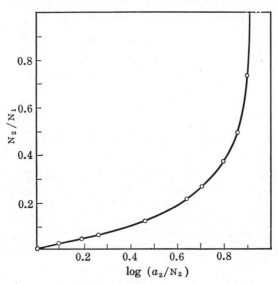

Fig. 26. Determination of activity of mercury in thallium amalgams

In this treatment it has been assumed that the activity of the solute, i.e., constituent 2, is known, and the procedure described was used to calculate the activity of the solvent. The general method may equally be employed to determine the activity of the solute if that of the solvent at a number of compositions is known. Actual use of this procedure will be made in §§ 39b, 39e, 39f.

38g. Analytical Procedure for Calculating Activities.—If the activity or activity coefficient of one constituent of a mixture could be expressed as a function of the mole fraction, the integration might be carried out analytically, instead of graphically. Actually the integration is not necessary for the relationship between the expressions for a_1/N_1 and a_2/N_2, that is, γ_1 and γ_2, respectively, are determined by the Gibbs-Duhem equation. If the pressure is 1 atm., the simplified form of the Margules equation (35.9) becomes

$$f_1 = \text{N}_1 \, f_1^0 e^{\frac{1}{2}\beta \text{N}_2^2},$$

where f_1^0, the fugacity in the standard state, i.e., pure liquid at 1 atm. pressure at the experimental temperature, replaces $f_1^\square$. Since the activity a_1 is equal to f_1/f_1^0,

[9] Lewis and Randall, ref. 6.

by equation (31.5), it is seen that

$$\frac{a_1}{N_1} = \gamma_1 = e^{\frac{1}{2}\beta N_2^2}$$

or

$$\log \frac{a_1}{N_1} = \frac{1}{2 \times 2.303} \beta N_2^2 = \beta' N_2^2. \qquad (38.34)$$

Similarly, for the second constituent,

$$\log \frac{a_2}{N_2} = \beta' N_1^2, \qquad (38.35)$$

where β', which is the same in both cases, is equal to 0.217β. If a/N for either constituent for any composition is known, the value of β can be determined, and hence a/N for the other constituent is readily found. The standard state in each case is the particular substance in the pure liquid state. By means of E.M.F. measurements, the activity coefficient a_2/N_2 of cadmium in an alloy with tin has been determined at a number of compositions at 430° C. The results are expressed to a close approximation by $\log (a_2/N_2) = 0.31 N_1^2$, so that the variation with composition of the activity coefficient of the tin in the same system is given by $\log (a_1/N_1) = 0.31 N_2^2$.

The foregoing method involves a single constant and can be used when the deviations from Raoult's law are not too great. Better results are obtained by using the van Laar equations (35.10). Thus, the activities of mercury (component 1) in liquid mixtures with tin (component 2) at 323° C, determined by vapor pressure measurements, can be expressed by

$$\log \frac{a_1}{N_1} = \frac{0.22 N_2^2}{(0.26 N_1 + N_2)^2}.$$

Hence, utilizing the relationship between the constants (§ 35d), it is found that

$$\log \frac{a_2}{N_2} = \frac{0.846 N_1^2}{(N_1 + 3.84 N_2)^2},$$

giving activity coefficients of the tin in amalgams of various compositions. The standard state in each case is the pure liquid metal at 323° C and 1 atm. pressure.[10]

Problem: Assuming the simplified Margules equation to be applicable to the fused LiBr-AgBr system, use the result obtained in the last problem in § 38e to derive general expressions for the activity coefficients of LiBr (a_1/N_1) and AgBr (a_2/N_2) as functions of the respective mole fractions, at 500° C.

In the problem mentioned, it was found that a_2/N_2 is 1.211 when N_2 is 0.5937; hence, if

$$\log \frac{a_2}{N_2} = \log \gamma_2 = \beta' N_1^2,$$

$$\log 1.211 = \beta' \times (1 - 0.5937)^2,$$

$$\beta' = 0.5035,$$

[10] Cf. J. H. Hildebrand, "Solubility of Non-electrolytes," 2nd ed., 1936, pp. 45 et seq.; see also, Lacher and Hunt, ref. 2; Lacher, Buck and Parry, ref. 2.

so that

$$\log \frac{a_2}{N_2} = \log \gamma_2 = 0.5035 N_1^2.$$

This gives the activity coefficient of the AgBr in the fused LiBr-AgBr system. By equations (38.34) and (38.35), the activity coefficient of the LiBr in the same system is given by the analogous expression

$$\log \frac{a_1}{N_1} = \log \gamma_1 = 0.5035 N_2^2.$$

38h. Osmotic Pressure and Activity.—The osmotic pressure of a solution is related to the activity of the solvent in it, and hence measurement of osmotic pressure should provide, in principle, a method for the determination of activities. Because of experimental difficulties this procedure has not found practical application, but the thermodynamic treatment is, nevertheless, of interest. The osmotic pressure may be defined as *the excess pressure which must be applied to a solution to prevent the passage into it of solvent when the two liquids are separated by a semipermeable membrane*, i.e., a membrane permitting the free passage of the molecules of solvent but not of solute. If a solution and the pure solvent, both at the same pressure, are separated by a semipermeable membrane, the system will not be in equilibrium, for there will be a tendency for the solvent to pass through the membrane into the solution. In order to establish equilibrium, the pressure on the solution must be increased above that exerted on the solvent. If P and P_0 are the pressures on the solution and solvent, respectively, when equilibrium is attained, then the difference $P - P_0$ is equal to Π, the osmotic pressure of the solution.

If μ_0 is the chemical potential of the pure solvent and μ is that in the solution, at the same pressure P_0, the two values will not be equal. However, as seen above, if the pressure on the solution is raised from P_0 to P, the chemical potential of the solvent in the solution at P will become equal to that of the pure solvent at P_0, for this is the condition of equilibrium.* It is therefore possible to write

$$\mu_0 = \mu + \int_{P_0}^{P} \left(\frac{\partial \mu}{\partial P}\right)_{T,N} dP, \qquad (38.36)$$

the left-hand side giving the chemical potential of the solvent at P_0, and the right-hand side that of the solvent in the solution when the pressure has been increased from P_0 to P. According to equation (26.26), $(\partial \mu / \partial P)_{T,N}$, for constant temperature and composition, is equal to the partial molar volume of the given constituent, i.e., the solvent in this case; thus equation (38.36) is equivalent to

$$\mu_0 = \mu + \int_{P_0}^{P} \bar{V}_1 dP, \qquad (38.37)$$

---

* In physical terms, increase of pressure from P_0 to P increases the vapor pressure (cf. § 27m) of the solvent in the solution, so that it becomes identical with that of the pure solvent at P_0.

where $\bar{V}_1$, is the partial molar volume of the solvent in the solution. The chemical potentials μ_0 and μ of the solvent in pure solvent and solution, respectively, may be expressed in terms of their respective fugacities [cf. equation (31.1)], viz.,

$$\mu_0 = \mu_1^* + RT \ln f_1^\square \quad \text{and} \quad \mu = \mu_1^* + RT \ln f_1, \quad (38.38)$$

where $f_1^\square$ and f_1 are the fugacities of the solvent in the pure state and in the solution, at the same pressure P_0. Making the substitution of the equations (38.38) into (38.37), the result is

$$RT \ln \frac{f_1^\square}{f_1} = \int_{P_0}^{P} \bar{V}_1 dP. \quad (38.39)$$

If the pressure P_0 is 1 atm., $f_1^\square$ is identical with the fugacity f_1^0 of the solvent in its standard state, and hence $f_1/f_1^\square$ may be replaced by a_1, the activity of the solvent in the solution, so that equation (38.39) becomes

$$-RT \ln a_1 = \int_{P_0}^{P} \bar{V}_1 dP. \quad (38.40)$$

In order to integrate equation (38.40) it is necessary to know how the partial molar volume $\bar{V}_1$ of the solvent in the solution varies with the external pressure. It may be supposed that the variation is linear, so that at any pressure P,

$$\bar{V}_1 = \bar{V}_0[1 - \alpha(P - P_0)],$$

where α is a constant and $\bar{V}_0$ is the partial molar volume of the solvent at the pressure P_0, i.e., 1 atm. If this expression is inserted into equation (38.40), the result is

$$-RT \ln a_1 = \bar{V}_0 \int_{P_0}^{P} [1 - \alpha(P - P_0)] dP$$
$$= \bar{V}_0(P - P_0)[1 - \tfrac{1}{2}\alpha(P - P_0)]. \quad (38.41)$$

As seen above, $P - P_0$ is equal to Π, the osmotic pressure of the solution, so that equation (38.41) becomes

$$-RT \ln a_1 = \bar{V}_0 \Pi (1 - \tfrac{1}{2}\alpha\Pi). \quad (38.42)$$

According to equation (38.42) it should be possible to determine the activity of the solvent in a solution by means of osmotic pressure measurements. Since such measurements are not easily made, the procedure is not convenient. Nevertheless, the accuracy of equation (38.42) may be tested by using the known activity of the solvent to calculate the osmotic pressure of the solution, and comparing the result with the experimental value. Provided the vapor pressures are not high, a_1, which is equal to f_1/f_1^0, may be replaced by $p_1/p_1^\square$, and so the osmotic pressure can be derived from vapor pressure measurements. For aqueous solutions, α is of the order of 4×10^{-5} atm.$^{-1}$ and may be neglected, except for solutions of high osmotic pressure. The results for sucrose solutions at 30° C obtained in this manner are given in Table XXIX.[11]

TABLE XXIX. OSMOTIC PRESSURES OF SUCROSE SOLUTIONS AT 30° C

Molality	Osmotic Pressure	
	Calc.	Obs.
0.1	2.47 atm.	2.47 atm.
1.0	27.0	27.2
2.0	58.5	58.4
3.0	96.2	95.2
4.0	138.5	139.0
6.0	231.0	232.3

EXERCISES

1. Compare the activity coefficients γ_N, γ_m and γ_c in an aqueous solution containing 0.1 mole of phenol per liter, taking the density of the solvent as 1.00 and that of the solution as 1.005 g. cc.$^{-1}$.

2. The partial pressures of chlorobenzene(1) at various mole fractions in mixtures with 1-nitropropane at 75° C are as follows (Lacher, Buck and Parry, ref. 2):

N_1	0.119	0.187	0.289	0.460	0.583	0.691	1.00
p_1	19.0	28.3	41.9	62.4	76.0	86.4	119 mm.

Express $\log \gamma$ for each component as a function of the mole fraction of the type $\log \gamma_1 = \beta' N_2^2$, etc. (Plot $\log \gamma_1$ against N_2^2 and determine the slope.)

3. Show that if the vapor cannot be treated as an ideal gas, equation (38.4) for the activity coefficient of a volatile solute would become $\gamma_N = (f_2/N_2)/(p_2^*/N_2^*)$, assuming the solvent to be nonvolatile so that p_2^* represents virtually zero total pressure; hence, suggest a possible procedure for determining γ_N in a case of this type.

4. The E.M.F. of the cell Pb(l) | PbCl$_2$ in LiCl-KCl fused | Pb-Bi(l) in which the mole fraction of bismuth is 0.770 is 0.5976 volt at 327° C (Strickler and Seltz, ref. 8). Calculate the activity and activity coefficient of the lead in the liquid alloy, indicating the standard state employed. Derive the constant β in the simple Margules equation.

5. The following data were obtained in a study of the vapor pressure of mercury (p_1) in equilibrium with bismuth amalgams at 321° C:

N_2(Bi)	0.1486	0.247	0.347	0.463	0.563	0.670	0.793	0.937
$p_1/p_1^\square$	0.908	0.840	0.765	0.650	0.542	0.432	0.278	0.092

[Hildebrand and Eastman, J. Am. Chem. Soc., 36, 2020 (1914)]. Determine the activity coefficients a_1/N_1 of the mercury in the various amalgams and by plotting N_1/N_2 against $\log (a_1/N_1)$ derive the activity coefficients of the bismuth at the mole fractions 0.1, 0.2, 0.4, 0.6, 0.8 and 0.9, by using the appropriate form of equation (38.33). Pure liquid bismuth should be taken as the standard state.

6. What would be the E.M.F. at 321° C of a cell consisting of two bismuth amalgam electrodes, with mole fractions 0.1 and 0.9 of bismuth, respectively, in the same electrolyte? (Use results of Exercise 5.)

[11] Vapor pressures from Berkeley, et al., Phil. Trans. Roy. Soc., A218, 295 (1919); osmotic pressures from Frazer and Myrick, J. Am. Chem. Soc., 38, 1920 (1916).

7. Utilize the results in Exercise 5 to express $\log \gamma_1$ (for the mercury) as a function of the mole fraction by means of the van Laar equation. Hence derive an expression for $\log \gamma_2$ (for the bismuth) and check the values obtained by the graphical integration.

8. The following results were obtained at 500° C for the E.M.F. of the cell Ag(s) | Fused AgCl(1) in LiCl(2) | Cl$_2$(g) at various mole fractions (N_1) of the silver chloride:

N_1	1.000	0.690	0.469	0.136
E	0.9001	0.9156	0.9249	0.9629 volt

[Salstrom, et al., J. Am. Chem. Soc., **58**, 1848 (1936)]. Calculate the activity and activity coefficient of the silver chloride at the various mole fractions. Determine the free energy of transfer of 1 mole of silver chloride from the pure fused state to the lithium chloride solution in which its mole fraction is 0.136, at 500° C.

9. "For a liquid system which does not deviate greatly from Raoult's law the activity coefficient of each of the two constituents is the same function of its respective mole fraction, at a given temperature, provided the pure liquid is taken as the standard state in each case." Justify this statement.

10. The vapor pressures of mercury in equilibrium with thallium amalgams containing various mole fractions (N_1) of mercury at 26° C were found to be as follows [Hirst and Olson, J. Am. Chem. Soc., **51**, 2398 (1929)]:

N_1	1.000	0.835	0.644	0.316
$p_1 \times 10^5$ cm.	20.1	15.15	9.6	6.9

Compare the activities of the mercury in the various amalgams with those obtained by the E.M.F. method at 20° C (cf. Table XXVIII).

11. The E.M.F. of a cell consisting of cadmium amalgam ($N_2 = 1.781 \times 10^{-3}$) as one electrode and a saturated amalgam as the other is $-$ 0.05294 volt at 25° C. What are the activity and activity coefficient of the cadmium in the dilute amalgam with reference to that in the saturated amalgam as the standard state?

The E.M.F. of a similar cell, in which the electrodes are pure solid cadmium and the saturated amalgam, is 0.05045 volt at 25° C. Determine the activity and activity coefficient of the cadmium in both amalgams referred to in the preceding paragraph, taking pure solid cadmium as the standard state.

What is the free energy of transfer of 1 g. atom of cadmium from the saturated to the more dilute amalgam at 25° C?

12. Show that for a nonideal liquid mixture, the free energy of mixing per mole is given by $\Delta F = N_1 RT \ln a_1 + N_2 RT \ln a_2$, the standard states being the pure liquids (cf. Exercise 24, Chapter XIV). Suggest how the free energy of mixing may be determined for a mixture of (i) two liquid metals, (ii) two volatile liquids.

13. Utilizing the results of the preceding exercise and of Exercise 24, Chapter XIV, justify the use of the term "excess free energy of mixing" for the quantity $\Delta F^E = N_1 RT \ln \gamma_1 + N_2 RT \ln \gamma_2$, which is a measure of the departure of the system from ideal behavior, for a given external pressure and composition. Show that the corresponding heat change ΔH^E is actually equal to the heat of mixing, at the specified composition [cf. Scatchard, et al., J. Phys. Chem., **43**, 119 (1939); J. Am. Chem. Soc., **60**, 1278 (1938); **61**, 3206 (1939); **62**, 712 (1940)].

The following partial vapor pressure data were obtained for a mixture containing equimolecular amounts of ethanol(1) and chloroform(2):

	35°	45°	55° C
p_1	59.3 (102.8)	101.9 (172.8)	166.3 (279.9) mm.
p_2	217.7 (295.1)	315.8 (433.5)	438.9 (617.8) mm.

The figures in parentheses in each case are the vapor pressures of the pure liquids. Assuming the vapors to behave as ideal gases, calculate (i) the free energy, (ii) the excess free energy, (iii) the heat, of mixing per mole of the equimolar mixture at 45° C. (Use a form of the Gibbs-Helmholtz equation to derive ΔH from ΔF; the total pressure may be supposed to be constant.)

14. Consider equations (38.33) and (38.24) and hence indicate how a_1/N_1 for mercury could be determined directly, by a graphical method, from the E.M.F. data on the thallium amalgam cells in Table XXVII.

15. The following E.M.F. results were obtained in measurements on cells with potassium amalgam electrodes at 25° C; in each case one electrode consisted of an amalgam in which the mole fraction of potassium (N_2') was 0.01984, while the composition (N_2) of the other was varied:

N_2	E	N_2	E
0.000472	0.13066 volt	0.01301	0.02359
0.000917	0.11267	0.01628	0.01191
0.002620	0.08281	0.01984	0
0.003272	0.07618	0.02252	− 0.00844
0.008113	0.04421	0.02530 (sat.)	− 0.01733

[Armbruster and Crenshaw, J. Am. Chem. Soc., 56, 2525 (1934)]. Determine the activity and activity coefficient of the potassium in each of these amalgams at 25° C. Utilize the result of the preceding exercise to determine the same quantities for the solvent (mercury). Indicate the standard state employed in each case.

16. Prove that if a solute is distributed between two immiscible solvents (I and II), the ratio of the activities in the two solvents, i.e., a_I/a_{II}, should be constant at constant temperature and pressure. Show that this result is the basis of the Nernst distribution law, i.e., c_I/c_{II} (or m_I/m_{II}) is constant for dilute solutions.

17. By using the results of the preceding exercise, propose a method for determining the activity coefficient in solvent I of a solute whose activity at various molalities is known in solvent II. Indicate the nature of the standard state employed.

18. The freezing points of three glycerol solutions in water are − 1.918° C for 1.0 molal, − 3.932° C for 2.0 molal and − 10.58° for 5.0 molal. Determine the activities and activity coefficients of the water in these solutions on the basis of the usual standard state, and consider the departure from Raoult's law. The vapor pressure of pure (supercooled) water at − 1.92° C is 3.980 mm.; what would be the aqueous vapor pressure of the 1.0 molal glycerol solution at this temperature?

19. Show that for a dilute solution, the vapor of the solvent behaving ideally,

$$\ln \frac{p_1^0}{p_1} = \frac{V_1^0 \Pi}{RT},$$

where V_1^0 is the molar volume of the pure solvent at temperature T and 1 atm. pressure. If the solution is dilute then show that $\Pi V_1^0 = (n_2/n_1)RT$, where n_1

and n_2 are the numbers of moles of solvent and solute, respectively (H. N. Morse, 1905). At 20° C, the vapor pressure of 0.9908 molal mannitol solution in water is 0.3096 mm. below that of pure water (17.535 mm.). Estimate the osmotic pressure of the solution, utilizing both the relationships just derived [Frazer, Lovelace and Rogers, *J. Am. Chem. Soc.*, **42**, 1793 (1920)]. The density of water at 20° C is 0.99823 g. cc.$^{-1}$.

20. From the results of Exercise 19, show that it is possible to derive the van't Hoff osmotic pressure equation $\Pi V = RT$ for a very dilute solution; V is the volume of solution containing 1 mole of solute.

21. Derive a general relationship between the osmotic pressure of a *dilute* solution and its freezing point depression. What form does it take for a very dilute solution?

22. Give the complete derivation of equation (37.8).

23. Show that the variation with temperature of a_2, the activity of the solute in terms of molality, is given by

$$\left(\frac{\partial \ln a_2}{\partial T}\right)_{P,N} = \frac{\bar{H}_2^0 - \bar{H}_2}{RT^2},$$

where $\bar{H}_2^0$ is equal to the partial molar heat content in the infinitely dilute solution.

24. Show that the partial molar entropy $\bar{S}_2^*$ of a solute at high dilution is related to the value in the standard state $\bar{S}_2^0$ by the expression $\bar{S}_2^* - \bar{S}_2^0 = -R \ln m^*$, where m^* is the molality of the dilute solution. Hence prove that the partial molar entropy of a solute approaches an infinitely large value at infinite dilution.

CHAPTER XVI

SOLUTIONS OF ELECTROLYTES

39. Activities and Activity Coefficients

39a. Mean Activities of Electrolytes.—The treatment of activities in the preceding chapter has referred in particular to solutions of nonelectrolytes, but since the activity concept has probably found its most useful application in connection with the study of electrolytes, the subject merits special consideration. Certain modifications, too, are desirable in order to allow for the presence of ions in the solution. For solutions of electrolytes, the standard state of *each ionic species* is chosen so that the ratio of its activity to its concentration becomes unity at infinite dilution, at 1 atm. pressure and the temperature of the solution.[1] The ionic concentration may be expressed in terms of the mole fraction, the molality or the molarity, just as for nonelectrolytes (cf. § 37b). The mole fraction is rarely used, almost the only case being that in which an attempt is made to calculate the corresponding activity coefficient of an ion from theoretical considerations (Chapter XVII). As a general rule, activities of ions are given in terms of molalities, and this standard state will be adopted for the present. The postulate is, therefore, that *the activity of an ion becomes equal to its molality at infinite dilution*.

Consider an electrolyte, represented by the formula $M_{\nu_+}A_{\nu_-}$, which ionizes in solution to yield the number ν_+ of positive ions M^{z+}, of valence z_+, and ν_- negative ions A^{z-}, of valence z_-; thus,

$$M_{\nu_+}A_{\nu_-} \rightleftharpoons \nu_+ M^{z+} + \nu_- A^{z-}.$$

The chemical potentials of each of these ions is given by the general equations [cf. equation (31.2)]

$$\mu_+ = \mu_+^0 + RT \ln a_+ \quad \text{and} \quad \mu_- = \mu_-^0 + RT \ln a_-, \qquad (39.1)$$

where μ_+ and μ_- are the chemical potentials, and a_+ and a_- are the activities of the ions M^{z+} and A^{z-}, respectively, in the given solution. The chemical potential of the electrolyte as a whole, represented by μ_2, is given by

$$\mu_2 = \mu_2^0 + RT \ln a_2, \qquad (39.2)$$

where a_2 is the activity of the solute. For a *strong electrolyte* the chemical potential μ_2 may be taken as equal to the sum of the chemical potentials of

---

[1] G. N. Lewis and M. Randall, "Thermodynamics and the Free Energy of Chemical Substances," 1923, pp. 326 *et seq.*; *J. Am. Chem. Soc.*, **43**, 1112 (1921); see also, Adams, *Chem. Rev.*, **19**, 1 (1936); Goranson, *J. Chem. Phys.*, **5**, 107 (1937).

the constituent ions; thus,

$$\mu_2 = \nu_+\mu_+ + \nu_-\mu_-. \tag{39.3}$$

Similarly, the standard state of the strong electrolyte as a whole may be chosen so that its chemical potential μ_2^0 in that state is equal to the sum of the chemical potentials of the ions in their respective standard states; hence,

$$\mu_2^0 = \nu_+\mu_+^0 + \nu_-\mu_-^0. \tag{39.4}$$

Substituting the values of μ_+, μ_- and μ_2, given by equations (39.1) and (39.2), into (39.3), and utilizing (39.4), it is found that

$$\nu_+RT \ln a_+ + \nu_-RT \ln a_- = RT \ln a_2,$$
$$a_+^{\nu_+} a_-^{\nu_-} = a_2. \tag{39.5}$$

This relationship is frequently employed to define the activity a_2 of a strong electrolyte in terms of the activities of the constituent ions.

If the total number of ions produced by one molecule of an electrolyte, i.e., $\nu_+ + \nu_-$, is represented by ν, then the **mean ionic activity** $a_\pm$ of the electrolyte is defined by

$$a_\pm^\nu = a_+^{\nu_+} a_-^{\nu_-}, \tag{39.6}$$

and hence, by equation (39.5),

$$a_2 = a_\pm^\nu. \tag{39.7}$$

The activity of each ion may be written as the product of its activity coefficient γ and its concentration; if the latter is expressed in terms of the molality m, it follows that

$$a_+ = \gamma_+ m_+ \quad \text{and} \quad a_- = \gamma_- m_-,$$

so that

$$\gamma_+ = \frac{a_+}{m_+} \quad \text{and} \quad \gamma_- = \frac{a_-}{m_-}, \tag{39.8}$$

it being understood that the activity coefficient γ is really γ_m, as defined in § 37b, III C. The **mean ionic activity coefficient** $\gamma_\pm$ of the electrolyte, given by

$$\gamma_\pm^\nu = \gamma_+^{\nu_+} \gamma_-^{\nu_-}, \tag{39.9}$$

can consequently be represented by

$$\gamma_\pm^\nu = \frac{a_+^{\nu_+} a_-^{\nu_-}}{m_+^{\nu_+} m_-^{\nu_-}} = \frac{a_\pm^\nu}{m_\pm^\nu},$$

or

$$\gamma_\pm = \frac{a_\pm}{m_\pm}, \tag{39.10}$$

where $m_\pm$, the **mean ionic molality** of the electrolyte, is defined by

$$m_\pm^\nu = m_+^{\nu_+} m_-^{\nu_-}. \tag{39.11}$$

For a strong electrolyte, which can be regarded as being completely ionized, a simple relationship may be derived between the mean ionic molality $m_\pm$ and the actual molality m of the solute. In this case m_+ is equal to $m\nu_+$ and m_- is equal to $m\nu_-$; hence,

$$m_\pm^\nu = m_+^{\nu_+} m_-^{\nu_-} = m^\nu(\nu_+^{\nu_+}\nu_-^{\nu_-}). \tag{39.12}$$

Problem: Determine the mean ionic molality of a 0.5 molal solution of sodium sulfate.

Since each molecule of Na_2SO_4 produces two Na^+ ions, i.e., $\nu_+ = 2$, and one SO_4^{--} ion, i.e., $\nu_- = 1$, the value of m_+ is $2m$, i.e., $2 \times 0.5 = 1$, and m_- is $1 \times m$, i.e., 0.5. Hence the mean ionic molality $m_\pm$ is given by

$$m_\pm^\nu = m_+^{\nu_+} m_-^{\nu_-} = 1^2 \times 0.5,$$

and since ν is $\nu_+ + \nu_- = 3$,

$$m_\pm^3 = 0.5, \qquad m_\pm = 0.794.$$

Relationships analogous to those given above may be derived in an exactly similar manner for the activities referred to mole fractions or molarities. As seen in § 37c, the activities for the various standard states, based on the ideal dilute solution, can be related to one another by equation (37.7). The result is, however, applicable to a single molecular species; the corresponding relationships between the *mean ionic activity coefficients* of a strong electrolyte, assumed to be completely ionized, are found to be

$$\gamma_N = \gamma_m(1 + 0.001\nu m M_1) = \gamma_c \frac{\rho + 0.001\nu c M_1 - 0.001 c M_2}{\rho_0}$$

$$\gamma_m = \gamma_c \frac{\rho - 0.001 c M_2}{\rho_0}, \tag{39.13}$$

where m is the molality and c is the molarity of the electrolyte; M_1 and M_2 are the molecular weights of the solvent and solute, respectively, and ν has the same significance as before, i.e., $\nu_+ + \nu_-$. At infinite dilution the three activity coefficients γ_N, γ_m and γ_c are, of course, identical, each being equal to unity, and for dilute solutions the values do not differ appreciably.

In the experimental determination of activity coefficients of strong electrolytes, by the methods described below, the molalities, etc., of the ions are taken as the stoichiometric values, that is, the total possible molality, etc., disregarding incomplete dissociation. For example, in the last problem, the molalities of the sodium and sulfate ions in the 0.5 molal solution of sodium sulfate were taken as exactly 1.0 and 0.5, respectively, without allowing for the possibility that the salt may be only partially dissociated at the specified concentration. The activity coefficients obtained in this manner are called **stoichiometric activity coefficients**; they allow for all variations from the postulated ideal behavior, including that due to incomplete dissociation. If the treatment is based on the *actual* ionic molalities, etc., in the given solution, as in the Debye-Hückel theory (Chapter XVII), there is obtained the true (or actual) activity coefficient. The ratio of the stoichiometric to the true activity coefficient is equal to the degree of dissociation of the electrolyte.

39b. Activities from Freezing Point Measurements.—The method of determining the activities of electrolytes based on measurements of the freezing points of solutions is capable of great accuracy over a considerable range of concentration. It has been used particularly for studying dilute solutions where other methods are less reliable. The procedure is based fundamentally on the determination of the activity of the solvent, which is then converted into that of the solute by a form of the Gibbs-Duhem equation, employing the same principles as were used in § 38f.

In § 38c, an expression of the form [cf. equation (38.11)]

$$-d \ln a_1 = \frac{1}{RT_0^2}(L_0 + b\theta + c\theta^2 + \cdots)d\theta, \qquad (39.14)$$

where b, c, etc., are constants, was derived for the activity of the solvent in a solution at its freezing point; this will now be utilized to obtain the activity of the solute. By the Gibbs-Duhem equation (38.28),*

$$d \ln a_2 = -\frac{N_1}{N_2} d \ln a_1$$
$$= -\frac{n_1}{n_2} d \ln a_1, \qquad (39.15)$$

where n_1 and n_2 are the numbers of moles of solvent and solute, respectively. If the molality, i.e., the number of moles of solute per 1000 g. solvent, is m, then n_2 may be set equal to m, and at the same time n_1 becomes $1000/M_1$, where M_1 is the molecular weight of the solvent.† Hence, equation (39.15) becomes

$$d \ln a_2 = -\frac{1000}{mM_1} d \ln a_1. \qquad (39.16)$$

Combining this result with equation (39.14) it is found that at the freezing point of the solution, at a pressure of 1 atm.,

$$d \ln a_2' = \frac{1000}{RT_0^2}\left(\frac{L_0}{M_1} + \frac{b\theta}{M_1} + \frac{c\theta^2}{M_1} + \cdots\right)\frac{d\theta}{m}. \qquad (39.17)$$

The primed symbol a_2' is used because the results refer to different temperatures for solutions of differing concentrations; subsequently (§ 39c), the results will be adjusted so as to apply to the same temperature for all concentrations.

* For an electrolyte solution, the exact form of this equation is $N_1 d \ln a_1 + \Sigma N_i d \ln a_i = 0$, where the summation is to be taken over every solute species i, ionic and nonionic, present in the solution. Using equation (39.5) for a strong electrolyte, this becomes identical with the expression used here.

† For aqueous solutions $1000/M_1$ is $1000/18.016$, i.e., 55.51 moles; this figure is frequently encountered in the literature of solutions.

Since L_0 is the molar heat of fusion of the pure solvent at its freezing point, at 1 atm., $RT_0^2 M_1/1000 L_0$ is exactly equivalent to the molal freezing point depression constant λ, as defined by equation (36.12); hence equation (39.17) becomes

$$d \ln a_2' = \frac{d\theta}{\lambda m} + \alpha \frac{\theta d\theta}{m}, \qquad (39.18)$$

where, for brevity, α is defined by

$$\alpha = \frac{1000}{RT_0^2 M_1} (b + c\theta + \cdots).$$

For dilute solutions, e.g., less than 1 molal, the terms $c\theta$, etc., may be neglected, so that α is a constant for the particular solvent. Utilizing the data for water given in § 38c, i.e., $L_0 = 1438$ cal. mole^{-1}, $T_0 = 293.16°$ K, and $\Delta C_P = 9$ cal. deg.$^{-1}$ mole^{-1}, α is found to be 5.7×10^{-4}.

Adopting the standard state for the electrolyte given in § 39a, the activity a_2 may be replaced by $a_\pm^\nu$ [equation (39.7)], so that $\ln a_2'$ is equal to $\nu \ln a_\pm'$; hence, from equation (39.18),

$$d \ln a_\pm' = \frac{d\theta}{\nu \lambda m} + \alpha \frac{\theta d\theta}{\nu m}, \qquad (39.19)$$

where ν is the total number of ions produced by one molecule of electrolyte in solution. The evaluation of the mean activity $a_\pm'$ of the electrolyte requires the integration of equation (39.19); the devices used for this purpose will now be described.

A function j is defined by

$$j = 1 - \frac{\theta}{\nu \lambda m}, \qquad (39.20)$$

which approaches unity as m approaches zero, i.e., with decreasing concentration.* Upon differentiation of this expression, recalling that ν and λ are constants, the result is

$$dj = \frac{\theta dm}{\nu \lambda m^2} - \frac{d\theta}{\nu \lambda m}$$

$$= (1 - j) \frac{dm}{m} - \frac{d\theta}{\nu \lambda m},$$

so that

$$\frac{d\theta}{\nu \lambda m} = (1 - j) \frac{dm}{m} - dj$$

$$= (1 - j) d \ln m - dj.$$

Combination with equation (39.19) then gives

$$d \ln a_\pm' = (1 - j) d \ln m - dj + \alpha \frac{\theta d\theta}{\nu m}. \qquad (39.21)$$

---

* For an electrolyte producing ν ions, equation (36.11) becomes $\theta = \nu \lambda m$ as infinite dilution is approached; hence $\theta/\nu \lambda m$ approaches unity and j becomes zero as m decreases toward zero.

By equations (39.10) and (39.12),
$$\ln \gamma_\pm = \ln a_\pm - \ln m - \ln (\nu_+^{\nu_+}\nu_-^{\nu_-})^{1/\nu},$$
so that
$$d \ln \gamma_\pm = d \ln a_\pm - d \ln m. \tag{39.22}$$

Hence equation (39.21) becomes
$$d \ln \gamma'_\pm = - jd \ln m - dj + \alpha \frac{\theta d\theta}{\nu m}. \tag{39.23}$$

Since $\gamma'_\pm$ is equal to unity at infinite dilution, i.e., when m is zero, in accordance with the choice of the ionic standard states, it follows upon integration, recalling that $j = 0$ when $m = 0$, that

$$\ln \gamma'_\pm = - \int_0^m jd \ln m - j + \frac{\alpha}{\nu} \int_0^m \frac{\theta}{m} d\theta. \tag{39.24}$$

The integrals in equation (39.24) may be evaluated by plotting j, obtained from freezing point measurements, against $\ln m$, for the first, and by plotting θ/m against θ, the freezing point depression, for the second, and determining the areas under the respective curves. For solutions less concentrated than about 0.1 molal, the second integral is usually negligible, and if the concentration is less than about 0.01 molal, the first integral may be evaluated in a simple manner by utilizing the empirical relationship between the function j and the molality m, viz.,[2]

$$j = Am^x, \tag{39.25}$$

where A and x are constants for a given electrolyte.* These constants are obtained from experimental values of the freezing point depression at various molalities, j being given by equation (39.20). Upon inserting equation (39.25) into (39.24), and neglecting the last term, it is found that

$$\ln \gamma'_\pm = - A \int_0^m m^x d \ln m - Am^x$$
$$= - \frac{A(x+1)}{x} m^x, \tag{39.26}$$

so that the activity coefficient at any molality less than 0.01 can be calculated.

The use of equation (39.25) to evaluate the first integral in (39.24) is not too accurate, and in any case it fails at concentrations exceeding 0.01 molal. A more reliable procedure is to determine the integral graphically. Instead of plotting j against $\ln m$, or j/m against j, it is preferable to change the

---

[2] Lewis and Linhart, *J. Am. Chem. Soc.*, **41**, 1951 (1919).

* According to the Debye-Hückel theory (Chapter XVII), A should be 0.375 for all uni-univalent strong electrolytes in aqueous solution at 0° C; x should be exactly 0.5 for all strong electrolytes.

variable from m to $m^{1/2}$; thus,

$$\int_0^m j d \ln m = 2 \int_0^{m^{1/2}} \frac{j}{m^{1/2}} dm^{1/2}, \qquad (39.27)$$

so that the required result is obtained by plotting $j/m^{1/2}$ against $m^{1/2}$.

39c. Corrections for More Concentrated Solutions.—As seen above, for solutions less concentrated than about 0.1 molal the second integral in equation (39.24) may be neglected. For more concentrated solutions, however, this integral must be determined graphically and its value included. For such solutions another factor must also be taken into consideration: since the freezing point varies with the concentration, the results do not all refer to the same temperature. It is necessary, therefore, to apply a correction involving the relative partial molar heat content of the solvent $\bar{L}_1$, i.e., the differential heat of dilution, for the given solution. If the concentration is less than approximately 0.1 molal, this correction is found to be very small, so that the values of the activity coefficient of the solute, as given by equation (39.24), will also represent the results at any temperature not too far removed from the freezing point of the solvent, e.g., 25° C for aqueous solutions.

For more concentrated solutions, allowance can be made for the variation of the activity coefficient with temperature, by utilizing equation (38.15), viz.,

$$\ln \frac{a_1''}{a_1'} = -\frac{1}{R} \int_{T'}^{T''} \frac{\bar{L}_1}{T^2} dT. \qquad (39.28)$$

By definition (§ 38c), $\bar{L}_1$ is equal to $\bar{H}_1 - H_{1l}^0$, where $\bar{H}_1$ is the partial molar heat content of the solvent in the solution, and H_{1l}^0 is the molar heat content of the pure liquid solvent, which has been chosen as the standard state for this constituent. Upon differentiation with respect to temperature, at constant (1 atm.) pressure, it is found that

$$\left(\frac{\partial \bar{L}_1}{\partial T}\right)_P = \bar{C}_{P1} - \bar{C}_{P1}^0, \qquad (39.29)$$

where $\bar{C}_{P1}$ is the partial molar heat capacity of the solvent in the solution and $\bar{C}_{P1}^0$ is the molar heat capacity of the pure solvent,* at 1 atm. pressure. If $\bar{C}_{P1} - \bar{C}_{P1}^0$ is approximately constant over a small temperature range, e.g., from T' to T''', it is possible to integrate equation (39.29) so as to yield

$$\bar{L}_1 - \bar{L}_1'' = (\bar{C}_{P1} - C_{P1}^0)(T - T''). \qquad (39.30)$$

If the value of $\bar{L}_1$ given by equation (39.30) is inserted in (39.28) and the integration carried out, the result is

$$\ln \frac{a_1''}{a_1'} = -\frac{\bar{L}_1''(T'' - T')}{RT'T''} + \frac{(\bar{C}_{P1} - \bar{C}_{P1}^0)}{R} \left(\frac{T'' - T'}{T'} - \ln \frac{T''}{T'}\right), \qquad (39.31)$$

---

* The partial molar notation is not strictly necessary in this case, but it is employed for the sake of consistency (cf. full treatment in § 44j).

where T' represents the freezing point of the solution, and T''' is the uniform temperature, e.g., 25° C, to which the activities are to be corrected. If the value of $\bar{L}_1$ and its variation with temperature, i.e., $\bar{C}_{P1} - \bar{C}_{P1}^0$, for the given solution are known, the right-hand side of equation (39.31) can be readily calculated; $\ln(a_1''/a_1')$ may, therefore, be regarded as known, and for simplicity the symbol X will be used to represent this quantity, viz.,

$$X = \ln \frac{a_1''}{a_1'}. \tag{39.32}$$

The transformation of equation (39.31) or (39.32) to give the activity of the solute is now made, as before (§ 39b), by means of the Gibbs-Duhem equation. If a_2 is the activity of the electrolyte *at the standard reference temperature* T''', then by equation (39.16),

$$d \ln a_2 = -\frac{1000}{mM_1} d \ln a_1'', \tag{39.33}$$

where M_1 is the molecular weight of the solvent. By definition [equation (39.32)], $\ln a_1'' - \ln a_1'$ is equal to X, and so it follows that

$$d \ln a_1'' - d \ln a_1' = dX,$$

and hence equation (39.33) can be written as

$$d \ln a_2 = -\frac{1000}{mM_1} d \ln a_1' - \frac{1000}{mM_1} dX. \tag{39.34}$$

Since a_1' here refers to the freezing point of the solution, it is equivalent to a_1 in equation (39.14), and hence, utilizing (39.16), the first term on the right-hand side of (39.34) may be replaced by (39.18), giving

$$d \ln a_2 = \frac{d\theta}{\lambda m} + \alpha \frac{\theta d\theta}{m} - \frac{1000}{mM_1} dX.$$

Recalling the postulate that $\ln a_2$ is equal to $\nu \ln a_\pm$ [equation (39.7)], it follows that

$$d \ln a_\pm = \frac{d\theta}{\nu \lambda m} + \alpha \frac{\theta d\theta}{\nu m} - \frac{1000}{\nu m M_1} dX$$

$$= (1 - j) d \ln m - dj + \alpha \frac{\theta d\theta}{\nu m} - \frac{1000}{\nu m M_1} dX. \tag{39.35}$$

Upon subtracting $d \ln m$ from both sides, it is evident that the left-hand side becomes $d \ln \gamma_\pm$ [equation (39.22)], whereas the right-hand side, with the exception of the last term, is equal to $d \ln \gamma_\pm'$, by equation (39.23). Hence, equation (39.35) may be written as

$$d \ln \gamma_\pm = d \ln \gamma_\pm' - \frac{1000}{\nu m M_1} dX.$$

Since the activity coefficient of the solute, at any temperature, is taken as unity at infinite dilution, when m is zero, it follows upon integration that

$$\ln \gamma_{\pm} = \ln \gamma'_{\pm} - \frac{1000}{\nu M_1} \int_0^m \frac{1}{m} dX, \qquad (39.36)$$

which is the expression for correcting the activity coefficient $\gamma'_{\pm}$, obtained from equation (39.24), to that, $\gamma_{\pm}$, at a standard temperature. The integral in equation (39.36) is evaluated graphically by plotting $1/m$ against the corresponding values of X, as obtained from equation (39.31), and measuring the area under the curve up to the point corresponding to a given molality. The results obtained for sodium chloride, at concentrations of 0.1 molal and greater, are reproduced in Table XXX.[3] It is seen that the correction term is negligible at 0.1 molal, but it increases as the concentration increases.

TABLE XXX. ACTIVITY COEFFICIENTS OF SODIUM CHLORIDE FROM FREEZING POINT MEASUREMENTS

m	γ'	γ (at 25° C)
0.1	0.798	0.798
0.2	0.750	0.752
0.5	0.682	0.689
1.0	0.630	0.650
2.0	0.613	0.661
3.0	0.627	0.704
4.0	0.657	0.765

If the value of $\gamma'_{\pm}$ given by equation (39.24) were strictly applicable at the freezing point of the solution, it would be a relatively simple matter to adjust the results to a standard temperature, e.g., 25°, by making use of an expression analogous to equation (39.28) involving a_2 and $\bar{L}_2$, i.e., $\bar{H}_2 - \bar{H}_2^0$, which is the partial molar heat content of the solute in the solution relative to the value at infinite dilution [cf. equation (44.31)]. Actually, the corrections obtained in this manner are a satisfactory approximation in many cases, provided the solutions are not more concentrated than about 1 molal. It should be noted, however, that although equation (39.23) gives $d \ln \gamma'_{\pm}$ at the freezing point of the particular solution, the value of $\ln \gamma'_{\pm}$ obtained after integration [equation (39.24)], is a kind of mean value for the temperature range from the freezing point of the solvent to that of the solution.* The treatment which resulted in equation (39.36) allows for this fact. An alternative method of applying the temperature correction is, therefore, to employ the latter equation, using graphical integration, to adjust the values of $\gamma'_{\pm}$ to give $\gamma_{\pm}$ at 0° C. The temperature range is quite small and the integral can be evaluated with greater accuracy than is possible when the standard temperature is taken as 25°. Having now obtained the activity coefficients for all the solutions at 0° C, it is a relatively simple matter to correct the values to 25° by making use of the appropriate form of equation (39.28) or (39.31),

[3] Lewis and Randall, ref. 1.

* It may be pointed out, for the sake of clarity, that this does not apply to $\ln a_1$ in equation (38.12), which gives the activity of the solvent at the actual freezing point of the solution; consequently, T' in equations (38.15), (39.29) and (39.31) represents this temperature.

with $\bar{L}_2$ and $\bar{C}_{P2}$. For work with concentrated solutions in which considerable accuracy is desired, this method is preferable to any other.[4]

The amount of reliable data available for the purpose of correcting activity coefficients obtained from freezing point measurements is not large. The freezing point method has thus been mainly used for the study of dilute solutions.

39d. Activities from Boiling Point Measurements.—Just as equation (39.24) for the mean activity coefficient of an ionic solute was obtained from (39.14), which gave the activity of the solvent in terms of the freezing point depression, so it is possible to derive an expression exactly analogous to equation (39.24) from (38.19) which relates the activity of the solvent to the rise of boiling point. The correction for temperature differences can be made in the same manner as described above, by using equation (39.36); the procedure is completely general for adjusting activity coefficients to a standard temperature. However, as indicated earlier, in actual practice the activity coefficients measured at several temperatures, by using boiling point elevations obtained at various external pressures, have been employed to calculate relative partial molar heat contents (cf. § 44g).[5]

39e. Activities of Electrolytes by the Isopiestic (Isotonic) Method.—One of the simplest methods for determining activities of electrolytes, provided the solutions are not too dilute, is based on a comparison of vapor pressures. If two (or more) solutions of different electrolytes in the same solvent are placed in an evacuated space, the solution of higher vapor pressure, i.e., higher fugacity (or activity) of the solvent, will distil over into that of lower vapor pressure until, when equilibrium is attained, the solutions will all have the same vapor pressure (and fugacity). Such solutions are said to be **isopiestic** or **isotonic**, *the solvent having the same activity in each.* Suppose one of these solutions contains a reference substance whose mean ionic activity coefficients at a number of molalities have been determined by a suitable method; it is then possible to calculate the activity coefficients in various solutions of other electrolytes.

For any experimental solute, equation (39.16), based on the Gibbs-Duhem equation, may be written as

$$\frac{mM_1}{1000} d \ln a_2 = - d \ln a_1,$$

and since, from equation (39.7), $d \ln a_2$ is equal to $\nu d \ln a_\pm$, it follows that

$$\frac{\nu m M_1}{1000} d \ln a_\pm = - d \ln a_1, \qquad (39.37)$$

where $a_\pm$ is the mean activity of the ions in the solution of molality m, and

---

[4] Lewis and Randall, ref. 1; see also, Randall and Young (L. E.), *J. Am. Chem. Soc.*, **50**, 989 (1928); Young (T. F.), *Chem. Rev.*, **13**, 103 (1933).

[5] Saxton and Smith, *J. Am. Chem. Soc.*, **54**, 2625 (1932); Smith, *et al.*, *ibid.*, **61**, 500, 1123 (1939); **63**, 1351 (1941).

a_1 is the activity of the solvent. Similarly, for a reference electrolyte, whose molality and mean ionic activity are indicated by the subscript R,

$$\frac{\nu_R m_R M_1}{1000} d \ln a_R = - d \ln a_1. \tag{39.38}$$

If the solutions of the experimental and reference substance are isopiestic, the activity a_1 of the solvent is the same in each case, and hence it is possible to equate (39.37) and (39.38); thus,

$$\nu m d \ln a_\pm = \nu_R m_R d \ln a_R.$$

According to equation (39.22), $d \ln a_\pm$ may be replaced by $d \ln m + d \ln \gamma_\pm$, i.e., by $d \ln m\gamma_\pm$, and similarly for $d \ln a_R$; hence,

$$\nu m d \ln m\gamma_\pm = \nu_R m_R d \ln m_R \gamma_R, \tag{39.39}$$

where γ_R is the mean activity coefficient of the ions in the reference solution.

The subsequent procedure is quite general, but in order to simplify the representation it will be supposed that the two electrolytes are of the same valence type, so that ν and ν_R are equal; equation (39.39) then becomes

$$m d \ln m\gamma_\pm = m_R d \ln m_R \gamma_R,$$

and upon rearrangement it is readily found that

$$d \ln \gamma_\pm = d \ln \gamma_R + d \ln \frac{m_R}{m} + \left(\frac{m_R}{m} - 1\right) d \ln m_R \gamma_R$$

$$= d \ln \gamma_R + d \ln r + (r - 1) d \ln m_R \gamma_R, \tag{39.40}$$

where r is written for the ratio of the molalities m_R/m of the two isotonic solutions. Further, since $m_R \gamma_R$ is equal to a_R,

$$d \ln m_R \gamma_R = d \ln a_R = 2 \frac{d a_R^{1/2}}{a_R^{1/2}},$$

and so equation (39.40) becomes upon integration

$$\ln \gamma_\pm = \ln \gamma_R + \ln r + 2 \int_0^{a_R^{1/2}} \frac{r - 1}{a_R^{1/2}} d a_R^{1/2}, \tag{39.41}$$

since $\gamma_\pm$ and γ_R are both unity, and hence $\ln \gamma_\pm$ and $\ln \gamma_R$ are both zero, at infinite dilution when m and m_R are zero. Since γ_R or a_R is supposed to be known at various molalities m_R, the integral in equation (39.41) can be evaluated graphically by plotting $(r - 1)/a_R^{1/2}$ against $a_R^{1/2}$. It is then possible to determine the area under the curve from $a_R^{1/2}$ equal to zero up to the point corresponding to the solution of molality m_R which is isopiestic with the experimental solution of molality m. All the information is then available for deriving $\ln \gamma_\pm$ for the experimental solution by means of equation (39.41).[6]

[6] Robinson and Sinclair, *J. Am. Chem. Soc.*, **56**, 1830 (1934).

39f. The Osmotic Coefficient and Activity Coefficient.—A method for determining the activity coefficient of electrolytes either directly from the vapor pressure of the solvent, or by the isopiestic comparison method, is based on the use of the **osmotic coefficient** of the solvent. The chemical potential of a solvent in a solution may be expressed in terms of its activity a_1 or its activity coefficient γ_1 by

$$\mu_1 = \mu_1^0 + RT \ln a_1 = \mu_1^0 + RT \ln N_1\gamma_1, \qquad (39.42)$$

where N_1 is the mole fraction of the solvent. The chemical potential may also be stated in terms of the **rational osmotic coefficient** g,* defined by

$$\mu_1 = \mu_1^0 + gRT \ln N_1, \qquad (39.43)$$

where g approaches unity at infinite dilution. For reasons which will appear shortly, another coefficient, known as the **practical osmotic coefficient** ϕ, is defined by

$$\mu_1 = \mu_1^0 - \phi RT \frac{M_1}{1000} \sum m_i, \qquad (39.44)$$

where $\sum m_i$ refers to the sum of the molalities of all the ions present in the solution. For a single electrolyte, one molecule of which yields ν ions in solution, $\sum m_i$ is equal to νm, where m is the molality of the electrolyte, and hence equation (39.44) becomes

$$\mu_1 = \mu_1^0 - \phi RT \frac{\nu m M_1}{1000}. \qquad (39.45)$$

Comparison with equation (39.42) then gives

$$\ln a_1 = -\phi \frac{\nu m M_1}{1000}, \qquad (39.46)$$

which relates the activity of the solvent to the practical osmotic coefficient.

The connection between the two osmotic coefficients, defined by equations (39.43) and (39.45), is

$$\phi = -\frac{1000}{\nu m M_1} g \ln N_1. \qquad (39.47)$$

If the solution is dilute, N_1 is close to unity, $\ln N_1$ is approximately equal to $N_1 - 1$ and hence to $-N_2$, where N_2 is here the *total* mole fraction of the ions present in the solution. It can be readily shown that this is related to the molality of the electrolyte [cf. equation (37.5)] by

$$N_2 = \frac{\nu m M_1}{\nu m M_1 + 1000}$$

* The term "osmotic coefficient" originates from the fact that it is virtually equivalent to the ratio of the actual to the ideal osmotic pressure of the given solution (see Exercise 17).

and hence
$$\ln N_1 \approx -\frac{\nu m M_1}{1000},$$

the last approximation being justified by the fact that the solution is dilute. If this result is substituted into equation (39.47), it is seen that ϕ and g are equal; in other words, the rational and practical osmotic coefficients are identical in dilute solutions.

The immediate importance of the practical osmotic coefficient of the solvent lies in its relationship to the mean activity coefficient of an electrolyte. From equation (39.37),
$$d \ln a_1 = -\frac{\nu m M_1}{1000} d \ln a_{\pm},$$

and, on the other hand, differentiation of (39.46) gives
$$d \ln a_1 = -\frac{\nu M_1}{1000}(\phi dm + m d\phi).$$

Combination of these two expressions for $d \ln a_1$ leads to the result
$$m d \ln a_{\pm} = \phi dm + m d\phi. \tag{39.48}$$

Again, as before, $d \ln a_{\pm}$ is equal to $d \ln m + d \ln \gamma_{\pm}$, by equation (39.22), and so (39.48) becomes
$$d \ln \gamma_{\pm} = (\phi - 1) d \ln m + d\phi, \tag{39.49}$$

which gives the mean ionic activity coefficient $\gamma_{\pm}$ at the molality m in terms of the osmotic coefficient.[7]

39g. Determination of Activity Coefficient from Osmotic Coefficient.— Although, in principle, equation (39.49) provides a method for determining activity coefficients, the details require consideration. The osmotic coefficients, in the first place, are determined from vapor pressure measurements. The activity a_1 of the solvent in a given solution is equal to f_1/f_1^0, by equation (31.5), or, approximately, to $p_1/p_1^{\square}$ [cf. equation (38.1)], where p_1 is the vapor pressure of the solvent over the solution and $p_1^{\square}$ is that of the pure solvent at the same temperature. Hence, by equation (39.46),
$$\ln \frac{p_1}{p_1^{\square}} \approx -\phi \frac{\nu m M_1}{1000},$$
$$\phi \approx -\frac{1000}{\nu m M_1} \ln \frac{p_1}{p_1^{\square}}, \tag{39.50}$$

so that the practical osmotic coefficient can be derived simply from vapor pressure measurements.

[7] Bjerrum, *Z. Elek.*, **24**, 321 (1918); *Z. phys. Chem.*, **104**, 406 (1923); see also, H. S. Harned and B. B. Owen, "The Physical Chemistry of Electrolytic Solutions," 1943, pp. 13, 287 *et seq.*

In order to determine $\gamma_\pm$ by combining equation (39.50) with (39.49), a function h is defined by

$$h = 1 - \phi, \tag{39.51}$$

which becomes zero at infinite dilution when ϕ is unity (see below). Hence, equation (39.49) may be written

$$d \ln \gamma_\pm = - h d \ln m - dh. \tag{39.52}$$

At infinite dilution, i.e., when m is zero, both h and $\ln \gamma_\pm$ are zero, so that integration of equation (39.52) gives

$$\ln \gamma_\pm = - \int_0^m h d \ln m - h. \tag{39.53}$$

This expression is seen to resemble the first two terms of equation (39.24) with h, defined by (39.51), replacing j, defined by (39.20). Although h and j become identical at infinite dilution, as will be shown below, there is an important difference between these two functions and hence between the activity coefficients derived from them. Whereas j applies to the freezing point of the solution, h refers to the particular temperature, e.g., 25° C, at which the osmotic coefficient is determined, e.g., from vapor pressure measurements.

As in the case of the analogous integral in equation (39.24), that in (39.53) is best evaluated by changing the variable to $m^{1/2}$ [cf. equation (39.27)], so that (39.53) becomes

$$\ln \gamma_\pm = - 2 \int_0^{m^{1/2}} \frac{h}{m^{1/2}} dm^{1/2} - h. \tag{39.54}$$

The values of $h/m^{1/2}$ are then plotted against $m^{1/2}$, and the area under the curve from zero to $m^{1/2}$ is determined, for any given molality m. Since ϕ, and hence h, can be obtained from actual vapor pressure measurements on a solution, it is possible by means of a series of such measurements, extending to high dilutions, to determine the mean ionic activity coefficient in any solution up to the highest concentration studied.[8]

In dilute solutions, the last term in equation (39.24) is negligible, and, as seen earlier, the results given by the first two terms are independent of temperature. It is thus possible to compare the first two terms of equation (39.24) with (39.53), which is applicable at all concentrations; it is seen that at high dilutions, therefore,

$$j = h = 1 - \phi,$$

so that h, like j, must become zero at infinite dilution. Upon introducing

[8] Randall and White, *J. Am. Chem. Soc.*, **48**, 2514 (1926). For an alternative method of applying the Gibbs-Duhem equation, see Randall and Longtin, *J. Phys. Chem.*, **44**, 306 (1940); Randall, Libby and Longtin, *ibid.*, **44**, 313 (1940).

the value of j given by equation (39.20), it follows that in dilute solutions

$$\phi = \frac{\theta}{\nu\lambda m}.$$

The osmotic coefficient, which is equal to the ratio of the actual osmotic pressure to the ideal value, is then equal to the ratio of the observed freezing point depression θ, to the ideal (infinite dilution) value $\nu\lambda m$.

In the foregoing method the mean ionic activity coefficient of the solute has been calculated from actual vapor pressure data. If the osmotic coefficients for a reference substance are known over a range of concentrations, the activity coefficients of another electrolyte can be derived from isopiestic measurements, without actually determining the vapor pressures. If m, ϕ and ν refer to an experimental electrolyte and m_R, ϕ_R and ν_R apply to a reference electrolyte which is isopiestic (isotonic) with the former, then by equation (39.46)

$$\phi\nu m = \phi_R \nu_R m_R,$$

since the activity of the solvent must be the same in the isopiestic solutions; hence,

$$\phi = \frac{\nu_R m_R}{\nu m}\phi_R. \tag{39.55}$$

As the osmotic coefficient ϕ_R for the reference substance at the molality m_R is known, the value of ϕ at the isopiestic molality m of the experimental electrolyte can be determined by means of equation (39.55). If the results are obtained for a series of concentrations of the latter, the activity coefficient at any molality can then be calculated from equation (39.54), as described above.[9]

39h. Activity Coefficients from E.M.F. Measurements.—The methods described so far for the determination of the activity coefficients of electrolytes are really based on measurements of the activity of the solvent, e.g., by freezing point depression, elevation of boiling point or vapor pressure, and their transformation by means of the Gibbs-Duhem equation. The procedures now to be considered give the mean ionic activity coefficients of the electrolyte directly. They depend on the use of suitable galvanic cells containing solutions of the substance being studied.[10] The first method requires the construction of a cell in which the reaction involves the formation of that substance at the activity in the solution. Consider, for example, the cell

$$\text{Pt, H}_2(1 \text{ atm.}) \mid \text{HCl}(m) \mid \text{AgCl}(s), \text{Ag},$$

[9] Cf. Scatchard, Hamer and Wood, *J. Am. Chem. Soc.*, **60**, 3061 (1938); Robinson and Harned, *Chem. Rev.*, **28**, 419 (1941).

[10] By using activity coefficients obtained from E.M.F. measurements, integration of the Gibbs-Duhem equation permits the evaluation of the activity of the solvent (water); cf., Newton and Tippetts, *J. Am. Chem. Soc.*, **58**, 280 (1936).

consisting of a hydrogen electrode, with gas at 1 atm. pressure, and a silver-silver chloride electrode immersed in a solution of hydrochloric acid at molality m. In the operation of this cell hydrogen gas dissolves at the left-hand electrode, to form hydrogen ions, while silver chloride at the right-hand electrode is reduced to form metallic silver, leaving chloride ions in solution; thus, the cell reaction is

$$\tfrac{1}{2}H_2(1 \text{ atm.}) + AgCl(s) = H^+(m) + Cl^-(m) + Ag(s),$$

the hydrogen and chloride ions being formed in a solution in which the molality of each is m.

It will be shown in § 45d that since the hydrogen gas, the silver chloride and the silver are in their respective standard states of unit activity, the E.M.F. of this cell depends only on the activities of the hydrogen and chloride ions in the hydrochloric acid solution. The actual value of the E.M.F. is given by

$$E = E^0 - \frac{RT}{F} \ln a_{H^+} a_{Cl^-}, \tag{39.56}$$

where R, T and F have their usual significance, and a_{H^+} and a_{Cl^-} are the activities of the ions in the solution contained in the cell; E^0 is the standard E.M.F. of the cell when the ionic activities in the solution are unity. Utilizing the standard state which makes the activity of each ion equal to its molality at infinite dilution (§ 39a), a_{H^+} may be set equal to $m_+\gamma_+$ and a_{Cl^-} to $m_-\gamma_-$, where m_+ and m_- are the molalities and γ_+ and γ_- the activity coefficients of the indicated ions. Hence equation (39.56) may be written as

$$E = E^0 - \frac{RT}{F} \ln m_+m_- - \frac{RT}{F} \ln \gamma_+\gamma_-. \tag{39.57}$$

In the present case, $\gamma_+\gamma_-$ is equal to $\gamma_\pm^2$ by equation (39.9), and m_+m_-, which is $m_\pm^2$, is in this case also equal to m^2, since ν_+ and ν_- are both unity [cf. equation (39.12)], so that (39.57) becomes, after rearrangement,

$$E + \frac{2RT}{F} \ln m - E^0 = -\frac{2RT}{F} \ln \gamma_\pm. \tag{39.58}$$

This expression provides a method for evaluating the mean ionic activity coefficient $\gamma_\pm$ in a hydrochloric acid solution of molality m from a measurement of the E.M.F., i.e., E, of the cell described above; it is necessary, however, to know the value of E^0, the standard E.M.F. For this purpose the data must be extrapolated to infinite dilution, and the reliability of the activity coefficients obtained from equation (39.58) depends upon the accuracy of this extrapolation. Two main procedures have been used for this purpose, but both are limited, to some extent, by the accuracy of E.M.F. measurements made with cells containing very dilute solutions.[11]

[11] Hitchcock, *J. Am. Chem. Soc.*, **50**, 2076 (1928); see also, Harned, *et al.*, *ibid.*, **54**, 1350 (1932); **55**, 2179 (1933); **58**, 989 (1936).

In order to describe the methods of extrapolation, it will be convenient to choose a definite temperature, viz., 25° C. Inserting the known value of R and F, converting the logarithms, and taking T as 298.16° K, equation (39.58) becomes

$$E + 0.1183 \log m - E^0 = - 0.1183 \log \gamma_{\pm}. \qquad (39.59)*$$

It will be seen in the next chapter, that by one form of the Debye-Hückel theory the variation with the molality of the mean activity coefficient of a uni-univalent electrolyte, such as hydrochloric acid, is given by

$$\log \gamma_{\pm} = - A\sqrt{m} + Cm, \qquad (39.60)$$

where A is a known constant, equal to 0.509 for water as solvent at 25° C, and C is another constant whose value is immaterial. Combination of this expression with equation (39.59) and rearrangement yields

$$E + 0.1183 \log m - 0.0602\sqrt{m} = E^0 - 0.1183Cm.$$

The left-hand side of this equation should consequently be a linear function of the molality, and extrapolation of the straight line plot to m equal zero should give the value E^0, which is required for the determination of the activity coefficients. Measurements have been made of the E.M.F.'s of the cell under consideration with hydrochloric acid at a number of different molalities m. The values of $E + 0.1183 \log m - 0.0602\sqrt{m}$ derived from these results plotted against m do not fall exactly on a straight line; nevertheless, a reasonably accurate extrapolation to m equal zero is possible, when it is found that E^0 is 0.2224 volt, within a probable accuracy of 0.0001 volt. Insertion of this value of E^0 in equation (39.59) gives

$$- 0.1183 \log \gamma_{\pm} = E + 0.1183 \log m - 0.2224,$$

from which the mean ionic activity coefficient of hydrochloric acid at any molality, for which E.M.F. measurements have been made, can be derived. Some of the results obtained in this manner are given in Table XXXI.[12]

TABLE XXXI. MEAN IONIC ACTIVITY COEFFICIENTS OF HYDROCHLORIC ACID FROM E.M.F. MEASUREMENTS AT 25° C

m	E	$E + 0.1183 \log m$	γ
0.1238	0.34199	0.23466	0.788
0.05391	0.38222	0.23218	0.827
0.02563	0.41824	0.22999	0.863
0.013407	0.44974	0.22820	0.893
0.009138	0.46860	0.22735	0.908
0.005619	0.49257	0.22636	0.926
0.003215	0.52053	0.22562	0.939

* With R in (int.) joules and F in (int.) coulombs, $R/2.3026F$ is 1.984×10^{-4}; hence $2RT/2.3026F$ at 25° C, i.e., 298.16° K, is 0.1183.

[12] Harned and Ehlers, *J. Am. Chem. Soc.*, **54**, 1350 (1932).

The second method of evaluating E^0 also makes use of the Debye-Hückel equation, but this time in a form which is more applicable to dilute solutions, viz.,

$$\log \gamma_{\pm} = - \frac{A\sqrt{m}}{1 + åB\sqrt{m}},$$

where A is identical with that in equation (39.60); $å$ and B are constants, the actual values of which are not required for the present purpose. If this expression is substituted in equation (39.59) the result is

$$E + 0.1183 \log m - E^0 = \frac{0.0602\sqrt{m}}{1 + åB\sqrt{m}}, \qquad (39.61)$$

taking A, as before, as equal to 0.509 for water as solvent at 25° C. Upon rearrangement, equation (39.61) becomes

$$E + 0.1183 \log m - 0.0602\sqrt{m} = E^0 - (E + 0.1183 \log m - E^0)åB\sqrt{m},$$

and hence the plot of the left-hand side against $(E + 0.1183 \log m - E^0)\sqrt{m}$ should be a straight line, the intercept for m equal zero giving E^0. It will be noted that for this purpose it is necessary to use a preliminary value of E^0. Employing the same E.M.F. measurements as before, E^0 was found to be 0.2225 volt. It is difficult to say whether this result is to be preferred over that obtained above, but the difference is not considerable; the effect on the activity coefficient does not exceed 0.002.[13]

Although the method described above has referred in particular to solutions of hydrochloric acid, it can be employed, in principle, to determine the activity coefficient of any suitable electrolyte. The essential requirement is a cell in which each of the two electrodes is reversible with respect to one of the ions of the electrolyte (cf. § 45a); for example, if the electrolyte is $M_{\nu_+}A_{\nu_-}$, then the cell can be represented formally by

$$M \,|\, M_{\nu_+}A_{\nu_-} \text{ solution } (m) \,|\, A.$$

The E.M.F. of this reversible cell is given by the expression (§ 45d)

$$E = E^0 - \frac{RT}{NF} \ln a_+^{\nu_+} a_-^{\nu_-}$$

$$= E^0 - \frac{\nu RT}{NF} \ln m_{\pm} - \frac{\nu RT}{NF} \ln \gamma_{\pm},$$

where ν is equal to $\nu_+ + \nu_-$, and N, the number of faradays associated with the formation of 1 mole of the solute in the cell, is equal to $\nu_+ z_+$ or to $\nu_- z_-$, where z_+ and z_- are the valences of the respective ions. This equation is analogous to (39.58) and, as before, determination of the mean ionic activity coefficient from the E.M.F. of the cell requires a knowledge of the standard

[13] D. A. MacInnes, "The Principles of Electrochemistry," 1939, p. 185; see also, Brown and MacInnes, J. Am. Chem. Soc., **57**, 1356 (1935).

E.M.F., i.e., E^0, which is obtained by an extrapolation procedure, as described above.

39i. Activities from Concentration Cells with Transference.

When two solutions of the same electrolyte, at different concentrations, are brought into contact and identical electrodes, which are reversible with respect to one or other of the ions of the electrolyte, are inserted in each solution, the result is a "concentration cell with transference." A typical cell of this kind is

$$\text{Pt, } H_2 \text{ (1 atm.)} \mid HCl(c') \vdots HCl(c) \mid H_2 \text{ (1 atm.), Pt,}$$

consisting of hydrogen gas electrodes with hydrochloric acid solutions of molarity c and c', respectively. The dotted line is used to indicate the "liquid junction" at which the two solutions are in actual contact.

When this cell produces 1 faraday of electricity, 1 g. atom of hydrogen gas dissolves at the left-hand electrode to yield 1 mole of hydrogen ions, and the same amount of these ions will be discharged to yield 1 g. atom of gas at the right-hand electrode. At the same time, t_+ mole of hydrogen ions will be transferred across the boundary between the two solutions in the direction of the current, i.e., from left to right, and t_- mole of chloride ions will move in the opposite direction; t_+ and t_- are the transference numbers of the positive and negative ions, respectively, their sum being equal to unity. The net result of the passage of 1 faraday of electricity is, therefore, the gain of $1 - t_+$, i.e., t_-, mole of hydrogen (positive) ions, and t_- mole of chloride (negative) ions in the solution at the right, and their loss from that at the left. The accompanying increase of free energy ΔF is then given by

$$\Delta F = t_-(\mu'_+ - \mu_+) + t_-(\mu'_- - \mu_-), \tag{39.62}$$

where the μ's are the chemical potentials of the indicated ions in the two solutions.

Since the transference numbers vary with composition, it is convenient to consider two solutions whose concentrations differ by an infinitesimal amount, viz., c and $c + dc$; under these conditions, equation (39.62) becomes

$$dF = - t_-(d\mu_+ + d\mu_-).$$

Utilizing the familiar relationship $\mu_i = \mu_i^0 + RT \ln a_i$ for the chemical potential of an ion in terms of its activity a_i, it is seen that

$$\begin{aligned} dF &= - t_-(RTd \ln a_+ + RTd \ln a_-) \\ &= - 2t_- RTd \ln a_\pm, \end{aligned} \tag{39.63}$$

where $a_\pm$, equal to $(a_+ a_-)^{1/2}$, is the mean activity of the hydrogen and chloride ions in the hydrochloric acid of molarity c, and t_- is the transference number of the chloride ion in this solution.

It may be noted that equation (39.63) is actually applicable to any concentration cell containing two different concentrations of a uni-univalent electrolyte in contact, provided the electrodes are reversible with respect to the positive ion. If the electrodes were reversible with respect to the negative ion, e.g., as in the Ag, AgCl electrode, the sign would be changed and t_- would be replaced by t_+.

If dE is the E.M.F. of the cell when the concentrations of the two solutions differ by an infinitesimal amount, then by equation (33.36) the free energy change dF, for the passage of one faraday, is equal to $- FdE$, where F is the faraday.

Upon equating this result to that given by equation (39.63), it is seen that

$$dE = 2t_- \frac{RT}{F} d \ln a_\pm. \tag{39.64}$$

The actual E.M.F. of the cell in which the concentrations of the two solutions differ by an appreciable amount can now be obtained by the integration of equation (39.64), but as this is not needed for the determination of activities, the procedure will not be given here. For the present requirement the mean activity may be replaced by the product of the mean ionic molarity, which in this case, i.e., uni-univalent electrolyte, is equal to the molarity c of the solution, and the mean ionic activity coefficient $\gamma_\pm$; thus,

$$dE = 2t_- \frac{RT}{F} (d \ln c + d \ln \gamma_\pm)$$

and hence,

$$\frac{dE}{t_-} = \frac{2RT}{F} (d \ln c + d \ln \gamma_\pm). \tag{39.65}$$

The activity is expressed here in terms of molarity, rather than molality, because the transference numbers, which are utilized in the calculations, are usually known as a function of the former. The procedure to be described below thus gives γ_c, but the results can be converted to γ_m, if required, by means of equation (39.13).

A quantity δ may be defined by

$$\frac{1}{t_-} = \frac{1}{t_R} + \delta, \tag{39.66}$$

where t_R is the transference number of the anion at a reference concentration c_R. If this result is inserted in (39.65), it is found upon rearrangement that

$$d \ln \gamma_\pm = \frac{F}{2t_R RT} dE - d \ln c + \frac{F}{2RT} \delta dE.$$

Integrating between the limits of c_R and c, the corresponding values of the mean ionic activity coefficients being γ_R and γ, it follows, after converting the logarithms, that

$$\log \frac{\gamma_\pm}{\gamma_R} = \frac{FE}{2.303 \times 2t_R RT} + \log \frac{c_R}{c} + \frac{F}{2.303 \times 2RT} \int_0^E \delta dE, \tag{39.67}$$

where E is the E.M.F. of a concentration cell in which one of the solutions has the constant molarity c_R, while the other has the variable concentration of c moles per liter.

The first two terms on the right-hand side of equation (39.67) may be evaluated directly from experimental data, after deciding upon the concentration c_R which is to represent the reference solution. The third term is obtained by graphical integration of δ against E, the value of δ being derived from the known variation of the transference number with concentration [equation (39.66)].

The procedure just described gives $\log (\gamma_\pm/\gamma_R)$, and hence the mean activity coefficient $\gamma_\pm$ in the solution of molarity c is known in terms of that (γ_R) at an arbitrary reference concentration c_R. It is desirable, however, to express the results in terms of one of the more common standard states; in this case the most convenient is that which makes the activity equal to the molarity at infinite dilution. For this purpose use is made of the Debye-Hückel expression analogous to that given above, viz.,

$$\log \gamma_{\pm(c)} = - \frac{A'\sqrt{c}}{1 + \mathring{a}B'\sqrt{c}},$$

where the constants A' and B' are related to A and B, but for dilute aqueous solutions they may be taken as identical. This may be written in the form

$$\log \frac{\gamma_\pm}{\gamma_R} + A'\sqrt{c} = -\log \gamma_R - \mathring{a}B'\left(\log \gamma_R + \log \frac{\gamma_\pm}{\gamma_R}\right)\sqrt{c}. \tag{39.68}$$

For solutions dilute enough for the Debye-Hückel equation to be applicable, the plot of $\log(\gamma_\pm/\gamma_R) + A'\sqrt{c}$ against $[\log \gamma_R + \log(\gamma_\pm/\gamma_R)]\sqrt{c}$ should be a straight line, the intercept for c equal zero giving the required value of $-\log \gamma_R$, by equation (39.68). The values of $\log(\gamma_\pm/\gamma_R)$ are obtained from equation (39.67), and $\log \gamma_R$, which is required for the purpose of the plot, is obtained by a short series of approximations. Once $\log \gamma_R$ has been determined, it is possible to derive $\log \gamma_\pm$ for any solution from the known value of $\log(\gamma_\pm/\gamma_R)$. The mean ionic activity coefficient of the given electrolyte can thus be evaluated from the E.M.F.'s of concentration cells with transference, provided the required transference number information is available.[14]

39j. Activities from Solubility Measurements.—The activity, or activity coefficient, of a sparingly soluble electrolyte can be determined in the presence of other electrolytes by means of solubility measurements. In a saturated solution the solid salt $M_{\nu_+}A_{\nu_-}$ will be in equilibrium, directly or indirectly, with $\nu_+ M^{z+}$ ions and $\nu_- A^{z-}$ ions in solution; thus,

$$M_{\nu_+}A_{\nu_-}(s) \rightleftharpoons \nu_+ M^{z+} + \nu_- A^{z-}.$$

The activity of the solid salt at atmospheric pressure is taken as unity, by convention, and hence the equilibrium constant is given by

$$K_s = a_+^{\nu_+} \times a_-^{\nu_-} \tag{39.69}$$

for a saturated solution at a particular temperature and 1 atm. pressure; a_+ and a_- are the activities of the positive and negative ions, respectively, in the saturated solution of the salt. This equation expresses the **solubility product principle** in its exact thermodynamic form; the constant K_s, as defined by equation (39.69), is the **activity solubility product** of the salt $M_{\nu_+}A_{\nu_-}$. The activity of each ion may be represented by the product of its molality and activity coefficient, so that

$$m_+^{\nu_+} m_-^{\nu_-} \times \gamma_+^{\nu_+}\gamma_-^{\nu_-} = K_s.$$

Referring to § 39a, it will be seen that this result may also be written as

$$(m_\pm \gamma_\pm)^\nu = K_s, \tag{39.70}$$

[14] MacInnes, et al., *J. Am. Chem. Soc.*, **57**, 1356 (1935); **58**, 1970 (1936); **59**, 503 (1937); **61**, 200 (1939); *Chem. Rev.*, **18**, 335 (1936).

where $m_\pm$ and $\gamma_\pm$ are the mean ionic molality and activity coefficient, respectively, in the saturated solution. It follows, then, from equation (39.70) that

$$\gamma_\pm = \frac{K_s^{1/\nu}}{m_\pm}. \qquad (39.71)$$

The mean activity coefficient of a sparingly soluble salt in any solution, containing other electrolytes, can thus be evaluated provided the solubility product (K_s) and the mean molality of the ions of the salt in the given solution are known. In order to obtain K_s the values of $m_\pm$ are determined from the experimentally observed solubilities of the sparingly soluble salt in the presence of various amounts of other electrolytes, and the results are extrapolated to infinite dilution (Fig. 27). In the latter case the activity coefficient is unity, in accordance with the chosen standard state, and hence, by equation (39.71), $K_s^{1/\nu}$ is equal to the extrapolated value of $m_\pm$.

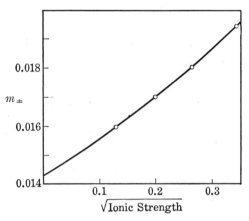

Fig. 27. Solubility of thallous chloride

Problem: By extrapolation of the experimental solubilities of TlCl in the presence of various electrolytes in water at 25° C, the limiting value of $m_\pm$ is found to be 0.01422. In the presence of 0.025 molal KCl, the solubility of the TlCl is 0.00869 molal; calculate the mean ionic activity coefficient of TlCl in this solution.

The number of ions ν produced by one molecule of TlCl is 2; hence, since $m_\pm$ is 0.01422 at infinite dilution, when $\gamma_\pm$ is 1, the value of $K_s^{1/2}$ is 0.01422.

In the KCl solution, the molality of the Tl⁺ ion, i.e., m_+, is 0.00869; the molality m_- of the Cl⁻ ion is, however, the sum of the values due to both KCl and TlCl, i.e., $0.025 + 0.00869 = 0.03369$. Hence the mean molality $m_\pm$ is $(0.00869 \times 0.03369)^{1/2} = 0.01711$, and by equation (39.71),

$$\gamma_\pm = \frac{0.01422}{0.01711} = 0.831.$$

The activity coefficients of thallous chloride, in the presence of various added electrolytes, determined from solubility measurements at 25°, are recorded in Table XXXII;[15] m is the *total* molality of the thallous chloride and the added substance. It will be observed that in the more dilute solutions the activity coefficient at a given molality is independent of the nature of the other electrolyte present in the solution. The significance of this fact will be considered below.

[15] Data from Lewis and Randall, ref. 1, p. 372.

TABLE XXXII. ACTIVITY COEFFICIENTS OF THALLOUS CHLORIDE IN PRESENCE OF VARIOUS ELECTROLYTES AT 25° C

Total Molality (m)	Added Electrolyte			
	KNO$_3$	KCl	HCl	TlNO$_3$
0.001	0.970	0.970	0.970	0.970
0.005	0.950	0.950	0.950	0.950
0.01	0.909	0.909	0.909	0.909
0.02	0.872	0.871	0.871	0.869
0.05	0.809	0.797	0.798	0.784
0.10	0.742	0.715	0.718	0.686
0.20	0.676	0.613	0.630	0.546

39k. The Ionic Strength Principle.—In order to systematize the data for the activity coefficients of simple salts, both in the pure state and in the presence of other electrolytes, G. N. Lewis and M. Randall (1921) introduced the concept of the **ionic strength**.[16] This property of a solution, represented by μ_m, is defined as *half the sum of the terms obtained by multiplying the molality of each ion present in the solution by the square of its valence;* thus,

$$\mu_m = \tfrac{1}{2}\sum m_j z_j^2, \qquad (39.72)$$

where m_j represents the molality of the jth ion and z_j is its valence, the summation being carried over all the ions present in the solution. In calculating the ionic strength it is necessary to use the *actual* molality of the ions, due allowance being made, especially with weak electrolytes, for incomplete dissociation. It is of interest to note that although the importance of the ionic strength was first realized from purely empirical considerations, it is now known to play an important part in the theory of electrolytes (Chapter XVII).

Problem: Compare the ionic strengths of solutions of (i) a uni-univalent, (ii) a uni-bi-(or bi-uni-)valent, and (iii) a bi-bivalent electrolyte, at the same molality m, assuming complete ionization.

(i) For a uni-univalent electrolyte, e.g., KCl, the molality of each ion is m and its valence unity, so that

$$\mu_m = \tfrac{1}{2}[(m \times 1^2) + (m \times 1^2)] = m.$$

(ii) For a uni-bivalent electrolyte, e.g., K$_2$SO$_4$, the molality of the positive ion is $2m$ and its valence unity, whereas the molality of the negative ion is m and its valence is two; hence,

$$\mu_m = \tfrac{1}{2}[(2m \times 1^2) + (m \times 2^2)] = 3m.$$

(iii) For a bi-bivalent electrolyte, e.g., ZnSO$_4$, the molality of each ion is m and its valence is two; thus,

$$\mu_m = \tfrac{1}{2}[(m \times 2^2) + (m \times 2^2)] = 4m.$$

The ratio of the ionic strengths is consequently 1 to 3 to 4. In general the ionic strength increases rapidly with the product z_+z_- of the ionic valences.

[16] Lewis and Randall, ref. 1, p. 373; *J. Am. Chem. Soc.*, **43**, 1112 (1921).

It was pointed out by Lewis and Randall that, in dilute solutions, *the activity coefficient of a given strong electrolyte is the same in all solutions of the same ionic strength*. The particular ionic strength may be due to the presence of other salts, but their nature does not appreciably affect the activity coefficient of the electrolyte under consideration. That this is the case for thallous chloride is shown by the data in Table XXXII; since the results are limited to uni-univalent salts, the molalities are identical with the ionic strengths. Another aspect of the ionic strength principle is that in dilute solutions electrolytes of the same valence type, e.g., KCl, NaCl and HCl, have been found to have equal activity coefficients at the same ionic strength. These generalizations, for which a theoretical basis will be given in Chapter XVII, hold only for solutions of relatively low ionic strength; as the concentration is increased the specific influence of the electrolytes present, especially if these yield ions of high valence, becomes evident.

391. Results of Activity Coefficient Measurements.—The mean ionic activity coefficients, in terms of molalities, of a number of strong electrolytes at 25° C are given in Table XXXIII, and some of the results are plotted in Fig. 28.[17] It will be seen that the values always decrease at first as the concentration of the solution is increased, but they frequently pass through a minimum and then increase again. The higher the product of the ionic valences, the sharper is the decrease of the activity coefficient as the molality is increased; it must be remembered, however, that when z_+z_- exceeds unity, the ionic strength is much greater than the molality. In the more dilute solutions, electrolytes of the same valence type, e.g., KCl and NaCl, CaCl$_2$ and ZnCl$_2$, etc., have almost identical activity coefficients at the same molality, but as the concentration is increased specific effects become apparent.

Measurements of the activity coefficients of electrolytes have been made in nonaqueous solutions and, in particular, in solvents consisting of mixtures of water and an organic liquid, e.g., methanol, ethanol or dioxane. The general nature of the results is similar to that for aqueous solutions. How-

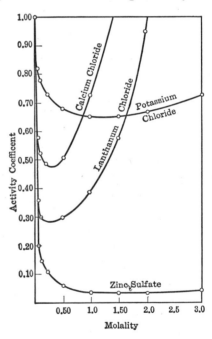

Fig. 28. Activity coefficients of electrolytes of different valence types

---

[17] For activity coefficient data and references, see W. M. Latimer, "The Oxidation States of the Elements, etc.," 1938; H. S. Harned and B. B. Owen, ref. 7; Harned and Robinson, *Chem. Rev.*, **28**, 419 (1941).

ever, the lower the dielectric constant of the solvent the more rapid is the fall in the activity coefficient of a given electrolyte with increasing ionic strength; further, a lower value is attained before the subsequent rise is observed.

TABLE XXXIII. MEAN IONIC ACTIVITY COEFFICIENTS IN AQUEOUS SOLUTION AT 25° C

Molality	0.001	0.005	0.01	0.02	0.05	0.1	0.2	0.5	1.0	2.0	3.0
HCl	0.966	0.928	0.905	0.875	0.830	0.796	0.767	0.757	0.809	1.009	1.316
HBr	0.966	0.929	0.906	0.879	0.838	0.805	0.782	0.790	0.871	1.17	1.67
HNO_3	0.965	0.927	0.902	0.871	0.823	0.785	0.748	0.715	0.720	0.783	0.876
H_2SO_4	0.830	0.643	0.545	0.455	0.341	0.266	0.210	0.155	0.131	0.125	0.142
NaOH	—	—	0.899	0.860	0.818	0.766	0.72	0.693	0.679	0.70	0.77
KOH	—	0.92	0.90	0.86	0.824	0.798	0.765	0.728	0.756	0.888	1.081
NaCl	0.966	0.929	0.904	0.875	0.823	0.778	0.732	0.679	0.656	0.670	0.719
NaBr	0.966	0.934	0.914	0.887	0.844	0.800	0.740	0.695	0.686	0.734	0.826
KCl	0.965	0.927	0.901	—	0.815	0.769	0.717	0.650	0.605	0.575	0.573
KBr	0.965	0.927	0.903	0.872	0.822	0.771	0.721	0.657	0.617	0.596	0.600
KI	0.965	0.927	0.905	0.88	0.84	0.776	0.731	0.675	0.646	0.641	0.657
$NaNO_3$	0.966	0.93	0.90	0.87	0.82	0.758	0.702	0.615	0.548	0.481	0.438
$CaCl_2$	0.888	0.789	0.732	0.669	0.584	0.531	0.482	0.457	0.509	0.807	1.55
$ZnCl_2$	0.88	0.789	0.731	0.667	0.628	0.575	0.459	0.394	0.337	0.282	—
Na_2SO_4	0.887	0.778	0.714	0.641	0.53	0.45	0.36	0.27	0.20	—	—
$MgSO_4$	—	—	0.40	0.32	0.22	0.18	0.13	0.088	0.064	0.055	0.064
$ZnSO_4$	0.734	0.477	0.387	0.298	0.202	0.148	0.104	0.063	0.044	0.035	0.041
$AlCl_3$	—	—	—	—	—	0.389	0.353	0.384	0.621	> 2	—
$LaCl_3$	0.853	0.716	0.637	0.552	0.417	0.356	0.298	0.303	0.387	0.954	—
$In_2(SO_4)_3$	—	0.16	0.11	0.08	0.035	0.025	0.021	0.014	—	—	—
$Ce_2(SO_4)_3$	—	—	0.171	0.112	0.063	0.041	—	—	—	—	—

A knowledge of activity coefficients is of importance in the study of equilibria involving ions and also in connection with the kinetics of ionic reactions. Some reference to the applications of the activity coefficients of electrolytes will be made in the succeeding chapters.

39m. Activity of Nonelectrolytes in Presence of Electrolytes.—Information concerning the influence of electrolytes on the activity coefficients of nonelectrolytes in aqueous solution is of some interest. Vapor pressure measurements can be utilized for this purpose, but a method based on solubility determinations is commonly employed. Consider an aqueous solution which is saturated, and hence in equilibrium, with a pure gaseous, liquid or solid nonelectrolyte solute at a given temperature. Provided this solute remains unchanged its activity is constant, and hence the activity in the saturated solution will not be affected by the addition of an electrolyte to the latter. Let m_0 be the solubility of the nonelectrolyte in pure water when its activity coefficient is γ_0, and suppose the solubility to be m in the presence of a certain amount of an added electrolyte when the activity coefficient of the nonelectrolyte is γ; since the standard states are the same, the activities in the saturated solutions will be equal, so that

$$m_0 \gamma_0 = m \gamma.$$

The activity coefficient γ, or its ratio to γ_0, i.e., γ/γ_0, can thus be readily determined from solubility measurements, since

$$\gamma = \frac{m_0}{m}\gamma_0. \qquad (39.73)$$

If the solubilities are expressed in terms of mole fractions or molarities, instead of molalities, alternative forms of the activity coefficient can be obtained.

From solubility determinations of various nonelectrolytes, such as helium, argon, hydrogen and other gases, ethyl acetate, diacetone alcohol, phenyl thiocarbamide, phenol and aniline, in the presence of electrolytes, it has been found that $\log(\gamma/\gamma_0)$ varies approximately in a linear manner with the ionic strength of the aqueous solution, viz.,

$$\log(\gamma/\gamma_0) = k\mu_m, \qquad (39.74)$$

where, for a given nonelectrolyte solute, k depends on the added electrolyte. The values of k also vary with the nature of the solute, but the order for a series of salts is very similar for different nonelectrolytes. In most cases k is positive, i.e., γ/γ_0 is greater than unity; according to equation (39.73) this means that m_0 is greater than m. In other words, the solubility of a nonelectrolyte in water is decreased by the addition of an electrolyte; this is the phenomenon known as the **salting-out effect**.

For most electrolytes k lies between zero and 0.1; hence, by equation (39.74), in a solution of ionic strength 0.1, γ/γ_0 for the nonelectrolyte varies from unity to 1.023. It is frequently assumed, therefore, that in dilute solutions of electrolytes, the activity coefficient of a nonelectrolyte, or of the undissociated portion of an electrolyte, is virtually unity.

There are two points in connection with equation (39.74) to which attention may be drawn. First, it is seen that the activity coefficient of the nonelectrolyte varies directly with the ionic strength; for electrolytes, in dilute solution, the activity coefficient depends on the square root of the ionic strength of the medium (see Chapter XVII). Second, the activity coefficient of the nonelectrolyte increases with the ionic strength whereas for an electrolyte the activity coefficient at first decreases. The subsequent increase observed in many cases is attributed to a type of salting-out effect.[18]

EXERCISES

1. Give the complete derivation of equation (39.13). The density of a 0.1 molar solution in ethanol of KI, which dissociates into two ions, is 0.8014; the density of the pure alcohol is 0.7919. Determine the ratio of the activity coefficients γ_N, γ_m and γ_c in this solution.

2. The densities of a number of aqueous sodium chloride solutions at 20° are as follows:

| Concn. | 10.05 | 20.25 | 41.07 | 62.48 | g. liter^{-1} |
| Density | 1.0053 | 1.0125 | 1.0268 | 1.0413 | g. cc.$^{-1}$ |

Consider to what extent (i) the molality, (ii) the molarity, of each ion is proportional to its mole fraction in the various solutions.

[18] Randall and Failey, *Chem. Rev.*, **4**, 285 (1927); Harned and Owen, ref. 7, pp. 397 *et seq.*

3. A 0.01 molal solution of cerous sulfate, i.e., $Ce_2(SO_4)_3$, has a mean ionic activity coefficient of 0.171 at 25° C. What are the values of m_+, m_-, $m_\pm$, $a_\pm$ and a_2? What is the ionic strength of the solution?

4. A solution contains 0.1 molal sodium sulfate, 0.3 molal copper sulfate and 0.05 molal zinc sulfate; assuming each salt to be completely ionized, what is the ionic strength?

5. Show that by setting ν equal to unity, equation (39.24) may be used to determine the activity coefficient of a nonelectrolyte solute; j is now defined as $1 - (\theta/\lambda m)$. For an aqueous solution, λ is 1.858 and α is 5.7×10^{-4}. The following values were obtained for the freezing point depressions of aqueous solutions of n-butanol at various molalities [Harkins and Wampler, *J. Am. Chem. Soc.*, **53**, 850 (1931)]:

m	θ	m	θ
0.0242	0.04452	0.21532	0.38634
0.04631	0.08470	0.32265	0.57504
0.06894	0.12554	0.54338	0.95622
0.09748	0.17680	0.95555	1.66128

Determine the activity coefficients of the solute at the molalities 0.1, 0.5 and 1.0 by graphical evaluation of the integrals. (Plot j/m against m for the first, and θ/m against θ for the second; when is the latter negligible?)

6. The freezing points of hydrochloric acid solutions, obtained by Randall and Young [*J. Am. Chem. Soc.*, **50**, 989 (1928)] by smoothing a large number of experimental results, are as follows:

m	θ	m	θ
0.001	0.003675	0.10	0.35209
0.002	0.007318	0.20	0.7064
0.005	0.018152	0.30	1.0689
0.01	0.036028	0.50	1.8225
0.02	0.07143	0.70	2.5801
0.05	0.17666	1.00	3.5172

Determine the mean ionic activity coefficients $\gamma'_\pm$ of the hydrochloric acid in the various solutions.

7. The relative partial molar heat contents ($\bar{L}_1$) in cal. mole^{-1} and the relative, partial molar heat capacities ($\bar{C}_{P1} - \bar{C}^0_{P1}$) in cal. deg.$^{-1}$ mole^{-1}, of the water in hydrochloric acid solutions are as follows:

m	0.01	0.04	0.10	0.25	0.5	0.75	1.00
$\bar{L}_1$	-0.0042	-0.031	-0.111	-0.44	-1.28	-2.60	-4.16
$\bar{C}_{P1} - \bar{C}^0_{P1}$	~ 0	~ 0	~ 0	-0.0056	-0.0159	-0.0293	-0.0450

Combine these data with the results obtained in Exercise 6 to determine the mean activity coefficient of the hydrochloric acid at molalities of 0.05, 0.1, 0.5 and 1.0 at 25° C.

8. From the values of the osmotic coefficient ϕ for potassium chloride solutions at 25° C given below, determine the activity coefficients at the various molalities by means of equation (39.54).

m	0.1	0.2	0.3	0.5	0.7	1.0	1.5
ϕ	0.926	0.913	0.906	0.900	0.898	0.899	0.906

9. The small proportion of undissociated molecules of hydrogen chloride in an aqueous solution will be in equilibrium with molecules in the vapor, on the

one hand, and with ions in solution, on the other hand. Hence, show that the fugacity (approximately, partial pressure) of hydrogen chloride over a solution of hydrochloric acid is proportional to $a_{\pm}^2$, where $a_{\pm}$ is the mean ionic activity in the solution.

The hydrogen chloride pressure in equilibrium with a 1.20 molal solution is 5.08×10^{-7} atm. and the mean ionic activity coefficient is known from E.M.F. measurements to be 0.842 at 25° C. Calculate the activity coefficients in the following solutions from the given hydrogen chloride pressures [Randall and Young, *J. Am. Chem. Soc.*, **50**, 989 (1928)]:

m	2.0	3.0	4.0	5.0	10.0
p	2.06×10^{-6}	7.78×10^{-6}	2.46×10^{-5}	7.01×10^{-5}	5.42×10^{-3} atm.

10. Show how E.M.F. measurements of the cell $Zn\,|\,ZnSO_4(m)\,|\,Hg_2SO_4(s)$, Hg, in which the reaction is $Zn + Hg_2SO_4(s) = Zn^{++}(m) + SO_4^{--}(m) + 2Hg$, might be used to determine the activity coefficients of zinc sulfate at various molalities.

11. The E.M.F.'s of the cell $H_2(1\text{ atm.})\,|\,HBr(m)\,|\,AgBr(s)$, Ag, in which the reaction is $\frac{1}{2}H_2 + AgBr(s) = H^+(m) + Br^-(m) + Ag$, with hydrobromic acid at various low molalities (m), at 25° C, were found to be as follows:

m	E	m	E
1.262×10^{-4}	0.53300	10.994×10^{-4}	0.42280
1.775	0.51618	18.50	0.39667
4.172	0.47211	37.18	0.36172

[Keston, *J. Am. Chem. Soc.*, **57**, 1671 (1935)]. Determine the standard E.M.F. E^0 of the cell. (The actual value is 0.0713 volt at 25° C.)

12. With more concentrated solutions of hydrobromic acid the following results were obtained for the E.M.F. of the cell referred to in the preceding exercise [Keston and Donelson, *J. Am. Chem. Soc.*, **58**, 989 (1936)]. Using the value of E^0 obtained there, determine the activity coefficients of hydrobromic acid at the various molalities.

m	E	m	E
0.001	0.42770	0.05	0.23396
0.05	0.34695	0.10	0.20043
0.01	0.31262	0.20	0.16625
0.02	0.27855	0.50	0.11880

13. The E.M.F.'s of the cell with transference Ag, $AgCl(s)\,|\,0.1\text{ N HCl}\,\vdots\,HCl(c)\,|\,AgCl(s)$, Ag at 25° C and the transference numbers of the hydrogen ion in the hydrochloric acid solution of concentration c mole liter^{-1}, are given below [Shedlovsky and MacInnes, *J. Am. Chem. Soc.*, **58**, 1970 (1935); Longsworth, *ibid.*, **54**, 2741 (1932)]. Utilize the data to calculate the activity coefficients of hydrochloric acid at the various concentrations.

$c \times 10^3$	E	t_{H^+}
3.4468	0.136264	0.8234
5.259	0.118815	0.8239
10.017	0.092529	0.8251
19.914	0.064730	0.8266
40.492	0.036214	0.8286
59.826	0.020600	0.8297
78.076	0.009948	0.8306
100.00	—	0.8314

14. The solubility of silver iodate ($AgIO_3$) in pure water at 25° C is 1.771×10^{-4} mole liter^{-1}; in the presence of various amounts of potassium nitrate the solubilities are as follows:

KNO_3 mole liter^{-1}	$AgIO_3$ mole liter^{-1}	KNO_3 mole liter^{-1}	$AgIO_3$ mole liter^{-1}
0.1301×10^{-2}	1.823×10^{-4}	1.410×10^{-2}	1.999×10^{-4}
0.3252	1.870	7.050	2.301
0.6503	1.914	19.98	2.665

[Kolthoff and Lingane, *J. Phys. Chem.*, **42**, 133 (1938)]. Calculate the activity coefficients of the silver iodate in the various solutions; state the total ionic strength in each case.

15. Show that the free energy of dilution of a solution from one molality (m) to another (m'), i.e., the difference in the partial molar free energy (μ) of the solute in the two solutions, is equal to

$$\Delta F = \nu RT \ln (a'/a) = \nu RT \ln (m'\gamma'/m\gamma),$$

where ν is the number of ions formed by one molecule of the electrolyte solute. The a's, m's and the γ's are the mean ionic activities, molalities and activity coefficients, respectively. How can the free energy of dilution be determined from E.M.F. measurements? Utilize the data in Table XXXIII to determine the free energy of dilution, in cal. mole^{-1} of solute, of a 1.0 molal solution of potassium chloride to 0.001 molal at 25° C.

16. Give the complete derivation of equation (39.15) for an electrolyte solution.

17. Making use of equations (38.42), (39.42) and (39.43), show that if the effect of pressure on the partial molar volume of the solvent is negligible, the rational osmotic coefficient is equal to the ratio of the actual osmotic pressure of a solution to the value it would have if it behaved ideally, i.e., the solvent obeyed Raoult's law.

CHAPTER XVII

THE DEBYE-HÜCKEL THEORY

40. Ionic Interaction in Solution

40a. Deviation from Ideal Behavior.—The activity or activity coefficient of an electrolyte has been regarded hitherto as a purely thermodynamic quantity which can be evaluated from observable properties of the solution. The treatment involves no theory, for the activity is defined in terms of the chemical potential by the expression $\mu = \mu^0 + RT \ln a$, and its experimental determination depends ultimately on this definition. The activity coefficient is related to the activity since it is equal to the latter divided by the measurable concentration of the substance in whatever units are chosen as convenient in connection with the specified standard state. It has been stated earlier (§ 1a) that theories dealing with the behavior of matter lie, strictly, outside the realm of pure thermodynamics. Nevertheless, it is desirable to refer briefly to the theory of P. Debye and E. Hückel (1923) which has made it possible to calculate the mean activity coefficients of dilute solutions of electrolytes without recourse to experiment. The theory thus bears the same relationship to thermodynamics as do those described earlier which permit of the estimation of other properties, such as heat capacity, entropy, etc. Although the treatment of Debye and Hückel has a number of limitations, it represents an important advance in the problem of accounting for the departure of dilute ionic solutions from ideal (Henry's law) behavior, as is evidenced by the deviations of the activity coefficients from unity even at very low concentrations.[1]

The first postulate of the **Debye-Hückel theory** is that if the ions of an electrolyte lost their charges and became neutral particles, the solution would behave like a dilute solution obeying Henry's law (§ 36a). The departure from ideal behavior is then attributed to the mutual interaction of the electrical charges carried by the ions. Hence, in writing the chemical potential (partial molar free energy) of an ion i in the form

$$\mu_i = \mu_i^0 + RT \ln a_i = \mu_i^0 + RT \ln \mathrm{N}_i + RT \ln \gamma_i, \qquad (40.1)$$

where the activity a_i of the ion is replaced by the product of its mole fraction N_i and its (rational) activity coefficient γ_i (cf. § 37b, III B), the term

---

[1] Debye and Hückel, *Physik. Z.*, **24**, 185, 334 (1923); **25**, 97 (1924); for reviews, see H. Falkenhagen, "Electrolytes," 1934 (translated by R. P. Bell); H. S. Harned and B. B. Owen, "The Physical Chemistry of Electrolytic Solutions," 1943; LaMer, *Trans. Electrochem. Soc.*, **51**, 507 (1927); Williams, *Chem. Rev.*, **8**, 303 (1931); Schingnitz, *Z. Elek.*, **36**, 861 (1930). For a critical examination of the Debye-Hückel treatment, see R. H. Fowler and E. A. Guggenheim, "Statistical Thermodynamics," 1939, Chapter IX.

$RT \ln \gamma_i$ is equal to the partial molar free energy $\mu_{i\text{el.}}$ of the given ions in the solution resulting from electrical interactions. If $F_{\text{el.}}$ represents the contribution to the free energy of the solution due to these electrical forces, then

$$\left(\frac{\partial F_{\text{el.}}}{\partial n_i}\right)_{T,P} = \mu_{i\text{el.}} = RT \ln \gamma_i, \tag{40.2}$$

where n_i is the number of moles of ions of the ith kind present in the solution.

40b. Electrical Potential of Ionic Atmosphere.—The problem is now to calculate the electrical free energy of the solution, and in this connection Debye and Hückel postulated that as a result of electrical interaction each ion in the solution is surrounded by an ionic atmosphere of opposite sign, arising somewhat in the following manner. Imagine a central positive ion situated at a particular point, and consider a small volume element at a short distance from it in the solution. As a result of the thermal movement of the ions, there will sometimes be an excess of positive and sometimes an excess of negative ions in this volume element. However, it will be expected to have a net negative charge as a consequence of the electrostatic attraction of the central positive ion for ions of opposite sign. In other words, *the probability of finding ions of opposite sign in the space surrounding a given ion is greater than the probability of finding ions of the same sign.* Every ion may thus be regarded as associated with an **ionic atmosphere** of opposite sign, the charge density of the atmosphere being greater in the immediate vicinity of any ion, and falling off with increasing distance. The net charge of the whole atmosphere extending through the solution is, of course, equal in magnitude, but opposite in sign, to that of the given ion.

By assuming that Boltzmann's law for the distribution of particles in a field of varying potential energy, which is based on statistical mechanics, is applicable to the distribution of the ions in the atmosphere of a central ion, Debye and Hückel were able to derive an expression relating the charge density at any point to the electrical potential at that point. By introducing Poisson's equation, which is based on Coulomb's inverse square law of force between electrostatic charges, another equation connecting the electrical charge density and the potential can be derived. From these two expressions the electrical potential ψ_i at an ion due to its surrounding atmosphere is found to be

$$\psi_i = -\frac{z_i \epsilon}{D} \cdot \frac{\kappa}{1 + \kappa \mathring{a}}, \tag{40.3}$$

where z_i is the number of charges carried by the ion, i.e., its valence, ϵ is the unit (electronic) charge, D is the dielectric constant of the medium, i.e., the solvent, which is assumed to be continuous in nature, $\mathring{a}$ is the mean distance of closest approach of the ions in the solution, and κ is defined by

$$\kappa^2 = \frac{4\pi \epsilon^2 N}{1000 D k T} \sum c_j z_j^2, \tag{40.4}$$

where N is the Avogadro number, k is the Boltzmann constant, i.e., the gas constant per single molecule, and T is the absolute temperature; c_j is the concentration in moles per liter and z_j is the valence of the ions, the summation being carried over all the kinds of ions present in the solution.

The value of the factor κ, as defined by equation (40.4), is seen to increase with the concentration of the ions in the solution, and for dilute solutions κa in the denominator of equation (40.3) may be neglected in comparison with unity, so that this equation becomes

$$\psi_i = -\frac{z_i \epsilon \kappa}{D}. \tag{40.5}$$

As seen above, the charge of the ionic atmosphere must be equal to the charge carried by the central ion, which in this case is $z_i\epsilon$. If a charge equal in magnitude and opposite in sign to that of the central ion itself were placed at a distance $1/\kappa$ from the ion, the potential produced at the ion in a medium of dielectric constant D is given by electrical theory as $-z_i\epsilon\kappa/D$. This is seen to be identical with the electrical potential, due to the ion atmosphere, given by equation (40.5). It follows, therefore, that *the effect of the ion atmosphere in a dilute solution is equivalent to that of a single charge, of the same magnitude, placed at a distance $1/\kappa$ from the central ion.* If the whole ionic atmosphere were concentrated at a point, or in a thin spherical shell, at a distance $1/\kappa$, the effect at the central ion would be the same as the actual ionic atmosphere. The quantity $1/\kappa$ can thus be regarded as a measure of the **radius of the ionic atmosphere** in the given solution. The more dilute the solution the smaller is the value of κ, and hence the larger the thickness of the atmosphere. Under these circumstances there is a greater justification in treating the medium, i.e., the solvent, as if it had a continuous, instead of a molecular, structure, as is done in connection with the dielectric constant.

40c. Electrical Free Energy and Activity Coefficient.—From a knowledge of the electrical potential at an ion due to its atmosphere, it is possible to calculate the work done in charging all the ions in a solution at a given concentration, and also at infinite dilution. The difference between these quantities, ignoring a possible small volume change in the charging process, is then identified with the electrical free energy of the solution. Utilizing equation (40.5) for the potential, the result for a dilute solution is found to be

$$F_{\text{el.}} = -\frac{N\epsilon^2\kappa}{3D}\sum n_j z_j^2, \tag{40.6}$$

where n_j represents the number of moles of an ionic species of valence z_j, the summation being carried over all the species present in the solution. The derivative of $F_{\text{el.}}$, at constant temperature and pressure, with respect to n_i, all the other n's being constant, is equal to $\mu_{i\text{el.}}$, and hence to $RT \ln \gamma_i$, by equation (40.2). Recalling that, according to equation (40.4), κ involves $c_j^{1/2}$, and hence $n_j^{1/2}$, it is found that

$$\left(\frac{\partial F_{\text{el.}}}{\partial n_i}\right)_{T,P} = RT \ln \gamma_i = -\frac{N\epsilon^2\kappa z_i^2}{2D}, \tag{40.7}$$

and hence

$$\ln \gamma_i = -\frac{N\epsilon^2\kappa z_i^2}{2DRT}. \tag{40.8}$$

An equation has thus been derived for the activity coefficient of any single ionic species in an electrolytic solution.

In order to put this result into a practical form, the expression for κ, i.e., equation (40.4), is introduced, and then the values of the universal constants N, R, k, ϵ and π are inserted. It will be observed that κ involves the factor $\sum c_j z_j^2$; this is very similar to twice the ionic strength defined by equation (39.72), except that the former contains the molarity, in place of the molality of each ion in the latter. If the ionic strength, in terms of molarity, is defined by analogy with equation (39.72) as

$$\mathfrak{u} = \tfrac{1}{2}\sum c_j z_j^2, \tag{40.9}$$

it is possible to write (40.4) as

$$\kappa^2 = \frac{8\pi\epsilon^2 N}{1000 DkT}\mathfrak{u}. \tag{40.10}$$

It should be noted that, as stated in § 39k, the value of c_j in equation (40.9) refers to the *actual ionic concentration*; in the event that one or more of the substances present in the solution is not completely dissociated, due allowance for this must be made when calculating the ionic strength. Upon combining equations (40.8) and (40.10), substituting the numerical values for the universal constants, and converting the logarithms, it is found that

$$\log \gamma_i = -\frac{1.824 \times 10^6}{(DT)^{3/2}} z_i^2 \sqrt{\mathfrak{u}}. \tag{40.11}$$

For a given solvent and temperature, D and T have definite values which may be inserted; equation (40.11) then takes the general form

$$\log \gamma_i = -Az_i^2\sqrt{\mathfrak{u}}, \tag{40.12}$$

where A is a constant for the solvent at the specified temperature. For water, the dielectric constant D is 88.15 at 0° C and 78.54 at 25° C; hence, for aqueous solutions A is 0.488 at 0° C, and 0.509 at 25° C.

40d. Mean Ionic Activity Coefficient.—No method is at present available for the experimental determination of the activity coefficient of a single ionic species, an expression for which has been derived above on the basis of the theoretical concepts of Debye and Hückel. All the methods described in Chapter XVI for the determination of activity coefficients give the *mean value for the two ions* of a particular electrolyte. In order to test the results of the Debye-Hückel treatment outlined above, it is desirable to derive from equation (40.12) a relationship for the mean ionic activity coefficient. According to equation (39.9)

$$\gamma_\pm^\nu = \gamma_+^{\nu_+}\gamma_-^{\nu_-},$$

where ν, equal to $\nu_+ + \nu_-$, is the total number of ions produced by one molecule of the given electrolyte; hence, taking logarithms,

$$\log \gamma_\pm = \frac{\nu_+ \log \gamma_+ + \nu_- \log \gamma_-}{\nu_+ + \nu_-}. \tag{40.13}$$

By equation (40.12), $\log \gamma_+$ is equal to $-Az_+^2\sqrt{\mu}$ and $\log \gamma_-$ to $-Az_-^2\sqrt{\mu}$; making use of the fact that for any electrolyte $\nu_+ z_+$ is equal to $\nu_- z_-$, where the valences z_+ and z_- are numerical only and do not include the sign, it is found from equation (40.13) that

$$\log \gamma_\pm = -Az_+z_-\sqrt{\mu}, \qquad (40.14)$$

where A is the same as before. For an aqueous solution at 25° C, the value of A, as seen above, is 0.509, so that

$$\log \gamma_\pm = -0.509 z_+ z_- \sqrt{\mu}. \qquad (40.15)$$

These equations give the mean ionic activity coefficient in terms of the ionic strength of the solution. It is important to remember that $\gamma_\pm$ refers to a particular electrolyte whose ions have the (numerical) valences of z_+ and z_-, but the ionic strength μ contains terms for *all* the ions present in the solution.

If the solution contains a single, strong, uni-univalent electrolyte the ionic strength μ is equal to the molarity c of the solution; at the same time z_+ and z_- are both unity, so that equation (40.14) takes the simple form

$$\log \gamma_\pm = -A\sqrt{c}, \qquad (40.16)$$

or, for such an aqueous solution at 25° C,

$$\log \gamma_\pm = -0.509\sqrt{c}. \qquad (40.17)$$

The various forms of equation (40.15), referred to as the **Debye-Hückel limiting law**, express the variation of the mean ionic activity coefficient of a solute with the ionic strength of the medium. It is called the "limiting law" because the approximations and assumptions made in its derivation are strictly applicable only at infinite dilution. The Debye-Hückel equation thus represents the behavior to which a solution of an electrolyte should approach as its concentration is diminished.

40e. Change of Units and Standard State.—Before proceeding to a comparison of the results derived from the Debye-Hückel treatment with those obtained experimentally, some consideration must be given to the matter of units and standard states. The ionic strength factor which appears in the Debye-Hückel equation involves ionic concentrations expressed in molarities, i.e., in moles per liter of solution. It has been seen, however, that in the study of activity coefficients, it is often more convenient to employ molalities, i.e., moles per 1000 g. of solvent; the ionic strength is then used in its original form as defined by equation (39.72). It will be apparent from equation (37.6), or it can be readily shown from general considerations, that in a dilute solution the molarity c and the molality m are related by $c/\rho_0 = m$, where ρ_0 is the density of the solvent. It follows, therefore, that

$$\mu = \mu_m \rho_0, \qquad (40.18)$$

for the concentration range in which the Debye-Hückel limiting law might be regarded as applicable. For aqueous solutions ρ_0 is so close to unity, that μ and μ_m may be taken as identical, but for other solvents it is necessary, when employing the Debye-Hückel equation, either to use μ involving molarities, or to apply the appropriate correction [equation (40.18)] to the molal ionic strength.

The activity coefficients given by the Debye-Hückel treatment presumably represent deviations from the dilute solution behavior, i.e., from Henry's law, and are consequently based on the standard state which makes the activity of an ion equal to its *mole fraction* at infinite dilution (§ 37b, III B). In the experimental determination of activity coefficients, however, it is almost invariably the practice to take the activity as equal to the molarity or the molality at infinite dilution. The requisite corrections can be made by means of equation (39.13), but this is unnecessary, for in solutions that are sufficiently dilute for the Debye-Hückel limiting law to be applicable, the difference between the various activity coefficients is negligible. The equations derived above may thus be regarded as being independent of the standard state chosen for the ions, provided only that the activity coefficients are defined as being unity at infinite dilution.

It should be noted that the Debye-Hückel theory yields true, and not stoichiometric, activity coefficients (§ 39a), since it is the behavior of the ions only, and not of the whole solute, which is taken into consideration. For strong electrolytes dissociation is virtually complete at all dilutions for which the limiting law may be expected to hold; for such solutes, therefore, the distinction between true and stoichiometric activity coefficients may be ignored.

41. Applications of the Debye-Hückel Equation

41a. Tests of the Debye-Hückel Equation: Qualitative.—Although the Debye-Hückel treatment is generally considered as applying to solutions of strong electrolytes, it should be emphasized that it is not restricted to such solutions. The results are of general applicability, but it should be noted, as mentioned previously, that in the calculation of the ionic strength the actual ionic concentration must be employed. For incompletely dissociated substances, such as weak and intermediate electrolytes, this involves a knowledge of the degree of dissociation which may not always be available with sufficient accuracy. It is for this reason that the Debye-Hückel limiting law equations are usually tested by means of data obtained with strong electrolytes, since they can be assumed to be completely dissociated at all concentrations for which the law should be valid.

It will be observed that the Debye-Hückel limiting equations contain no reference, apart from the valence, to the specific properties of the electrolytes that may be present in the solution. The mean ionic activity coefficient of a given solute, in a particular solvent at a definite temperature, should thus depend only on the ionic strength of the solution, the actual nature of the electrolytes present being immaterial. This conclusion is identical with the empirical result stated in § 39k. Further, for different electrolytes of the same valence type the Debye-Hückel theory requires the activity coefficients to be the same in solutions of equal ionic strength; this expectation is again in general agreement with experiment, provided the solutions are dilute (§§ 39k, 39l).

It is evident from equation (40.15) that, at a definite ionic strength in a given solvent at constant temperature, the deviation of the mean ionic activity coefficient of an electrolyte from unity should be greater the higher the valences of the ions constituting the electrolyte. This is in harmony with the results given in Table XXXIII.

41b. Tests of the Debye-Hückel Equation: Quantitative.—In essence, the Debye-Hückel limiting law, e.g., equation (40.14), states that the plot of $\log \gamma_{\pm}$ of a given electrolyte against the square root of the ionic strength, i.e., $\sqrt{\mu}$, should approach, with increasing dilution, a straight line of slope equal to $-Az_+z_-$, where A is a constant whose value depends on the dielectric constant of the solvent and its temperature. For aqueous solutions at 25° C, A should be 0.509, and consequently the limiting slope of $\log \gamma_{\pm}$ against $\sqrt{\mu}$ should be $-0.509z_+z_-$. How far this theoretical expectation is fulfilled is shown by the results in Fig. 29 for three salts of different valence

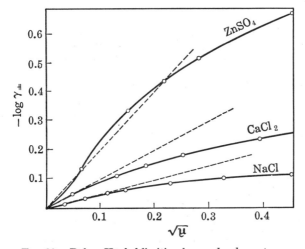

Fig. 29. Debye-Hückel limiting law and valence type

types, viz., sodium chloride, $z_+z_- = 1$; calcium chloride, $z_+z_- = 2$; zinc sulfate, $z_+z_- = 4$. The theoretical slope in each case is indicated by a broken straight line. It is evident that the experimental results approach the values required by the Debye-Hückel limiting law with increasing dilution, i.e., with decreasing ionic strength. It is not absolutely certain that the limiting slopes are exactly constant for salts of the same valence type, and that the values are always precisely equal to $-0.509z_+z_-$. There is a possibility that there may be slight variations from the theoretical figure according to the nature of the salt, apart from its valence type, but the activity coefficient data for dilute solutions are not sufficiently accurate for a clear decision to be made in this connection. It may be concluded, therefore, that as a first approximation, at least, the Debye-Hückel law represents the limiting behavior of electrolytes in aqueous solutions at 25° C.

According to equation (40.11) it appears that for electrolytes of the same valence type, or for a given electrolyte, the limiting slope of the plot of log $\gamma_{\pm}$ against $\sqrt{\mu}$ at constant temperature should be inversely proportional to $D^{3/2}$, where D is the dielectric constant of the solvent. Experimental data, obtained by E.M.F. measurements, on the activity coefficients of hydrochloric acid in methyl and ethyl alcohols, and in a number of dioxane-water mixtures, with dielectric constants varying from 9.5 to 78.6, are in satisfactory agreement with the theoretical requirement. It may be noted that the lower the dielectric constant the smaller the concentration at which the limiting slope is attained (Fig. 30); there are theoretical reasons for this, but as they lie outside the scope of thermodynamics they will not be discussed.[2]

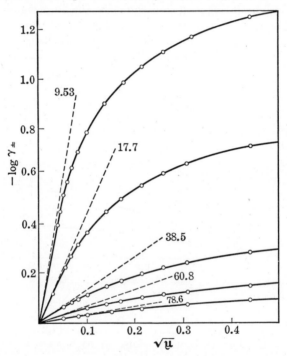

FIG. 30. Debye-Hückel limiting law and dielectric constant

The influence of one other variable, namely, the temperature, remains to be considered. It is not an easy matter to vary the temperature without changing the dielectric constant, and so these factors are taken together. From equation (40.11) it is evident that the limiting Debye-Hückel slope should vary as $1/(DT)^{3/2}$, where T is the absolute temperature at which the activity coefficients are measured. The experimental results obtained under a wide variety of conditions, e.g., from liquid ammonia at $-75°$ C to water at the normal boiling point, $100°$ C, are generally in satisfactory agreement

[2] Harned, et al., J. Am. Chem. Soc., **61**, 49 (1939); for references to numerous tests of the Debye-Hückel theory, see ref. 1; also, S. Glasstone, "The Electrochemistry of Solutions," 2nd ed., 1937, Chapter VII.

with the theoretical requirements. Where discrepancies have been observed they can probably be explained by incomplete dissociation of the electrolyte in media of low dielectric constant.

41c. Solubility and the Debye-Hückel Theory.—An interesting application of the Debye-Hückel equation is to be found in the study of the influence of "inert" electrolytes on the solubility of a sparingly soluble salt. By equation (39.70), in a saturated solution

$$(m_\pm \gamma_\pm)^\nu = K_s, \qquad (41.1)$$

where $m_\pm$ is the mean ionic molality of the salt at saturation and K_s is a constant for the salt, viz., the solubility product; $\gamma_\pm$ is the mean ionic activity coefficient of the salt in the given solution, which may contain added electrolytes. If S is the saturation solubility of the salt, expressed in molalities, then by equation (39.12)

$$m_\pm^\nu = m^\nu(\nu_+^{\nu_+}\nu_-^{\nu_-}) = S^\nu(\nu_+^{\nu_+}\nu_-^{\nu_-}), \qquad (41.2)$$

the molality m being identified with S, the solubility. Substitution of this result into equation (41.1), and simplifying gives

$$S\gamma_\pm = \text{constant}. \qquad (41.3)$$

This equation accounts for the fact, so frequently observed, that the addition of an inert salt, not having an ion in common, will increase the solubility of a sparingly soluble salt. The inert salt increases the ionic strength of the medium, and hence the activity coefficient $\gamma_\pm$ decreases. In order for the product $S\gamma_\pm$ to remain constant the solubility S must increase correspondingly.

If S_0 is the solubility of the salt in pure water and S is that in the presence of another electrolyte which has no ion in common with the saturating salt, and γ_0 and γ are the corresponding mean ionic activity coefficients, then by equation (41.3)

$$\frac{S}{S_0} = \frac{\gamma_0}{\gamma},$$

or

$$\log \frac{S}{S_0} = \log \gamma_0 - \log \gamma.$$

Introducing the values of γ_0 and γ as obtained from the Debye-Hückel limiting law equation (40.14), it follows, provided the ionic strength is low enough for the equation to be applicable, that

$$\log \frac{S}{S_0} = Az_+z_-(\sqrt{\mu} - \sqrt{\mu_0}), \qquad (41.4)$$

where μ_0 and μ are the ionic strengths of the solutions containing the sparingly soluble salt only, and of that to which other electrolytes have been added, respectively. Since μ_0 is a constant for a given saturating salt, it is apparent from equation (41.4) that the plot of $\log (S/S_0)$ against $\sqrt{\mu}$ should be a straight line of slope Az_+z_-, where z_+ and z_- are the (numerical) valences of the ions of the saturating salt.

A large number of solubility determinations have been made of sparingly soluble salts of various valence types, with values of z_+z_- from 1 to 9, in solvents of dielectric constant from 10.4, as in ethylene chloride, to 78.6, in water at 25° C. The results have been found in nearly all cases to be in very fair agreement with the requirements of the Debye-Hückel limiting law.[3] Appreciable discrepancies have been observed, however, when the saturating salt is of a high valence type, especially in the presence of added ions of high valence. The effect is particularly large when the high valence ions of the saturating and added salts have opposite signs; this suggests that there are electrostatic factors which have not been allowed for in the Debye-Hückel treatment.[4]

Problem: The solubility of silver iodate in pure water at 25° C is 1.771×10^{-4} mole liter^{-1}; calculate the solubility in the presence of 0.3252×10^{-2} mole liter^{-1} of potassium nitrate.

In this case, S_0 is 1.771×10^{-4} mole liter^{-1}, and since silver iodate is a uni-univalent salt, this is also the value of μ_0. In the potassium nitrate solution, μ is, as a first approximation, equal to $(0.3252 \times 10^{-2}) + (1.771 \times 10^{-4}) = 0.343 \times 10^{-2}$; hence, by equation (41.4)

$$\log \frac{S}{1.771 \times 10^{-4}} = 0.509[(0.343 \times 10^{-2})^{1/2} - (1.77 \times 10^{-4})^{1/2}],$$

taking A as 0.509 for water at 25° C, with z_+ and z_- both equal to unity. Solving this equation, the solubility S in the potassium nitrate solution is found to be 1.868×10^{-4} mole (AgIO$_3$) liter^{-1}. (The experimental value is 1.870×10^{-4} mole liter^{-1}.)

41d. Debye-Hückel Theory and the Osmotic Coefficient.—Multiplication of equation (39.49), i.e., $d \ln \gamma_\pm = (\phi - 1)d \ln m + d\phi$, by m throughout gives

$$\begin{aligned} md \ln \gamma_\pm &= (\phi - 1)dm + md\phi \\ &= d[m(\phi - 1)] \end{aligned} \qquad (41.5)$$

for the relationship between the osmotic coefficient ϕ of the solvent and the mean ionic activity coefficient $\gamma_\pm$ of the solute. Upon integration of equation (41.5) between the molality limits of zero and m the result is

$$\int_0^m md \ln \gamma_\pm = m(\phi - 1),$$

and hence,

$$\phi = 1 + \frac{1}{m} \int_0^m md \ln \gamma_\pm. \qquad (41.6)$$

---

[3] Cf., Brønsted and LaMer, *J. Am. Chem. Soc.*, **46**, 555 (1924); LaMer, King and Mason, *ibid.*, **49**, 363 (1927).

[4] LaMer and Cook, *J. Am. Chem. Soc.*, **51**, 2622 (1929); LaMer and Goldman, *ibid.*, **51**, 2632 (1929); Neuman, *ibid.*, **54**, 2195 (1932).

41d APPLICATIONS OF THE DEBYE-HÜCKEL EQUATION

In dilute solution the molality and molarity are approximately proportional, so that

$$\phi = 1 + \frac{1}{c}\int_0^c cd\ln\gamma_\pm, \qquad (41.7)$$

where c is the molarity corresponding to the molality m at which the osmotic coefficient ϕ is determined. Utilizing the Debye-Hückel limiting law expression for $\ln\gamma_\pm$, i.e., $2.303\log\gamma_\pm$, and noting that $\sqrt{\mu}$ involves $c^{1/2}$, it is readily found that

$$\phi = 1 + \tfrac{1}{3}\ln\gamma_\pm = 1 + \frac{2.303}{3}\log\gamma_\pm,$$

so that, by equation (40.14),

$$1 - \phi = \frac{2.303}{3}Az_+z_-\sqrt{\mu}. \qquad (41.8)$$

This is the limiting law for the osmotic coefficient as a function of the ionic strength of the solution. At high dilutions, when this equation may be expected to be applicable, $1 - \phi$ is equal to j (§ 39g), so that, by equation (39.20),

$$\phi = 1 - j = \frac{\theta}{\lambda\nu m}, \qquad (41.9)$$

where θ is the observed freezing point lowering of the solution, λ is the molal lowering for infinite dilution, ν is the number of ions produced by one molecule of the solute, and m is the molality of the solution. The Debye-Hückel theory thus provides a direct (limiting) relationship between the freezing point and the

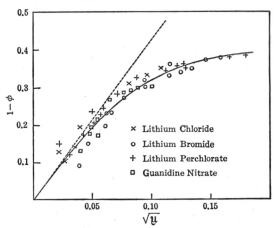

FIG. 31. Debye-Hückel limiting law and freezing point depression

ionic strength in dilute solutions. An illustration of the test of equation (41.8), with the values of ϕ being determined by means of (41.9) from freezing point measurements in cyclohexanol as solvent, dielectric constant 15.0 at 23.6° C, is provided by the results in Fig. 31.[5] The full curve is drawn through the values of $1 - \phi$ for a number of uni-univalent electrolytes, while the broken line shows the limiting slope required by equation (41.8).

[5] Schreiner and Frivold, Z. phys. Chem., 124, 1 (1926).

41e. The Debye-Hückel Theory in More Concentrated Solutions.—

Even a superficial comparison of the Debye-Hückel limiting law equation with the activity coefficient data in Table XXXIII (or Fig. 28) shows that the equation requires considerable modification if it is to be valid in solutions of appreciable concentration. According to equation (40.14), the activity coefficient should decrease steadily with increasing ionic strength, whereas the actual values pass through a minimum and then increase. Further, according to the Debye-Hückel treatment the activity coefficient of a particular electrolyte at a given ionic strength should depend only on its valence type, and not on the specific nature of the ions. The experimental results, however, show that this is true only for very dilute solutions. At appreciable concentrations, e.g., greater than 0.02 molal, the activity coefficients of HCl, NaCl and KCl differ at the same molality, indicating the existence of specific ionic effects.

It will be recalled that in § 40b the solution under consideration was supposed to be so dilute that $\kappa \mathring{a}$ was negligible in comparison with unity. If, however, the term $1 + \kappa \mathring{a}$ is retained in the denominator of the expression [equation (40.3)] for the electrical potential, it will also appear in the equations for the electrical free energy and the activity coefficient. In this event, equation (40.14) becomes

$$\log \gamma_\pm = - \frac{A z_+ z_- \sqrt{\mu}}{1 + \mathring{a}\kappa}, \tag{41.10}$$

where $\mathring{a}$ is the **mean distance of closest approach** of the ions in the solution; this varies with the nature of the ions, and thus a specific property of the electrolyte is introduced. It is seen from equation (40.10) that κ is equivalent to $B\mu^{1/2}$, where B involves universal constants together with the dielectric constant and the temperature. Hence equation (41.10) may be written as

$$\log \gamma_\pm = - \frac{A z_+ z_- \sqrt{\mu}}{1 + \mathring{a} B \sqrt{\mu}}, \tag{41.11}$$

for a given solvent and temperature; for water as solvent, B is 0.325×10^8 at $0°$ C and 0.329×10^8 at $25°$ C.

In order to utilize the modified form of the Debye-Hückel equation, it is necessary to know the value of $\mathring{a}$ for each electrolyte; although this should be of the order of the diameter of the ions, i.e., about 10^{-8} cm., there is, unfortunately, no independent method of assessing its exact value in any particular case. The general correctness of equation (41.11) may be tested by determining the value of $\mathring{a}$ required to make this equation harmonize with the experimental activity coefficients, and seeing if the results are reasonable.

By rearranging equation (41.11) it can be put in the form

$$- \frac{A z_+ z_- \sqrt{\mu}}{\log \gamma_\pm} = 1 + \mathring{a} B \sqrt{\mu}, \tag{41.12}$$

so that if the left-hand side of this equation, obtained from *experimental values* of $\gamma_\pm$, is plotted against $\sqrt{\mu}$, the result should be a straight line of slope $åB$. Since B is known, the mean distance of closest approach $å$ of the ions can thus be determined. It has been found that for solutions of moderate ionic strength, e.g., up to about 0.1 molal, this plot is, in fact, linear, and the values of $å$ for several electrolytes derived from the corresponding slopes are about 3 to 5×10^{-8} cm. (Table XXXIV).[6] It is seen, therefore, that the

TABLE XXXIV. MEAN DISTANCE OF CLOSEST APPROACH OF IONS

Electrolyte	$å$	Electrolyte	$å$
HCl	5.3×10^{-8} cm.	CaCl$_2$	5.2×10^{-8} cm.
NaCl	4.4	MgSO$_4$	3.4
KCl	4.1	K$_2$SO$_4$	3.0
CsNO$_3$	3.0	La$_2$(SO$_4$)$_3$	3.0

use of reasonable values for $å$ in equation (41.11) frequently makes it possible to represent the variation of the activity coefficient with the ionic strength up to appreciable concentrations. However, it must be pointed out that in some cases the experimental activity coefficients lead to values of $å$ which, although of the correct order of magnitude, viz., 10^{-8} cm., cannot represent the distance of closest approach of the ions. For example, with potassium nitrate it is necessary to postulate a value of 0.43×10^{-8} cm. for $å$ in order that equation (41.11) may reproduce the dependence of the activity coefficient upon the ionic strength in aqueous solution at 25° C. Salts of high valence type in particular behave abnormally in this respect. These discrepancies provide further support for the view that the Debye-Hückel treatment is not quite complete, although it provides a very close approximation in many instances.

41f. The Hückel and Brønsted Extensions.—It can be seen from equation (41.11) that as the ionic strength increases, the value of $\log \gamma_\pm$ should approach a constant limiting value equal to $-Az_+z_-/åB$. Actually, however, it is known that $\log \gamma_\pm$ passes through a minimum and then increases. It is evidently necessary to include an additional term in equation (41.11), and various lines of argument lead to the view that it should be proportional to the ionic strength. The inclusion of such a term, e.g., $C'\mu$, sometimes called the "salting-out" term [cf. equation (39.74)], leads to what is known as the **Hückel equation**, viz.,

$$\log \gamma_\pm = -\frac{Az_+z_-\sqrt{\mu}}{1 + åB\sqrt{\mu}} + C'\mu, \qquad (41.13)$$

where C' is a constant which must be derived from experimental data. An equation of this type, with properly selected values of the two constants $å$ and C', has been found to represent the behavior of many electrolytes up to concentrations as high as three or more molal. It should be noted that over this range the value of the activity coefficient depends on the particular standard state chosen for the activities; equation (41.13), as already seen, gives the rational activity coefficient

[6] For further data, see Harned and Owen, ref. 1, p. 381.

γ_N, and this must be corrected to yield γ_m or γ_c, as required, by means of equation (39.13).[7]

Upon dividing the numerator by the denominator of the fraction in equation (41.13) and neglecting all terms in the power series beyond that involving μ, the result is

$$\log \gamma_{\pm} = - Az_+z_-\sqrt{\mu} + åABz_+z_-\mu + C'\mu$$
$$= - Az_+z_-\sqrt{\mu} + C\mu, \qquad (41.14)$$

where C is a constant for the given electrolyte. This relationship is of the same form as an empirical equation proposed by J. N. Brønsted (1922), and hence is in general agreement with experiment.[8]

41g. Uses of the Debye-Hückel Equations.—The various equations derived from the Debye-Hückel theory have found a number of uses in thermodynamic problems. For example, the limiting law equation may be employed to calculate activity coefficients in dilute solutions when experimental values are not available. For a uni-univalent electrolyte the results obtained in a solvent of high dielectric constant, such as water, are reliable up to an ionic strength of about 0.01. By assuming an average value of about 3×10^{-8} cm. for $å$, the product $åB$ is approximately unity, since B is 0.33×10^8 in water at 25° C, and equation (41.11) takes the simple form

$$\log \gamma_{\pm} = - \frac{Az_+z_-\sqrt{\mu}}{1 + \sqrt{\mu}}, \qquad (41.15)$$

which may be used to give rough activity coefficients, especially of uni-univalent electrolytes, in aqueous solutions up to about 0.1 molal.

Problem: Calculate the approximate mean ionic activity coefficient of a 0.1 molal uni-univalent electrolyte in water at 25° C.

Since A is 0.509 for water at 25° C, and z_+ and z_- are both unity, equation (41.15) gives

$$\log \gamma_{\pm} = - \frac{0.509\sqrt{0.1}}{1 + \sqrt{0.1}} = - 0.122,$$
$$\gamma_{\pm} = 0.755.$$

(Some experimental values are 0.766 for NaOH, 0.769 for KCl, and 0.778 for NaCl.)

The Hückel equation (41.13), appropriately adjusted to give γ_m, has been frequently employed for the analytical representation of activity coefficient values as a function of the ionic strength of the solution, and various forms of the Debye-Hückel and Brønsted equations have been used for the purpose of extrapolating experimental results. Some instances of such applications have been given earlier (§§ 39h, 39i), and another is described in the next section.

[7] Hückel, *Physik. Z.*, **26**, 93 (1925); see also, Butler, *J. Phys. Chem.*, **33**, 1015 (1929); Scatchard, *Physik. Z.*, **33**, 22 (1932); for applications, see Harned and Owen, ref. 1.

[8] Brønsted, *J. Am. Chem. Soc.*, **44**, 938 (1922); Brønsted and LaMer, *ibid.*, **46**, 555 (1924).

41h. Debye-Hückel Equation and Equilibrium Constants.—When any electrolyte, such as an acid or a base, is dissolved in a suitable solvent an equilibrium is established between the free (solvated) ions and the undissociated portion of the solute. In the simple case of a uni-univalent electrolyte MA, such as a monobasic acid, the equilibrium may be represented by

$$MA \rightleftharpoons M^+ + A^-,$$

and the equilibrium constant is given by

$$K = \frac{a_{M^+} a_{A^-}}{a_{MA}}, \tag{41.16}$$

where the a terms are the activities of the indicated species, a_{MA} being that of the undissociated molecules. The constant K is referred to as the **ionization constant** or, better, as the **dissociation constant** of the substance MA in the given solvent at a definite temperature. Writing the activity of each species as the product of its molar concentration c_i and activity coefficient γ_i, in this case γ_c, equation (41.16) becomes

$$K = \frac{c_{M^+} c_{A^-}}{c_{MA}} \cdot \frac{\gamma_{M^+} \gamma_{A^-}}{\gamma_{MA}}$$

$$= K' \frac{\gamma_{M^+} \gamma_{A^-}}{\gamma_{MA}}. \tag{41.17}$$

It should be understood that c_{M^+} and c_{A^-} refer to the *actual concentrations* of the ions, allowing for possible incomplete dissociation; γ_+ and γ_- are then the "true" activity coefficients, e.g., as given by the Debye-Hückel theory. The equilibrium function K', which is equal to $c_{M^+} c_{A^-}/c_{MA}$, is in general not constant,* but it becomes equal to the true equilibrium (dissociation) constant when the activity coefficient factor is unity, i.e., at infinite dilution.

Before describing the method of extrapolating the results to infinite dilution, there are some qualitative aspects of equation (41.17) that are of interest. The following considerations are quite general, but they are particularly important in connection with the dissociation of weak electrolytes, e.g., weak acids and weak bases. If the ionic strength of the solution is increased, by the addition of "neutral" salts, the activity coefficient factor in equation (41.17) decreases. As a result, in order to maintain the equilibrium constant, K' must increase; this means that the dissociation of the weak acid or base will increase, thus increasing $c_{M^+} c_{A^-}$ while decreasing c_{MA}. At high salt concentrations, the activity coefficients will increase, after passing through a minimum, and hence K' must then decrease. Consequently, in the presence of large amounts of salts the extent of dissociation of a weak electrolyte will decrease, after having passed through a maximum value. This variation of the degree of dissociation with the ionic strength of the

---

* If the ionic strength of the medium is maintained constant, e.g., by the addition of "inert" electrolytes, the equilibrium function K' will remain virtually constant, although it will, in general, differ from the true dissociation constant.

medium is the basis of what is called the "secondary kinetic salt effect" in homogeneous catalysis. If a particular reaction is catalyzed by hydrogen ions, for example, the concentration of these ions, and hence their catalytic effect, can be changed merely by altering the salt concentration of the solution. The results, in dilute solution, can often be accounted for quantitatively by utilizing the Debye-Hückel limiting law to express the ionic activity coefficients.[9]

The particular application of the Debye-Hückel equation to be described here refers to the determination of the true equilibrium constant K from values of the equilibrium function K' at several ionic strengths; the necessary data for weak acids and bases can often be obtained from conductance measurements. If the solution of the electrolyte MA is sufficiently dilute for the limiting law to be applicable, it follows from equation (40.12), for the activity coefficient of a single ionic species, that

$$\log \gamma_{M^+} = \log \gamma_{A^-} = - A \sqrt{\mu}, \qquad (41.18)$$

since z_+ and z_- are both unity. Assuming no ions other than M^+ and A^-, derived from MA, to be present in the solution, it follows that

$$\mu = \tfrac{1}{2}\sum c_j z_j^2 = \tfrac{1}{2}[(c_{M^+} \times 1^2) + (c_{A^-} \times 1^2)]. \qquad (41.19)$$

If c is the stoichiometric, i.e., total, concentration of the solute, in moles per liter, and α is the degree of dissociation, the concentrations c_{M^+} and c_{A^-} of the respective ions are both equal to αc; hence equation (41.19) becomes

$$\mu = \tfrac{1}{2}(\alpha c + \alpha c) = \alpha c,$$

so that, by (41.18),

$$\log \gamma_{M^+} = \log \gamma_{A^-} = - A \sqrt{\alpha c}. \qquad (41.20)$$

In the region in which the Debye-Hückel limiting law is applicable, the activity coefficient of the molecules of undissociated MA is probably very close to unity, as may be inferred from the known variation of the activity coefficient of a neutral molecule in the presence of added electrolytes (cf. § 39m). It follows, therefore, upon taking logarithms of equation (41.17) that

$$\log K = \log K' + \log \gamma_{M^+} + \log \gamma_{A^-}$$
$$= \log K' - 2A \sqrt{\alpha c},$$

or

$$\log K' = \log K + 2A \sqrt{\alpha c}. \qquad (41.21)$$

It is seen from this equation that if the value of $\log K'$ in any solution is plotted against $\sqrt{\alpha c}$, where c is the concentration (molarity) and α is the degree of dissociation of the electrolyte MA in that solution, the result should be a straight line with intercept equal to $\log K$. For concentrations that are too high for the limiting law to hold, the values of $\log K'$ do not fall on the straight line, but they approach it with decreasing concentration

[9] See, for example, S. Glasstone, "Textbook of Physical Chemistry," 2nd ed., 1946, p. 1139; see also, pp. 1115–18.

(Fig. 32). This procedure has been utilized for the determination of dissociation constants of weak acids from conductance measurements.[10]

Another application of the Debye-Hückel equations, involving an equilibrium between polyvalent ions, is worthy of mention because of its interest in another connection (§ 45h). If a solution of ferric perchlorate, containing

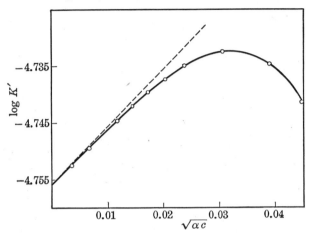

Fig. 32. Dissociation constant of acetic acid

free perchloric acid to repress hydrolysis, is shaken with finely divided silver, the equilibrium

$$Ag(s) + Fe(ClO_4)_3 \rightleftharpoons AgClO_4 + Fe(ClO_4)_2$$

is attained. Since the perchlorates are strong electrolytes, they may be regarded as completely dissociated, provided the concentrations are not too high; the equilibrium may thus be represented by

$$Ag(s) + Fe^{+++} \rightleftharpoons Ag^+ + Fe^{++},$$

where the ionic concentrations are equal to the total concentrations of the respective salts. The equilibrium constant is then given by

$$K = \frac{a_{Ag^+} a_{Fe^{++}}}{a_{Fe^{+++}}} = \frac{c_{Ag^+} c_{Fe^{++}}}{c_{Fe^{+++}}} \cdot \frac{\gamma_{Ag^+} \gamma_{Fe^{++}}}{\gamma_{Fe^{+++}}}$$

$$= K' \frac{\gamma_{Ag^+} \gamma_{Fe^{++}}}{\gamma_{Fe^{+++}}}, \qquad (41.22)$$

the activity of the metallic (solid) silver being taken as unity; the value of the function K' in any solution can be determined by analyzing the solution for ferrous, ferric and silver ions when equilibrium is established.

[10] MacInnes and Shedlovsky, *J. Am. Chem. Soc.*, **54**, 1429 (1932); MacInnes, *J. Frank. Inst.*, **225**, 661 (1938); D. A. MacInnes, "The Principles of Electrochemistry," 1939, Chapter 19.

Upon taking logarithms of equation (41.22), and utilizing the extended Debye-Hückel equation in the form of (41.14), adapted to a single ion, viz.,

$$\log \gamma_i = - A z_i^2 \sqrt{\mu} + C_i \mu,$$

it is found that

$$\begin{aligned}\log K &= \log K' + \log \gamma_{Ag^+} + \log \gamma_{Fe^{++}} - \log \gamma_{Fe^{+++}} \\ &= \log K' - (z_{Ag^+}^2 + z_{Fe^{++}}^2 - z_{Fe^{+++}}^2) A \sqrt{\mu} + C \mu \\ &= \log K' + 4A \sqrt{\mu} + C \mu, \end{aligned} \quad (41.23)$$

where A is equal to 0.509 for aqueous solutions at 25° C, and C is a composite constant made up of the various C_i's. By rearranging equation (41.23), it is seen that

$$\log K' + 4A \sqrt{\mu} = \log K - C \mu,$$

so that the plot of the left-hand side of this expression, which is available from the experimental determinations of K' in solutions of various ionic strengths, against the ionic strength should be a straight line; the intercept for μ equal to zero, i.e., $\log K$, should give the value of K, the true equilibrium constant. Even if the plot is not exactly linear, it should provide a convenient method for extrapolating the data to infinite dilution. From measurements made in solutions of ionic strengths varying from about 1.4 to 0.1, the equilibrium constant has been estimated as 0.53 at 25° C.[11] It appears from other evidence (§ 45h), however, that this result is somewhat too high, the correct value being probably about 0.35. The error is undoubtedly due to the somewhat lengthy extrapolation necessary to obtain K from the experimental data. If accurate measurements could be made in dilute solutions, the final result obtained by the procedure described above would be more reliable.

EXERCISES

1. Show that the expression for the mean ionic activity coefficient, derived from equation (40.11), can be written in the form

$$\log \gamma_\pm = - \frac{1.824 \times 10^6}{\nu(DT)^{3/2}} \sum \nu_i z_i^2 \sqrt{\mu}.$$

2. Plot the logarithms of the activity coefficients of silver iodate obtained in Exercise 14, Chapter XVI against the square roots of the corresponding (total) ionic strengths, and hence test the Debye-Hückel limiting law.

3. Calculate the effective radius of the ionic atmosphere $(1/\kappa)$ in a 0.1 molal aqueous solution of potassium sulfate at 25° C. The dielectric constant of water at 25° C is 78.54.

4. Show that for electrolytes of the same valence type, the Debye-Hückel-Brønsted equation (41.14), as applied to moderately concentrated solutions, leads to the result

$$\log (\gamma / \gamma_R) = kc,$$

[11] Schumb and Sweetser, *J. Am. Chem. Soc.*, **57**, 871 (1935).

where γ/γ_R is the ratio of the mean ionic activity coefficient of any electrolyte to that of a reference substance of the same valence type at the same concentration c; k is a constant for the given electrolyte. The plot of log (γ/γ_R) against c should thus be a straight line [Åkerlöf and Thomas, *J. Am. Chem. Soc.*, **56**, 593 (1934)]. Test this by the data for hydrochloric acid and potassium chloride in Table XXXIII at molalities of 0.1 and higher. (Molalities may be used to represent the concentration.)

5. Evaluate the Debye-Hückel constants A and B for methanol as solvent at 25° C, its dielectric constant being 31.5. Hence, write an expression for the mean ionic activity coefficient of zinc chloride in moderately dilute methanol solutions as a function of the ionic strengths μ and μ_m at 25° C. (The density of methanol is 0.790.)

6. Compare the mean ionic activity coefficient of a 0.1 molal solution of (i) a uni-univalent, (ii) a bi-bivalent, electrolyte in water and methanol as solvents respectively, at 25° C. The mean ionic diameter $å$ may be taken as 3Å in each case.

7. Utilize the mean ionic activity coefficients for solutions up to 0.05 molal in Table XXXIII to determine by a graphical procedure the mean ionic diameter $å$ of hydrochloric acid.

8. Utilize the following data for the solubility of the uni-bivalent salt $[Co(NH_3)_4(C_2O_4)]_2S_2O_6$ in the presence of various amounts of added electrolytes [Brønsted and LaMer, *J. Am. Chem. Soc.*, **45**, 555 (1924)] to test the Debye-Hückel limiting law at 25° C. The solubility in pure water is 1.545×10^{-4} mole liter^{-1}.

Added Salt (mole liter^{-1})	Solubility (mole liter^{-1})	Added Salt (mole liter^{-1})	Solubility (mole liter^{-1})
0.001 NaCl	1.597×10^{-4}	0.01 KNO$_3$	1.880×10^{-4}
0.002 NaCl	1.645	0.005 MgSO$_4$	1.866
0.005 NaCl	1.737	0.010 BaCl$_2$	2.032

Plot log (S/S_0) against $\sqrt{\mu} - \sqrt{\mu_0}$; determine the slope and compare with the calculated value for a uni-bivalent electrolyte.

9. Show that if $\gamma_{1,1}$ is the mean ionic activity coefficient of a uni-univalent electrolyte, and γ is that of any other electrolyte with ions of valence z_+ and z_-, at the same ionic strength, then the Debye-Hückel limiting law requires that $\gamma = \gamma_{1,1}^{z_+ z_-}$. What general conclusions concerning the effect of valence on the activity coefficient can be drawn from this result?

10. Using the values of A and B for water given in the text, plot $-\log \gamma_\pm$ for a uni-univalent electrolyte against $\sqrt{\mu}$ for ionic strengths 0.01, 0.1, 0.5 and 1.0, taking $å$ equal to (i) zero, (ii) 2Å, (iii) 4Å, (iv) 8Å. Hence, draw general conclusions as to the effect of increasing the ionic size. Investigate qualitatively the result of increasing the valence of the ions.

11. By means of the Debye-Hückel theory, calculate the activity coefficients of silver iodate in the various potassium nitrate solutions mentioned in Exercise 14, Chapter XVI. Compare the values with those derived from the observed solubilities.

12. Account for the following observations: (i) the addition of increasing amounts of an "inert" electrolyte causes the solubility of a sparingly soluble salt first to increase to a maximum and then to decrease steadily; (ii) the addition of a salt with an ion in common with the sparingly soluble salt causes the solubility of

the latter to decrease and then to increase. (Complex ion formation should be disregarded.)

13. Justify, by means of the Debye-Hückel limiting law, the statement in the footnote in § 41h.

14. The freezing point depressions (θ) of dilute aqueous solutions of potassium chloride are given below:

c (mole liter^{-1})	θ	c (mole liter^{-1})	θ
0.001596	0.00586°	0.02867	0.1011°
0.002143	0.00784°	0.04861	0.1702°
0.008523	0.03063°	0.1047	0.3606°

[Lange, Z. phys. Chem., **A186**, 147 (1934)]. Plot $1 - \phi$ against $\sqrt{\mu}$ and determine the limiting slope; compare the value with that given by the Debye-Hückel theory. (The constant A is 0.488 for water at 0° C.)

15. The dielectric constant of cyclohexanol is 15.0 and its freezing point is 23.6° C. Calculate the limiting slope of $1 - \phi$ against $\sqrt{\mu}$, according to the Debye-Hückel theory, and compare the result with that in Fig. 31.

16. The following data were derived from conductance measurements of solutions of picric acid in methanol at 25° C [D. A. MacInnes, "The Principles of Electrochemistry," 1939, p. 362]:

c (mole liter^{-1})	α	K'
1.00×10^{-1}	0.05871	3.662×10^{-4}
5.00×10^{-2}	0.07823	3.320
2.50×10^{-2}	0.1037	2.996
1.25×10^{-2}	0.1379	2.758
6.25×10^{-3}	0.1820	2.532
3.125×10^{-3}	0.2408	2.384

Use the Debye-Hückel equation (for A, see Exercise 5), in conjunction with a graphical procedure, to evaluate the true dissociation constant from the K' values. (The result is 1.845×10^{-4}.)

17. The solubility of barium sulfate in pure water at 25° C is 9.57×10^{-6} mole liter^{-1}. Estimate the solubility in the presence of (i) 0.01 molar sodium chloride, (ii) 0.01 molar sodium sulfate. Use the Debye-Hückel equation (41.11), assuming a value of 3Å for the mean distance of closest approach of the ions.

18. The mean ionic activity coefficients of hydrochloric acid in water and in dioxane-water mixtures, at an ionic strength of 0.001, together with the corresponding temperatures and dielectric constants are given below. Use the data to test the Debye-Hückel limiting law, with particular reference to the effect of temperature and dielectric constant.

$t°$ C	D	$\gamma_\pm$
0°	88.15	0.967
25°	9.53	0.398
50°	15.37	0.675

[Harned and Owen, ref. 1].

CHAPTER XVIII

PARTIAL MOLAR PROPERTIES

42. Determination of Partial Molar Properties

42a. Thermodynamic Significance.—The importance of partial molar properties will be evident from the many uses of these quantities that have been made in the preceding sections of this book. In the present chapter it is proposed to consider certain of these properties, such as the partial molar volume and heat content, and related matters, in somewhat greater detail. It may be pointed out that the treatment of activities given in Chapters XV and XVI is essentially a convenient method for studying partial molar free energies or, more correctly, partial molar free energies relative to the value in an arbitrary standard state. This may be seen by writing the equation $\mu_i = \mu_i^0 + RT \ln a_i$ in the form

$$RT \ln a_i = \mu_i - \mu_i^0 = \bar{F}_i - \bar{F}_i^0,$$

since the chemical potential μ is equivalent to the partial molar free energy $\bar{F}$. The determination of the activity a_i is thus equivalent to the evaluation of the partial molar free energy $\bar{F}_i$ of the constituent i relative to that ($\bar{F}_i^0$) in the chosen standard state.

One significant aspect of partial molar properties is that represented by equation (26.6). If $\bar{G}_i$ is the partial molar value of any property in a system containing n_i moles of the constituent i, then the total value G for the system is given by the sum of all the $n_i \bar{G}_i$ terms. For a system consisting of a single, pure substance the partial molar property is identical with the ordinary molar value. This result has often been used in the earlier treatment.

Another aspect of partial molar volumes and heat contents, in particular, arises from the thermodynamic requirement that for an ideal gas mixture or for an ideal liquid solution, as defined for example in § 30a and § 34a, respectively, there is no change of volume or of heat content upon mixing the components. This means that the partial molar volume and heat content of each substance in the mixture are equal to the respective molar values for the pure constituents. Any deviation of the partial molar quantity from the molar value then gives an indication of departure from ideal behavior; this information is useful in connection with the study of solutions.

42b. Apparent Molar Properties.—Although not of direct thermodynamic significance, the **apparent molar property** [1] is related to the corresponding partial molar property, as will be seen below. The importance of

[1] Cf. G. N. Lewis and M. Randall, "Thermodynamics and the Free Energy of Chemical Substances," 1923, p. 35; *J. Am. Chem. Soc.*, 43, 233 (1921).

the apparent molar properties lies in the fact that they are usually capable of direct experimental determination in cases when the partial molar properties are not. For this reason, apparent molar quantities are frequently employed, particularly for the study of the thermodynamic properties of systems of two components. In the subsequent treatment it will be assumed throughout that the systems are of this kind, the two constituents being represented by the subscripts 1 and 2; when a distinction is possible, the solvent will be regarded as constituent 1 and the solute as constituent 2.

If G is the value of a particular property for a mixture consisting of n_1 moles of constituent 1 and n_2 moles of constituent 2, and G_1 is the value of the property *per mole of pure constituent 1*, then the apparent molar value, represented by ϕ_G, of the given property for the component 2 is given by

$$\phi_G = \frac{G - n_1 G_1}{n_2}. \tag{42.1}$$

In order to indicate the fact that the value of ϕ_G as given by equation (42.1) applies to the constituent 2, i.e., the solute, a subscript 2 is sometimes included. However, this is usually omitted, for in the great majority of cases it is understood that *the apparent molar property refers to the solute*. It is seen from equation (42.1) that ϕ_G is the apparent contribution of 1 mole of the component 2 to the property G of the mixture. If the particular property were strictly additive for the two components, e.g., volume and heat content for ideal gas and liquid solutions, the value of ϕ_G would be equal to the actual molar contribution, and hence also to the partial molar value. For nonideal systems, however, the quantities are all different.

42c. Determination of Partial Molar Quantities: I. Direct Method.[2]—In view of the definition of the partial molar property $\bar{G}_i$ as

$$\bar{G}_i = \left(\frac{\partial G}{\partial n_i}\right)_{T, P, n_1, \ldots}, \tag{42.2}$$

an obvious method for its determination is to plot the value of the extensive property G, at constant temperature and pressure, for various mixtures of the two components against the number of moles, e.g., n_2, of one of them, the value of n_1 being kept constant. The slope of the curve at any particular composition, which may be determined by drawing a tangent to the curve, gives the value of $\bar{G}_2$ at that composition.[3] Since the molality of a solution represents the number of moles of solute associated with a constant mass, and hence a constant number of moles, of solvent, the plot of the property G against the molality can be used for the evaluation of the partial molar property of the solute. Once $\bar{G}_2$ at any composition has been determined, the corresponding value of $\bar{G}_1$ is readily derived by means of the relationship $G = n_1 \bar{G}_1 + n_2 \bar{G}_2$.

[2] For methods of determining partial molar properties, see ref. 1; see also, Young and Vogel, *J. Am. Chem. Soc.*, **54**, 3025 (1932).

[3] See Latshaw, *J. Am. Chem. Soc.*, **47**, 793 (1925); Gucker and Brennen, *ibid.*, **54**, 886 (1932).

In view of the difficulty of determining the exact slope of the curve at all points, it is preferable to use an analytical procedure instead of the graphical one just described. The property G is then expressed as a function of the number of moles of one component, e.g., the molality, associated with a constant amount of the other component. Upon differentiation with respect to n, i.e., the molality, an expression for the partial molar property is obtained.

Problem: At concentrations exceeding 0.25 molal, the volume of a NaCl solution, per 1000 g. of water, at 25° C is given by

$$V = 1002.9 + 16.40m + 2.5m^2 - 1.2m^3 \text{ ml.}$$

The molar volume of pure water at 25° C is 18.069 ml. mole^{-1}. Derive general expressions for the partial molar volume and the apparent molar volume of sodium chloride in aqueous solutions, and compare the values for a 1 molal solution.

The partial molar volume is given by

$$\bar{V}_2 = \left(\frac{\partial V}{\partial n_2}\right)_{T,P,n_1} = \left(\frac{\partial V}{\partial m}\right)_{T,P},$$

and hence, utilizing the expression for V,

$$\bar{V}_2 = 16.40 + 5.0m - 3.6m^2 \text{ ml. mole}^{-1}.$$

In a 1 molal solution, therefore, $\bar{V}_2$ is 17.8 ml. mole^{-1}.

The apparent molar volume ϕ_V is given by equation (42.1) as

$$\phi_V = \frac{V - n_1 V_1}{n_2} = \frac{V - (1000/18.016)V_1}{m},$$

since n_1 is equal to $1000/M_1$, where M_1, equal to 18.016, is the molecular weight of the solvent, i.e., water, and n_2 is m, the molality. Consequently, utilizing the expression for V and taking V_1 as 18.069, it follows that

$$\phi_V = 16.40 + 2.5m - 1.2m^2 \text{ ml. mole}^{-1},$$

and hence in a 1 molal solution the apparent molar volume is 17.7 ml. mole^{-1}.

II. From Apparent Molar Properties.—A method that is often more convenient and accurate than that described above, makes use of the apparent molar property. It is seen from equation (42.1) that

$$G - n_1 G_1 = n_2 \phi_G,$$

and if n_1 is maintained constant, so that $n_1 G_1$ is constant, differentiation with respect to n_2, *constant temperature and pressure being understood*, gives

$$\bar{G}_2 = \left(\frac{\partial G}{\partial n_2}\right)_{n_1} = n_2 \left(\frac{\partial \phi_G}{\partial n_2}\right)_{n_1} + \phi_G \quad (42.3)$$

or

$$\bar{G}_2 = \left(\frac{\partial \phi_G}{\partial \ln n_2}\right)_{n_1} + \phi_G. \quad (42.4)$$

Since the molality m is equivalent to n_2, with n_1 constant, equations (42.3) and (42.4) may be written as

$$\bar{G}_2 = m \frac{d\phi_G}{dm} + \phi_G \tag{42.5}$$

and

$$\bar{G}_2 = \frac{d\phi_G}{d \ln m} + \phi_G, \tag{42.6}$$

respectively. If the apparent molar property ϕ_G is determined for various values of n_2, with n_1 constant, or at various molalities, the partial molar property $\bar{G}_2$ can be calculated from the slope, at any given composition, of the plot of ϕ_G against n_2 (or m) or against $\log n_2$ (or $\log m$). The method based on the use of equation (42.3) or (42.5) is usually more accurate than that involving the logarithmic plot, since it does not give undue importance to results obtained in dilute solutions. An analytical method can, of course, be used in place of the graphical procedure if ϕ_G can be expressed as a function of n_2 or of the molality.

For use in a later connection, an alternative form of equation (42.5) is required and it will be derived here. The right-hand side of this equation is equivalent to $d(m\phi_G)/dm$, that is,

$$\frac{d(m\phi_G)}{dm} = \bar{G}_2,$$

and upon integration, m varying between the limits of zero and m, and $m\phi_G$ between zero and $m\phi_G$, it is found that

$$m\phi_G = \int_0^m \bar{G}_2 dm,$$

$$\phi_G = \frac{1}{m} \int_0^m \bar{G}_2 dm.$$

For dilute solutions, the molality is proportional to the molar concentration c, and hence it is permissible to put this result in the form

$$\phi_G = \frac{1}{c} \int_0^c \bar{G}_2 dc. \tag{42.7}$$

III. Method of Intercepts.—The method of intercepts is useful in many instances, especially as it gives simultaneously the partial molar properties of both constituents of a binary mixture of any composition. Let G represent the mean value of a particular extensive property *per mole* of mixture, so that the observed value of the property G for the system is given by

$$G = (n_1 + n_2)\text{G}.$$

Differentiation with respect to n_2, at constant n_1, *constant temperature and pressure being understood*, then gives

$$\bar{G}_2 = \left(\frac{\partial G}{\partial n_2}\right)_{n_1} = G + (n_1 + n_2)\left(\frac{\partial G}{\partial n_2}\right)_{n_1}. \quad (42.8)$$

The mole fraction N_1 of the component 1 is defined by

$$N_1 = \frac{n_1}{n_1 + n_2},$$

and differentiation of this expression, keeping n_1 constant, gives

$$dN_1 = -\frac{n_1 dn_2}{(n_1 + n_2)^2} = -\frac{N_1 dn_2}{n_1 + n_2},$$

that is,

$$\frac{n_1 + n_2}{dn_2} = -\frac{N_1}{dN_1}.$$

Utilizing this result,

$$(n_1 + n_2)\left(\frac{\partial G}{\partial n_2}\right)_{n_1} = -N_1 \left(\frac{\partial G}{\partial N_1}\right)_{n_1},$$

and upon insertion in equation (42.8), it follows that

$$\bar{G}_2 = G - N_1 \left(\frac{\partial G}{\partial N_1}\right)_{n_1}.$$

The value of dG/dN_1 is independent of the method whereby N_1 is varied, and hence it is unnecessary to postulate n_1 constant; consequently, it is possible to write

$$\bar{G}_2 = G - N_1 \frac{dG}{dN_1}. \quad (42.9)$$

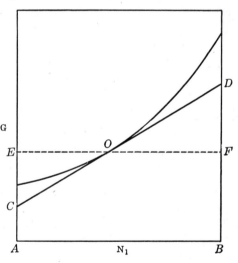

Fig. 33. Partial molar quantities by method of intercepts

The values of the mean molar property G for mixtures of various compositions are plotted against the mole fraction N_1, as shown in Fig. 33. Let O be the point at which the partial molar property is to be determined; at O draw the tangent CD and the horizontal line EF, parallel to the base line AB. The slope of CD is dG/dN_1, and so CE is equal to $N_1(dG/dN_1)$ at O. Since AE is the value of G at that point, it is evident from equation (42.9) that the distance AC gives the partial molar property $\bar{G}_2$. In an exactly similar manner it can be shown that BD is equal to $\bar{G}_1$ for the mixture whose composition is represented by the point O; hence, both $\bar{G}_1$ and $\bar{G}_2$ are obtained at the same time.

IV. General Methods.—In the methods described above for the determination of partial molar quantities, it has been tacitly assumed that the property G is one which is capable of experimental determination. Such is the case, for example, if G represents the volume or the heat capacity. However, if the property under consideration is the heat content then, like the free energy, it cannot be determined directly. In cases of this kind modified methods, which involve measurements of changes in the property, rather than of the property itself, can be used. It should be pointed out that the procedures are quite general and they are frequently adopted for the study of properties susceptible of direct measurement, as well as of those which are not.

If n_1 moles of constituent 1 and n_2 moles of constituent 2 are mixed, the observable change ΔG in the property G is given by

$$\Delta G = G - (n_1 G_1 + n_2 G_2), \qquad (42.10)$$

where G is the value of the property for the mixture, and G_1 and G_2 are the molar properties for the pure substances. The change ΔG upon mixing can be determined experimentally, even though G cannot. For a series of mixtures, in which n_1 is maintained constant while n_2 is varied, the change in the property per mole of constituent 2 is $\Delta G/n_2$, and equation (42.10) may be written as

$$n_2(\Delta G/n_2) = G - (n_1 G_1 + n_2 G_2).$$

Upon differentiating with respect to n_2, with n_1 constant, the constancy of temperature and pressure being understood, the result is

$$\bar{G}_2 - G_2 = n_2 \left[\frac{\partial (\Delta G/n_2)}{\partial n_2} \right]_{n_1} + \frac{\Delta G}{n_2}. \qquad (42.11)$$

This equation is seen to be similar to (42.3), and the values of $\bar{G}_2 - G_2$ can be derived in an analogous graphical manner from the plot of $\Delta G/n_2$ against n_2, with n_1 constant. An equivalent relationship can be employed to obtain $\bar{G}_1 - G_1$.

An alternative procedure is to divide both sides of equation (42.10) by $n_1 + n_2$; thus,

$$\Delta \text{g} = \text{g} - (\text{n}_1 G_1 + \text{n}_2 G_2), \qquad (42.12)$$

where Δg is the change in the property upon mixing the constituents, *per mole of mixture*. As before, g is the value of the property itself per mole, and n_1 and n_2 are the mole fractions. By differentiating equation (42.12) with respect to n_1, with temperature and pressure constant, and bearing in mind that n_2 is equal to $1 - \text{n}_1$, it is possible to derive an expression for $\text{n}_1 (d\text{g}/d\text{n}_1)$ which, upon insertion into equation (42.9), gives

$$\bar{G}_2 - G_2 = \Delta \text{g} - \text{n}_1 \frac{d(\Delta \text{g})}{d\text{n}_1}. \qquad (42.13)$$

This expression is seen to be similar to equation (42.9), and hence the method of intercepts can be applied to the plot of Δg against n_1, for various mixtures, in order to obtain $\bar{G}_2 - G_2$ and $\bar{G}_1 - G_1$ for any composition. If G_1 and G_2 are known, as would be the case if G represented the volume, then $\bar{G}_1$ and $\bar{G}_2$ can also be determined.

When N_1 is zero, i.e., pure 2, and N_2 is unity, i.e., pure 1, ΔG must be zero; the plot of ΔG against N_1 is thus of the form shown in Fig. 34 if ΔG is positive, e.g., heat of mixing of ethyl iodide and ethyl acetate. If ΔG is negative, the curve is similar in form but it lies on the other side of the N_1-axis. Another possibility is that ΔG is positive when one constituent is present in excess, and negative when the other is in larger proportion.[4]

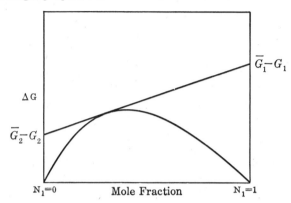

FIG. 34. Generalized method of intercepts

It is frequently more convenient to determine the partial specific properties, defined in terms of grams instead of moles, of the constituents of a solution, and then to multiply the results by the respective molecular weights to yield the partial molar properties. Any of the methods described above may be adapted for this purpose. The value of the property G or of ΔG per mole is replaced by the value per gram, and N or n, the mole fraction or number of moles, is replaced by the corresponding gram fraction or number of grams, respectively.

43. Partial Molar Volumes

43a. Partial Molar Volumes from Density Measurements.—A convenient form of method I for determining partial molar volumes in liquid solutions is based on density measurements at a number of different concentrations. Consider a solution containing n_1 moles of solvent of molecular weight M_1 to 1 mole of solute of molecular weight M_2 in a total volume of V liters; then its density ρ is given by

$$\rho = \frac{n_1 M_1 + M_2}{1000 V},$$

so that

$$n_1 = \frac{1000 \rho V - M_2}{M_1}.$$

[4] Sosnick, *J. Am. Chem. Soc.*, **49**, 2255 (1927); R. F. Newton, private communication.

Upon differentiating, at constant temperature and pressure, the result is

$$dn_1 = \frac{1000(\rho dV + V d\rho)}{M_1},$$

and hence

$$\frac{dV}{dn_1} = \frac{M_1}{1000[\rho + V(d\rho/dV)]}. \tag{43.1}$$

Since the value of dV/dn_1 refers to a constant amount, viz., 1 mole, of the constituent 2, it is equivalent to $(\partial V/\partial n_1)_{T,P,n_2}$, and hence to $\bar{V}_1$, so that equation (43.1) gives the value of the partial molar volume, in liters, of constituent 1. In order to use this expression, V on the right-hand side is replaced by $1/c$, where c is the concentration in moles per liter, so that

$$\bar{V}_1 = \frac{M_1}{1000[\rho - c(d\rho/dc)]}, \tag{43.2}$$

since $V(d\rho/dV)$ is equal to $-c(d\rho/dc)$. To evaluate $\bar{V}_1$ it is necessary, therefore, to determine the density of the solution at various concentrations, and to plot the results. The slope at any point gives the required value of $d\rho/dc$, and this, together with the density, can then be inserted in equation (43.2), so as to obtain the partial molar volume of the constituent 1. An expression for $\bar{V}_2$, similar to equation (43.2), and involving $d\rho/dc$, can be derived in an analogous manner.

43b. Liquid Mixtures.—When two liquids which form an ideal solution are mixed there will be no change of volume, and the partial molar volume of each constituent will be equal to its ordinary molar volume, as indicated earlier. If a solution exhibits positive deviations from Raoult's law (§ 35a), there is usually an increase of volume upon mixing, and the partial molar volume of each substance is greater than its molar volume in the pure state. This may be attributed to the mean attractive force between the molecules in the mixture being smaller than for the constituents separately. In fact, the same underlying cause is responsible for the increased vapor pressure, i.e., positive deviations, and the volume change. For a system which shows negative deviation from Raoult's law, the attractive force in the mixture is greater because of the net attraction of the two molecular species; this results in a lowering of the vapor pressure and a decrease of volume. Negative deviations from ideal behavior are thus, in general, associated with partial molar volumes that are less than the respective molar

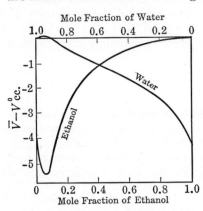

FIG. 35. Partial molar volumes in water-ethanol mixtures

volumes. An illustration of a system of this type is provided by the results in Fig. 35, which give the partial molar volumes of water and ethanol in mixtures of these two liquids.[5] The results for the alcohol are of particular interest, since they show a steady decrease, followed by an increase in dilute solution, when the mole fraction is less than 0.1. The interpretation of these facts is of considerable interest although it lies outside the immediate scope of thermodynamics.

43c. Solutions of Electrolytes.—Because of the thermodynamic relationship between the partial molar volume and the partial molar free energy (chemical potential) it is possible to develop a theoretical treatment of partial molar volumes of electrolytes by means of the Debye-Hückel concept.[6] According to the arguments in Chapter XVII, the mean activity coefficient given by the Debye-Hückel limiting law, e.g., equation (40.14), represented by $\gamma_{D.H.}$, is determined solely by the electrostatic forces between the ions in the solution at constant temperature and pressure. Strictly speaking, the complete mean activity coefficient $\gamma_{\pm}$ should contain another factor, designated $\gamma_{n.e.}$, to allow for nonelectrical forces in the solution; this factor may be close to unity, but it will not necessarily be independent of temperature and pressure. Consequently, it is possible to write, in general, $\gamma_{\pm} = \gamma_{D.H.} \times \gamma_{n.e.}$ or

$$\ln \gamma_{\pm} = \ln \gamma_{D.H.} + \ln \gamma_{n.e.}. \tag{43.3}$$

By equation (39.2),

$$\mu_2 - \mu_2^0 = RT \ln a_2,$$

where μ_2^0 is the chemical potential of the electrolyte in its standard state, which is defined in terms of a reference state at 1 atm. pressure, and is consequently independent of the pressure. Replacing a_2 by $(m_{\pm}\gamma_{\pm})^\nu$, in accordance with the treatment in § 39a, it is seen that

$$\mu_2 - \mu_2^0 = \nu RT \ln m_{\pm} + \nu RT \ln \gamma_{\pm}, \tag{43.4}$$

ν being the total number of ions produced by the dissociation of one molecule of solute. If equation (43.3) is substituted into (43.4), and the resulting expression differentiated with respect to pressure, at constant temperature and composition, recalling that, by equation (26.26), $(\partial \mu_2/\partial P)_{T,N}$ is equal to $\bar{V}_2$, whereas μ_2^0 is independent of pressure, it is found that

$$\nu RT \left(\frac{\partial \ln \gamma_{D.H.}}{\partial P} \right)_{T,N} = \bar{V}_2 - \nu RT \left(\frac{\partial \ln \gamma_{n.e.}}{\partial P} \right)_{T,N}. \tag{43.5}$$

As the solution approaches infinite dilution and the electrostatic forces become negligible, $\ln \gamma_{D.H.}$ approaches zero irrespective of the pressure; that is to say, as m tends to zero, the left-hand side of equation (43.5) becomes zero, and hence

$$\nu RT \left(\frac{\partial \ln \gamma_{n.e.}}{\partial P} \right)_{T,N} = \bar{V}_2^* \quad \text{as} \quad m \to 0,$$

where $\bar{V}_2^*$ is the partial molar volume of the solute at infinite dilution. Although this result is strictly applicable only at infinite dilution, it may be assumed to

---

[5] Adapted from Lewis and Randall, ref. 1, p. 40.
[6] Redlich and Rosenfeld, Z. phys. Chem., **A155**, 65 (1931); R. F. Newton, private communication.

hold in the same region as does the Debye-Hückel limiting law, and hence substitution in equation (43.5) gives

$$\bar{V}_2 - \bar{V}_2^* = \nu RT \left(\frac{\partial \ln \gamma_{\text{D.H.}}}{\partial P} \right)_{T,N} = 2.303 \nu RT \left(\frac{\partial \log \gamma_{\text{D.H.}}}{\partial P} \right)_{T,N}. \quad (43.6)$$

The Debye-Hückel equation (40.11) for $\log \gamma_i$, for a single ion, may be combined with (40.13) to give an alternative expression for the mean ionic activity coefficient represented here by the symbol $\gamma_{\text{D.H.}}$; thus (cf. Exercise 1, Chapter XVII),

$$\log \gamma_{\text{D.H.}} = - \frac{1.824 \times 10^6}{\nu(DT)^{3/2}} \sum \nu_i z_i^2 \sqrt{\mu}, \quad (43.7)$$

where ν_i is the number of ions of valence z_i formed from one molecule of the solute in the solution. For the present purpose, that is, the evaluation of the partial molar volume of the solute, equation (43.7) is restricted to a solution containing a *single* completely dissociated electrolyte. In this case, the concentration c_i of any ion is equal to $\nu_i c$, where c is the concentration of the solute in moles per liter, and hence

$$\mu = \tfrac{1}{2} \sum c_i z_i^2 = \tfrac{1}{2} c \sum \nu_i z_i^2,$$

so that equation (43.7) becomes

$$\log \gamma_{\text{D.H.}} = - \frac{1.824 \times 10^6}{\sqrt{2} \nu (DT)^{3/2}} (\sum \nu_i z_i^2)^{3/2} \sqrt{c}. \quad (43.8)$$

This result may now be inserted into equation (43.6) and the differentiation carried out, at constant temperature and composition; taking R as 8.314×10^7 ergs deg.$^{-1}$ mole^{-1}, it is found that

$$\bar{V}_2 - \bar{V}_2^* = \frac{2.469 \times 10^{14}}{D^{3/2} T^{1/2}} (\sum \nu_i z_i^2)^{3/2} \frac{1}{2} \left(\frac{3}{D} \cdot \frac{\partial D}{\partial P} + \frac{1}{V} \cdot \frac{\partial V}{\partial P} \right)_T \sqrt{c}. \quad (43.9)$$

Attention may be called to the presence of the term $(\partial V/\partial P)_T/V$; this arises because the concentration is inversely proportional to the volume, and the latter varies with the pressure, at constant temperature. Actually, $-(\partial V/\partial P)_T/V$ is the compressibility of the solution, but since equation (43.9) can apply only to dilute solutions, it may be regarded as the compressibility of the solvent, and hence is known. The dependence of dielectric constant upon pressure, which is required for the value of $(\partial D/\partial P)_T$, has not been studied to any extent except for water as solvent, and even in this case the results are not too reliable. Taking the compressibility of water as 45×10^{-12} (dyne cm.$^{-2}$)$^{-1}$ and $(\partial D/\partial P)_T/D$ as 47×10^{-12} (dyne cm.$^{-2}$)$^{-1}$ at 25 °C, with D equal to 78.54, equation (43.9) reduces to

$$\bar{V}_2 - \bar{V}_2^* = 0.99 (\sum \nu_i z_i^2)^{3/2} \sqrt{c} \text{ cc. mole}^{-1}, \quad (43.10)$$

for dilute aqueous solutions at 25° C; hence,

$$\bar{V}_2 = \bar{V}_2^* + 0.99 (\sum \nu_i z_i^2)^{3/2} \sqrt{c} \text{ cc. mole}^{-1}. \quad (43.11)$$

Problem: Derive an expression for the partial molar volume of BaCl$_2$ in dilute aqueous solution at 25° C.

For BaCl$_2$, $\nu_+ = 1$, $z_+ = 2$, and $\nu_- = 2$, $z_- = 1$; hence $\sum \nu_i z_i^2 = [(1 \times 2^2) + (2 \times 1^2)] = 6$, so that $(\sum \nu_i z_i^2)^{3/2} = 6^{3/2} = 14.7$. From equation (43.11),

therefore,
$$\bar{V}_2 = \bar{V}_2^* + 0.99 \times 14.7\sqrt{c}$$
$$= \bar{V}_2^* + 14.6\sqrt{c} \text{ cc. mole}^{-1}.$$

The accuracy of equation (43.11) may be tested by plotting the experimental values of $\bar{V}_2$, the partial molar volume of the solute, at various concentrations against $\sqrt{c}$; the results should approach a straight line of slope 0.99 $(\sum \nu_i z_i^2)^{3/2}$ in dilute aqueous solution at 25° C. A simpler method, however, is to derive an analogous relationship involving the apparent molar volume, for this can be obtained more readily from the experimental data. The general equation (42.7), for dilute solutions, in the present case takes the form

$$\phi_V = \frac{1}{c}\int_0^c \bar{V}_2 dc. \tag{43.12}$$

If the expression for $\bar{V}_2$ given by equation (43.11) is now introduced, it follows directly that
$$\phi_V = \phi_V^* + \tfrac{2}{3} \times 0.99(\sum \nu_i z_i^2)^{3/2}\sqrt{c}$$
$$= \phi_V^* + 0.66(\sum \nu_i z_i^2)^{3/2}\sqrt{c} \text{ cc. mole}^{-1}. \tag{43.13}$$

The verification of this equation requires accurate measurements of density of aqueous solutions at high dilutions, and so not many data are available for the purpose. From such information as is available, however, it appears that the limiting slope of the plot of the apparent molar volume of an electrolyte solute against the square root of the molar concentration is approximately equal to $0.66(\sum \nu_i z_i^2)^{3/2}$, as required by equation (43.13).[7]

Attention may be called to the interesting fact that for a large number of salts the apparent molar volume in aqueous solution is a linear function of the square root of the molarity at concentrations exceeding about 0.25 molar up to quite high values. The slope of this line, however, is specific for each electrolyte, whereas, according to equation (43.13), in dilute solution, it should be dependent only on the valence type of the solute. In general, the slope at high concentrations differs from the theoretical limiting Debye-Hückel slope as given by equation (43.13).[8]

44. Partial Molar Thermal Properties

44a. Relative Partial Molar Heat Contents.—The partial molar thermal properties, namely, heat content and heat capacity, are of particular interest, as well as of practical importance, as will be seen from some of the examples to be given below. In accordance with the general definition (§ 26a), the partial molar heat content of any constituent of a solution is represented by

$$\bar{H}_i = \left(\frac{\partial H}{\partial n_i}\right)_{T, P, n_1, \ldots},$$

---

[7] Redlich, *J. Phys. Chem.*, **44**, 619 (1940); Redlich and Bigeleisen, *Chem. Rev.*, **30**, 171 (1942).

[8] For review, see H. S. Harned and B. B. Owen, "The Physical Chemistry of Electrolytic Solutions," 1943, pp. 250 *et seq.*; also, Gucker, *Chem. Rev.*, **13**, 111 (1933).

the amounts of all the constituents, except n_i, being constant. As indicated earlier, since it is not possible to determine absolute values of the heat content H of any system, it is the practice to consider values relative to the reference state chosen in connection with the definition of activity (§ 37b), viz., pure liquid at 1 atm. pressure for the solvent (constituent 1) and the infinitely dilute solution at 1 atm. for the solute (constituent 2).

For the solvent, the standard state and the reference state are identical, on the basis of the usual convention, and consequently the molar heat content in the reference state may be represented by H_1^0. The partial molar heat content of the solvent in any solution relative to the heat content in the reference state is then $\bar{H}_1 - H_1^0$; this quantity is called the **relative partial molar heat content** of the solvent, and is represented by the symbol $\bar{L}_1$, so that in any solution

$$\bar{L}_1 = \bar{H}_1 - H_1^0. \tag{44.1}$$

For the sake of consistency, a slight modification is sometimes made in this expression. The partial molar heat content of the pure solvent, i.e., $\bar{H}_1^0$, which is the same as in a solution at infinite dilution, is, of course, identical with the molar heat content of the pure solvent, i.e., $\bar{H}_1^0 = H_1^0$; hence, equation (44.1) may be written

$$\bar{L}_1 = \bar{H}_1 - \bar{H}_1^0. \tag{44.2}$$

For the solute, the reference state is the infinitely dilute solution, and although this is not the same as the standard state, the partial molar heat contents are the same in both cases (§ 37d). The reference value, which is the partial molar heat content of the solute at infinite dilution, can then be represented by the symbol $\bar{H}_2^0$. The relative partial molar heat content $\bar{L}_2$ in any solution is thus given by

$$\bar{L}_2 = \bar{H}_2 - \bar{H}_2^0. \tag{44.3}$$

It may be noted that with the reference states chosen above, $\bar{H}_1$ becomes identical with $\bar{H}_1^0$, and $\bar{H}_2$ becomes identical with $\bar{H}_2^0$, at infinite dilution. Hence, in an infinitely dilute solution, $\bar{L}_1$ and $\bar{L}_2$ are both zero.

In accordance with the usual properties of partial molar quantities [cf. equation (26.6)], the heat content of a system consisting of n_1 moles of constituent 1, and n_2 moles of constituent 2, is

$$H = n_1 \bar{H}_1 + n_2 \bar{H}_2, \tag{44.4}$$

and hence, by equations (44.2) and (44.3),

$$H - (n_1 \bar{H}_1^0 + n_2 \bar{H}_2^0) = n_1 \bar{L}_1 + n_2 \bar{L}_2. \tag{44.5}$$

The quantity $n_1 \bar{H}_1^0 + n_2 \bar{H}_2^0$ may be taken as the reference value of the total heat content of the solution, so that the left-hand side of equation (44.5) gives the total relative heat content, for which the symbol L is used. It follows, therefore, that

$$L = n_1 \bar{L}_1 + n_2 \bar{L}_2, \tag{44.6}$$

which is consistent with the general property of partial molar quantities.

44b. Heats of Solution and Dilution.

The partial molar heat contents are related to such experimental quantities as heats of solution and dilution, and hence these will be considered. If H is the heat content of a solution, as described above, and H_1 and H_2 are the molar heat contents of the pure solvent and solute,* respectively, at the same temperature and pressure, the change of heat content upon mixing the constituents of the solution is then

$$\Delta H = H - (n_1 H_1 + n_2 H_2). \tag{44.7}$$

This quantity, divided by n_2 so as to give the value of ΔH per mole of solute, is the **total** or **integral heat of solution** at the given concentration (§ 12l).

If equation (44.7) is differentiated with respect to n_2, while the temperature, pressure and n_1 are maintained constant, it is found that

$$\left[\frac{\partial(\Delta H)}{\partial n_2}\right]_{T, P, n_1} = \left(\frac{\partial H}{\partial n_2}\right)_{T, P, n_1} - H_2$$
$$= \bar{H}_2 - H_2, \tag{44.8}$$

since, by definition, the first term on the right-hand side is equal to $\bar{H}_2$. The quantity obtained in this manner is the **partial** or **differential heat of solution** of the solute; it is seen from equation (44.8) to be the *increase of heat content per mole of solute when it is dissolved in a large volume of solution at a particular concentration*, so that there is no appreciable change in the latter.

Another expression for the differential heat of solution of the solute may be obtained by adding and subtracting the term $\bar{H}_2^0$ at the right-hand side of equation (44.8); thus,

$$\bar{H}_2 - H_2 = (\bar{H}_2 - \bar{H}_2^0) - (H_2 - \bar{H}_2^0)$$
$$= \bar{L}_2 - L_2, \tag{44.9}$$

where $\bar{L}_2$ is the relative partial molar heat content of the solute in the given solution, and L_2 is the relative (partial) molar heat content of the pure solute, the reference state in each case being the partial molar heat content at infinite dilution. Since $-L_2$ is equal to $\bar{H}_2^0 - H_2$, it is evident from equation (44.8) that it represents the *differential heat of solution of the solute at infinite dilution*.

By differentiation of equation (44.7) with reference to n_1, keeping T, P and n_2 constant, the result, analogous to (44.8), is

$$\left[\frac{\partial(\Delta H)}{\partial n_1}\right]_{T, P, n_2} = \bar{H}_1 - H_1$$
$$= \bar{H}_1 - H_1^0 \quad \text{or} \quad \bar{H}_1 - \bar{H}_1^0, \tag{44.10}$$

since in this case H_1, for the pure solvent, is identical with H_1^0 and $\bar{H}_1^0$. The left-hand side of (44.10), which is seen to be equal to the relative partial molar heat content $\bar{L}_1$ of the solvent in the given solution, is called the **partial**

---

* The quantity H_1 is identical with H_1^0 and $\bar{H}_1^0$, but the same relationship does not apply to H_2; the latter refers to pure solute, whereas $\bar{H}_2^0$ applies to infinite dilution.

or **differential heat of dilution.*** It may be defined as *the change of heat content per mole of solvent when it is added to a large volume of the solution at the given concentration.*

The **integral heat of dilution*** is *the change of heat content, per mole of solute, when a solution is diluted from one specified concentration to another.* By Hess's law, i.e., by the first law of thermodynamics, the integral heat of dilution must be equal to the difference of the integral heats of solution in the initial and final states. If the dilution is carried out by means of an infinite amount of solvent, so that the final solution is infinitely dilute, the heat change is referred to as the **integral heat of dilution to infinite dilution.**

44c. Relative Apparent Molar Heat Contents.—If ϕ_H is used to represent the apparent molar heat content of the solute, then by equation (42.1),

$$\phi_H = \frac{H - n_1 \bar{H}_1^0}{n_2}, \tag{44.11}$$

where $\bar{H}_1^0$ is used, as before, in place of H_1^0, for the molar heat content of the pure solvent; hence,

$$H = n_1 \bar{H}_1^0 + n_2 \phi_H. \tag{44.12}$$

At infinite dilution, equation (44.4) becomes

$$H^0 = n_1 \bar{H}_1^0 + n_2 \bar{H}_2^0, \tag{44.13}$$

and at the same time the general equation (44.12) takes the form

$$H^0 = n_1 \bar{H}_1^0 + n_2 \phi_H^0, \tag{44.14}$$

where ϕ_H^0 is the apparent molar heat content of the solute at *infinite dilution.* Comparison of equations (44.13) and (44.14) then shows that

$$\phi_H^0 = \bar{H}_2^0, \tag{44.15}$$

so that the apparent and the partial molar heat contents of the solute are identical at infinite dilution.†

By equation (44.12), $H - n_1 \bar{H}_1^0$ is equal to $n_2 \phi_H$, and since, by (44.15), $n_2 \phi_H^0$ is equal to $n_2 \bar{H}_2^0$, it follows that

$$H - (n_1 \bar{H}_1^0 + n_2 \bar{H}_2^0) = n_2 \phi_H - n_2 \phi_H^0,$$

and hence, by (44.5),

$$n_1 \bar{L}_1 + n_2 \bar{L}_2 = n_2(\phi_H - \phi_H^0)$$
$$= n_2 \phi_L, \tag{44.16}$$

where the **relative apparent molar heat content** of the solute, i.e., $\phi_H - \phi_H^0$, is represented by ϕ_L. This quantity, as will be seen below, is equal in

---

* Heats of dilution are sometimes considered as heats of solution of the *solvent*, to indicate that there is no essential difference between the solvent and solute.

† This result is of general applicability; thus $\phi_G = \bar{G}_2$ for infinite dilution at any constant pressure. In equation (44.15) the implied pressure is that of the reference state, i.e., 1 atm. (cf. § 44a).

magnitude but opposite in sign to the (integral) heat of dilution to infinite dilution per mole of solute, as defined in § 44b, for a solution consisting of n_1 moles of solvent and n_2 moles of solute.

44d. Heats of Dilution to Infinite Dilution.—When n_2 moles of solute are dissolved in an infinite amount, i.e., ∞ moles, of solvent, the heat content of the infinitely dilute solution, i.e., H^0, is given by equation (44.4) as

$$H^0 = \infty \bar{H}_1^0 + n_2 \bar{H}_2^0, \qquad (44.17)$$

the partial molar heat contents being $\bar{H}_1^0$ and $\bar{H}_2^0$, for solvent and solute, respectively. For a solution consisting of n_1 moles of solvent and n_2 moles of solute, the heat content H is [cf. equation (44.4)]

$$H = n_1 \bar{H}_1 + n_2 \bar{H}_2. \qquad (44.18)$$

If $\infty - n_1$ moles of solvent are added to this solution, it will become identical with the one just considered, i.e., n_2 moles of solute in an infinite amount of solvent; the change of heat content is then the heat of dilution to infinite dilution. The heat content of $\infty - n_1$ moles of solvent is $(\infty - n_1)\bar{H}_1^0$, and so the sum of the heat contents of the given solution and the infinite amount of solvent, before mixing, is $n_1\bar{H}_1 + n_2\bar{H}_2 + (\infty - n_1)\bar{H}_1^0$. The heat content, after mixing, is given by equation (44.17), and hence the heat of dilution to infinite dilution, represented by $\Delta H_{c \to 0}$, is

$$\Delta H_{c \to 0} = (\infty \bar{H}_1^0 + n_2 \bar{H}_2^0) - [n_1 \bar{H}_1 + n_2 \bar{H}_2 + (\infty - n_1)\bar{H}_1^0]$$
$$= n_1(\bar{H}_1^0 - \bar{H}_1) + n_2(\bar{H}_2^0 - \bar{H}_2). \qquad (44.19)$$

By the definitions of equations (44.2) and (44.3), this becomes

$$\Delta H_{c \to 0} = - n_1 \bar{L}_1 - n_2 \bar{L}_2,$$

and hence, by (44.16),

$$\Delta H_{c \to 0} = - n_2 \phi_L. \qquad (44.20)$$

The value of $\Delta H_{c \to 0}$ is for the infinite dilution of a solution containing n_2 moles of solute, and consequently the integral heat of dilution to infinite dilution, *per mole of solute*, is equal $- \phi_L$, as stated above.

The integral heat of dilution of any solution to infinite dilution can be determined experimentally. To measure the heat change accompanying the actual dilution of a given solution by an infinite amount of solvent is, of course, not practicable. The procedure employed is to dilute the given solution in stages by using finite amounts of pure solvent, or by adding a more dilute to a more concentrated solution. The heat content changes in the various states are additive, and from the results it is possible to derive the heats of dilution to increasing extents. The data obtained in this manner for the dilution of 0.1 molar sodium chloride solution to various final molarities c, at 25° C, are given in Table XXXV.[9] The values of $\Delta H_{0.1 \to c}$ represent the heats of dilution, *per mole of solute*, of the 0.1 molar solution to the final concentration of c mole per liter. Extrapolation of these

[9] Robinson, *J. Am. Chem. Soc.*, **54**, 1311 (1932); Harned and Owen, ref. 8, p. 225.

values to infinite dilution, preferably by plotting $\Delta H_{0.1 \to c}$ against $\sqrt{c}$, gives -83.0 cal. mole^{-1} for $\Delta H_{0.1 \to 0}$, the integral heat of dilution to infinite dilution. Hence for 0.1 molar sodium chloride at 25° C, the value of ϕ_L, i.e., $\phi_H - \phi_H^0$, is 83.0 cal. mole^{-1}.

TABLE XXXV. HEAT OF DILUTION OF 0.1N SODIUM CHLORIDE SOLUTION AT 25° C

c	$\Delta H_{0.1 \to c}$	$\Delta H_{c \to 0}$
0.05 mole liter^{-1}	-12.8 cal. mole^{-1}	-70.2 cal. mole^{-1}
0.025	-27.8	-55.2
0.0125	-42.9	-40.1
0.00605	-52.9	-30.1
0.00305	-61.5	-21.5
0.00153	-67.4	-15.6
0.00076	-73.1	-9.9
0.00039	-75.7	-7.3
0	(-83.0)	0

Once the integral heat of diluting a given solution, e.g., 0.1 molar, to infinite dilution is known, the values of ϕ_L for other solutions can be determined much more easily. All that is necessary is to measure the total (integral) heat change accompanying the dilution of any solution to 0.1 molar; upon adding the value of ϕ_L for 0.1 molar, the result is ϕ_L for the former solution. For solutions more dilute than 0.1 molar, the values of ϕ_L, i.e., $-\Delta H_{c \to 0}$ per mole of solute, can be obtained by subtracting $\Delta H_{0.1 \to c}$, which is determined experimentally, from $\Delta H_{0.1 \to 0}$, which may be regarded as known.

The relative apparent molar heat contents of sulfuric acid (solute) in mixtures with water (solvent), for various compositions at 25° C, are recorded in Table XXXVI.*[10] It may be noted that when ϕ_L is positive, $\Delta H_{c \to 0}$ is negative, and the dilution process is accompanied by the evolution of heat. It is seen, therefore, from Table XXXVI that heat is evolved upon the infinite dilution of sulfuric acid solutions at all concentrations down to the lowest studied, viz., 0.00108 molal.

44e. Integral Heat of Finite Dilution.—The relative apparent molar heat content values in Table XXXVI give not only the heat changes accompanying infinite dilution of a particular solution, but also the change in heat content for a finite dilution. This is, of course, merely a reversal of the procedure described above for determining heats of dilution to infinite dilution. If a solution is diluted from molality m' to molality m, the change of heat content per mole of solute is represented by

$$\Delta H_{m' \to m} = \Delta H_{m' \to 0} - \Delta H_{m \to 0}$$
$$= -(\phi_L' - \phi_L), \qquad (44.21)$$

where ϕ_L' and ϕ_L are the relative apparent molar heat contents of the solute at the molalities m' and m, respectively.

[10] Calculated by Craig and Vinal, *J. Res. Nat. Bur. Stand.*, **24**, 475 (1940).

* The relative partial molar heat contents $\bar{L}_1$ and $\bar{L}_2$ are also included in the table; they are derived from the ϕ_L values, as explained in § 44f.

TABLE XXXVI. HEAT CONTENTS OF AQUEOUS SOLUTIONS OF SULFURIC ACID AT 25° C

Molality m	$\dfrac{\text{Moles } H_2O}{\text{Moles } H_2SO_4}$	ϕ_L cal. mole^{-1}	$\dfrac{d\phi_L}{dm^{1/2}}$	$\bar{L}_1$ cal. mole^{-1}	$\bar{L}_2$ cal. mole^{-1}
0	∞	0	33.0 × 10^3	0	0
0.00108	51,200	1,160	30.8	0.01	1,665
0.00217	25,600	1,540	29.8	0.03	2,234
0.00434	12,800	2,080	27.4	0.06	2,889
0.00867	6,400	2,900	17.74	− 0.13	3,725
0.01734	3,200	3,480	12.74	− 0.26	4,318
0.03469	1,600	4,040	8.28	− 0.48	4,811
0.06938	800	4,550	5.24	− 0.91	5,280
0.1388	400	5,020	3.32	− 1.54	5,638
0.2775	200	5,410	1.64	− 2.16	5,842
0.5551	100	5,620	0.72	− 2.68	5,888
1.1101	50	5,780	0.54	− 5.70	6,065
2.2202	25	6,070	0.82	− 24.4	6,681
3.7004	15	6,550	1.40	− 89.8	7,896
5.551	10	7,300	1.98	− 233	9,632
6.938	8	7,880	2.20	− 362	10,777
9.251	6	8,800	2.25	− 570	12,222
13.876	4	10,410	2.24	− 1,043	14,581
18.502	3	11,660	2.06	− 1,477	16,089
27.75	2	13,580	1.76	− 2,318	18,216
55.51	1	16,720	1.27	− 4,731	21,451
111.01	0.5	19,730	0.64	− 6,743	23,102
∞	0	(23,540)	(0.00)		(23,540) = $\bar{L}_2$

Problem: What is the net amount of heat required to remove half of the water from a solution containing 1 mole of H_2SO_4 to 400 moles H_2O at 25° C?

The final solution will contain 1 mole H_2SO_4 to 200 moles H_2O, and the heat which must be supplied is equal in magnitude but opposite in sign to the heat evolved upon diluting this solution to one containing 1 mole H_2SO_4 to 400 moles H_2O. The change of heat content accompanying the dilution in this case is given by equation (44.21) as $-[\phi_L \text{ (for } 200H_2O/1H_2SO_4) - \phi_L \text{ (for } 400H_2O/1H_2SO_4)]$. Hence, from Table XXXVI,

$$\Delta H = -(5{,}410 - 5{,}020) = -390 \text{ cal.}$$

The heat required to remove the water is thus 390 cal.

Problem: How much heat is evolved when 1 mole of H_2SO_4 is added to 200 moles H_2O at 25° C?

The process is here equivalent to dilution from an initial solution containing $0H_2O/1H_2SO_4$ to a final solution of $200H_2O/1H_2SO_4$. Hence, by Table XXXVI, the value of ΔH is $-[\phi_L \text{ (for } 0H_2O/1H_2SO_4) - \phi_L \text{ (for } 200H_2O/1H_2SO_4)]$; thus,

$$\Delta H = -(23{,}540 - 5{,}410) = -18{,}130 \text{ cal.}$$

The amount of heat evolved is 18,130 cal. This is the integral heat of solution of the sulfuric acid in the solution.

Although it is mainly of theoretical interest, there is a simple relationship between the integral heat of finite dilution $\Delta H_{m' \to m}$ and the corresponding apparent molar heat contents ϕ'_H and ϕ_H of the solute in the initial and final solutions. Thus, by definition [cf. equation (44.16)],

$$\phi'_L = \phi'_H - \phi^0_H \quad \text{and} \quad \phi_L = \phi_H - \phi^0_H,$$

the same reference value ϕ^0_H being applicable in both cases. Inserting these expressions into equation (44.21), it is seen that

$$\Delta H_{m' \to m} = -(\phi'_H - \phi_H). \tag{44.22}$$

44f. Determination of Relative Partial Molar Heat Contents.—A number of methods are available for evaluating the relative partial molar heat contents $\bar{L}_1$ and $\bar{L}_2$ of the constituents of a solution. One of these makes use of the apparent quantities, such as those given in Table XXXVI. Since

$$L = n_1 \bar{L}_1 + n_2 \bar{L}_2 = n_2 \phi_L, \tag{44.23}$$

by equations (44.6) and (44.16), it follows, upon differentiation with respect to n_2, the temperature, pressure and number of moles n_1 of solvent being constant, that

$$\left(\frac{\partial L}{\partial n_2}\right)_{T, P, n_1} = \bar{L}_2 = \phi_L + n_2 \left(\frac{\partial \phi_L}{\partial n_2}\right)_{T, P, n_1}. \tag{44.24}$$

This result is identical with the general equation (42.3), and could, of course, have been obtained directly from it. Since n_2 in equation (44.24) may be replaced by the molality m, for this satisfies the condition that n_1 is constant, it follows that, *at constant temperature and pressure,*

$$\bar{L}_2 = \phi_L + m \frac{d\phi_L}{dm} = \phi_L + \frac{d\phi_L}{d \ln m}. \tag{44.25}$$

By plotting ϕ_L against m (or log m), and determining the slope $d\phi_L/dm$ (or $d\phi_L/d \log m$) at any required molality, the value of $\bar{L}_2$ can be found, as described in § 42c, II.

For electrolytes, more accurate results are obtained by plotting ϕ_L against $m^{1/2}$; thus, utilizing the relationship

$$dm = 2m^{1/2} dm^{1/2},$$

it follows from equation (44.25) that

$$\bar{L}_2 = \phi_L + \frac{m^{1/2}}{2} \cdot \frac{d\phi_L}{dm^{1/2}}. \tag{44.26}$$

If the value of $\bar{L}_2$ given by equation (44.24) is substituted in (44.23), it is found that

$$\bar{L}_1 = -\frac{n_2^2}{n_1}\left(\frac{\partial \phi_L}{\partial n_2}\right)_{T, P, n_1}. \tag{44.27}$$

Further, if n_2 is replaced by the molality, then n_1 is equal to $1000/M_1$, where

M_1 is the molecular weight of the solvent; hence equation (44.27) becomes

$$\bar{L}_1 = -\frac{M_1 m^2}{1000} \cdot \frac{d\phi_L}{dm}, \qquad (44.28)$$

at constant temperature and pressure. For a solution of an electrolyte, this is preferably written in the form

$$\bar{L}_1 = -\frac{M_1 m^{3/2}}{2000} \cdot \frac{d\phi_L}{dm^{1/2}}. \qquad (44.29)$$

It is thus possible to derive both $\bar{L}_1$ and $\bar{L}_2$ from the slope of the plot of ϕ_L against m for a nonelectrolyte, or against $m^{1/2}$ for an electrolyte.* The results obtained in this manner, for aqueous solutions of sulfuric acid, are included in Table XXXVI, above.

In the procedure just described, both $\bar{L}_1$ and $\bar{L}_2$ are obtained from the same data, viz., the relative apparent molar heat contents; other methods, however, give only $\bar{L}_1$ or $\bar{L}_2$, as will be seen shortly. If one of these quantities is known, the other can be calculated by using a form of the general equation (26.8); thus, in the present case, *at constant temperature and pressure,*

$$n_1 d\bar{L}_1 + n_2 d\bar{L}_2 = 0,$$

so that

$$d\bar{L}_1 = -\frac{n_2}{n_1} d\bar{L}_2$$

$$= -\frac{mM_1}{1000} d\bar{L}_2. \qquad (44.30)$$

Consequently, if n_2/n_1 (or m) is plotted against $\bar{L}_2$ for various compositions, it is possible to evaluate $\bar{L}_1$ by graphical integration (cf. § 38f).

According to the results derived in § 44b, it is seen that $\bar{L}_1$ is equal to the differential heat of dilution of the given solution. This may be obtained experimentally by determining the heat changes upon mixing a definite quantity of solute with varying amounts of solvent. If the results are plotted against the number of moles of solvent, the slope, i.e., $[\partial(\Delta H)/\partial n_1]_{T,P,n_2}$, at any concentration gives the differential heat of dilution, and hence $\bar{L}_1$, at that concentration [cf. equation (44.10)]; this is equivalent to method I (§ 42c). If the solute is a solid, this procedure is not convenient. The method that can then be employed is to take a solution of definite concentration and determine the heat change *per mole of solvent*, i.e., $\Delta H/n_1$, upon dilution with varying amounts of solvent. If these data are extrapolated to zero added solvent, the result is equivalent to the heat change accompanying the dilution of the solution by a mole of solvent under such conditions that the composition remains virtually constant; this is evidently the differential heat of dilution of the given solution. An approximate value of this

---

* It will be evident that if either $\bar{L}_1$ or $\bar{L}_2$ is determined in this manner, the other could be obtained directly from equation (44.23).

quantity may be determined directly by adding a small amount of solvent to a solution, and assuming that the heat change per mole of solvent is equal to $\bar{L}_1$ at a mean concentration between the initial and final compositions.[11]

Problem: Solve the first problem in § 44e by using the relative partial molar heat contents $\bar{L}_1$ and $\bar{L}_2$ in Table XXXVI.

The relative heat content L in the initial state is $400\bar{L}_1 + \bar{L}_2$ for the system $400H_2O/1H_2SO_4$; this is $(400 \times -1.54) + 5{,}638 = 5{,}022$ cal. The final solution consists of 200 moles H_2O and 1 mole H_2SO_4, and in addition there are 200 moles of pure H_2O. Since pure solvent is the reference state, $\bar{L}_1$ for the latter will be zero (cf. Table XXXVI), and hence L for the final state is $200\bar{L}_1 + \bar{L}_2$ for the system $200H_2O/1H_2SO_4$, i.e., $(200 \times -2.16) + 5{,}842 = 5{,}410$ cal. The change of heat content is thus $5{,}410 - 5{,}022 = 388$ cal. Since the final state has a larger heat content than the initial state, 388 cal. must be supplied. (The small difference from the result in § 44e is due to the use of two decimal places only in $\bar{L}_1$. It should be noted that since L is equal to $n_2\phi_L$, by equation (44.23), the method used here is identical in principle with that employed in the previous solution of the problem. The use of relative heat contents is, however, more fundamental.)

Problem: Calculate the heat change when 1 mole of H_2SO_4 is added to a solution of 1 mole H_2SO_4 in 400 moles H_2O, using Table XXXVI.

The relative heat content L in the initial state is equal to $400\bar{L}_1 + \bar{L}_2$ for the solution $400H_2O/1H_2SO_4$, plus L_2 for the pure H_2SO_4; this is $(400 \times -1.54) + 5{,}638 + 23{,}540 = 28{,}562$ cal. In the final state, L is $400\bar{L}_1 + 2\bar{L}_2$ for the solution $200H_2O/1H_2SO_4$, i.e., $(400 \times -2.16) + (2 \times 5{,}842) = 10{,}820$ cal. The change of heat content is thus $10{,}820 - 28{,}562 = -17{,}742$ cal. (This problem could also be solved by using the ϕ_L values in Table XXXVI.)

44g. Activity Coefficients and Relative Partial Molar Heat Contents.—It was seen in § 39c that the variation of the activity coefficient with temperature is dependent upon $\bar{L}_2$; this variation can be utilized, both directly and indirectly, to evaluate relative partial molar heat contents. By equation (26.25), $[\partial(\mu_i/T)/\partial T]_{P,N} = -\bar{H}_i/T^2$ for any constituent of a solution; hence, utilizing (43.4), for an electrolyte, in the form

$$\frac{\mu_2}{T} - \frac{\mu_2^0}{T} = \nu R \ln m_\pm \gamma_\pm,$$

it follows, upon differentiation with respect to temperature, at constant pressure and composition, that

$$-\frac{\bar{H}_2 - \bar{H}_2^0}{T^2} = \nu R \left(\frac{\partial \ln \gamma_\pm}{\partial T}\right)_{P,N}.$$

Since $\bar{H}_2 - \bar{H}_2^0$ is equal to $\bar{L}_2$, this is equivalent to

$$\bar{L}_2 = -\nu R T^2 \left(\frac{\partial \ln \gamma_\pm}{\partial T}\right)_{P,N} \tag{44.31}$$

[11] Randall and Bisson, *J. Am. Chem. Soc.*, **42**, 347 (1920); Randall and Rossini, *ibid.*, **51**, 323 (1929); Rossini, *J. Res. Nat. Bur. Stand.*, **6**, 791 (1930); **9**, 679 (1931); Gucker, *J. Am. Chem. Soc.*, **61**, 459 (1939).

or
$$\bar{L}_2 = -\nu R \left(\frac{\partial \ln \gamma_\pm}{\partial (1/T)} \right)_{P,N}. \tag{44.32}$$

By determining the mean ionic activity coefficient at several temperatures, e.g., from the elevation of boiling point at different pressures,* the values of $\bar{L}_2$ at various concentrations may be calculated from the plot of $\ln \gamma_\pm$ (or $\log \gamma_\pm$) against either T or $1/T$. The relative partial molar heat contents of a few salts have been determined in this manner.[12]

The reverse of this procedure provides one of the main applications of the relative partial molar heat contents. If these are determined by means of thermal measurements, as described in § 44f, they can be employed to evaluate the activity coefficients at one temperature if those at another temperature are known (§§ 38c, 39c).

Problem: Assuming $\bar{L}_2$ to remain constant, calculate the relative change in the mean ionic activity coefficient of 1 molal sulfuric acid solution from 0° to 25° C.

Integration of equation (44.31) with $\bar{L}_2$ constant gives

$$\log \frac{\gamma}{\gamma'} = \frac{\bar{L}_2}{4.576 \nu} \left(\frac{1}{T} - \frac{1}{T'} \right),$$

where γ and γ' are the mean ionic activity coefficients at T and T', respectively. For H_2SO_4, ν is 3, i.e., $2H^+ + SO_4^{--}$, and by Table XXXVI, $\bar{L}_2$ for a 1 molal solution is close to 6,000 cal. Hence,

$$\log \frac{\gamma_{273}}{\gamma_{298}} = \frac{6,000}{4.576 \times 3} \left(\frac{298 - 273}{273 \times 298} \right) = 0.134,$$

so that $\gamma_{273}/\gamma_{298} = 1.36$. (If allowance were made for the variation of $\bar{L}_2$ with temperature, using partial molar heat capacities given later (Table XXXVIII), a slightly smaller ratio, about 1.34, is obtained. The experimental value, derived from E.M.F. measurements, is 1.33. The variation of $\gamma_\pm$ with temperature is unusually large, because $\bar{L}_2$ for sulfuric acid is considerably greater than for most electrolytes.)

An indirect method of using equation (44.31) is based on the employment of E.M.F. measurements. For example, if the solute is a suitable electrolyte $M_{\nu_+}A_{\nu_-}$, the E.M.F. of a cell of the type $M\,|\,$Solution of $M_{\nu_+}A_{\nu_-}\,|\,A$, referred to in § 39h, is, in the general case,

$$E = E^0 - \frac{RT}{NF} \ln a_+^{\nu_+} a_-^{\nu_-}$$

$$= E^0 - \frac{\nu RT}{NF} \ln m_\pm \gamma_\pm,$$

where N is the number of faradays associated with the formation of 1 mole of the solute $M_{\nu_+}A_{\nu_-}$ in the cell. Upon rearranging this result, and differentiating with

---

* The variation of the activity coefficient with pressure is negligible.
[12] Smith, et al., J. Am. Chem. Soc., **61**, 1123 (1939); **63**, 1351 (1941).

respect to temperature, at constant pressure and composition, it is seen that

$$\left(\frac{\partial (E - E^0)/T}{\partial T}\right)_{P,N} = -\frac{\nu R}{NF}\left(\frac{\partial \ln \gamma_{\pm}}{\partial T}\right)_{P,N}. \qquad (44.33)$$

Utilizing equation (44.31), it follows that

$$\bar{L}_2 = NFT^2\left(\frac{\partial (E - E^0)/T}{\partial T}\right)_{P,N} \qquad (44.34)$$

$$= NFT\left(\frac{\partial (E - E^0)}{\partial T}\right)_{P,N} - NF(E - E^0), \qquad (44.35)$$

which is a form of the Gibbs-Helmholtz equation (25.31), since $\bar{L}_2$ is equal to $\bar{H}_2 - \bar{H}_2^0$ [cf. also, equation (45.3)].

By means of equation (44.35), it is evidently possible to determine the relative partial molar heat content of an electrolyte by measuring the E.M.F. of a cell of the type mentioned above at several temperatures, so that the temperature coefficient can be evaluated. This gives the first term on the right-hand side of equation (44.35), and the second term is obtained from the actual E.M.F.'s.[13]

The most convenient procedure, however, is to express the E.M.F. of the cell, with electrolyte at a given molality or molarity, as a power series function of the temperature; thus,

$$E = a + bT + cT^2 + dT^3 + \cdots, \qquad (44.36)$$

where a, b, c, etc., are empirical constants. Similarly, the standard E.M.F. can be represented by an analogous expression

$$E^0 = a^0 + b^0T + c^0T^2 + d^0T^3 + \cdots. \qquad (44.37)$$

If the temperature range is not too large, it is not necessary to proceed beyond three terms; hence,

$$\frac{E - E^0}{T} = \frac{(a - a^0)}{T} + (b - b^0) + (c - c^0)T.$$

Upon differentiating with respect to temperature and inserting the result in equation (44.34), it follows that

$$\bar{L}_2 = -NF[(a - a^0) - (c - c^0)T^2]. \qquad (44.38)$$

The relative partial molar heat content of the solute can thus be determined at any temperature within the range in which equations (44.36) and (44.37) are applicable.

The relative partial molar heat contents of the constituents of a liquid metallic alloy can be obtained in an analogous manner by making use of cells of the type described in § 38e, viz., Metal A(liquid) | Fused salt of A | Solution of A in metal B(liquid). In this case the standard state is taken as pure liquid A, and, consequently, E^0, the standard E.M.F., is zero; equation (44.34) then becomes

$$NFT^2\left(\frac{\partial (E/T)}{\partial T}\right)_{P,N} = \bar{L}_A.$$

[13] Harned, et al., J. Am. Chem. Soc., **54**, 423 (1932); **55**, 2179, 4838 (1933); LaMer, et al., ibid., **55**, 1004, 4343 (1933); Åkerlöf, et al., ibid., **59**, 1855 (1937); **62**, 620 (1940); see also, Harned and Owen, ref. 8, Chapters 10, 11, 12 and 13.

The values of $\bar{L}_A$ for a series of compositions can thus be obtained by changing the proportion of A in the alloy electrode; the corresponding values of $\bar{L}_B$ can then be calculated by a suitable application of the equation (26.8), i.e., $n_A d\bar{L}_A + n_B d\bar{L}_B = 0$, as seen in § 44f.[14]

44h. Determination of Heat of Solution of Solute.—In some cases, particularly, when the solute is a liquid which is completely miscible with the solvent, e.g., sulfuric acid and water, or a gas that is very soluble, e.g., hydrogen chloride in water, it is possible to extrapolate the relative partial molar heat content $\bar{L}_2$, i.e., $\bar{H}_2 - \bar{H}_2^0$, to the value for the pure solute. This quantity is $H_2 - \bar{H}_2^0$, where H_2 refers to the pure solute, and hence is identical with L_2 defined in § 44b, and employed in equation (44.9). Incidentally, for the pure solute n_1 is zero, and hence, by equation (44.16) the corresponding value of $\bar{L}_2$, in this case L_2, is equal to ϕ_L. The figures on the last line of Table XXXVI were obtained by extrapolation of the ϕ_L values. Once L_2 is known, the differential heat of solution of the solute at any composition can be calculated, since it is equal to $\bar{L}_2 - L_2$ [cf. equation (44.9)]; $-L_2$ is itself equal to the differential heat of solution at infinite dilution, as seen in § 44b.

An expression for the integral heat of solution can be derived in various ways; perhaps the simplest is to utilize equation (44.21). In this case the initial state m' is the pure solute, so that ϕ_L' is identical with L_2, as found above; hence the integral heat of solution, to form a solution of molality m, is $-(L_2 - \phi_L)$ or $\phi_L - L_2$, where ϕ_L refers to the final molality. This procedure was actually used in the second problem in § 44e.

When the solute is a solid having a limited solubility, then it is not possible to obtain L_2 at all accurately by extrapolation of the ϕ_L values, and another procedure is used. The solute, e.g., a salt, can invariably be obtained in the pure state, and the integral heat of solution to a particular concentration can then be determined experimentally. By reversing the procedure just described, it is thus possible to evaluate L_2, provided ϕ_L is known for the same concentration.

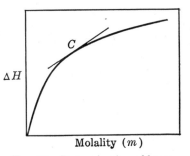

Fig. 36. Determination of heats of solution

Alternatively, the differential heat of solution $[\partial(\Delta H)/\partial n_2]_{n_1}$, which is equal to $\bar{L}_2 - L_2$, may be obtained if the observed heat change ΔH, when m moles of solute (equivalent to n_2) are added to a definite quantity, e.g., 1000 g. of solvent (equivalent to constant n_1), is plotted against m, and the slope determined at any required composition (e.g., C, Fig. 36). If $\bar{L}_2$ at this composition is available, then L_2 may be derived immediately.

---

[14] See, for example, Strickler and Seltz, *J. Am. Chem. Soc.*, **58**, 2084 (1936); Seltz, *Trans. Electrochem. Soc.*, **77**, 233 (1940).

It may be remarked that in dilute solution the heat of solution ΔH is usually a linear function of the molality (see Fig. 36); the integral heat of solution per mole, i.e., $\Delta H/m$, is then equal to the differential heat of solution $d(\Delta H)/dm$. Provided the solution is sufficiently dilute, therefore, the experimental value of the former may be identified with $-\bar{L}_2$, since this is equivalent to the differential heat of solution at infinite dilution.

Problem: The integral heat of solution at infinite dilution per mole of KCl was found by extrapolation to be 4,120 cal. at 25° C. The integral heat of solution of 1 mole of KCl in 55.5 moles of water, to make a 1 molal solution, is 4,012 cal. From the slope of the plot of the heat of solution per mole of solute against the number of moles of solvent, the differential heat of dilution of 1 molal KCl solution is found to be 3.57 cal. mole^{-1}. What is the differential heat of solution of KCl in this solution?

The integral heat of solution at infinite dilution may be identified with the differential heat of solution under these conditions, and hence $-L_2 = 4,120$ cal. mole^{-1}. Further, for a finite solution, the integral heat of solution per mole of solute is $\phi_L - L_2$ (or $L - L_2$); this is 4,012 cal. mole^{-1} for the 1 molal solution, so that ϕ_L (or L) is -108 cal. mole^{-1}. The differential heat of dilution, 3.57 cal. mole^{-1}, is $\bar{L}_1$; since $L = n_1\bar{L}_1 + n_2\bar{L}_2$, where n_1 is 55.5 and n_2 is 1 mole, $\bar{L}_2 = -306$ cal. mole^{-1}. The differential heat of solution of the solute is $\bar{L}_2 - L_2$, and hence in the 1 molal solution this is $-306 + 4,120 = 3,814$ cal. mole^{-1} of KCl.

For a system consisting of two similar, miscible liquids, e.g., alcohol and water or benzene and toluene, it is neither necessary nor desirable to distinguish between solvent and solute. The treatment for deriving heats of solution may be applied to either constituent. The fact that the pure liquid is chosen as the reference state for each substance is immaterial, for the results must be independent of the particular reference state.

44i. Thermal Properties and the Debye-Hückel Theory.—Since equation (44.31) relates the relative partial molar heat content of an electrolyte to the mean ionic activity coefficient, there is clearly a possibility of introducing the Debye-Hückel treatment.[15] Thus, inserting equation (43.8) * into (44.31) and carrying out the differentiation, allowing for the fact that c is inversely proportional to the volume, the result is

$$\bar{L}_2 = -\frac{2.469 \times 10^{14}}{D^{3/2}T^{1/2}} (\sum \nu_i z_i^2)^{3/2} \frac{3}{2} \left(1 + \frac{T}{D} \cdot \frac{\partial D}{\partial T} + \frac{T}{3V} \cdot \frac{\partial V}{\partial T}\right)_P \sqrt{c} \text{ ergs mole}^{-1}.$$

(44.39)

For water at 25° C, $(\partial D/\partial T)_P$, at atmospheric pressure, is -0.3613 deg.$^{-1}$, the dielectric constant D is 78.54, and the coefficient of cubical expansion $(\partial V/\partial T)_P/V$ is 2.58×10^{-4} deg.$^{-1}$; the quantity in the large parentheses is thus -0.346.

[15] Bjerrum, *Z. phys. Chem.*, **119**, 145 (1926); Scatchard, *J. Am. Chem. Soc.*, **53**, 2037 (1931); H. Falkenhagen, "Electrolytes," 1934 (translated by R. P. Bell); Harned and Owen, ref. 8, p. 48.

* By following the procedure used in § 43c it can be shown that in dilute solutions, at least, the nonelectrical factor in the activity coefficient is independent of temperature; hence, $\gamma_{D.H.}$ may be used for $\gamma_\pm$ when differentiating with respect to temperature.

Consequently, dividing by 4.184×10^7 to convert the result into calories, equation (44.39) becomes

$$\bar{L}_2 = 255(\sum \nu_i z_i^2)^{3/2} \sqrt{c} \text{ cal. mole}^{-1}. \tag{44.40}$$

The quantity most easily determined experimentally is the integral heat of infinite dilution $\Delta H_{c \to 0}$; as seen in § 44d, the value per mole of solute is equal to $-\phi_L$, the relative apparent molar heat content of the solute at the concentration c. By utilizing a similar procedure to that employed in deriving equation (43.12), it is found that

$$\Delta H_{c \to 0} = -\phi_L = -\frac{1}{c}\int_0^c \bar{L}_2 dc,$$

and hence, by (44.40),

$$\Delta H_{c \to 0} = -\tfrac{2}{3} \times 255(\sum \nu_i z_i^2)^{3/2} \sqrt{c}$$
$$= -170(\sum \nu_i z_i^2)^{3/2} \sqrt{c} \text{ cal. mole}^{-1}. \tag{44.41}$$

The heat of dilution to infinite dilution, per mole of solute, should thus be proportional to the square root of the concentration, at least for dilute solutions. The slope of the plot of $\Delta H_{c \to 0}$ against $\sqrt{c}$ should approach the limiting value of $-170(\sum \nu_i z_i^2)^{3/2}$ for aqueous solutions at 25° C. The experimental results appear to be in general agreement with equation (44.41). Although the data are probably not precise enough to permit the exact verification of the theoretical slope, they show that equation (44.41) probably constitutes a satisfactory limiting approximation (Fig. 37).[16]

It is of interest to observe that the negative sign of $\Delta H_{c \to 0}$ means that the infinite dilution of a dilute solution of an electrolyte is accompanied by an evolution of heat. At higher concentrations, however, when the limiting equation (44.41) is no longer, even approximately, applicable, the sign of the heat of dilution for many salts is reversed (cf. Fig. 37); heat is then absorbed when the solution is diluted to infinite dilution.

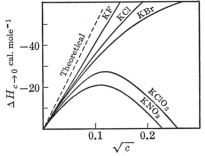

FIG. 37. Heats of dilution and the Debye-Hückel limiting law

Attention may be called to the fact that equation (44.41) can be derived from the Debye-Hückel treatment in an alternative manner, which is based on the same fundamental principles as that just described. In § 40a the deviation from ideal behavior, as represented by the activity coefficient, was attributed to the interaction of the ions, and the heat of dilution to infinite dilution may be ascribed to the same cause. The quantity $\Delta H_{c \to 0}$ can thus be identified with $-H_{el.}$ corresponding to $F_{el.}$ evaluated in § 40c; these quantities are then related by a form of the Gibbs-Helmholtz equation [cf. (25.28)],

$$-\Delta H_{c \to 0} = H_{el.} = -T^2\left[\frac{\partial(F_{el.}/T)}{\partial T}\right]_P.$$

[16] For reviews, see Lange and Robinson, *Chem. Rev.*, **9**, 89 (1931); Robinson and Wallace, *ibid.*, **30**, 195 (1942); Harned and Owen, ref. 8, pp. 223 *et seq.*

If equation (40.6) for $F_{\text{el.}}$ is inserted in this expression, and the differentiation performed, the result obtained, after making appropriate substitutions for $(\partial D/\partial T)_P$, $(\partial V/\partial T)_P/V$, etc., is found to be identical with (44.41).

44j. Partial Molar Heat Capacities.—The heat capacity at constant pressure C_P of a solution containing n_1 moles of solvent and n_2 moles of solute is given by

$$C_P = \left(\frac{\partial H}{\partial T}\right)_{P,N},$$

the pressure and composition being constant. Upon differentiation with respect to n_1, maintaining n_2 constant, it follows that

$$\bar{C}_{P1} = \left(\frac{\partial C_P}{\partial n_1}\right)_{T,P,n_2} = \frac{\partial H}{\partial T \partial n_1}, \tag{44.42}$$

where $\bar{C}_{P1}$ is the **partial molar heat capacity**, at constant pressure, of the constituent 1 of the given solution. The partial molar heat content $\bar{H}_1$ of this constituent is defined by

$$\bar{H}_1 = \left(\frac{\partial H}{\partial n_1}\right)_{T,P,n_2},$$

and hence differentiation with respect to temperature gives

$$\left(\frac{\partial \bar{H}_1}{\partial T}\right)_{P,N} = \frac{\partial H}{\partial n_1 \partial T} = \bar{C}_{P1}, \tag{44.43}$$

the result being identical with $\bar{C}_{P1}$, by equation (44.42). The partial molar heat capacity of the solvent in any particular solution may thus be defined by either equation (44.42) or (44.43). Similarly, for the solute, i.e., constituent 2,

$$\bar{C}_{P2} = \left(\frac{\partial C_P}{\partial n_2}\right)_{T,P,n_1} = \left(\frac{\partial \bar{H}_2}{\partial T}\right)_{P,N}. \tag{44.44}$$

If equation (44.2) for the relative partial molar heat content, i.e.,

$$\bar{L}_1 = \bar{H}_1 - \bar{H}_1^0,$$

is differentiated with respect to temperature, at constant pressure and composition, it follows that

$$\left(\frac{\partial \bar{L}_1}{\partial T}\right)_{P,N} = \left(\frac{\partial \bar{H}_1}{\partial T}\right)_{P,N} - \left(\frac{\partial \bar{H}_1^0}{\partial T}\right)_{P,N}$$

$$= \bar{C}_{P1} - \bar{C}_{P1}^0, \tag{44.45}$$

where $\bar{C}_{P1}^0$, identical with C_{P1} or C_{P1}^0, is the molar heat capacity of the pure solvent or the partial molar heat capacity of the solvent in a solution at infinite dilution. Thus, $\bar{C}_{P1}^0$ may be regarded as an experimental quantity, and if the variation of the relative partial molar heat content of the solvent with temperature, i.e., $(\partial \bar{L}_1/\partial T)_{P,N}$, is known, it is possible to determine $\bar{C}_{P1}$ at the corresponding composition of the solution. The necessary data are rarely available from direct thermal measurements of $\bar{L}_1$, such as those described in § 44f, at several temperatures, but the information can often be obtained, although not very accurately, from E.M.F. measurements.

By differentiating the expression for the relative partial molar heat content of the solute [equation (44.3)] it is found, in an exactly similar manner to that used above, that

$$\left(\frac{\partial \bar{L}_2}{\partial T}\right)_{P,N} = \left(\frac{\partial \bar{H}_2}{\partial T}\right)_{P,N} - \left(\frac{\partial \bar{H}_0^2}{\partial T}\right)_{P,N}$$
$$= \bar{C}_{P2} - \bar{C}_{P2}^0. \qquad (44.46)$$

In this expression $\bar{C}_{P2}^0$ is the partial molar heat capacity of the solute in the infinitely dilute solution. Although the experimental significance of this quantity is not immediately obvious, it will be shown below to be capable of direct determination. Thus, from a knowledge of the variation of $\bar{L}_2$, the partial molar heat content of the solute, with temperature, it should be possible to derive, with the aid of equation (44.46), the partial molar heat capacity of the solute $\bar{C}_{P2}$ at the given composition.

It was seen in § 44g that it is sometimes possible to express $\bar{L}_2$ as a function of temperature by means of E.M.F. measurements. In this event, the differentiation can be carried out so as to give $\bar{C}_{P2} - \bar{C}_{P2}^0$ directly. Thus, utilizing equation (44.38), it follows from (44.46) that

$$\bar{C}_{P2} - \bar{C}_{P2}^0 = 2NF(c - c^0)T,$$

which is applicable at the same concentration as are (44.36) and (44.37). In Table XXXVII a comparison is made of the relative partial molar heat capacities of hydrochloric acid in various solutions as obtained from E.M.F. and from thermal measurements; the corresponding values of the relative partial molar heat contents are also given.[17]

TABLE XXXVII. THERMAL PROPERTIES OF HYDROCHLORIC ACID IN AQUEOUS SOLUTION AT 25° C

Molality	$\bar{L}_2$		$\bar{C}_{P2} - \bar{C}_{P2}^0$	
	E.M.F.	Thermal	E.M.F.	Thermal
0.02	87	100 cal. mole^{-1}	1.5	— cal. deg.$^{-1}$ mole^{-1}
0.05	132	150	2.2	—
0.10	181	203	2.9	2.4
0.20	245	274	3.7	3.4
0.50	396	431	5.3	5.3
1.0	606	645	7.6	7.5
2.0	1055	1056	11.2	10.6
3.0	1506	1486	14.0	13.0

It will be recalled that in connection with partial molar heat contents it was not possible to determine actual values, but only the values relative to

[17] See ref. 13.

a chosen reference state. However, with partial molar heat capacities, actual values can be obtained from experimental measurements at constant pressure (or volume), just as can the total heat capacity of the system. In fact, by determining the latter at various compositions, the partial molar heat capacity of each constituent can be evaluated by the method of intercepts (§ 42c, III and IV). For solutes of limited solubility, e.g., electrolytes, when heat capacity measurements cannot be made over the whole range of composition from pure 1 to pure 2, only one intercept, namely, the one which gives $\bar{C}_{P1}$, can be obtained. The corresponding $\bar{C}_{P2}$ could then be estimated by the use of equation (26.8) and graphical integration. However, because of the relatively low concentrations, in terms of mole fractions, the method of intercepts is not very accurate for solutions of salts.

44k. Apparent Molar Heat Capacities.—Another approach to the problem of partial molar heat capacities is through a consideration of the respective apparent quantities. In accordance with the usual definition [equation (42.1)], the apparent molar heat capacity of the solute is given by

$$\phi_C = \frac{C_P - n_1 \bar{C}_{P1}^0}{n_2}, \qquad (44.47)$$

where $\bar{C}_{P1}^0$ has been written, for consistency, in place of C_{P1}^0, the heat capacity of the pure solvent; C_P is the experimental heat capacity of the solution containing n_1 moles of solvent and n_2 moles of solute. In accordance with the general equation (42.3), therefore,

$$\bar{C}_{P2} = \phi_C + n_2 \left(\frac{\partial \phi_C}{\partial n_2}\right)_{n_1}, \qquad (44.48)$$

constant temperature and pressure being understood. Utilizing the molality m in place of n_2, in the usual manner, this becomes

$$\bar{C}_{P2} = \phi_C + m \frac{d\phi_C}{dm}. \qquad (44.49)$$

By plotting the observed values of the apparent molar heat capacity against the molality, it is possible to determine $d\phi_C/dm$ at any molality, and hence $\bar{C}_{P2}$ at that molality can be obtained from equation (44.49). This procedure is satisfactory for nonelectrolytes, but for electrolytes it is preferable to plot ϕ_C as a function of $m^{1/2}$, as in § 44f; thus, equation (44.49) may be written

$$\bar{C}_{P2} = \phi_C + \frac{m^{1/2}}{2} \cdot \frac{d\phi_C}{dm^{1/2}}. \qquad (44.50)$$

For solutions of electrolytes containing large proportions of solute, e.g., concentrated solutions of sulfuric acid, the plot of ϕ_C against $m^{1/2}$ gives less accurate results than does an alternative method. Utilizing the fact that the mole fractions N_1 and N_2 are equal to $n_1/(n_1 + n_2)$ and $n_2/(n_1 + n_2)$, respectively, it can be readily shown that equation (44.48) can be put in the form

$$\bar{C}_{P2} = \phi_C + \mathrm{N}_1\mathrm{N}_2 \frac{d\phi_C}{d\mathrm{N}_2} \qquad (44.51)$$

at constant temperature and pressure. The values of $d\phi_C/d\mathrm{N}_2$ are then obtained from the plot of ϕ_C against N_2, the mole fraction of the solute; thus, $\bar{C}_{P2}$ can be obtained for any composition by means of equation (44.51).

It will be observed from equation (44.49) that at infinite dilution, when m is zero but $d\phi_C/dm$ is finite, $\bar{C}_{P2}$ must become equal to ϕ_C; hence [cf. equation (44.15)],

$$\bar{C}_{P2}^0 = \phi_C^0.$$

Since the value of the apparent molar heat capacity at infinite dilution may be obtained by the extrapolation of experimental data, it is thus possible, as indicated earlier, to determine $\bar{C}_{P2}^0$ the partial molar heat capacity of the solute at infinite dilution.

The foregoing treatment permits the calculation of the partial molar heat capacities of the solute; it will now be shown how the values for the solvent may be obtained from the same data. By combining equations (44.47) and (44.48) with the general relationship for partial molar quantities, viz.,

$$C_P = n_1 \bar{C}_{P1} + n_2 \bar{C}_{P2},$$

where C_P is, as before, the observed heat capacity of the solution of the specified composition, and $\bar{C}_{P1}$ and $\bar{C}_{P2}$ are the partial molar heat capacities of solvent and solute in that solution, it is found that, at constant temperature and pressure,

$$\bar{C}_{P1} = \bar{C}_{P1}^0 - \frac{n_2^2}{n_1}\left(\frac{\partial \phi_C}{\partial n_2}\right)_{n_1}, \qquad (44.52)$$

which is analogous to equation (44.27). As before, replacing n_2 by the molality, and n_1 by $1000/M_1$, where M_1 is the molecular weight of the solvent, equation (44.52) becomes

$$\bar{C}_{P1} = \bar{C}_{P1}^0 - \frac{M_1 m^2}{1000} \cdot \frac{d\phi_C}{dm},$$

and hence

$$\bar{C}_{P1} = \bar{C}_{P1}^0 - \frac{M_1 m^{3/2}}{2000} \cdot \frac{d\phi_C}{dm^{1/2}}.$$

The former of these two equations may be used for nonelectrolytes and the latter for electrolytes. Since $\bar{C}_{P1}^0$ is actually equal to the molar heat capacity of the pure solvent, its value may be taken as known. For solutions of high molality, better results are obtained by utilizing the equivalent form

$$\bar{C}_{P1} = \bar{C}_{P1}^0 - \mathrm{N}_2^2 \frac{d\phi_C}{d\mathrm{N}_2},$$

where N_2 is the mole fraction of the solute.[18]

[18] Randall and Ramage, J. Am. Chem. Soc., 49, 93 (1927); Randall and Rossini, ibid., 52, 323 (1930); Rossini, J. Res. Nat. Bur. Stand., 4, 313 (1930); 6, 791 (1931); 7, 47 (1931); 9, 679 (1932).

The partial molar heat capacities of water (1) and sulfuric acid (2) at 25° C, obtained as described above, are recorded in Table XXXVIII.[19]

TABLE XXXVIII. HEAT CAPACITIES OF AQUEOUS SOLUTIONS OF SULFURIC ACID AT 25° C

Molality m	$\dfrac{\text{Moles } H_2O}{\text{Moles } H_2SO_4}$	ϕ_C	$\bar{C}_{P1}$ cal. deg.$^{-1}$ mole^{-1}	$\bar{C}_{P2}$
0.0347	1,600	6.00	18.01	7.06
0.0694	800	6.85	18.01	8.37
0.1388	400	8.10	18.01	10.22
0.2775	200	9.82	18.00	12.84
0.5551	100	12.32	17.97	16.55
1.110	50	15.85	17.90	21.78
1.850	30	19.35	17.77	26.80
3.700	15	22.40	18.04	22.01
5.551	10	21.22	18.57	15.68
6.938	8	19.85	18.88	12.92
9.251	6	18.77	18.02	18.77
13.876	4	20.15	16.46	26.39
18.502	3	22.20	15.33	30.26
27.753	2	26.03	12.51	37.05
44.406	1.25	31.52	9.92	41.64
55.51	1.00	33.16	15.09	36.08
85.62	0.6483	32.40	22.94	29.21
186.51	0.2976	31.40	17.96	31.42
388.70	0.1428	31.68	14.23	32.22
∞	0	32.88	2.94	32.88

Attention may be called to the remarkable variations in these quantities with the composition, particularly the sharp drop of $\bar{C}_{P1}$ from about 18 to 9.9, followed by a rapid increase to over 20 cal. deg.$^{-1}$ mole^{-1}, and the corresponding rise of $\bar{C}_{P2}$ up to 41.6, followed by the decrease to 29 cal. deg.$^{-1}$ mole^{-1}, in the region of 50 molal. This composition corresponds closely to equimolecular amounts of water and sulfuric acid, and suggests the formation of a strong complex, viz., $H_2SO_4 \cdot H_2O$. The other, minor, variations must also have a definite chemical significance, but their consideration lies outside the scope of this book. It may be mentioned, however, that these variations are not evident in the total heat capacity of the mixture; they become apparent only when the results are treated thermodynamically so as to give the respective partial molar quantities.

Problem: A solution of 2 moles of H_2SO_4 to 100 moles H_2O at 30° C is mixed with one containing 1 mole of H_2SO_4 to 200 moles H_2O at 25° C. Assuming the heat capacities to remain constant, and ignoring losses due to radiation, etc., what will be the final temperature?

The problem is here somewhat similar to that involved in the determination of flame temperatures in §§ 13a, 13c. The simplest procedure is to bring the initial solutions to the reference temperature, 25° C, and allow them to mix at this temperature; from the net heat change, the temperature of the final solution can be calculated.

[19] Calculations by Craig and Vinal, ref. 10 from data that are probably not too reliable.

For the solution of 2 moles H_2SO_4 to 100 moles H_2O, i.e., $50H_2O/1H_2SO_4$, C_P is $100\bar{C}_{P1} + 2\bar{C}_{P2} = 100 \times 17.90 + 2 \times 21.78 = 1{,}833.6$ cal. deg.$^{-1}$ (from Table XXXVIII). Hence, if this is cooled from 30° to 25° C, ΔH is $-5 \times 1{,}833.6 = -9{,}168$ cal., assuming C_P to remain constant. The other solution is already at 25° C, and for this ΔH is zero.

Since L is equal to $n_2\phi_L$, by equation (44.23), the relative heat contents of the initial solutions at 25° C are $2\phi_L$ for $50H_2O/1H_2SO_4$ plus ϕ_L for $200H_2O/1H_2SO_4$, i.e., $2 \times 5{,}780 + 5{,}410 = 16{,}970$ cal. (from Table XXXVI). The final solution contains 3 moles of H_2SO_4 and 300 moles H_2O, so that the relative heat content is $3\phi_L$ for $100H_2O/1H_2SO_4$, i.e., $3 \times 5{,}620 = 16{,}860$ cal. Hence ΔH for mixing is $16{,}860 - 16{,}970 = -110$ cal. The total ΔH for cooling and mixing is thus $-9{,}168 - 110 = -9{,}278$ cal.

This value of ΔH must be equal in magnitude but opposite in sign to that required to heat the final solution to the final temperature. The heat capacity of the solution is $300 \times 17.97 + 3 \times 16.55 = 5{,}441$ cal. deg.$^{-1}$. Hence, the increase of temperature, above 25° C, is $9{,}278/5{,}441 = 1.71°$; the final temperature is, therefore, 26.71° C.

441. Heat Capacities and the Debye-Hückel Theory.—By combining the general equation (44.46) with the expression for $\bar{L}_2$ [equation (44.39)] derived from the Debye-Hückel theory, it is found that for a solution containing a single strong electrolyte,

$$\bar{C}_{P2} - \bar{C}_{P2}^0 = \frac{2.469 \times 10^{14}}{(DT)^{3/2}} (\textstyle\sum \nu_i z_i^2)^{3/2} [f(D, V, T)]_P \sqrt{c} \text{ ergs deg.}^{-1} \text{ mole}^{-1}, \quad (44.53)$$

where $f(D, V, T)$ is a somewhat complicated function involving the first and second derivatives of D and V with respect to temperature, at constant pressure. These quantities are known only approximately, but for aqueous solutions at 25° C, equation (44.53) becomes, after conversion to calories,

$$\bar{C}_{P2} - \bar{C}_{P2}^0 = 4.7(\textstyle\sum \nu_i z_i^2)^{3/2} \sqrt{c} \text{ cal. deg.}^{-1} \text{ mole}^{-1}. \quad (44.54)$$

From this result, it is possible to obtain, by means of an equation of the form of (42.7), an expression for the apparent molar heat capacity; thus,

$$\phi_C = \phi_C^0 + \tfrac{2}{3} \times 4.7(\textstyle\sum \nu_i z_i^2)^{3/2} \sqrt{c} \text{ cal. deg.}^{-1} \text{ mole}^{-1}. \quad (44.55)$$

It is doubtful if heat capacity measurements of sufficient reliability have yet been made in solutions dilute enough for equations (44.53) to (44.55) to be applicable. It cannot be said, therefore, that the Debye-Hückel limiting law for heat capacities of electrolytes has been adequately tested. Such a test, involving as it does two successive differentiations with respect to temperature of the Debye-Hückel equation for the activity coefficient (or electrical free energy), would be extremely stringent.[20] It may be mentioned that, as in the case of the apparent molar volume, both the apparent and relative molar heat capacities have been found to vary in a linear manner with the square root of the concentration in the range from about 0.2 to 3.0 molar. The slope of the plot of the heat capacity function against $\sqrt{c}$ is specific, varying even among electrolytes of the same valence

---

[20] Cf. Young and Machin, *J. Am. Chem. Soc.*, **58**, 2254 (1936); Wallace and Robinson, *ibid.*, **63**, 958 (1941).

type. The slopes are, of course, quite different from the limiting values required by the Debye-Hückel equations.[21]

EXERCISES

1. Derive an expression for $\bar{V}_2$ in terms of $d\rho/dc$, analogous to equation (43.2). The variation of the density of aqueous ammonium nitrate solutions with the molarity (c) at 25° C is given by

$$\rho = 0.99708 + 3.263 \times 10^{-2}c - 9.63 \times 10^{-4}c^{3/2} - 4.73 \times 10^{-5}c^2 \text{ g. ml.}^{-1}$$

[Gucker, J. Phys. Chem., **38**, 307 (1934)]. Determine the partial molar volume of the salt in a 1.0 molar solution.

2. Show that if V is the volume of a solution consisting of m moles of solute in 1,000 g. of solvent (i.e., molality m), then

$$\bar{V}_2 = \frac{1}{\rho}\left(M_2 - V\frac{d\rho}{dm}\right),$$

where M_2 is the molecular weight of the solute and $\bar{V}_2$ is its partial molar volume in the given solution.

3. The apparent molar heat capacity of sucrose in water is given as a function of the molality by the expression

$$\phi_C = 151.50 + 1.130m - 0.0466m^2 \text{ cal. deg.}^{-1} \text{ mole}^{-1}$$

[Gucker, Pickard and Planck, J. Am. Chem. Soc., **61**, 459 (1939)]. Derive expressions for $\bar{C}_{P1}$ and $\bar{C}_{P2}$; $\bar{C}_{P1}^0$ is 18.02 cal. deg.$^{-1}$ mole^{-1}. Calculate the heat capacity of a solution consisting of 1 mole of sucrose and 1,000 g. water.

4. Give in full the derivation of equation (42.13).

5. Prove that, in general, $\phi_G^0 = \bar{G}_2^0$, so that the apparent molar value of any property of a solution at infinite dilution is equal to the partial molar property of the solute in the same solution.

6. When various numbers of moles (n_1) of water were added to 9,500 g. of a 2.9 molar solution of strontium chloride at 25° C, the observed changes of heat content were as follows:

n_1	37.2	30.0	27.7	20.0	10.0 moles
ΔH	$-$ 129.8	$-$ 113.3	$-$ 109.2	$-$ 86.3	$-$ 48.5 cal.

[Stearn and Smith, J. Am. Chem. Soc., **42**, 18 (1920)]. Evaluate the differential heat of dilution, i.e., $\bar{L}_1$, of the 2.9 molar strontium chloride solution.

7. The mean ionic activity coefficient of 1 molal sodium chloride was found to be 0.641 at 80°, 0.632 at 90° and 0.622 at 100° C [Smith and Hirtle, J. Am. Chem. Soc., **61**, 1123 (1939)]. Determine the value of $\bar{L}_2$ at about 90° C.

8. Using the values for $\bar{L}_2$ for hydrochloric acid solutions at various molalities at 25° C, calculate by a graphical method the values of $\bar{L}_1$ at 0.1, 0.5, 1.0, 2.0 and 3.0 molal.

m	$\bar{L}_2$	m	$\bar{L}_2$	m	$\bar{L}_2$
0.0001	7.2 cal.	0.04	136 cal.	1.0	645 cal.
0.0016	28.6	0.10	202	2.0	1,055
0.01	71	0.50	430	3.0	1,484

---

[21] Randall and Rossini, ref 18; Rossini, ref. 18; Gucker, et al., J. Am. Chem. Soc., **54**, 1358 (1932); **55**, 1013 (1933); **57**, 78 (1935). For nonelectrolytes, see Gucker, et al., ibid., **59**, 447, 2152 (1937); **61**, 459 (1939).

EXERCISES 459

9. The values of $\bar{C}_{P2} - \bar{C}^0_{P2}$ for hydrochloric acid in the temperature range from 0° to 25° may be taken to be constant, viz., 2.4 cal. deg.$^{-1}$ for 0.1 molal, 7.5 cal. deg.$^{-1}$ for 1.0 molal, and 13.0 cal. deg.$^{-1}$ for 3.0 molal. Use these results together with the data in Exercise 8 to calculate the ratio of the mean ionic activity coefficients of hydrochloric acid at the temperatures 0° and 25° C for 0.1, 1.0 and 3.0 molal solutions. What error is involved in neglecting the variation of $\bar{L}_2$ with temperature, i.e., in taking $\bar{C}_{P2} - \bar{C}^0_{P2}$ to be zero?

10. Making use of the known heats of formation of lead dioxide, pure sulfuric acid, lead sulfate and water, and of the data in Table XXXVI, determine the change of heat content for the reaction

$$\text{Pb} + \text{PbO}_2 + 2\text{H}_2\text{SO}_4(25\ aq) = 2\text{PbSO}_4 + 2\text{H}_2\text{O}(\text{N}_1/\text{N}_2 = 25.0)$$

which takes place in the lead storage battery. (Compare Exercise 4, Chapter V which is for a slightly different concentration of acid.)

11. Upon mixing chloroform and the dimethyl ether of ethylene glycol, the following changes of heat content, per mole of mixture, were observed for various mole fractions (N) of chloroform at 3° C:

N	0.1	0.2	0.4	0.5	0.6	0.8	0.9
ΔH	−290	−560	−1010	−1180	−1180	−720	−400 cal.

[Adapted from Zellhofer and Copley, *J. Am. Chem. Soc.*, **60**, 1343 (1938).] Determine $\bar{H} - H^0$ for each component by a graphical method at the mole fractions 0.2, 0.4, 0.6 and 0.8, and plot the results. Does the system exhibit positive or negative deviations from ideal behavior?

12. Pure acetone has a vapor pressure of 185.2 mm. and ether one of 443.5 mm. at 20° C. For a liquid mixture containing 0.457 mole fraction of the former, the respective partial vapor pressures are 105.2 and 281.8 mm., respectively. Assuming ideal behavior of the vapors, determine the partial molar free energy $\bar{F} - F^0$ of each constituent of the mixture, relative to that of the pure liquid.

13. The E.M.F. of the $\text{H}_2 | \text{HCl}(m) | \text{AgCl}(s)$, Ag cell can be expressed as a function of the temperature of the form

$$E_t = E + a(t - 25) + b(t - 25)^2,$$

t being on the centigrade scale. The following values of a and b were obtained for various molalities of hydrochloric acid, and for the standard (hypothetical 1 molal ideal) solution:

m	0.001	0.01	0.1	1.0	Standard
$a \times 10^6$	560.1	177.9	−180.5	−512.1	−639.64
$b \times 10^6$	−3.149	−3.041	−2.885	−2.541	−3.181.

Determine $\bar{L}_2$ and $\bar{C}_{P2} - \bar{C}^0_{P2}$ for each solution at 25° C [Harned and Ehlers, *J. Am. Chem. Soc.*, **54**, 2179 (1933)].

14. The apparent relative molar heat contents ϕ_L of potassium chloride in aqueous solution at 25° C at various molalities are as follows:

m	ϕ_L	m	ϕ_L	m	ϕ_L
0.01	38 cal. mole^{-1}	0.50	48	2.00	−169
0.05	69	0.70	18	3.00	−300
0.10	78	1.00	−26	4.00	−405
0.20	81	1.50	−99	4.50	−448

Plot ϕ_L against $m^{1/2}$ and hence verify the values of $\bar{L}_2$ given below.

m	$\bar{L}_2$	m	$\bar{L}_2$
0.10	91 cal. mole^{-1}	2.00	-446
0.20	72	3.00	-648
0.50	-21	4.00	-782
1.00	-176	4.82 (sat.)	-840

(Data from Harned and Owen, ref. 8.)

15. The heat of solution of solid potassium chloride to form an infinitely dilute solution at 25° C is $+$ 4,120 cal. mole^{-1}. Use this result and the data in the preceding exercise to determine (i) the magnitude and sign of the heat change when 1 mole of KCl is dissolved in sufficient water to make a saturated solution (4.82 molal); (ii) the integral heat of dilution of a 4.0 molal solution to 0.1 molal, per mole KCl; (iii) the heat change per mole KCl, apart from heat of vaporization, required to concentrate a 0.01 molal solution to the saturation point; (iv) the heat change when 1 mole of KCl separates from a large amount of saturated solution; (v) the heat change when 1 mole of KCl is added to a solution consisting of 1 mole of KCl and 1000 g. water; (vi) the heat change when 1 mole of KCl is added to a very large amount of a 1 molal solution; and (vii) the heat change, apart from heat of vaporization, for the removal of 1 mole of water from a large amount of a 1 molal solution. Give the sign of ΔH in each case.

16. Verify equation (44.41) by using the expression for $F_{\text{el.}}$ given by (40.6).

17. Show that for a solution consisting of n_1 moles of solvent and n_2 moles of solute the following relationship holds: n_2 (integral heat of solution $-$ differential heat of solution) $= n_1$ (differential heat of dilution).

18. The following thermal properties have been recorded for sodium hydroxide solutions at 20° C [adapted from Åkerlöf and Kegeles, *J. Am. Chem. Soc.*, **62**, 620 (1940)]:

m	$\bar{L}_1$	$\bar{L}_2$	$\bar{C}_{P1}$	$\bar{C}_{P2}$
	cal. mole^{-1}		cal. deg.$^{-1}$ mole^{-1}	
0.5	0.6	-14	17.96	-11.3
1.0	2.4	-162	17.81	-1.6
1.5	4.4	-250	17.64	4.4
2.0	5.3	-275	17.46	7.8
3.0	0.3	-170	17.16	9.7
4.0	-16.1	88	16.94	7.7

Assuming the heat capacities to be independent of temperature, determine the final temperature when 1 mole of solid NaOH is dissolved in 1,000 g. water, given that the heat of solution to form an infinitely dilute solution is $-$ 10,100 cal.

Determine the final temperature in each of the following cases: (i) 1,000 g. water are added to a solution consisting of 4 moles NaOH and 1,000 g. water; (ii) 2 moles NaOH are added to a solution containing 2 moles NaOH and 1,000 g. water; (iii) a solution consisting of 1 mole NaOH and 500 g. water is mixed with one consisting of 0.5 mole NaOH and 1,000 g. water.

19. Show that if the vapor behaves as an ideal gas, the relative partial molar heat content $\bar{L}_1$ of the solvent in a solution is given by

$$L_1 = RT^2 \left(\frac{\partial \ln (p_1^0/p_1)}{\partial T} \right)_{P,N},$$

where p_1 is the vapor pressure in equilibrium with the solution, of constant composition, and p_1^0 is that of the pure solvent; the external (total) pressure P is supposed to be 1 atm. Hence suggest possible methods for the determination of $\bar{L}_1$ of a given solution.

20. Show that in very dilute solution the nonelectrical factor $\gamma_{n.e.}$ in the activity coefficient of an electrolyte, defined in § 43c, is independent of temperature, at constant pressure and composition.

CHAPTER XIX

E.M.F.'S AND THE THERMODYNAMICS OF IONS

45. E.M.F.'s AND ELECTRODE POTENTIALS

45a. Reversible Cell and Reversible Electrodes.—It was seen in Chapter XIII that the E.M.F. of a reversible cell is a measure of the free energy change of the process taking place in the cell, and this fact has been already utilized to determine activities and activity coefficients. There are, however, many other useful applications of reversible cells, and some of these will be described in the succeeding sections. It is desirable, in this connection, to consider first the fundamental structural basis of reversible galvanic cells. Every reversible cell consists of at least two reversible electrodes, and several types of such electrodes are known. The *first type* of reversible electrode consists of *an element in contact with a solution of its own ions*, e.g., zinc in zinc sulfate solution or silver in silver nitrate solution. Although electrodes of this kind usually involve metals, certain nonmetals yield reversible electrodes, at least in principle; these are hydrogen, in particular, and also oxygen and the halogens, the corresponding ions being hydrogen, hydroxyl and halogen ions respectively.* Since the electrode material is a nonconductor and often gaseous when the element is a nonmetal, a sheet of platinum, or other inert conductor, coated with a finely divided layer of this metal, is employed for the purpose of maintaining electrical contact.

If the reversible element is a metal or hydrogen, which is in equilibrium with positive ions (cations), the reaction occurring at the electrode when it forms part of a cell may be formulated as

$$M \rightleftharpoons M^{z+} + z\epsilon,$$

where ϵ represents an electron; the valence of the cation in this case is z, so that it carries z positive charges. The direction of the reaction indicated depends on the direction of flow of current through the cell of which the electrode under consideration is part. If the electrode element is a nonmetal A, the corresponding reactions are

$$A^{z-} \rightleftharpoons A + z\epsilon,$$

where A^{z-} is the negative ion (anion) of valence z, e.g., OH^-, Cl^-. With an oxygen electrode, which is theoretically reversible with respect to hydroxyl

---

* Although no method has yet been found for establishing a reversible oxygen gas-hydroxyl ion solution electrode, this electrode is possible in principle; other electrodes, e.g., nickel in a nickel salt solution, are somewhat similar. Such electrodes will be referred to as "theoretically reversible" or as "reversible in principle."

ions, the reaction may be written

$$2\text{OH}^- \rightleftharpoons \tfrac{1}{2}\text{O}_2 + \text{H}_2\text{O} + 2\epsilon.$$

Electrodes of the *second type* consist of *a metal and a sparingly soluble salt of this metal in contact with a solution of a soluble salt of the same anion.* An example of this form of electrode consists of silver, solid silver chloride and a solution of a soluble chloride, e.g., hydrochloric acid, viz.,

$$\text{Ag, AgCl}(s) \,|\, \text{HCl soln.}$$

The reaction in this electrode consists, first, of the passage of silver ions into solution, or the reverse, i.e.,

$$\text{Ag}(s) \rightleftharpoons \text{Ag}^+ + \epsilon,$$

and this is followed by interaction of the silver ions with the chloride ions in the solution, i.e.,

$$\text{Ag}^+ + \text{Cl}^- \rightleftharpoons \text{AgCl}(s)$$

to form the slightly soluble salt silver chloride. The net reaction is then

$$\text{Ag}(s) + \text{Cl}^- \rightleftharpoons \text{AgCl}(s) + \epsilon,$$

which is essentially the same as that of a chlorine electrode,

$$\text{Cl}^- \rightleftharpoons \tfrac{1}{2}\text{Cl}_2 + \epsilon,$$

except that the silver chloride is to be regarded as the source of the chlorine. Electrodes of the type under consideration are thus reversible with respect to the common anion, namely, the chloride ion in this case. The Ag, AgCl electrode is thermodynamically equivalent to a chlorine gas electrode with the gas at a pressure equal to that in equilibrium with silver chloride dissociating at the experimental temperature (cf. § 45j) according to the reaction

$$\text{AgCl}(s) = \text{Ag}(s) + \tfrac{1}{2}\text{Cl}_2(g).$$

Electrodes of this kind have been made with other similar halides, e.g., silver bromide and iodide, and mercurous chloride and bromide, and also with various sparingly soluble sulfates, oxalates, etc.

The *third type* of reversible electrode consists of *an inert metal, e.g., gold or platinum, immersed in a solution containing both oxidized and reduced states of an oxidation-reduction system*, e.g., Fe^{++} and Fe^{+++}, Fe(CN)_6^{---} and Fe(CN)_6^{----}, and $\text{Mn}^{++} + 4\text{H}_2\text{O}$ and $\text{MnO}_4^- + 8\text{H}^+$. The purpose of the inert metal is to act as a conductor for making electrical contact, just as with a gas electrode. The oxidized and reduced states may consist of more than a single species, as in the case of the manganous ion-permanganate system, just given, and the substances involved are not necessarily ionic. For example, in an important type of reversible electrode the oxidized state is a quinone, together with hydrogen ions, while the corresponding hydroquinone is the reduced state. Electrodes of the kind being considered are sometimes called **oxidation-reduction electrodes** or **redox electrodes**; the chemical reactions occurring at these electrodes are either oxidation of the reduced state or reduction of the oxidized state, e.g.,

$$Fe^{++} \rightleftharpoons Fe^{+++} + \epsilon,$$
$$Mn^{++} + 4H_2O \rightleftharpoons MnO_4^- + 8H^+ + 5\epsilon,$$

depending on the direction of the current through the cell. In order that the electrode may behave reversibly, it is essential that both oxidized and reduced states of the given system should be present.

The three types of reversible electrodes described above differ formally, as far as their construction is concerned; nevertheless, *they are all based on the same fundamental principle. A reversible electrode always involves an oxidized and a reduced state,* using the terms "oxidized" and "reduced" in their broadest sense; thus, oxidation refers to the loss of electrons, and reduction means gain of electrons. If the electrode consists of a metal M and its ions M^+ the former is the reduced state and the latter is the oxidized state. Similarly, for an anion electrode, the A^- ions are the reduced state and A represents the oxidized state. In the Ag, AgCl electrode, the metallic silver and the chloride ions together form the reduced state of the system, while the silver chloride is the oxidized state. It can be seen, therefore, that all three types of reversible electrodes are made up from the reduced and oxidized states of a given system, and in every case the electrode reaction may be written in the general form

$$\text{Reduced state} \rightleftharpoons \text{Oxidized state} + N\epsilon,$$

where N is the number of electrons by which the oxidized and reduced states differ.[1]

45b. Reactions in Reversible Cells.—The reaction occurring at a reversible electrode is either oxidation, i.e.,

$$\text{Reduced state} \rightarrow \text{Oxidized state} + N\epsilon,$$

or reduction, i.e.,

$$\text{Oxidized state} + N\epsilon \rightarrow \text{Reduced state}.$$

It can be seen, therefore, that in a reversible cell consisting of two reversible electrodes, there is a continuous flow of electrons, and hence of current, if oxidation occurs at one electrode, where electrons are set free, and reduction occurs at the other electrode, where the electrons are taken up. According to the convention widely adopted, the E.M.F. *of the cell is positive when oxidation takes place at the left-hand electrode of the cell as written, reduction occurring at the right-hand electrode, as a result of spontaneous operation.* If the E.M.F. is positive, there is thus a tendency for electrons to pass from left to right through the wire connecting the electrodes outside the cell. If the reverse is the case, so that reduction is taking place at the left-hand electrode and oxidation at the right-hand electrode, the E.M.F. of the cell as written is

[1] See S. Glasstone, "An Introduction to Electrochemistry," 1942, Chapter VI.

negative. If the cell is written in the opposite direction, the sign of its E.M.F. would be reversed; it is important, therefore, to pay special attention to the actual representation of the cell.

If the sign of the E.M.F. of a cell is known, there is no uncertainty concerning the direction of the reaction, but if it is not available, the reaction is written as if oxidation takes place at the left-hand electrode. By utilizing the general information concerning electrode reactions, it is usually a simple matter to determine the process taking place in a particular cell. Consider, for example, the cell

$$\text{Pt, } H_2 \text{ (1 atm.)} \mid \text{HCl}(m) \mid \text{AgCl}(s), \text{Ag,}$$

referred to in § 39h; assuming the process at the left-hand electrode to be oxidation, i.e., liberation of electrons, it may be written as

$$\tfrac{1}{2}H_2 \text{ (1 atm.)} = H^+ + \epsilon,$$

whereas at the right-hand electrode, the reduction process is

$$\text{AgCl}(s) + \epsilon = \text{Ag}(s) + \text{Cl}^-,$$

so that the complete cell reaction is

$$\tfrac{1}{2}H_2 \text{ (1 atm.)} + \text{AgCl}(s) = H^+(m) + \text{Cl}^-(m) + \text{Ag}(s),$$

the hydrogen and chloride ions,* representing hydrochloric acid, being formed at the molality m existing in the cell. The complete reaction would occur for the passage of 1 faraday of electricity, for 1 electron only is involved in each individual electrode process as written.

If the cell had been written in the reverse direction, viz.,

$$\text{Ag, AgCl}(s) \mid \text{HCl}(m) \mid H_2 \text{ (1 atm.), Pt,}$$

the reaction would be exactly opposite to that given above; thus, for 1 faraday,

$$\text{Ag}(s) + H^+(m) + \text{Cl}^-(m) = \text{AgCl}(s) + \tfrac{1}{2}H_2 \text{ (1 atm.).}$$

In this case the silver is oxidized to silver chloride at the left-hand electrode, while hydrogen ions are reduced to hydrogen gas at the right-hand electrode. At moderate concentrations of hydrochloric acid the cell in this form has a negative E.M.F., and so the spontaneous cell reaction is the reverse of that given here, coinciding with the direction given above.

Another example of a reversible cell for which the reaction is of interest is

$$\text{Ag} \mid \text{AgClO}_4 \text{ soln.} \quad \text{Fe(ClO}_4)_2, \text{Fe(ClO}_4)_3 \text{ soln.} \mid \text{Pt,}$$

equivalent to

$$\text{Ag} \mid \text{Ag}^+ \quad \text{Fe}^{++}, \text{Fe}^{+++} \mid \text{Pt.}$$

* The ions are, of course, hydrated in solution; this may be understood without actually indicating the inclusion of water molecules.

Assuming oxidation to take place at the left-hand electrode, the process is
$$Ag(s) \rightarrow Ag^+ + \epsilon,$$
whereas at the right-hand electrode the reduction process is
$$Fe^{+++} + \epsilon \rightarrow Fe^{++},$$
so that the complete cell reaction, for the passage of 1 faraday, is
$$Ag(s) + Fe^{+++} = Ag^+ + Fe^{++}$$
with the ions in solution, i.e.,
$$Ag(s) + Fe(ClO_4)_3 = AgClO_4 + Fe(ClO_4)_2.$$

45c. Change of Heat Content in Cell Reaction.—The E.M.F. of a reversible cell is related to the free energy change ΔF of the reaction occurring within it, at constant temperature and pressure, by equation (33.36), viz.,

$$\Delta F = - NFE, \qquad (45.1)$$

where E is the E.M.F., and N is the number of faradays (F) required for the cell reaction. If the entropy change for the process were known it would be possible to determine the change in heat content ΔH, i.e., the heat of reaction at constant pressure, at the given temperature. Alternatively, ΔS for the reaction may be determined, as indicated in § 33m, from the temperature coefficient of the E.M.F. of the cell. As given in that section, the treatment referred in particular to the *standard* entropy change, but the results are applicable to any conditions existing in the cell.

The matter may be treated quite generally by utilizing the Gibbs-Helmholtz equation (25.31), viz.,

$$\Delta F = \Delta H + T \left[\frac{\partial (\Delta F)}{\partial T} \right]_P, \qquad (45.2)$$

so that combination with (45.1) gives

$$NFE = - \Delta H + NFT \left(\frac{\partial E}{\partial T} \right)_P$$

or

$$\Delta H = - NF \left[E - T \left(\frac{\partial E}{\partial T} \right)_P \right]. \qquad (45.3)$$

From this equation it is possible to calculate ΔH for any cell reaction, irrespective of whether the substances involved are present in their standard states or not. It will be evident that the term $NF(\partial E/\partial T)_P$ is equivalent to ΔS, the entropy change for the cell reaction.

Problem: The E.M.F. of the cell

$$Zn \mid ZnCl_2(1.0m) \mid AgCl(s), Ag$$

is 1.015 volts at 0° C and 1.005 volts at 25° C. Assuming the temperature coefficient to be approximately constant in this vicinity, calculate the heat change of the cell reaction at 25° C.

First, consider the actual cell reaction. Since the E.M.F. is positive, oxidation takes place at the left-hand electrode; thus,

$$\text{Zn}(s) \rightarrow \text{Zn}^{++} + 2\epsilon.$$

The simultaneous reduction reaction at the right-hand electrode is

$$2\text{AgCl}(s) + 2\epsilon \rightarrow 2\text{Ag}(s) + 2\text{Cl}^-,$$

so that the complete cell reaction is

$$\text{Zn}(s) + 2\text{AgCl}(s) = \text{Zn}^{++}(1m) + 2\text{Cl}^-(2m) + 2\text{Ag}(s),$$

or

$$\text{Zn}(s) + 2\text{AgCl}(s) = \text{ZnCl}_2(1m) + 2\text{Ag}(s),$$

for the passage of 2 faradays, since 2 electrons are involved in each electrode process.* It will be noted that the $\text{Zn}^{++} + 2\text{Cl}^-$ is equivalent to zinc chloride in the 1 molal solution existing in the cell.

The heat of the reaction as written may be obtained from equation (45.3). By taking F, the faraday, as 23,070 cal. volt^{-1} g. equiv.$^{-1}$ (Table 1, Appendix) with E in volts and $(\partial E/\partial T)_P$ in volt deg.$^{-1}$, the value of ΔH will be in calories. At 25° C, E is 1.005 volts and $(\partial E/\partial T)_P$ may be taken as equal to $\Delta E/\Delta T$, i.e., $(1.005 - 1.015)/25 = -4.0 \times 10^{-4}$ volt deg.$^{-1}$. Hence, since N is 2 equiv. in this case, it follows that at 25° C, i.e., 298.2° K,

$$\Delta H = -2 \times 23,070[1.005 + (298.2 \times 4.0 \times 10^{-4})]$$
$$= -51,830 \text{ cal.} = -51.83 \text{ kcal.}$$

It may be noted that since the zinc chloride solution is 1 molal, the ionic activities are somewhat different from unity and vary with temperature; hence ΔH as calculated is not quite the same as the standard value, although the difference would probably be less than the experimental error.

45d. General Expression for E.M.F.'s of Reversible Cells.—Suppose that the general reaction

$$a\text{A} + b\text{B} + \cdots = l\text{L} + m\text{M} + \cdots$$

can take place in a reversible cell of E.M.F. equal to E, for the passage of N faradays; for this reaction the free energy change ΔF is $-NFE$, with the substances A, B, ..., L, M, etc., at the activities actually existing in the cell. If these substances were all in their respective chosen standard states, the E.M.F. would be the standard value E^0, and the standard free energy change ΔF^0 would be $-NFE^0$.

---

* It would be quite permissible to write the reaction as

$$\tfrac{1}{2}\text{Zn}(s) + \text{AgCl}(s) = \tfrac{1}{2}\text{ZnCl}_2(1m) + \text{Ag}(s)$$

for the passage of *one* faraday.

According to the reaction isotherm derived in § 33a, the free energy change of the process is represented by

$$\Delta F = \Delta F^0 + RT \ln \frac{a_L^l \times a_M^m \times \cdots}{a_A^a \times a_B^b \times \cdots} \qquad (45.4)$$

$$= \Delta F^0 + RT \ln J_a, \qquad (45.5)$$

where, as in § 32c, the symbol J_a indicates a function of the same form as the equilibrium constant involving, in this instance, the activities of the various substances in the given cell. Replacing ΔF by $-NFE$ and ΔF^0 by $-NFE^0$, the result is

$$E = E^0 - \frac{RT}{NF} \ln J_a, \qquad (45.6)$$

which is the general equation for the E.M.F. of any reversible cell.

In order to employ equation (45.6), the values may be inserted for the constants R and F; since E is almost invariably expressed in volts, F may be conveniently used in coulombs g. equiv.$^{-1}$, and R in int. joules (i.e., volt-coulombs) deg.$^{-1}$ mole^{-1} (see Table 1, Appendix). Upon converting the logarithms, therefore, equation (45.6) becomes

$$E = E^0 - \frac{2.3026 \times 8.313}{96,500} \cdot \frac{T}{N} \log J_a$$

$$= E^0 - 1.984 \times 10^{-4} \frac{T}{N} \log J_a \text{ volts.} \qquad (45.7)$$

At 25° C, which is the temperature commonly employed for E.M.F. measurements, T is 298.16° K, and hence

$$E = E^0 - \frac{0.05914}{N} \log J_a \text{ volts.} \qquad (45.8)$$

The application of the foregoing equations may be illustrated with reference to the cells considered in § 45b. Thus for the cell Pt, H_2 (1 atm.) | HCl(m) | AgCl(s), Ag, the reaction for 1 faraday ($N = 1$) is

$$\tfrac{1}{2}H_2 \text{ (1 atm.)} + \text{AgCl}(s) = H^+ + Cl^- + \text{Ag}(s),$$

and hence, by equation (45.6),

$$E = E^0 - \frac{RT}{F} \ln \frac{a_{H^+} a_{Cl^-} a_{Ag}}{a_{H_2}^{\frac{1}{2}} a_{AgCl}}. \qquad (45.9)$$

Since the silver and silver chloride are present as solids, they are in their conventional standard states;* hence a_{Ag} and a_{AgCl} may be set equal to unity. Further, the hydrogen gas is at a pressure of 1 atm. and although

---

* Unless there is a specific statement to the contrary, it may be assumed that the external pressure is 1 atm., so that pure solids taking part in cell reactions are in their standard states.

this does not represent unit fugacity, the departure from ideal behavior is so small that the difference is quite negligible; hence the hydrogen gas may be regarded as being in its standard state, so that a_{H_2} is also unity. Consequently, equation (45.9) becomes

$$E = E^0 - \frac{RT}{F} \ln a_{H^+} a_{Cl^-}. \tag{45.10}$$

Another convenient form of this equation is obtained by replacing $a_{H^+}a_{Cl^-}$ by $(a_{HCl})^2$ in accordance with (39.6); hence,

$$E = E^0 - \frac{2RT}{F} \ln a_{HCl}, \tag{45.11}$$

where a_{HCl} is the mean activity, i.e., $a_{\pm}$, of the hydrochloric acid in the cell solution. The standard E.M.F. of the cell E^0 is seen to be its E.M.F. when the product of the activities of the hydrogen and chloride ions, or the activity of the hydrochloric acid, in the cell is unity. Actually this standard E.M.F. was determined by the extrapolation procedure described in § 39h, its value being 0.2224 volt at 25° C.

For the other cell considered in § 45b, involving silver and iron perchlorates, the reaction for 1 faraday is

$$Ag(s) + Fe^{+++} = Ag^+ + Fe^{++},$$

so that the E.M.F. is given by

$$E = E^0 - \frac{RT}{F} \ln \frac{a_{Ag^+} a_{Fe^{++}}}{a_{Fe^{+++}}}, \tag{45.12}$$

the activity of the solid silver being omitted, since this is unity. In general, the activity factors for substances which are present in the cell in their standard states may be omitted from the expression for the E.M.F.

45e. Single Electrode Potentials.—There is at present no method available for the experimental determination of the potential of a single electrode; it is only the E.M.F. of a cell, made by combining two or more electrodes, which can be actually measured. In fact, it is doubtful if absolute single electrode potentials, like single ionic activities, have any real thermodynamic significance. However, by choosing an arbitrary zero of potential, it is possible to express the potentials of individual electrodes on an arbitrary reference scale. The zero of potential used in thermodynamic studies is the potential of a reversible hydrogen electrode with gas at 1 atm. pressure (or ideally, unit fugacity) in a solution containing hydrogen ions at unit activity. This particular electrode, i.e., H_2 (1 atm.), $H^+(a_{H^+} = 1)$, is the **standard hydrogen electrode**, since both the hydrogen gas and the hydrogen ions are in their respective standard states of unit activity. The convention given above, therefore, *is to take the potential of the standard hydrogen electrode as zero at all temperatures;* electrode potentials based on this scale are said to refer to the **hydrogen scale**.

If any electrode, e.g., one consisting of the metal M in contact with a solution of its ions M^{z+}, is combined with the standard hydrogen electrode to make the complete cell

$$M\,|\,M^{z+}(a_+) \qquad H^+(a_{H^+} = 1)\,|\,H_2 \text{ (1 atm.)},$$

the potential of the left-hand electrode, on the hydrogen scale, would be equal to the measured E.M.F. of the cell, since the potential of the right-hand electrode is zero, by convention.* The reaction taking place in this cell for the passage of z_+ faradays is

$$M(s) + z_+H^+(a_{H^+} = 1) = M^{z+}(a_+) + \tfrac{1}{2}z_+H_2 \text{ (1 atm.)},$$

and since the solid metal M, the hydrogen ions and the hydrogen gas are all in their standard states, it follows immediately, from equation (45.6), that the E.M.F. of the cell or the potential of the electrode is given by

$$E_M = E_M^0 - \frac{RT}{z_+ F} \ln a_+, \qquad (45.13)$$

where a_+ is the activity of the M^{z+} ions in the electrode solution. The standard electrode potential E_M^0 is seen from equation (45.13) to be that when the M^{z+} ions in the solution are at unit activity.

If the electrode is reversible with respect to an anion of valence z_-, so that the cell obtained by combining it with a standard hydrogen electrode is

$$A\,|\,A^{z-}(a_-) \qquad H^+(a_{H^+} = 1)\,|\,H_2 \text{ (1 atm.)},$$

the cell reaction for z_- faradays is

$$A^{z-}(a_-) + z_-H^+(a_{H^+} = 1) = A + \tfrac{1}{2}z_-H_2 \text{ (1 atm.)}.$$

Here again, the electrode potential is equal to the E.M.F. of the cell, given by equation (45.6) as

$$\begin{aligned}E_A &= E_A^0 - \frac{RT}{z_- F} \ln \frac{1}{a_-} \\ &= E_A^0 + \frac{RT}{z_- F} \ln a_-,\end{aligned} \qquad (45.14)$$

where a_- is the activity of the A^{z-} ions in the electrode solution; the nonmetal A is assumed to be in its standard state. As before, the standard potential E_A^0 of the electrode is that when the reversible ions, i.e., A^{z-}, are present at unit activity.

The results just derived may be put in a general form applicable to electrodes of all types. When any reversible electrode is combined with a standard hydrogen electrode, as depicted above, an oxidation reaction takes place at the former, while at the latter the hydrogen ions are reduced to

* It is tacitly assumed here, and subsequently, that liquid-junction potentials, i.e., potentials existing at the junction between two different solutions, do not arise or else have been completely eliminated.

hydrogen gas. Writing the electrode process in the general form given at the end of § 45a, viz.,

$$\text{Reduced state} = \text{Oxidized state} + N\epsilon,$$

and the corresponding hydrogen electrode reaction as

$$NH^+(a_{H^+} = 1) + N\epsilon = \tfrac{1}{2}NH_2 \text{ (1 atm.)},$$

the complete cell reaction for the passage of N faradays is

$$\text{Reduced state} + NH^+(a_{H^+} = 1) = \text{Oxidized state} + \tfrac{1}{2}NH_2 \text{ (1 atm.)}.$$

The E.M.F. of the cell, which is equal to the potential of the reversible electrode under consideration, is then obtained from equation (45.6) as

$$E_{\text{el.}} = E^0_{\text{el.}} - \frac{RT}{NF} \ln \frac{(\text{Oxidized state})}{(\text{Reduced state})}, \qquad (45.15)$$

the activities of the hydrogen gas and hydrogen ions being omitted since they are both unity in the standard hydrogen electrode. This equation, in which the parentheses are used to indicate the product of the activities of the substances concerned, is completely general, and expresses the potential of any electrode on the hydrogen scale. The standard potential $E^0_{\text{el.}}$ is that for the electrode when all the substances concerned are present in their respective standard states of unit activity.

In the case of an electrode consisting of a metal M (reduced state) and its ions M^{z+} (oxidized state), the activity of the reduced state, i.e., the metal, is unity; hence, the general equation (45.15) becomes equivalent to (45.13), since N is now equal to z_+. On the other hand, for an electrode involving anions A^{z-} (reduced state) and A (oxidized state), the activity of the latter is unity, and equation (45.15) becomes identical with (45.14).

For an electrode containing a conventional oxidation-reduction system, terms are included in both numerator and denominator of equation (45.15). For example, if the system is manganous-permanganate ions, as given in § 45a, the expression for the electrode potential is

$$E_{\text{el.}} = E^0_{\text{el.}} - \frac{RT}{5F} \ln \frac{a_{\text{MnO}_4^-} a_{H^+}^8}{a_{\text{Mn}^{++}}},$$

the solution being assumed to be dilute enough for the activity of the water to be taken as unity (cf. § 38c, problem).

45f. Sign of Electrode Potential.—It will be recalled that in deriving equation (45.15) for the potential of an electrode on the hydrogen scale, the electrode was placed at the left in the hypothetical cell in which it was combined with a hydrogen electrode. The reaction taking place in the cell, as seen above, involved the conversion of the reduced state to the oxidized state of the (left-hand) electrode system. For this reason, the potentials given by equation (45.15), and also by the related equations (45.13) and (45.14), are called **oxidation potentials**. If the electrode had been written to the right of the cell, the reaction taking place within it would be reduction, and the potential would be equal to the value given by equation (45.15) *but with the sign reversed.* The **reduction potential** of any electrode is thus equal in magnitude, but of *opposite sign,* to the corresponding oxidation potential.

In order to indicate whether the process in a particular electrode is to be regarded as oxidation or reduction, a simple convention is employed. If the electrode material is a metal, an oxidation electrode is represented by writing the reduced state of the system to the left and the oxidized state to the right; the following, for example, are all oxidation electrodes:

$$Cu, Cu^{++} \text{ (e.g., } CuSO_4 \text{ soln.)}; \qquad Ag, Ag^+ \text{ (e.g., } AgNO_3 \text{ soln.)};$$
$$Ag, AgCl(s), Cl^- \text{ (e.g., KCl soln.)}.$$

The potentials of these electrodes are given by the appropriate forms of equation (45.15).

If the electrodes are written in the reverse manner, with the oxidized state to the left and the reduced state to the right, viz.,

$$Cu^{++}, Cu; \qquad Ag^+, Ag; \qquad Cl^-, AgCl(s), Ag,$$

it is implied that the electrode process is reduction, and the potential is given by equation (45.15) with the sign reversed.

When the electrode material is nonmetallic, and an inert metal must be used to act as an electrical conductor, the convention is to write the symbol of the inert metal, e.g., Pt, to the left when an oxidation electrode is implied and to the right when a reduction process is to be inferred. The order of writing the components of the solution is immaterial, although if one is a gas or a sparingly soluble liquid or solid it is usually written next to the inert metal. The following are oxidation electrodes:

$$Pt, Cl_2, Cl^-; \qquad Pt, O_2, OH^-; \qquad Pt, I_2, I^-; \qquad Pt, Fe^{++}, Fe^{+++},$$

and the corresponding reduction electrodes are:

$$Cl^-, Cl_2, Pt; \qquad OH^-, O_2, Pt; \qquad I^-, I_2, Pt; \qquad Fe^{++}, Fe^{+++}, Pt.$$

If these conventions are rigidly adopted, many of the complexities which are to be found in the literature become unnecessary.[2]

If any two reversible electrodes are combined to form a reversible cell, e.g.,

$$Pt, Cl_2(g) \,|\, Cl^- \,|\, AgCl(s), Ag,$$

then in accordance with the conventions given above the process at the left-hand electrode is presumed to be oxidation, while at the right-hand electrode a reduction is supposed to take place when the cell operates spontaneously, upon closing the external circuit. It will be noted that this is in agreement with the convention adopted in § 45b. *The E.M.F. of a complete cell is then*

---

[2] G. N. Lewis and M. Randall, "Thermodynamics and the Free Energy of Chemical Substances," 1923, p. 402.

equal to the algebraic sum of the potentials of the two electrodes, one being an oxidation potential and the other a reduction potential. Assuming, for simplicity, that in the cell depicted above the chlorine gas is at 1 atm. pressure and the activity of the chloride ions is unity, the (oxidation) potential of the left-hand electrode is found experimentally to be -1.358 volt, and the (reduction) potential of the right-hand electrode is $+0.2224$ volt, at 25° C. The E.M.F. of the cell as written is therefore $-1.358 + 0.222 = -1.136$ volt. The negative sign shows immediately that if this cell were to be constructed, then upon closing the circuit there would actually be reduction at the chlorine gas electrode and oxidation at the silver electrode. By writing the cell in the opposite direction, i.e.,

$$\text{Ag, AgCl}(s) \,|\, \text{Cl}^- \,|\, \text{Cl}_2(g), \text{Pt},$$

the E.M.F. would be 1.136 volt, indicating that the spontaneous cell reaction involves oxidation at the left-hand electrode, viz.,

$$\text{Ag}(s) + \text{Cl}^- = \text{AgCl}(s) + \epsilon,$$

and reduction at the right-hand electrode, viz.,

$$\tfrac{1}{2}\text{Cl}_2 + \epsilon = \text{Cl}^-.$$

The net reaction in the cell when the circuit is closed is therefore

$$\text{Ag}(s) + \tfrac{1}{2}\text{Cl}_2(g) = \text{AgCl}(s),$$

that is, the formation of solid silver chloride from solid silver and chlorine gas; this is in agreement with expectation, for the reverse reaction would certainly not be spontaneous at ordinary temperature and pressure.

Attention may be called to a matter of interest in connection with the E.M.F. of a cell. As seen above, the latter is the sum of an oxidation and a reduction potential, and this is equivalent to *the difference of two oxidation potentials*. As a consequence, the E.M.F. of a cell is independent of the arbitrary zero chosen for the representation of single electrode potentials; the actual value of the zero of the scale, whatever it may be, cancels out when taking the difference of two potentials on the same scale.

45g. Standard Electrode Potentials.—For the purpose of recording electrode potentials the important quantity is the **standard potential** for any given system. The actual value of this potential depends on the chosen standard states; as usual, pure solids and liquids at atmospheric pressure are taken as being in their standard states, and the standard state of an ion is generally chosen so that the ratio of the activity to the molality becomes unity in very dilute solution (§§ 37b, 39a). The activities of ions are thus expressed in terms of their respective molalities. Once its standard potential is known the potential of an electrode containing arbitrary activities (concentrations) of the various substances concerned can be readily calculated by means of the appropriate form of equation (45.15). The obvious method of determining the standard potential of any electrode is to set up this electrode with the components at known activities, to combine it with a standard hydrogen electrode, and then to measure the E.M.F. of the resulting cell. This will give the potential of the electrode on the hydrogen scale. By inserting the value for E in equation (45.15) and using the known activities, the standard potential E^0 can be evaluated. Actually this procedure is not simple: the standard hydrogen electrode is not always a practical device, the activities of the cell substances are not necessarily known, and there is always the possibility of a junction between two liquids in the cell which introduces an uncertain "liquid junction" potential.

One way whereby the difficulties are overcome is to use a reference electrode, of a form more convenient than the hydrogen electrode, as a secondary standard. One such reference electrode is the silver-silver chloride electrode. As already seen, the standard E.M.F. of the cell formed by combining a hydrogen electrode with this electrode is 0.2224 volt at 25° C (§§ 39h, 45d); thus, the E.M.F. of the cell

$$\text{Pt, H}_2 \text{ (1 atm.)} | \text{H}^+(a_{\text{H}^+} = 1) \quad \text{Cl}^-(a_{\text{Cl}^-} = 1) | \text{AgCl}(s), \text{Ag}$$

is 0.2224 volt. The left-hand electrode is the standard hydrogen electrode, and so its potential, by convention, is zero; thus the standard reduction potential of the silver chloride electrode, viz.,

$$\text{Cl}^-(a_{\text{Cl}^-} = 1) | \text{AgCl}(s), \text{Ag},$$

is 0.2224 volt at 25° C.

In order to determine the standard potential of a metal forming a soluble, highly dissociated chloride, e.g., zinc, the measurements are made on cells of the type

$$\text{Zn} | \text{ZnCl}_2(m) | \text{AgCl}(s), \text{Ag}.$$

The cell reaction for the passage of 2 faradays is

$$\text{Zn}(s) + 2\text{AgCl}(s) = 2\text{Ag}(s) + \text{Zn}^{++} + 2\text{Cl}^-,$$

and since the solid zinc, silver chloride and silver are in their respective standard states, the E.M.F. is given by equation (45.6) as

$$E = E^0 - \frac{RT}{2F} \ln a_{\text{Zn}^{++}} a_{\text{Cl}^-}^2. \tag{45.16}$$

The activity of each ionic species may be represented as the product of its molality and its (stoichiometric) activity coefficient; hence $a_{\text{Zn}^{++}}$ is equal to $m\gamma_+$ and a_{Cl^-} to $2m\gamma_-$, where m is the molality of the zinc chloride solution in the cell. By equation (39.9), $\gamma_+\gamma_-^2$ is equal to $\gamma_\pm^3$, where $\gamma_\pm$ is the mean ionic activity coefficient; hence equation (45.16) becomes

$$E = E^0 - \frac{RT}{2F} \ln [m \times (2m)^2] \gamma_\pm^3$$

$$= E^0 - \frac{RT}{2F} \ln 4m^3 - \frac{3RT}{2F} \ln \gamma_\pm,$$

which, upon rearrangement, gives

$$E + \frac{RT}{2F} \ln 4m^3 = E^0 - \frac{3RT}{2F} \ln \gamma_\pm. \quad (45.17)$$

The problem of evaluating E^0 is now similar to that in § 39h. The simplest procedure is to determine the E.M.F. of the cell for various molalities of zinc chloride, and to plot the values of $E + (RT/2F) \ln 4m^3$ as ordinate against a function of the molality, e.g., $\sqrt{m}$, as abscissa. At infinite dilution, i.e., when m is zero, the right-hand side of equation (45.17) is equal to E^0, since $\gamma_\pm$ is then unity and $\ln \gamma_\pm$ is zero. The extrapolated value of the ordinate for $m = 0$ is thus equal to E^0. Alternatively, the method of extrapolation utilizing the Debye-Hückel limiting law, as in § 39h, may be employed.

The value of E^0 for the cell was found to be 0.9834 volt at 25° C, and since the standard (reduction) potential of the right-hand i.e., silver chloride, electrode was recorded above as 0.2224 volt, it follows that the standard (oxidation) potential E_{Zn}^0 of the zinc electrode is given by $E_{Zn}^0 + 0.2224 = 0.9834$ volt, so that E_{Zn}^0 is 0.7610 volt at 25°.

If the chloride of the metal is not suitable, for one reason or another, the sulfate may prove satisfactory. In this event a sulfate reference electrode, e.g., SO_4^{--}, $Hg_2SO_4(s)$, Hg may be employed in place of the silver chloride electrode. The standard (reduction) potential of this electrode has been found to be 0.6141 volt at 25° C, by means of measurements similar to those described in connection with the silver chloride cell in § 39h.

In some cases, e.g., the silver-silver ion electrode, neither the chloride nor the sulfate is satisfactory because of their sparing solubility. It is then necessary to construct cells containing solutions of the nitrate, e.g., silver nitrate. There is, however, no nitrate reference electrode, and a chloride electrode is employed, the two solutions being separated by a solution of an inert electrolyte, constituting a "salt bridge." Due allowance must then be made for liquid junction potentials, but as the values of these potentials are somewhat uncertain, the results are not always highly accurate.[3]

45h. Standard Potentials from Equilibrium Constants and Free Energy Data.—When direct measurement of electrode potentials is not convenient or when the allowance for the activity coefficients is uncertain, it is often possible to determine the standard electrode potential by indirect procedures. One of these is based upon the relationship $\Delta F^0 = - NFE^0$, where E^0 is the standard E.M.F. of a cell, and ΔF^0 is the standard free energy change of the reaction taking place within it. As seen in § 33b, ΔF^0 is related to the equilibrium constant K of the reaction, i.e., $\Delta F^0 = - RT \ln K$ [cf. equation (33.6)], and hence it follows that

$$NFE^0 = RT \ln K,$$

$$E^0 = \frac{RT}{NF} \ln K, \quad (45.18)$$

---

[3] For descriptions of determination of standard potentials, see Lewis and Randall, ref. 2, Chapter XXX; see also, D. A. MacInnes, "The Principles of Electrochemistry," 1939, Chapters 10 and 16; Glasstone, ref. 1, Chapters VII and VIII.

or at 25° C,

$$E^0 = \frac{0.05914}{N} \log K. \tag{45.19}$$

It is seen, therefore, that if the equilibrium constant of a reaction could be determined experimentally, the standard E.M.F. of the cell in which that reaction takes place can be calculated. The constant K is, of course, the true (thermodynamic) equilibrium constant, and in its determination allowance should be made for departure of the solution from ideal behavior, either by including activity coefficients or by extrapolation to infinite dilution.

The method may be illustrated by reference to the Sn, Sn^{++} electrode. If combined with a Pb, Pb^{++} electrode there would be obtained the cell

$$Sn | Sn^{++} \quad Pb^{++} | Pb,$$

in which the reaction

$$Sn(s) + Pb^{++} = Sn^{++} + Pb(s)$$

would take place for the passage of 2 faradays. The equilibrium constant K is given by

$$K = \frac{a_{Sn^{++}}}{a_{Pb^{++}}},$$

the activities of the solid tin and lead being unity. By shaking finely divided metallic tin and lead with a solution containing the perchlorates of these metals, and analyzing the solution after the equilibrium

$$Sn(s) + PbClO_4(aq) \rightleftharpoons SnClO_4(aq) + Pb(s)$$

was established, the value of K was found to be 2.98 at 25° C. Upon insertion of this result into equation (45.19), the standard E.M.F. of the tin-lead cell, depicted above, is found to be

$$E^0 = \frac{0.05914}{2} \log 2.98 = 0.014.$$

This is equal to the standard (oxidation) potential of the Sn, Sn^{++} electrode minus that of the Pb, Pb^{++} electrode; the latter is known to be 0.126 volt, and so the standard (oxidation) potential of the Sn, Sn^{++} electrode is 0.140 volt at 25° C.[4]

The same method has been used to determine the standard potential of the ferrous-ferric electrode, by considering the cell

$$Ag | Ag^+ \quad Fe^{++}, Fe^{+++} | Pt,$$

in which the reaction, for the passage of 1 faraday, is

$$Ag(s) + Fe^{+++} = Ag^+ + Fe^{++}.$$

The value of K for this equilibrium was given as 0.53 at 25° C in § 41h, and

[4] Noyes and Toabe, *J. Am. Chem. Soc.*, **39**, 1537 (1917).

hence the standard E.M.F. of the cell is found to be -0.016, by equation (45.19). The standard (oxidation) potential of the Ag, Ag$^+$ electrode is known to be -0.799 volt, and hence that of the Pt, Fe^{++}, Fe^{+++} electrode is -0.783 volt at 25° C. It appears from various direct E.M.F. measurements that the correct standard potential is about -0.771 volt, the discrepancy being probably due to uncertainty in the extrapolation of the equilibrium constant values to infinite dilution, as indicated in § 41h.

The procedure just described essentially utilizes free energy values to calculate E.M.F.'s, and another example, based on the same principle, is the determination of the standard potential of the oxygen electrode. The problem is to evaluate the E.M.F. of the cell

Pt, H$_2$ (1 atm.) | H$^+$(a_{H^+} = 1), OH$^-$(a_{OH^-} = 1) | O$_2$ (1 atm.), Pt,

for the potential of the left-hand electrode is zero, by convention, and the right-hand electrode represents the standard (reduction) potential of the oxygen electrode. The reaction taking place in this cell is

$$H_2 \text{ (1 atm.)} + \tfrac{1}{2}O_2 \text{ (1 atm.)} = H_2O(l),$$

for the passage of 2 faradays. The standard free energy change of this reaction is known to be -56.70 kcal. at 25° C (see Table XXIV), and this is equivalent to $-NFE^0$, where N is here equal to 2 equiv. Hence, taking F as 23,070 cal. volt^{-1} g. equiv.$^{-1}$, it is seen that

$$E^0 = \frac{56.70 \times 1,000}{2 \times 23,070} = 1.229 \text{ volts}$$

This is the standard E.M.F. of the cell

Pt, H$_2$ (1 atm.) | Water | O$_2$ (1 atm.), Pt,

in which both hydrogen and oxygen electrodes are in contact with the *same solution*, the latter having the same activity as pure water, at 25° C. If the hydrogen ion activity in this solution is taken as unity, the potential of the left-hand electrode is then zero, and hence 1.229 volts is the potential of the electrode

H$_2$O(l), H$^+$(a_{H^+} = 1) | O$_2$ (1 atm.), Pt.

It is known from other investigations that in pure water the product of the activities of hydrogen and hydroxyl ions, i.e., $a_{H^+}a_{OH^-}$, is always equal to 1.008×10^{-14} at 25° C (§ 45k). Hence the electrode just indicated can also be represented as

OH$^-$(a_{OH^-} = 1.008×10^{-14}) | O$_2$ (1 atm.), Pt,

or reversing the electrode, so as to imply oxidation, the potential of the electrode

Pt, O$_2$ (1 atm.) | OH$^-$(a_{OH^-} = 1.008×10^{-14})

is -1.229 volts at 25° C. This result may be inserted into equation (45.14), which gives the oxidation potential of an electrode yielding negative ions; thus, at 25° C, since the valence of the OH$^-$ ions is unity,

$$-1.229 = E^0 + 0.05914 \log (1.008 \times 10^{-14}),$$
$$E^0 = -0.401 \text{ volt.}$$

The standard (oxidation) potential of the oxygen gas electrode is thus -0.401 volt at 25° C.

Another method for calculating standard potentials from free energy data will be described in § 46c; this employs heat content and entropy values, so that the free energies are obtained indirectly.

45i. Standard Electrode Potentials and their Applications.—By the use of methods, such as those described above, involving either E.M.F. measurements or free energy and related calculations, the standard potentials of a large number of electrodes have been determined. Some of the results on the hydrogen scale for a temperature of 25° C are quoted in Table XXXIX;[5] as noted above the standard state for any ion is taken as the (hypothetical) ideal solution of that ion at unit molality. The potentials in Table XXXIX are all oxidation potentials, as indicated by the symbols (cf. § 45f) and by the expression for the electrode process in each case. The corresponding standard reduction potentials are obtained by reversing the signs. The standard E.M.F. of any cell can be obtained by adding the standard oxidation potential of the left-hand electrode to the standard reduction potential of the right-hand electrode. The same result is obtained, of course, by subtracting the standard oxidation potential of the right-hand electrode from that of the left-hand electrode.

Numerous applications of standard electrode potentials have been made in various aspects of electrochemistry and analytical chemistry, as well as in thermodynamics. Some of these applications will be considered here, and others will be mentioned later. Just as standard potentials which cannot be determined directly can be calculated from equilibrium constant and free energy data, so the procedure can be reversed and electrode potentials used for the evaluation, for example, of equilibrium constants which do not permit of direct experimental study. Some of the results are of analytical interest, as may be shown by the following illustration. Stannous salts have been employed for the reduction of ferric ions to ferrous ions in acid solution, and it is of interest to know how far this process goes toward completion. Although the solutions undoubtedly contain complex ions, particularly those involving tin, the reaction may be represented, approximately, by

$$Sn^{++} + 2Fe^{+++} = Sn^{++++} + 2Fe^{++}.$$

A knowledge of the equilibrium constant would provide an indication of the extent to which the reaction proceeds from left to right; if it were large, then the process could be regarded as being virtually complete. The reaction between stannous and ferric ions can, in principle, be made to take place in the reversible cell

$$Pt\,|\,Sn^{++}, Sn^{++++} \quad Fe^{++}, Fe^{+++}\,|\,Pt,$$

---

[5] Data mainly from W. M. Latimer, "The Oxidation States of the Elements and Their Potentials in Aqueous Solutions," 1938, where a very complete collection of electrode potentials will be found.

TABLE XXXIX. STANDARD (OXIDATION) POTENTIALS AT 25° C *

Electrode	Reaction	Potential
Li, Li$^+$	Li $\rightarrow$ Li$^+$ + ϵ	3.024 volt
K, K$^+$	K $\rightarrow$ K$^+$ + ϵ	2.924
Na, Na$^+$	Na $\rightarrow$ Na$^+$ + ϵ	2.712
Pt, H$_2$, OH$^-$	$\frac{1}{2}$H$_2$ + OH$^-$ $\rightarrow$ H$_2$O + ϵ	0.828
Zn, Zn^{++}	Zn $\rightarrow$ Zn^{++} + 2ϵ	0.762
Fe, Fe^{++}	Fe $\rightarrow$ Fe^{++} + 2ϵ	0.440
Pt, Cr^{++}, Cr^{+++}	Cr^{++} $\rightarrow$ Cr^{+++} + ϵ	0.41
Cd, Cd^{++}	Cd $\rightarrow$ Cd^{++} + 2ϵ	0.402
Co, Co^{++}	Co $\rightarrow$ Co^{++} + 2ϵ	0.283
Ni, Ni^{++}	Ni $\rightarrow$ Ni^{++} + 2ϵ	0.236
Ag, AgI(s), I$^-$	Ag + I$^-$ $\rightarrow$ AgI + ϵ	0.1522
Sn, Sn^{++}	Sn $\rightarrow$ Sn^{++} + 2ϵ	0.140
Pb, Pb^{++}	Pb $\rightarrow$ Pb^{++} + 2ϵ	0.126
H$_2$, H$^+$	$\frac{1}{2}$H$_2$ $\rightarrow$ H$^+$ + ϵ	0.000
Pt, Ti^{+++}, Ti^{++++}	Ti^{+++} $\rightarrow$ Ti^{++++} + ϵ	$-$ 0.06
Ag, AgBr(s), Br$^-$	Ag + Br$^-$ $\rightarrow$ AgBr + ϵ	$-$ 0.0711
Pt, Sn^{++}, Sn^{++++}	Sn^{++} $\rightarrow$ Sn^{++++} + 2ϵ	$-$ 0.15
Pt, Cu$^+$, Cu^{++}	Cu$^+$ $\rightarrow$ Cu^{++} + ϵ	$-$ 0.167
Ag, AgCl(s), Cl$^-$	Ag + Cl$^-$ $\rightarrow$ AgCl + ϵ	$-$ 0.2224
Hg, Hg$_2$Cl$_2$(s), Cl$^-$	Hg + Cl$^-$ $\rightarrow$ $\frac{1}{2}$Hg$_2$Cl$_2$ + ϵ	$-$ 0.2680
Cu, Cu^{++}	Cu $\rightarrow$ Cu^{++} + 2ϵ	$-$ 0.340
Pt, Fe(CN)$_6^{----}$, Fe(CN)$_6^{---}$	Fe(CN)$_6^{----}$ $\rightarrow$ Fe(CN)$_6^{---}$ + ϵ	$-$ 0.356
Pt, O$_2$, OH$^-$	2OH$^-$ $\rightarrow$ $\frac{1}{2}$O$_2$ + H$_2$O + 2ϵ	$-$ 0.401
Pt, I$_2$(s), I$^-$	I$^-$ $\rightarrow$ $\frac{1}{2}$I$_2$ + ϵ	$-$ 0.536
Pt, MnO$_4^{--}$, MnO$_4^-$	MnO$_4^{--}$ $\rightarrow$ MnO$_4^-$ + ϵ	$-$ 0.54
Hg, Hg$_2$SO$_4$(s), SO$_4^{--}$	2Hg + SO$_4^{--}$ $\rightarrow$ Hg$_2$SO$_4$ + 2ϵ	$-$ 0.6141
Pt, Fe^{++}, Fe^{+++}	Fe^{++} $\rightarrow$ Fe^{+++} + ϵ	$-$ 0.771
Hg, Hg$_2^{++}$	2Hg $\rightarrow$ Hg$_2^{++}$ + 2ϵ	$-$ 0.7986
Ag, Ag$^+$	Ag $\rightarrow$ Ag$^+$ + ϵ	$-$ 0.7995
Pt, Hg$_2^{++}$, Hg^{++}	Hg$_2^{++}$ $\rightarrow$ 2Hg^{++} + 2ϵ	$-$ 0.906
Pt, Br$_2$(l), Br$^-$	Br$^-$ $\rightarrow$ $\frac{1}{2}$Br$_2$ + ϵ	$-$ 1.066
Pt, Tl$^+$, Tl^{+++}	Tl$^+$ $\rightarrow$ Tl^{+++} + 2ϵ	$-$ 1.22
Pt, Cl$_2$(g), Cl$^-$	Cl$^-$ $\rightarrow$ $\frac{1}{2}$Cl$_2$ + ϵ	$-$ 1.358
Pt, Mn^{++}, MnO$_4^-$, H$^+$	Mn^{++} + 4H$_2$O $\rightarrow$ MnO$_4^-$ + 8H$^+$ + 5ϵ	$-$ 1.52
Pt, Ce^{+++}, Ce^{++++}	Ce^{+++} $\rightarrow$ Ce^{++++} + ϵ	$-$ 1.61
Pt, PbSO$_4$(s), PbO$_2$(s), SO$_4^{--}$	PbSO$_4$ + 2H$_2$O $\rightarrow$ PbO$_2$ + 4H$^+$ + SO$_4^{--}$ + 2ϵ	$-$ 1.685
Pt, Pb^{++}, Pb^{++++}	Pb^{++} $\rightarrow$ Pb^{++++} + 2ϵ	$-$ 1.7
Pt, Co^{++}, Co^{+++}	Co^{++} $\rightarrow$ Co^{+++} + ϵ	$-$ 1.8

* Numerous other standard potentials can be calculated from the standard free energy values in Table 5 at the end of the book.

for the passage of 2 faradays. The standard E.M.F. is equal to the standard (oxidation) potential of the stannous-stannic electrode, i.e., $-$ 0.15 volt, minus the standard (oxidation) potential of the ferrous-ferric electrode, i.e., $-$ 0.771 volt; hence, E^0 is $-$ 0.15 $-$ ($-$ 0.771) = 0.62 volt at 25° C. By equation (45.19), therefore, since N is 2,

$$0.62 = \frac{0.05914}{2} \log K,$$
$$K = 9.3 \times 10^{20}.$$

The large value of the equilibrium constant shows that the reaction proceeds to virtual completion from left to right, within the limits of analytical accuracy.

An examination of these calculations shows that the large value of the equilibrium constant is determined by the large difference between the standard potentials of the Sn^{++}, Sn^{++++} and Fe^{++}, Fe^{+++} systems. In general, the greater the difference in the standard potentials, the more completely will the system with the lower (algebraic) oxidation potential be reduced, and that with the higher oxidation potential be oxidized. This conclusion is applicable to oxidation and reduction in the general sense, although the example given above refers to conventional oxidation and reduction.

For the sake of completeness, attention may be called to the fact that the calculation of an equilibrium constant from standard potentials is equivalent to the determination of the standard free energy of the corresponding reaction. Since ΔF^0 is equal $-RT \ln K$ the former can be readily obtained if K is known. Alternatively, of course, ΔF^0 could be calculated directly from $-NFE^0$, without the use of the equilibrium constant.

Another important use of standard potentials is for the determination of solubility products, for these are essentially equilibrium constants (§ 39j). If $M_{\nu_+}A_{\nu_-}$ is a *sparingly soluble salt*, a knowledge of the standard potentials of the electrodes M, $M_{\nu_+}A_{\nu_-}(s)$, A^{z-} and M, M^{z+} permits the solubility product to be evaluated. A simple example is provided by silver chloride for which the standard (oxidation) potential of the Ag, $AgCl(s)$, Cl^- electrode is known to be -0.2224 volt at 25° C. The activity of the chloride ion in the standard electrode is unity, and hence the silver ion activity must be equal to the solubility product of silver chloride. The value of a_{Ag^+} may be derived from equation (45.13), utilizing the standard potential of silver; thus E_M is -0.2224 volt, E_M^0 for silver is -0.799, and z is 1, so that at 25° C,

$$-0.2224 = -0.799 - 0.05914 \log a_{Ag^+},$$

$$a_{Ag^+} = 1.78 \times 10^{-10},$$

so that the solubility product $a_{Ag^+} \times a_{Cl^-}$ of silver chloride is 1.78×10^{-10} at 25° C, in terms of molalities.

Since the solubility product, e.g., of silver chloride, is the equilibrium constant of the process

$$AgCl(s) = Ag^+ + Cl^-,$$

the ions being in solution, the standard free energy change ΔF^0 is equal to $-RT \ln K_s$, where K_s is the solubility product. Hence, the standard free energy change for the solution of solid silver chloride in water at 25° C is equal to $-2.303 \times 1.987 \times 298.2 \log 1.78 \times 10^{-10}$, i.e., 13,300 cal. It will be observed that this value is positive, and hence (cf. § 25f) solid silver chloride will not dissolve to form silver and chloride ions *at unit activity* in terms of molality. Because of the low solubility of silver chloride in water it is obvious that the reverse process, namely the precipitation of this salt from a solution in which the silver and chloride ions have activities, or (approximately) concentrations, of 1 molal, will be spontaneous. For this process the standard free energy change is $-$ 13,300 cal. at 25° C.

45j. Dissociation Pressures.—In § 45a it was stated that the Ag, AgCl(s), Cl$^-$ electrode could be regarded as a chlorine electrode with the gas at a pressure equal to the dissociation pressure of silver chloride at the experimental temperature. That this must be the case can be seen in the following manner. Consider the reaction between solid silver chloride and hydrogen taking place in the hydrogen-silver chloride cell; this may be written in two stages, viz., first, an equilibrium between solid silver chloride, on the one hand, and solid silver and chlorine gas at its dissociation pressure p_{Cl_2} on the other hand,

$$AgCl(s) \rightleftharpoons Ag(s) + \tfrac{1}{2}Cl_2(p_{Cl_2}),$$

and second, reaction between the chlorine and hydrogen in the presence of water, to form hydrochloric acid, i.e., hydrogen and chloride ions in solution,

$$\tfrac{1}{2}Cl_2(p_{Cl_2}) + \tfrac{1}{2}H_2 \,(1 \text{ atm.}) = H^+ + Cl^-.$$

The free energy change for the dissociation of solid silver chloride to yield chlorine gas at the pressure p_{Cl_2} is zero, since the system is in equilibrium; hence, the free energy change of the second reaction is identical with that for the over-all process. An electrode consisting of chlorine gas at the dissociation pressure p_{Cl_2} would thus have the same potential, in a given chloride ion solution, as would the Ag, AgCl(s) electrode.

The conclusion just reached may be utilized to determine the dissociation pressure of chlorine gas in equilibrium with solid silver chloride; for this purpose it is necessary to consider the problem of a gas electrode in which the pressure differs from the standard value of 1 atm. For such electrodes, the general equation (45.15) is still applicable, but allowance must be made for the activity, i.e., approximately the pressure, of the chlorine gas, as well as of the reversible ions. For a chlorine electrode, the gas is the oxidized state and the chloride ion the reduced state, and hence the electrode process may be written as

$$2Cl^- = Cl_2(g) + 2\epsilon,$$

for the passage of 2 faradays of electricity. By equation (45.15), therefore, the oxidation potential is given by

$$E_{Cl_2} = E^0_{Cl_2} - \frac{RT}{2F} \ln \frac{p_{Cl_2}}{a^2_{Cl^-}}$$

$$= E^0_{Cl_2} - \frac{RT}{2F} \ln p_{Cl_2} + \frac{RT}{F} \ln a_{Cl^-}. \qquad (45.20)^*$$

If the pressure (or fugacity) of the chlorine gas were 1 atm., this would reduce to exactly the same form as equation (45.14).

At 25° C, the value of $E^0_{Cl_2}$ is $-$ 1.358 (Table XXXIX), and if the activity a_{Cl^-} of the chloride ions is taken as unity, the oxidation potential of the chlorine

---

* It would be permissible to write the electrode reaction as $Cl^- = \tfrac{1}{2}Cl_2(g) + \epsilon$, so that N is unity; in this case, however, the activity of the chlorine gas would be $p^{\frac{1}{2}}_{Cl_2}$ and that of the ions a_{Cl^-}, thus leading to the same result as in equation (45.20).

gas electrode would be

$$E_{Cl_2} = -1.358 - \frac{0.05914}{2}\log p_{Cl_2},$$

from equation (45.20). Further, if p_{Cl_2} is the dissociation pressure of silver chloride at 25° C, this potential must be identical with that of the Ag, AgCl(s), Cl$^-$(a_{Cl^-} = 1) electrode, which is $-$ 0.2224 volt; hence,

$$-0.2224 = -1.358 - \frac{0.05914}{2}\log p_{Cl_2}$$
$$p_{Cl_2} = 3.98 \times 10^{-39} \text{ atm.}$$

The dissociation pressure of solid silver chloride at 25° C is thus 4×10^{-39} atm.; if the necessary thermal data were available, it would be possible to calculate the dissociation pressure at other temperatures (cf. §§ 33h, 33i).

Although the preceding discussion has dealt in particular with silver chloride, the general conclusions are applicable quite generally, not only to the chlorides of other metals, but also to bromides, iodides and oxides. Thus, the potential of an electrode M, MO(s), OH$^-$ is exactly equivalent to that of the oxygen electrode Pt, O$_2$(p_{O_2}), OH$^-$, in the same hydroxyl ion solution, with oxygen gas at the dissociation pressure of the oxide MO, at the experimental temperature. By utilizing this relationship, together with heat of reaction and heat capacity data, estimates have been made of the temperatures at which various metallic oxides dissociate freely in air, the dissociation pressure being then equal to the partial pressure of oxygen in the atmosphere, i.e., 0.21 atm.

Problem: The standard potential of the Ag, Ag$_2$O(s), OH$^-$ electrode is $-$ 0.344 volt at 25° C. The heat of formation of silver oxide is $-$ 7,300 cal. at 25° and ΔC_P is about 1.0 cal. deg.$^{-1}$ mole^{-1}. Estimate the temperature at which silver oxide will dissociate freely in air.

Treating the electrode as an oxygen electrode in a solution in which a_{OH^-} is unity, the potential is given by equation (45.15) as

$$E = E^0 - \frac{RT}{4F}\ln p_{O_2},$$

since 4 faradays are required to convert an oxygen molecule into OH$^-$ ions. In this case, E is $-$ 0.344, and E^0 for the oxygen electrode is $-$ 0.401 volt (Table XXXIX); hence, at 25° C,

$$-0.344 = -0.401 - \frac{0.05914}{4}\log p_{O_2},$$

so that p_{O_2} is 1.40×10^{-4} atm. This may now be taken as the equilibrium constant for the reaction

$$2\text{Ag}_2\text{O}(s) = 4\text{Ag}(s) + \text{O}_2(g),$$

two molecules of silver oxide being necessary to produce one molecule of oxygen, so that K_p is equivalent to p_{O_2}. For this reaction ΔH at 298° K is $-$ 2ΔH for the *formation* of the oxide, i.e., 14,600 cal., and ΔC_P is, similarly, $-$ 2.0 cal. deg.$^{-1}$; hence, by equation (12.16) or (33.29),

$$14,600 = \Delta H_0^0 - 2.0 \times 298; \quad \Delta H_0^0 = 15,200 \text{ cal.}$$

Utilizing equation (33.30), with K equal to 1.40×10^{-4} atm. at 298° K,

$$\log (1.40 \times 10^{-4}) = -\frac{15{,}200}{2.303RT} - \frac{2.0}{R}\log T + \frac{I'}{2.303}.$$

Taking R as 2 cal. deg.$^{-1}$ mole^{-1}, $I'/2.303$ is found to be 9.76; hence, the general expression for $\log K$ becomes

$$\log K = -\frac{15{,}200}{4.576T} - 1.0 \log T + 9.76.$$

The oxide will dissociate freely in air when p_{O_2}, i.e., K, is 0.21 atm.; inserting this value in the expression for $\log K$, and solving by successive approximations, T is found to be 435° K, i.e., 162° C. (The experimental value is about 150° C.)

45k. The Dissociation (Ionization) Constant of Water.—In pure water, and in any aqueous solution, an equilibrium exists between the undissociated molecules of water, on the one hand, and hydrogen and hydroxyl ions, on the other hand; thus,

$$H_2O(l) \rightleftharpoons H^+(aq) + OH^-(aq).$$

The equilibrium (dissociation) constant of this reaction is given by

$$K = \frac{a_{H^+} a_{OH^-}}{a_{H_2O}},$$

but for pure water, or for dilute aqueous solutions of electrolytes, the activity of the water, i.e., a_{H_2O}, may be taken as unity; the result, referred to as the **dissociation** (or **ionization**) **constant** or, sometimes, as the **ionic product**, of water, is then

$$K_w = a_{H^+} a_{OH^-}. \tag{45.21}$$

One of the most accurate methods for determining the dissociation constant of water is based on E.M.F. measurements of cells of the type

$$\text{Pt, H}_2 \text{ (1 atm.)} \,|\, \text{MOH}(m_1) \quad \text{MCl}(m_2) \,|\, \text{AgCl}(s), \text{Ag},$$

where M is an alkali metal, e.g., lithium, sodium or potassium. The E.M.F. of this cell, like that for other hydrogen-silver chloride cells (§ 45d) is given by

$$E = E^0 - \frac{RT}{F} \ln a_{H^+} a_{Cl^-}. \tag{45.22}$$

Since, by equation (45.21), $a_{H^+} a_{OH^-}$ is equal to K_w, it follows that

$$E = E^0 - \frac{RT}{F} \ln K_w - \frac{RT}{F} \ln \frac{a_{Cl^-}}{a_{OH^-}}$$

$$= E^0 - \frac{RT}{F} \ln K_w - \frac{RT}{F} \ln \frac{m_{Cl^-}}{m_{OH^-}} - \frac{RT}{F} \ln \frac{\gamma_{Cl^-}}{\gamma_{OH^-}}, \tag{45.23}$$

where, in equation (45.23), the activities have been replaced by the products of the respective molalities and activity coefficients. Upon rearrangement, equation (45.23) gives

$$E - E^0 + \frac{RT}{F} \ln \frac{m_{Cl^-}}{m_{OH^-}} = -\frac{RT}{F} \ln K_w - \frac{RT}{F} \ln \frac{\gamma_{Cl^-}}{\gamma_{OH^-}},$$

$$\frac{F(E - E^0)}{2.303RT} + \log \frac{m_{Cl^-}}{m_{OH^-}} = -\log K_w - \log \frac{\gamma_{Cl^-}}{\gamma_{OH^-}}. \quad (45.24)$$

At infinite dilution the activity coefficient fraction $\gamma_{Cl^-}/\gamma_{OH^-}$ becomes equal to unity, so that $\log(\gamma_{Cl^-}/\gamma_{OH^-})$ is then zero; under these conditions, therefore, the right-hand side of equation (45.24) becomes equal to $-\log K_w$. It follows, therefore, that if the left-hand side of this equation, for various molalities of MOH and MCl, is plotted against some function of the concentration, e.g., the ionic strength, the intercept for infinite dilution gives the value of $-\log K_w$. At 25° C, E^0 for the cell under consideration is known to be 0.2224 volt, and the actual E.M.F. of the cell, i.e., E, can be measured; by assuming m_{Cl^-} to be equal to m_2 (for MCl) and m_{OH^-} to be equal to m_1 (for MOH), the left-hand side of equation (45.24) can be readily evaluated. From results of this kind, with various chlorides and hydroxides, K_w has been found to be 1.008×10^{-14} at 25° C. Accurate measurements of the same type have been made in the temperature range from 0° to 50° C, with cells containing a number of different chlorides.

The E.M.F. of the hydrogen-silver chloride cell

$$\text{Pt, H}_2 \text{ (1 atm.)} \,|\, \text{HCl}(m_1') \text{ MCl}(m_2') \,|\, \text{AgCl}(s), \text{Ag},$$

containing hydrochloric acid and a metallic chloride solution, is

$$E' = E^0 - \frac{RT}{F} \ln a'_{H^+} a'_{Cl^-}. \quad (45.25)$$

If this result is combined with equation (45.22) for the E.M.F. of the similar cell containing the hydroxide (and chloride) of the metal M, it is seen that

$$E - E' = \frac{RT}{F} \ln \frac{a'_{H^+} a'_{Cl^-}}{a_{H^+} a_{Cl^-}}$$

$$= \frac{RT}{F} \ln \frac{m'_{H^+} m'_{Cl^-}}{m_{H^+} m_{Cl^-}} + \frac{RT}{F} \ln \frac{\gamma'_{H^+} \gamma'_{Cl^-}}{\gamma_{H^+} \gamma_{Cl^-}}, \quad (45.26)$$

the standard potentials E^0 being the same in equations (45.22) and (45.25). If the ionic strengths in the two cells are kept equal, then provided the solutions are relatively dilute, the activity coefficient factor in equation (45.26) is unity and its logarithm is zero; equation (45.26) then becomes

$$E - E' = \frac{RT}{F} \ln \frac{m'_{H^+} m'_{Cl^-}}{m_{H^+} m_{Cl^-}}. \quad (45.27)$$

By introducing the term $(RT/F) \ln m_{OH^-}$, equation (45.27) can be rearranged to give

$$-\frac{RT}{F} \ln m_{H^+} m_{OH^-} = E - E' - \frac{RT}{F} \ln \frac{m'_{H^+} m'_{Cl^-} m_{OH^-}}{m_{Cl^-}}, \quad (45.28)$$

where, as before, the primed quantities refer to the cell containing hydrochloric acid, and those without primes apply to the alkali hydroxide cell. The particular interest of equation (45.28) lies in the fact that it permits of the determination of $m_{H^+} m_{OH^-}$, known as the **molal ionization product** of water, in various halide solutions. In a pure halide solution the molalities of the hydrogen and hydroxyl-ions are equal; hence, the square root of $m_{H^+} m_{OH^-}$ gives the actual molalities of the hydrogen and hydroxyl ions, produced by the ionization of water, in the halide solution.[6]

451. The Dissociation (Ionization) Constant of a Weak Acid.—The dissociation constants of weak acids, as defined in § 41h, may also be determined by means of E.M.F. measurements. For this purpose another form of the hydrogen-silver chloride cell, viz.,

$$\text{Pt, } H_2 \text{ (1 atm.)} | HA(m_1) NaA(m_2) NaCl(m_3) | AgCl(s), \text{ Ag,}$$

where HA is the weak acid and NaA is its sodium salt, is employed. The E.M.F. of this cell is given, as usual, by

$$E = E^0 - \frac{RT}{F} \ln a_{H^+} a_{Cl^-}$$

$$= E^0 - \frac{RT}{F} \ln m_{H^+} m_{Cl^-} - \frac{RT}{F} \ln \gamma_{H^+} \gamma_{Cl^-}. \quad (45.29)$$

The dissociation constant K of the acid may be expressed in the form

$$K = \frac{a_{H^+} a_{A^-}}{a_{HA}} = \frac{m_{H^+} m_{A^-}}{m_{HA}} \cdot \frac{\gamma_{H^+} \gamma_{A^-}}{\gamma_{HA}},$$

and combination of this result with equation (45.29) gives

$$E = E^0 - \frac{RT}{F} \ln \frac{m_{HA} m_{Cl^-}}{m_{A^-}} - \frac{RT}{F} \ln \frac{\gamma_{HA} \gamma_{Cl^-}}{\gamma_{A^-}} - \frac{RT}{F} \ln K,$$

$$\frac{F(E - E^0)}{2.303 RT} + \log \frac{m_{HA} m_{Cl^-}}{m_{A^-}} = -\log \frac{\gamma_{HA} \gamma_{Cl^-}}{\gamma_{A^-}} - \log K. \quad (45.30)$$

If the left-hand side of this equation, for various molalities of HA, NaA and NaCl, is plotted against the ionic strength, the value extrapolated to infinite dilution gives $-\log K$, since the first term on the right-hand side of equation (45.30) is then zero. As a first approximation, m_{HA} is taken as equal to m_1, m_{Cl^-} to m_3 and m_{A^-} to m_2; this neglects the dissociation of the acid HA in the presence of the salt. Allowance for the dissociation can be made, if necessary, by utilizing a preliminary value of K for the given acid.

[6] Harned and Hamer, *J. Am. Chem. Soc.*, **55**, 2194 (1933); for reviews, see H. S. Harned and B. B. Owen, "The Physical Chemistry of Electrolytic Solutions," 1943, Chapter 15; *Chem. Rev.*, **25**, 31 (1939).

The procedure just described has been used for the determination of the dissociation constants of a number of weak monobasic acids. It has also been extended to the study of polybasic acids and amino-acids.

Dissociation constants of weak acids can also be derived from measurements on "unbuffered" cells of the type

$$\text{Pt, H}_2 \text{ (1 atm.)} | \text{HA}(m_1) \quad \text{NaCl}(m_3) | \text{AgCl}(s), \text{Ag}.$$

These are similar to those described above except for the omission of the salt NaA from the electrolyte. By means of these cells it is also possible to determine the quantity $m_{\text{H}^+} m_{\text{A}^-} / m_{\text{HA}}$, which gives the actual extent of dissociation of the weak acid, in solutions containing various amounts of sodium chloride or other halides.[7]

46. Thermodynamics of Ions in Solution

46a. Free Energies of Formation of Ions.—In Chapter XIII the standard free energies of formation of various neutral molecules were given; it will now be shown how corresponding values may be obtained for many ions. The results may then be combined to determine the standard free energy changes of reactions involving ions as well as neutral molecules. In the cell

$$\text{M} | \text{M}^{z+}(a_+ = 1) \quad \text{H}^+(a_{\text{H}^+} = 1) | \text{H}_2 \text{ (1 atm.)},$$

the E.M.F. of which is equal to the standard potential E_{M}^0 of the M, M^{z+} electrode, the reaction for z_+ faradays is (cf. § 45e).

$$\text{M}(s) + z_+ \text{H}^+(a_{\text{H}^+} = 1) = \text{M}^{z+}(a_+ = 1) + \tfrac{1}{2} z_+ \text{H}_2 \text{ (1 atm.)}.$$

Hence the standard free energy change of this reaction is $-z_+ F E_{\text{M}}^0$. However, the convention of taking the standard potential of the hydrogen electrode as zero is equivalent to stating that the standard free energy change for the transfer of hydrogen gas to hydrogen ions, or the reverse, is arbitrarily set equal to zero. Consequently, on the basis of this convention, $-z_+ F E_{\text{M}}^0$ is the standard free energy change accompanying the formation of M^{z+} ions in solution, viz.,

$$\text{M}(s) = \text{M}^{z+}(a_+ = 1) + z_+ \epsilon.$$

By adopting the foregoing convention, therefore, it is possible to define the standard free energy of formation of a positive ion M^{z+}, from the element M, as $-z_+ F E_{\text{M}}^0$, where E_{M}^0 is the standard (oxidation) potential on the hydrogen scale.

Similarly, for an electrode involving negative ions, the reaction in the cell

$$\text{A} | \text{A}^{z-}(a_- = 1) \quad \text{H}^+(a_{\text{H}^+} = 1) | \text{H}_2 \text{ (1 atm.)}$$

is

$$\text{A}^{z-}(a_- = 1) + z_- \text{H}^+(a_{\text{H}^+} = 1) = \text{A} + \tfrac{1}{2} z_- \text{H}_2 \text{ (1 atm.)}.$$

[7] Harned and Ehlers, *J. Am. Chem. Soc.*, **54**, 1350 (1932); Nims, *ibid.*, **55**, 1946 (1933); Harned and Owen, ref. 6.

for z_- faradays. Hence, adopting the same convention as above, viz., that the standard free energy change of formation of hydrogen ions is zero, $-z_-FE_A^0$ is the standard free energy of the process

$$A^{z-}(a_- = 1) = A + z_-\epsilon.$$

This is the reverse of the standard free energy of formation of the A^{z-} ion, and consequently the latter is equal to $z_-FE_A^0$ where E_A^0 is the standard (oxidation) potential of the A, A^{z-} electrode.*

Provided the potential of an electrode of the type M, M^{z+} or A, A^{z-}, where M and A are elements, can be measured, the standard free energy of formation of the M^{z+} and A^{z-} ions, respectively, can be readily determined. For this purpose the data in Table XXXIX may be used. If the ion is one which cannot form a stable electrode of the simple type indicated, indirect methods must be adopted; sometimes electrode potential or equilibrium data can be employed, but in other cases recourse must be had to heat content and entropy values. For example, the potential of the Fe, Fe^{+++} electrode cannot be determined experimentally because the system is unstable; however, the potentials of the Fe, Fe^{++} and the Pt, Fe^{++}, Fe^{+++} systems are known, viz., $+0.441$ and -0.771 volt at 25°, respectively, and from these the free energy of formation of the ferric ion can be determined. Thus, for the process

$$Fe = Fe^{++} + 2\epsilon,$$

the standard free energy change is $-2F \times 0.441$, i.e., $-0.882F$, and for

$$Fe^{++} = Fe^{+++} + \epsilon,$$

it is $-F \times (-0.771)$, i.e., $0.771F$. The total standard free energy change for the formation of ferric ions,

$$Fe = Fe^{+++} + 3\epsilon,$$

is thus $(-0.882 + 0.771)F$ i.e., $-0.111F$. To express the free energy change in calories, F is taken as 23,070 cal. volt^{-1}, so that the free energy of formation of ferric ions is $-2,560$ cal. g. ion^{-1}.

The calculation of free energies of formation of ions from equilibrium data may be illustrated by reference to the iodate (IO_3^-) ion. At 25° C the equilibrium constant K for the system

$$3I_2(s) + 3H_2O(l) = IO_3^- + 5I^- + 6H^+$$

is approximately 2×10^{-56}, so that the standard free energy change is given by

$$\Delta F^0 = -RT \ln K = -4.576 \times 298.2 \log K$$
$$= 76{,}000 \text{ cal.} = 76.0 \text{ kcal.}$$

* In general, the standard free energy of formation of any ion in solution is equal to $-zFE^0$, where z is the valence of the ion *including its sign*.

By convention (cf. § 33k) the free energy of formation of every element in its standard state is taken as zero, and, as seen above, the same applies to the hydrogen *ion*. The standard free energy of formation of liquid water at 25° C is -56.70 kcal. mole^{-1} (Table XXIV), and that of the iodide ion is derived from its standard potential (Table XXXIX) as $1 \times 23{,}070 \times (-0.536)$ cal., i.e., -12.37 kcal. g. ion^{-1}. It follows, therefore, that

$$[\Delta F^0(\mathrm{IO}_3^-) + (5 \times -12.37) + (6 \times 0)] - [(3 \times 0) + (3 \times -56.70)] = 76.0,$$
$$\Delta F^0(\mathrm{IO}_3^-) = -32.3 \text{ kcal. g. ion}^{-1},$$

so that the standard free energy of formation of the iodate ion at 25° C is -32.3 kcal. g. ion^{-1}.

A similar result has been obtained from measurements of the potential of the electrode Pt, $\mathrm{IO}_3^- + 6\mathrm{H}^+, \tfrac{1}{2}\mathrm{I}_2(s)$, for which the reaction is

$$\tfrac{1}{2}\mathrm{I}_2(s) + 3\mathrm{H}_2\mathrm{O}(l) = \mathrm{IO}_3^- + 6\mathrm{H}^+ + 5\epsilon.$$

The standard potential at 25° C is -1.195 volt, and so the standard free energy change is $-5 \times 23{,}070 \times (-1.195)$ cal., or 137.8 kcal. The standard free energies of formation of solid iodine and of hydrogen ions are taken as zero, and since that for liquid water is -56.70 kcal. mole^{-1} at 25° C, it follows that

$$[\Delta F^0(\mathrm{IO}_3^-) + (6 \times 0)] - [(\tfrac{1}{2} \times 0) + (3 \times -56.70)] = 137.8,$$
$$\Delta F^0(\mathrm{IO}_3^-) = -32.3 \text{ kcal. g. ion}^{-1},$$

in agreement with the result obtained above.

46b. Standard Entropies of Ions.—The evaluation of the standard entropies of ions has provided a method for the calculation of standard free energies of formation, and hence the standard electrode potentials, of ions which are not susceptible of direct experimental study. The data may also be used in conjunction with other entropy values, such as those given in Chapter IX, to determine the entropy changes of reactions involving both molecules and ions. Several methods are available for determining the standard entropies of ions, the particular one used depending on the circumstances.

For ions which are reversible with respect to a metallic element that dissolves in dilute acid solution, the following procedure is the simplest. Consider the general reaction accompanying the solution of the metal M, viz.,

$$\mathrm{M} + z_+\mathrm{H}^+ = \mathrm{M}^{z+} + \tfrac{1}{2}z_+\mathrm{H}_2;$$

the standard free energy change ΔF^0 is $-z_+ F E^0_\mathrm{M}$, and this can be determined if the standard potential E^0_M is known. The standard heat content change ΔH^0 of the same reaction is virtually identical with the experimental heat of solution of 1 g. atom of the metal in a dilute acid solution. By utilizing the familiar thermodynamic relationship $\Delta S^0 = (\Delta H^0 - \Delta F^0)/T$, it is thus possible to determine the standard entropy change of the reaction between the metal and hydrogen ions in solution. Considering the species involved in the reaction it is seen that

$$\Delta S^0 = (S^0_{M^+} + \tfrac{1}{2}z_+ S_{H_2}) - (S_M + z_+ S^0_{H^+}), \tag{46.1}$$

where $S^0_{M^+}$ and $S^0_{H^+}$ are the standard entropies of the M^{z+} and H^+ ions, respectively, and S_{H_2} and S_M are the entropies of the hydrogen gas at 1 atm. and of the metal M, respectively. These two latter quantities are presumably known, from the data in Tables XV and XIX, and if by convention *the standard entropy of the hydrogen ion is taken as zero,** it is possible to calculate $S^0_{M^+}$ for the given ion by means of equation (46.1).[8]

Problem: The standard heat of solution of 1 g. atom of potassium in dilute acid is found to be -60.15 kcal. at 25° C. The standard oxidation potential of this metal is 2.924 volt at the same temperature; the standard entropy of solid potassium is 15.2 E.U. g. atom^{-1} and of gaseous hydrogen it is 31.21 E.U. mole^{-1}. Calculate the standard entropy of the potassium ion in solution at 25° C.

For the solution reaction

$$K(s) + H^+ = K^+ + \tfrac{1}{2}H_2(g),$$

ΔH^0 is about -60.15 kcal., i.e., $-60,150$ cal., and ΔF^0 is obtained from the standard potential as $-1 \times 23,070 \times 2.924$, i.e., $-67,450$ cal. Hence,

$$\Delta S^0 = [-60,150 - (-67,450)]/298.2 = 24.5 \text{ E.U.},$$

and by equation (46.1)

$$24.5 = (S^0_{K^+} + \tfrac{1}{2} \times 31.2) - (15.2 + 0),$$

$$S^0_{K^+} = 24.1 \text{ E.U. g. ion}^{-1}.$$

If the standard potential of an electrode is known at two or more temperatures, the entropy of the reversible ion can be determined by using a form of the Gibbs-Helmholtz relationship. As seen in § 33l, the standard entropy change may be expressed in terms of the standard electrode potential, viz.,

$$\Delta S^0 = NF \frac{dE^0}{dT}, \tag{46.2}$$

and this is also the value of ΔS^0 given by equation (46.1). The subsequent procedure for obtaining $S^0_{M^+}$ is then similar to that described above.

---

* Since it has been postulated that the standard potential of the hydrogen electrode is zero at all temperatures, it follows that ΔS^0 for the reaction $\tfrac{1}{2}H_2(g) = H^+ + \epsilon$ must also be zero. An alternative convention would therefore be to take the standard entropy of the hydrogen ion as equal to that of $\tfrac{1}{2}H_2(g)$; this is equivalent to postulating zero entropy for the electrons. It is immaterial, however, which convention is employed, provided it is adhered to strictly. Since the convention that S^0 for the hydrogen ion is zero is widely used in the literature, it will be adopted here.

[8] Latimer and Buffington, *J. Am. Chem. Soc.*, **48**, 2297 (1926); Latimer, *Chem. Rev.*, **18**, 349 (1936), and numerous papers by Latimer, *et al.*, in *J. Am. Chem. Soc.*; see also, Reinhardt and Crockford, *J. Phys. Chem.*, **46**, 473 (1942).

Problem: From measurements of the standard potential of zinc, the value of dE^0/dT was found to be -1.00×10^{-4} volt deg.$^{-1}$ at 25° C. The standard entropy of solid zinc is 9.95 E.U. g. atom^{-1} and of hydrogen gas it is 31.21 E.U. mole^{-1} at 25° C. Calculate the standard entropy of the zinc ion.[9]

In this case, the reaction with hydrogen ions in solution is

$$Zn(s) + 2H^+ = Zn^{++} + H_2(g),$$

and the standard entropy change is $2 \times 23{,}070 \times (-1.00 \times 10^{-4})$ cal. deg.$^{-1}$; hence ΔS^0 is found to be -4.61 E.U. By equation (46.1), therefore,

$$-4.61 = (S^0_{Zn^{++}} + 31.21) - (9.95 + 0),$$
$$S^0_{Zn^{++}} = -25.9 \text{ E.U. g. ion}^{-1}.$$

For anions, in particular, an entirely different procedure is adopted to determine the standard entropy of the ion in solution. Consider the solid salt $M_{\nu_+}A_{\nu_-}$ in equilibrium with its ions M^+ and A^- in a saturated solution; thus,

$$M_{\nu_+}A_{\nu_-}(s) \rightleftharpoons \nu_+ M^+ + \nu_- A^-.$$

The equilibrium constant is equal to the solubility product K_s, at 1 atm. pressure, and hence the standard free energy change ΔF^0 of the indicated process is given by $-RT \ln K_s$ (cf. §45i).

The significance of ΔH^0 for the reaction may be seen in the following manner. The total heat content of the products (right-hand side) in their standard states is $\nu_+ \bar{H}^0_+ + \nu_- \bar{H}^0_-$, and this is equivalent to $\bar{H}^0_2$, the standard partial molar heat content of the solute $M_{\nu_+}A_{\nu_-}$ in solution. The standard heat content of the initial state (left-hand side), which consists of pure solid solute at 1 atm. pressure, is identical with the molar heat content H_S of the solid salt. It is seen, therefore, that

$$\Delta H^0 = \bar{H}^0_2 - H_S. \tag{46.3}$$

Since $\bar{H}^0_2$ is equal to the partial molar heat content of the solute at infinite dilution, it follows from equation (44.8) that ΔH^0 in this case is equal to the differential heat of solution of the solid salt in the infinitely dilute solution. In dilute solution the total heat of solution usually varies in a linear manner with the molality, and so the differential heat of solution is then equal to the integral heat of solution per mole (cf. § 44h).

For a sparingly soluble salt the differential heat of solution in the saturated solution is virtually the same as that at infinite dilution; the former can be derived from the solubility of the salt at two temperatures (§ 36f), and the resulting value can be used for ΔH^0 of the process under consideration. Alternatively, the experimentally determined heat of precipitation of a sparingly soluble salt may be taken as approximately equal, but of opposite sign, to the differential heat of solution.

[9] Bates, *J. Am. Chem. Soc.*, **61**, 522 (1939).

By means of arguments exactly similar to those given above, it can be shown that
$$\Delta F^0 = \bar{F}_2^0 - F_S = \mu_2^0 - F_S, \tag{46.4}$$
and by combining equations (46.3) and (46.4) with the thermodynamic relationship $\Delta F^0 = \Delta H^0 - T\Delta S^0$, it is found that for the present process
$$\Delta S^0 = \frac{\Delta H^0 - \Delta F^0}{T} = \bar{S}_2^0 - S_S, \tag{46.5}$$
where $\bar{S}_2^0$ is the standard partial molar entropy of the salt in solution and S_S is the molar entropy of the solid salt. The standard partial molar entropy of the solute in solution may be taken as the sum of the molar entropies of the constituent ions, i.e.,
$$\bar{S}_2^0 = \nu_+ S_+^0 + \nu_- S_-^0, \tag{46.6}$$
so that by equation (46.5)
$$\Delta S^0 = \nu_+ S_+^0 + \nu_- S_-^0 - S_S. \tag{46.7}$$
If the standard entropy of either of the ions and that of the solid salt is known, the entropy of the other ion may thus be calculated.

As indicated previously, ΔH^0 can be determined in various ways, and S_S for the solid salt can be obtained from heat capacity measurements, based on the third law of thermodynamics. There remains the evaluation of ΔF^0 to be considered. If the solubility product K_s is known in terms of activities, either by extrapolation of solubility measurements (§ 39j) or from electrode potentials (§ 45i), ΔF^0 can be obtained directly, since it is equal to $-RT \ln K_s$, as stated above. When the activity solubility product is not available, use may be made of equation (39.70) for the solubility product, i.e.,
$$K_s = (m_\pm \gamma_\pm)^\nu.$$
As seen in § 41c, $m_\pm^\nu$ may be replaced by $m^\nu(\nu_+^{\nu_+}\nu_-^{\nu_-})$, and m may be identified with the saturation solubility of the given salt; hence, it follows that
$$K_s = (m\gamma_\pm)^\nu \nu_+^{\nu_+}\nu_-^{\nu_-}. \tag{46.8}$$
The value of m may be obtained from the solubility of the salt at the specified temperature and $\gamma_\pm$ can be derived from the Debye-Hückel equation, if the solution is sufficiently dilute. For more concentrated solutions, $\gamma_\pm$ may be estimated by extrapolation from activity coefficients determined in more dilute solution, or by comparison with that of a similar salt at the same concentration.

Problem: The solubility of barium sulfate in water is 9.57×10^{-6} molal at 25°; the heat of solution is 5,970 cal. mole^{-1}. The standard entropy of the barium ion is 2.3 E.U. g. ion^{-1} and that of solid barium sulfate is 31.5 E.U. mole^{-1}. Determine the standard entropy of the sulfate ion in solution.

For BaSO$_4$, both ν_+ and ν_- are unity and ν is 2, so that by equation (46.8)
$$\Delta F^0 = -RT \ln K_s = -2RT \ln m\gamma_\pm.$$

The value of m in the saturated solution is given as 9.57×10^{-6} and $\gamma_\pm$ for this concentration may be estimated from the Debye-Hückel limiting law. Since both ions are bivalent, the ionic strength is equal to $4 \times 9.57 \times 10^{-6}$, and hence by equation (40.15), $-\log \gamma_\pm = 0.509 \times 4 \times \sqrt{4 \times 9.57 \times 10^{-6}} = 0.0126$, so that $\gamma_\pm = 0.971$. Consequently,

$$\Delta F^0 = -2 \times 4.576 \times 298.2 \log (9.57 \times 10^{-6} \times 0.971)$$
$$= 13,740 \text{ cal.}$$

Since ΔH^0 is 5,970 cal., ΔS^0 of solution is $(5,970 - 13,740)/298.2 = -26.0$ E.U. Hence, by equation (46.7),

$$-26.0 = S^0_{Ba^{++}} + S^0_{SO_4^{--}} - S^0_{BaSO_4}$$
$$= 2.3 + S^0_{SO_4^{--}} - 31.5,$$
$$S^0_{SO_4^{--}} = 3.2 \text{ E.U. g. ion}^{-1} \text{ at } 25° \text{ C.}$$

(An average value from various data is given as 4.4 E.U. g. ion^{-1}.)

The standard free energies of formation of a number of ions and the corresponding entropies at 25° C, based on the conventions stated in the preceding sections, are recorded in Table XL.[10] The ΔH^0 values for the

TABLE XL. STANDARD FREE ENERGIES AND HEATS OF FORMATION (IN KCAL. G. ION^{-1}) AND ENTROPIES (IN CAL. DEG.$^{-1}$ G. ION^{-1}) OF IONS AT 25° C *

Ion	ΔF^0 kcal.	ΔH^0 kcal.	S^0 E.U.	Ion	ΔF^0 kcal.	ΔH^0 kcal.	S^0 E.U.
Al^{+++}	-115.5	-126.3	-76	PO$_4^{---}$	-241.0	-229	-45
Fe^{+++}	-2.53	-9.3	-61	AsO$_4^{---}$	-153.4	-215	
Ca^{++}	-132.7	-129.5	-11.4	SO$_4^{--}$	-176.1	-216.3	4.4
Mg^{++}	-107.8	-110.2	-31.6	NO$_3^-$	-26.25	-49.5	35.0
Zn^{++}	-35.18	-36.3	-25.7	CO$_3^{--}$	-126.4	-160.5	-13.0
Fe^{++}	-20.3	-20.6	-25.9	I$^-$	-12.33	-13.6	25.3
Hg^{++}	39.42	41.6	-6.5	Br$^-$	-24.58	-28.7	19.7
K$^+$	-67.43	-60.3	24.2	Cl$^-$	-31.33	-39.9	13.50
Na$^+$	-62.59	-57.5	14.0	OH$^-$	-37.59	-54.8	-2.49
Ag$^+$	18.44	25.2	17.54	CN$^-$	39.14	34.9	25

* For further values, see Table 5 at end of book.

formation of various ions are also given; these may be derived from the ΔF^0 and $T\Delta S^0$ data, or they may be obtained by an alternative procedure described below (§ 46d).

46c. Application of Standard Entropies.—If the heat of solution of a metal M in dilute acid is known, and the entropy of the M^{z+} ion is available, it is obviously a simple matter, by reversing the calculations described above, e.g., for the potassium ion, to calculate the free energy of formation of the M^{z+} ion, and hence the standard potential of the M, M^{z+} electrode. The standard potentials of Mg, Mg^{++} and Al, Al^{+++}, for example, which cannot be obtained by direct measurement, have been calculated in this manner.

[10] Entropy and free energy data mainly from Latimer, ref. 5.

To calculate the standard potential of an anion electrode two methods have been used. The first is possible if the heat of formation from its elements of the corresponding acid in dilute aqueous solution is available. In order to illustrate the procedure the standard potential of the chlorine electrode will be calculated. For the reaction

$$\tfrac{1}{2}H_2(g) + \tfrac{1}{2}Cl_2(g) + aq = H^+(aq) + Cl^-(aq),$$

ΔH^0 is found to be $- 39.94$ kcal. at 25° C. The standard entropies of hydrogen and chlorine gases are 31.21 and 53.31 E.U. mole^{-1} and that of the chloride ion is 13.5 E.U. g. ion^{-1}; since the standard entropy of the hydrogen ion, by convention, is zero,

$$\Delta S^0 = (0 + 13.5) - \tfrac{1}{2}(31.21 + 53.31) = - 28.76 \text{ E.U.}$$

Consequently, the standard free energy change ΔF^0 of the reaction, i.e., $\Delta H^0 - T\Delta S^0$, is $- 39{,}940 - (298.2 \times - 28.76)$, i.e., $- 31{,}360$ cal. By convention, the standard free energies of formation of the hydrogen and chlorine gases and the hydrogen ion are all zero; hence the standard free energy of formation of the chloride ion is $- 31{,}360$ cal. at 25° C. As seen earlier, this is equal to $z\_FE_A^0$, where E_A^0 is the standard potential of the Pt, $Cl_2(g)$, Cl^- electrode, so that

$$23{,}070 \times E_A^0 = - 31{,}360 \text{ cal.}$$
$$E_A^0 = - 1.359 \text{ volts.}$$

The direct experimental value is $- 1.358$ volts at 25° C.

Another method makes use of the heat of formation of a dilute solution of a salt of the anion. Once again the procedure will be explained by reference to the chloride ion. The heat of formation ΔH^0 from its elements of sodium chloride in dilute solution, i.e.,

$$Na(s) + \tfrac{1}{2}Cl_2(g) + aq = Na^+(aq) + Cl^-(aq),$$

is $- 97.4$ kcal. at 25° C. The standard entropy of solid sodium is 12.2 E.U. g. atom^{-1}, that of chlorine gas is 53.31 E.U. mole^{-1}, and the values for the sodium and chloride ions in solution are 14.0 and 13.5 E.U. g. ion^{-1}, respectively. Hence, for the reaction given,

$$\Delta S^0 = (14.0 + 13.5) - [12.2 + (\tfrac{1}{2} \times 53.3)] = - 11.35 \text{ E.U.},$$

and consequently

$$\Delta F^0 = \Delta H^0 - T\Delta S^0 = - 97{,}400 - (298.2 \times - 11.35)$$
$$= - 94{,}000 \text{ cal.}$$

Since the standard free energies of formation of the solid sodium and the chlorine gas are both zero, this result represents the sum of the free energies of formation of the sodium and chloride ions. The standard potential of sodium is 2.714 volt, and so the standard free energy of formation of the sodium ion is $- 1 \times 23{,}070 \times 2.714$, i.e., $- 62{,}600$ cal. The standard free energy of formation of the chloride ion is therefore, $- 94{,}000 - (- 62{,}600)$, i.e., $- 31{,}400$ cal., and the corresponding potential would be $- 1.36$ volts at 25° C. This method has been used for determining the standard potential of the Pt, $F_2(g)$, F^- electrode, for which direct measurements have not been made.

46d. Standard Heats of Formation of Ions.—The standard free energies of formation of ions are based on the convention that for the hydrogen ion the value is zero (§ 46a); this is consistent with the postulate that the standard potential of the hydrogen electrode is zero (§ 45e). Since it has been stipulated that this potential be taken as zero at all temperatures, it follows, from equation (45.3) for example, that ΔH^0, the standard heat of formation of the hydrogen ion, must also be zero. This is the convention adopted in the determination of the standard heats of formation of other ions; two general methods have been employed for the purpose.

The first method, which uses free energy and entropy data, may be illustrated by reference to the heat of formation of the SO_4^{--} ion. It is seen from Table XL that the standard free energy of formation of this ion is -176.1 kcal., so that for the reaction

$$H_2(g) + S(s) + 2O_2(g) + aq = 2H^+(aq) + SO_4^{--}(aq)$$

the value of ΔF^0 is $-176,100$ cal. at 25° C. From Tables XV, XIX and XL, ΔS^0 is given by [0 (for $2H^+$) + 4.4 (for SO_4^{--})] − [31.21 (for H_2) + 7.62 (for S) + 2 × 49.00 (for $2O_2$)] = -132.5 E.U. Consequently ΔH^0 for the formation of the sulfate ion, taking that for the hydrogen ion as zero, is equal to $\Delta F^0 + T\Delta S^0$, i.e., $-176,100 + (298.2 \times -132.5) = -215,600$ cal., or -215.6 kcal.

If, now, the heat of formation of a metallic sulfate at infinite dilution is known, e.g.,

$$2Na(s) + S(s) + 2O_2(g) + aq = 2Na^+(aq) + SO_4^{--}(aq)$$
$$\Delta H^0 = -330.5 \text{ kcal.,}$$

the heat of formation of the associated cation, i.e., the Na^+ ion, can be calculated. In the present case, the standard heat of formation of the sodium ion is $\frac{1}{2}[-330.5 - (-215.6)] = -57.5$ kcal. From this result and the heat of formation of sodium chloride at infinite dilution, viz., -97.4 kcal., the standard heat of formation of the chloride ion, i.e., $-97.4 - (-57.5) = -39.9$ kcal., can be obtained. Proceeding in this manner, the values for other ions can be derived. Instead of starting with the SO_4^{--} ion, as described above, ΔH^0 for the formation of the chloride ion, or of another ion, could have been evaluated from the appropriate ΔF^0 and S^0 data and made the basis of the calculations.

The second general method makes use of thermal data only. At 25° C, the heat of formation of gaseous hydrogen chloride is -22.06 kcal., and the heat of solution at infinite dilution is -17.88 kcal., so that for the reaction

$$\tfrac{1}{2}H_2(g) + \tfrac{1}{2}Cl_2(g) + aq = H^+(aq) + Cl^-(aq)$$

ΔH^0 is -39.94 kcal. By convention, the heats of formation of the hydrogen gas, the chlorine gas and the hydrogen ion are all zero, so that the standard heat of formation of the chloride ion is -39.94 kcal. From the known heats of formation of various chlorides at infinite dilution, the heats of formation of the associated cations can be evaluated.

Another approach to the subject is through a consideration of the heat of neutralization of a strong acid and strong base at infinite dilution; this is equivalent to the standard heat change of the reaction

$$H^+(aq) + OH^-(aq) = H_2O(l) \qquad \Delta H^0 = -13.50 \text{ kcal.,}$$

at 25° C. By combining this with the heat of formation of liquid water, viz.,

$$H_2(g) + \tfrac{1}{2}O_2(g) = H_2O(l) \qquad \Delta H^0 = -68.32 \text{ kcal.,}$$

it is seen that

$$H_2(g) + \tfrac{1}{2}O_2(g) = H^+(aq) + OH^-(aq) \qquad \Delta H^0 = -54.82 \text{ kcal.}$$

On the basis of the usual convention, the standard heat of formation of the OH^- ion is -54.8 kcal. at 25° C. Utilizing the known heat of formation of sodium hydroxide in dilute solution, i.e., -112.2 kcal., the heat of formation of the sodium ion is found to be $-112.2 - (-54.8) = -57.4$ kcal., in satisfactory agreement with the result obtained previously.

Instead of starting with an anion in the calculation of the heats of formation of ions, it is possible to use a cation. For this purpose it is necessary to know the heat of solution of a metal in very dilute acid solution, e.g.,

$$Al(s) + 3HCl(aq) = Al^{+++}(aq) + 3Cl^-(aq) + \tfrac{3}{2}H_2(g).$$

Since the dilute acid may be regarded as completely ionized, this reaction is equivalent to

$$Al(s) + 3H^+(aq) = Al^{+++}(aq) + \tfrac{3}{2}H_2(g),$$

and hence the heat change, i.e., -126.3 kcal. at 25° C, in this case, is equal to the standard heat of formation of the Al^{+++} ion. The values for other cations can be derived in the same manner, and hence the heats of formation of anions can be calculated from the heats of formation of dilute solutions of the appropriate salts.

Problem: Utilize the standard heats of formation of the ions to determine the heat of formation of a dilute solution of calcium nitrate at 25° C.

The required result is obtained by adding ΔH^0 for the Ca^{++} ion to the value for $2NO_3^-$ ions; from Table XL, this is seen to be $-129.5 + 2 \times -49.5$, i.e., -228.5 kcal. at 25° C. (This is very close to the experimental value.)

The standard heats of formation of ions, as recorded in Table XL, may be combined with the free energies of formation and the entropies, in the same table, to calculate thermodynamic quantities for a variety of ionic reactions. Further, these data may be utilized in conjunction with those given in Tables V, XV, XIX and XXIV to determine standard free energy, entropy and heat content changes for reactions involving both ions and neutral molecules.

Problem: Calculate the values at 25° C of ΔF^0, ΔS^0 and ΔH^0 for the reaction

$$2NO(g) + \tfrac{3}{2}O_2(g) + H_2O(l) + aq = 2H^+(aq) + 2NO_3^-(aq).$$

The free energies and heats of formation and the entropies of the various species are as follows (see Tables V, XV, XIX, XXIV and XL):

	$NO(g)$	$O_2(g)$	$H_2O(l)$	H^+	NO_3^-
Free Energy (kcal.)	20.66	0	− 56.70	0	− 26.25
Heat Content (kcal.)	21.6	0	− 68.32	0	− 49.5
Entropy (E.U.)	50.34	49.00	16.75	0	35.0

Hence,

$\Delta F^0 = (0 - 2 \times 26.25) - [(2 \times 20.66) + 0 - 56.70] = -37.12$ kcal.
$\Delta H^0 = (0 - 2 \times 49.5) - [(2 \times 21.6) + 0 - 68.32] = -73.9$ kcal.
$\Delta S^0 = (0 + 2 \times 35.0) - [(2 \times 50.34) + (\tfrac{3}{2} \times 49.00) + 16.75] = -120.9$ E.U.

From the ΔF^0 and ΔH^0 values, ΔS^0 should be $(-73{,}900 + 37{,}120)/298.2 = -123.4$ E.U., the small difference being due to the fact that some of the data are not quite exact.

EXERCISES

1. Determine the chemical reactions taking place in the following (theoretically) reversible cells:

    (i) $Cd \mid CdSO_4$ soln. $\mid PbSO_4(s)$, Pb
   (ii) $K(Hg) \mid KCl$ soln. $\mid AgCl(s)$, Ag
  (iii) Pt, $H_2(g) \mid HI$ soln. $\mid I_2(s)$, Pt
  (iv) Fe $\mid FeCl_2$ soln. $\vdots$ $FeCl_2$, $FeCl_3$ soln. $\mid$ Pt
   (v) Zn, $Zn(OH)_2(s) \mid NaOH$ soln. $\mid HgO(s)$, Hg
  (vi) Hg, $Hg_2Cl_2(s) \mid KCl$ soln. $\vdots$ K_2SO_4 soln. $\mid Hg_2SO_4(s)$, Hg
 (vii) Pt, $Cl_2(g) \mid KCl$ soln. $\vdots$ KBr soln. $\mid Br_2(l)$, Pt
(viii) Pt, $H_2(g) \mid HCl$ soln. $\vdots$ NaOH soln. $\mid H_2(g)$, Pt.

2. Write equations for the E.M.F.'s of the cells given in Exercise 1; liquid junction potentials may be ignored.

3. Devise (theoretically) reversible cells in which the over-all reactions are:

    (i) $PbO(s) + H_2(g) = Pb(s) + H_2O(l)$
   (ii) $H_2(g) + \tfrac{1}{2}O_2(g) = H_2O(l)$
  (iii) $Zn(s) + 2AgBr(s) = ZnBr_2$ soln. $+ 2Ag(s)$
  (iv) $\tfrac{1}{2}H_2(g) + \tfrac{1}{2}Cl_2(g) = HCl$ soln.
   (v) $5Fe^{++} + MnO_4^- + 8H^+ = 5Fe^{+++} + Mn^{++} + 4H_2O$
  (vi) $2Hg(l) + Cl_2(g) = Hg_2Cl_2(s)$
 (vii) $Ce^{++++} + Ag(s) = Ce^{+++} + Ag^+$
(viii) $2Ag(s) + \tfrac{1}{2}O_2(g) = Ag_2O(s)$

Suggest possible uses for some of these cells.

4. Propose cells, without liquid junction, which might be used for determining the activities (or activity coefficients) of (i) H_2SO_4, (ii) KCl, (iii) NaOH, (iv) $CdBr_2$, in solution. Why is it not possible to determine the activity of a nitrate in this manner?

5. The E.M.F. of the cell H_2 (1 atm.) | $HCl(a = 1)$ | $AgCl(s)$, Ag is 0.22551 volt at 20°, 0.22239 volt at 25° and 0.21912 volt at 30° C [Harned and Ehlers, *J. Am. Chem. Soc.*, **54**, 1350 (1932)]. Write out the cell reaction and determine the standard change of (i) free energy, (ii) heat content, (iii) entropy, at 25° C. Compare the results with those obtained from tabulated data.

6. The variation with temperature of the E.M.F. of the cell H_2 (1 atm.) | $HBr(a = 1)$ | $Hg_2Br_2(s)$, Hg is given by Larson [*J. Am. Chem. Soc.*, **62**, 764 (1940)] as

$$E = 0.13970 - 8.1 \times 10^{-5}(t - 25) - 3.6 \times 10^{-6}(t - 25)^2,$$

where t is the temperature on the centigrade scale. Write down the cell reaction and determine the standard change of (i) free energy, (ii) heat content (iii) entropy, at 25° C. Given the entropies of $H_2(g)$, $Hg_2Br_2(s)$ and $Hg(l)$, determine that of the Br^- ion.

7. Suggest methods for the experimental determination of the standard potentials of the following electrodes, using cells without liquid junctions: (i) Pb, Pb^{++}, (ii) Cu, Cu^{++}, (iii) Hg, $Hg_2SO_4(s)$, SO_4^{--}, (iv) Ag, $Ag_2O(s)$, OH^-.

8. The E.M.F. of the cell Zn | $ZnCl_2(m)$ | $AgCl(s)$, Ag, with various molalities m of zinc chloride, was found to be as follows at 25° C:

m	E	m	E
0.002941	1.1983	0.04242	1.10897
0.007814	1.16502	0.09048	1.08435
0.01236	1.14951	0.2211	1.05559
0.02144	1.13101	0.4499	1.03279

[Scatchard and Tefft, *J. Am. Chem. Soc.*, **52**, 2272 (1930)]. Determine the standard potential of the Zn, Zn^{++} electrode.

9. By utilizing the standard potentials given in Table XXXIX, calculate the true equilibrium constant and standard free energy change of the reaction $Ag(s) + Fe^{+++} = Ag^+ + Fe^{++}$ at 25° C (cf. § 41h).

10. The equilibrium constant for the reaction $CuCl(s) + AgCl(s) + aq = Cu^{++} + 2Cl^- + Ag(s)$ was found to be 1.86×10^{-6} at 25° C [Edgar and Cannon, *J. Am. Chem. Soc.*, **44**, 2842 (1922)]. Using the known standard potentials of the Ag, $AgCl(s)$, Cl^- and Cu, Cu^{++} electrodes, calculate that of the Cu, $CuCl(s)$, Cl^- electrode.

11. Account quantitatively for the fact that metallic mercury is able to reduce ferric chloride to ferrous chloride, to virtual completion within the limits of analytical accuracy.

12. The equilibrium constant of the reaction $\frac{1}{2}HgO(s) + \frac{1}{2}Hg(l) + \frac{1}{2}H_2O(l) + Br^- = \frac{1}{2}Hg_2Br_2(s) + OH^-$ is 0.204 at 25° C [Newton and Bolinger, *J. Am. Chem. Soc.*, **52**, 921 (1930)]. The E.M.F. of the cell H_2 (1 atm.) | $HBr(a = 1)$ | $Hg_2Br_2(s)$, Hg is 0.1397 volt and that of the cell H_2 (1 atm.) | $NaOH(aq)$ | $HgO(s)$, Hg is 0.9264 volt at 25° C. Calculate the standard free energy change of the reaction $H_2O(l) = H^+ + OH^-$, and the dissociation constant (K_w) of water.

13. Using tabulated free energy and entropy values, calculate the standard E.M.F., and the temperature coefficient at 25° C, of the cell Pt, H_2 | HBr soln. | $AgBr(s)$, Ag. What would be the E.M.F. if the pressure (fugacity) of the hydrogen gas was 0.1 atm. and the molality of the hydrobromic acid solution was 0.1?

14. Using the standard potential of the Pt, $Cl_2(g)$, Cl^- electrode, and taking the partial pressure of hydrogen chloride in equilibrium with a hydrochloric acid solution of unit activity as 4.97×10^{-7} atm. at 25° C (cf. Exercise 9, Chapter XVI), together with the known heat capacities of hydrogen, chlorine, and hydrogen chloride, derive a general expression for the standard free energy of formation of hydrogen chloride gas as a function of the absolute temperature.

15. The solubility of silver iodate in water is 1.771×10^{-4} mole liter^{-1} at 25° C. Calculate the standard free energy change of the process $AgIO_3(s) = Ag^+ + IO_3^-$. Using the standard potential of the Ag, Ag^+ electrode, calculate that of the Ag, $AgIO_3(s)$, IO_3^- electrode at 25° C.

16. The standard potential of the Hg, Hg_2^{++} electrode is -0.7986 and that of the Pt, Hg_2^{++}, Hg^{++} electrode is -0.910 volt at 25° C. Calculate (i) the standard potential of the Hg, Hg^{++} electrode, (ii) the ratio of Hg_2^{++} to Hg^{++} ion activities in a solution of mercurous and mercuric ions in equilibrium with metallic mercury, at 25° C.

17. The temperature coefficient of the standard potential of the Ag, Ag^+ electrode ($E^0 = -0.7995$ volt) is 0.967×10^{-3} volt deg.$^{-1}$ in the vicinity of 25° C [Lingane and Larson, *J. Am. Chem. Soc.*, **59**, 2271 (1937)]. Determine the entropy and the heat of formation of the Ag^+ ion.

18. The solubility of silver chloride in pure water is 1.314×10^{-5} molal, and the mean ionic activity coefficient is then 0.9985 [Neuman, *J. Am. Chem. Soc.*, **54**, 2195 (1932)]. The heat of solution of the salt is 15,740 cal. mole^{-1}. Taking the entropy of solid silver chloride as 22.97 E.U. mole^{-1}, and using the results of the preceding exercise, calculate the standard free energy and heat of formation and the entropy of the Cl^- ion at 25° C.

19. Show that the E.M.F. of the cell $M \mid M^+(a = 1) \vdots A^-(a = 1) \mid MA(s)$, M is given by $E = (RT/F) \ln K_s$, where K_s is the solubility product of the sparingly soluble salt MA. Propose a cell for determining the solubility product of silver iodide, and use the data in Table XXXIX to calculate the value at 25° C.

20. The standard potential of the silver azide electrode, i.e., Ag, $AgN_3(s)$, N_3^-, is -0.2919 volt at 25° C [Harned and Nims, *J. Am. Chem. Soc.*, **60**, 262 (1938)]. If the solubility of silver chloride is 1.314×10^{-5} molal, calculate that of silver azide at 25° C. (Complete dissociation may be assumed in the saturated solution in each case.)

21. The heat of solution (ΔH^0) of cadmium in dilute acid is -17.1 kcal. g. atom^{-1} at 25° C [Richards, *et al.*, *J. Am. Chem. Soc.*, **44**, 1051, 1060 (1922)]. Using the known standard Cd, Cd^{++} potential, and the entropy of Cd and $H_2(g)$, determine the standard entropy of the Cd^{++} ion.

22. The heat of formation of a dilute solution of sodium fluoride is -136.00 kcal. at 25° C. Using the standard entropies of $Na(s)$, $F_2(g)$, Na^+ and F^-, together with the known standard potential Na, Na^+, calculate the standard potential of the Pt, $F_2(g)$, F^- electrode.

23. The heat of solution of aluminum in dilute acid is -126.3 kcal. at 25° C. Using the standard entropies of $Al(s)$, $H_2(g)$ and Al^{+++}, calculate the standard potential of the Al, Al^{+++} electrode.

24. From the standard free energies of formation of $PbSO_4(s)$, Pb^{++} and SO_4^{--}, calculate the solubility product of lead sulfate at 25° C. Assuming complete dissociation, estimate the solubility in water, in terms of molality.

25. From the standard free energies of formation of OH^- ions and of $H_2O(l)$, calculate the dissociation product (K_w) of water at 25° C.

26. The standard potential of the Ag, Ag$_2$O(s), OH$^-$ electrode is $-$ 0.344 volt and that of the Pt, AgO(s), Ag$_2$O(s), OH$^-$ electrode is $-$ 0.57 volt at 25° C. Using the standard O$_2$(g), OH$^-$ potential, determine the oxygen dissociation pressure of (i) Ag$_2$O, (ii) AgO, at 25° C. What would be the dissociation pressure in the latter case if the reaction AgO = Ag + $\tfrac{1}{2}$O$_2$(g) occurred?

27. Derive the general expression

$$\left(\frac{\partial E}{\partial P}\right)_T = -\frac{\Delta V}{NF}$$

for the influence of pressure on the E.M.F. of any cell, at constant temperature. Show that for the cell H$_2$(P) | HCl(m) | Hg$_2$Cl$_2$(s), Hg(l),

$$\Delta V = 2(\bar{V}_{\text{HCl}(m)} + V_{\text{Hg}}) - (V_{\text{H}_2(P)} + V_{\text{Hg}_2\text{Cl}_2})$$

where $\bar{V}_{\text{HCl}(m)}$ is the partial molar volume of the hydrochloric acid, i.e., $\bar{V}_2$, in the solution of molality m; $V_{\text{H}_2(P)}$ is the molar volume of hydrogen at the pressure P in the cell; V_{Hg} and $V_{\text{Hg}_2\text{Cl}_2}$ refer to the indicated substances in the pure states. The terms $\bar{V}_{\text{HCl}}$, V_{Hg} and $V_{\text{Hg}_2\text{Cl}_2}$ are relatively small and their resultant is virtually zero; hence, show that in this case

$$E_2 - E_1 = \frac{2RT}{NF}\ln\frac{f_2}{f_1},$$

where E_1 and E_2 are the E.M.F.'s of the cells with hydrogen gas at fugacities f_1 and f_2, respectively. Compare this result with that obtained from the general equation (45.6).

28. With 0.1 molar hydrochloric acid at 25° C, E_1 is 0.3990 volt when the hydrogen pressure (or fugacity) is 1 atm., and E_2 is 0.4850 volt when the pressure of the gas in 568.8 atm. Calculate the fugacity of hydrogen at the latter pressure using the equation derived in the preceding exercise. Compare the result with that obtained from the compressibility equation

$$PV = RT(1 + 5.37 \times 10^{-4}P + 3.5 \times 10^{-8}P^2).$$

[Hainsworth, Rowley and MacInnes, *J. Am. Chem. Soc.*, **46**, 1437 (1924).]

29. The potential of the tetrathionate-thiosulfate electrode (Pt, S$_4$O$_6^{--}$, 2S$_2$O$_3^{--}$), involving the process 2S$_2$O$_3^{--}$ = S$_4$O$_6^{--}$ + 2ϵ, cannot be determined directly. Estimate its value, given that the standard entropies of the S$_2$O$_3^{--}$ and S$_4$O$_6^{--}$ ions are 8 and 35 E.U., respectively, and that ΔH^0 for the reaction 2S$_2$O$_3^{--}$ + I$_2$(s) = S$_4$O$_6^{--}$ + 2I$^-$ in solution is $-$ 7.76 kcal. at 25° C. Any other data required may be obtained from tables in the book.

30. The E.M.F. of a lead storage battery containing 2.75 molal sulfuric acid was found to be 2.005 volt at 25° C. The aqueous vapor pressure of the acid solution at this temperature is about 20.4 mm., while that of pure water is 23.8 mm. The mean ionic activity coefficient of the sulfuric acid is 0.136. Calculate the standard free energy change of the cell reaction at 25° C and check the values from tabulated free energy data.

31. Using tabulated standard free energies and the known activity coefficients of nitric acid solutions, calculate the free energy change of the reaction

$$3\text{NO}_2(g) + \text{H}_2\text{O}(l) = 2\text{HNO}_3 \text{ (soln.)} + \text{NO}(g)$$

at 25° C, (i) for the standard state, (ii) for 2 molal nitric acid.

32. Assuming the heats of formation to remain approximately constant in the range from 25° to 30° C, calculate the ionic product of water at these two temperatures, using standard free energy and heat of formation data only.

33. Show that the heat of solution of an electrolyte at infinite dilution is equal to the sum of the standard heats of formation of its constituent ions minus the standard heat of formation of the pure solute. From the data in Tables V and XL calculate the heats of solution at infinite dilution per mole of (i) NaOH, (ii) H_2SO_4, (iii) AgCl, (iv) K_2SO_4.

APPENDIX

TABLE 1. CONSTANTS AND CONVERSION FACTORS *

1 liter	1000.028 cm.^{3}
1 atm	1.01325×10^6 dynes cm.$^{-2}$
1 int. joule	1.00017 abs. joule
1 (defined) cal	4.1833 int. joules
	4.1833 int. volt-coulombs
	4.1840 abs. joules
	0.041292 liter-atm.
	41.293 cc.-atm.
1 liter-atm	1.0133×10^9 ergs
	1.0131×10^2 int. joules
	24.218 cal.
1 cc.-atm	0.024212 cal.
Molar volume of ideal gas at 0° C and 1 atm	22.4140 liter mole^{-1}
Ice Point	273.16° K
Molar gas constant (R)	8.3144 abs. joules deg.$^{-1}$ mole^{-1}
	8.3130 int. joules deg.$^{-1}$ mole^{-1}
	1.9872 cal. deg.$^{-1}$ mole^{-1}
	0.082054 liter-atm. deg.$^{-1}$ mole^{-1}
	82.057 cc.-atm. deg.$^{-1}$ mole^{-1}
Avogadro number (N)	6.0228×10^{23} mole^{-1}
Boltzmann constant ($k = R/N$)	1.3805×10^{-16} erg deg.$^{-1}$
Planck constant (h)	6.6242×10^{-27} erg sec.
Velocity of light (c)	2.99776×10^{10} cm. sec.$^{-1}$
hc/k	1.4385 cm. deg.
Faraday (F)	96,500 int. coulombs g. equiv.$^{-1}$
	23,070 cal. volt^{-1} g. equiv.$^{-1}$

* Mainly from publications of the National Bureau of Standards, cf., *J. Res. Nat. Bur. Stand.*, **34**, 143 (1945).

TABLE 2. PROPERTIES OF GASES AND LIQUIDS *

Substance	T_c	P_c	B. P.	ΔH_v at B. P.
H_2	33.2° K	12.8 atm.	20.3° K	216 cal. mole^{-1}
N_2	126.0	33.5	77.3	1,360
O_2	154.3	49.7	90.2	1,610
Cl_2	417	76	239.5	4,800
CO	134.4	34.6	81.1	1,410
CO_2	304.1	72.9	—	—
N_2O	309.6	71.7	183.7	—
NO	179	65	121.4	—
NO_2	431	99	—	—
HCl	324.5	81.6	189.5	3,600
HBr	363	84	206.2	3,950
HI	424	82	237	4,300
H_2O	647.3	218.2	373.2	9,717
H_2S	373.5	88.9	213.4	4,490
SO_2	430.3	77.7	263.2	6,080
SO_3	491.4	83.8	317.8	9,500
NH_3	405.5	111.5	239.8	5,560
HCN	458.6	56.9	299.2	5,700
CH_4	191.1	45.8	111.7	1,950
C_2H_6	305.2	48.8	185.9	3,800
C_3H_8	369.9	42.0	231.0	4,500
n-C_4H_{10}	426.0	36.0	272.6	5,300
iso-C_4H_{10}	407.1	37.0	263.0	5,080
C_2H_4	282.8	50.7	169.3	—
C_3H_6	364.8	45.0	226.2	—
C_2H_2	309.1	61.7	184.7	—
CH_3OH	513.2	98.7	337.8	8,410
C_2H_5OH	516.2	63.1	351.6	9,400
C_6H_6	561.6	47.7	353.3	7,400

* Data mainly adapted from International Critical Tables. For empirical formulae for estimating critical temperatures and heats of vaporization, see Watson, *Ind. Eng. Chem.*, **23**, 360 (1931); Othmer, *ibid.*, **32**, 841 (1940); **34**, 1072 (1942); Meissner, *ibid.*, **33**, 1440 (1941); **34**, 521 (1942). When no other information is available, the Guldberg-Guye rule may be used, i.e., $T_c \approx 1.6 \times$ B. P. (in °K), to obtain the critical temperature. Heats of vaporization may be estimated from Trouton's rule.

TABLE 3. HEAT CAPACITIES OF GASES AT 1 ATM. PRESSURE *
$$C_P = \alpha + \beta T + \gamma T^3 + \delta T^4 \text{ cal. deg.}^{-1} \text{ mole}^{-1}$$

Gas	α	$\beta \times 10^3$	$\gamma \times 10^6$	$\delta \times 10^9$
H_2	6.947	− 0.200	0.4808	—
D_2	6.830	0.210	0.468	—
O_2	6.095	3.253	− 1.017	—
	6.148	3.102	− 0.923	—
N_2	6.449	1.413	− 0.0807	—
	6.524	1.250	− 0.001	—
Cl_2	7.576	2.424	− 0.965	—
Br_2	8.423	0.974	− 0.3555	—
CO	6.342	1.836	− 0.2801	—
	6.420	1.665	− 0.196	—
HCl	6.732	0.4325	0.3697	—
HBr	6.578	0.955	0.1581	—
H_2O	7.219	2.374	0.267	—
	7.256	2.298	0.283	—
CO_2	6.396	10.100	− 3.405	—
	6.214	10.396	− 3.545	—
	† 5.152	15.224	− 9.681	2.313
H_2S	5.974	10.208	− 4.317	—
N_2O	6.529	10.515	− 3.571	—
SO_2	6.147	13.844	− 9.103	2.057
NH_3	6.189	7.787	− 0.728	—
SO_3	† 3.603	36.310	− 28.828	8.649
	6.077	23.537	− 9.687	—
CH_4	† 4.171	14.450	0.267	− 1.722
	3.381	18.044	− 4.300	—
C_2H_6	† 1.279	42.464	− 16.420	2.035
	2.195	38.282	− 11.001	—
C_3H_8	† − 1.209	73.734	− 38.666	7.961
	0.410	64.710	− 22.582	—
n-C_4H_{10}	† − 0.012	92.506	− 47.998	9.706
	4.357	72.552	− 22.145	—
C_2H_4	2.706	29.160	− 9.059	—
C_2H_2	11.942	4.387	$(− 0.232 \times 10^6 T^{-2})$	—
C_6H_6	− 9.478	119.930	− 80.702	—
CH_3OH	4.398	24.274	− 6.855	—
C_2H_5OH	3.578	49.847	− 16.991	—
CH_3COCH_3	† 2.024	64.401	− 34.285	7.082
	5.371	49.227	− 15.182	—

* Spencer, et al., *J. Am. Chem. Soc.*, **56**, 2311 (1934); **64**, 250 (1942); **67**, 1859 (1945); see also, Bryant, *Ind. Eng. Chem.*, **25**, 820 (1933). Most of the data are applicable from about 300° to 1500° K.

† More accurate values, involving four constants.

TABLE 4. HEAT CAPACITIES OF SOLIDS AND LIQUIDS AT 1 ATM. PRESSURE *

$$C_P = a + bT + cT^{-2} \text{ cal. deg.}^{-1} \text{ mole}^{-1}$$

Substance	a	$b \times 10^3$	$c \times 10^{-6}$
Al	4.80	3.22	—
Al_2O_3	22.08	8.971	0.5225
B	1.54	4.40	—
Cd	5.46	2.466	—
CaO	10.0	4.84	0.1080
$CaCO_3$	19.68	11.89	0.3076
Graphite	2.673	2.617	0.1169
Cu	5.44	1.462	—
CuO	10.87	3.576	0.1506
Pb	5.77	2.02	—
PbO	10.33	3.18	—
PbS	10.63	4.01	—
Hg(l)	6.61	—	—
Ag	5.60	1.50	—
AgCl	9.60	9.29	—
Zn	5.25	2.70	—
ZnO	11.40	1.45	0.1824
ZnS	12.81	0.95	0.1946

* Kelley, *U. S. Bur. Mines Bull.*, 371 (1934). Except for mercury, the values are applicable from 0° C to the melting point or to about 1000° C, whichever is the lower.

TABLE 5.* STANDARD FREE ENERGIES AND HEATS OF FORMATION (IN KCAL. MOLE^{-1}) AND ENTROPIES (IN CAL. DEG.$^{-1}$ MOLE^{-1}) AT 25° C †

Substance	ΔF^0 kcal.	ΔH^0 kcal.	S^0 E.U.	Substance	ΔF^0 kcal.	ΔH^0 kcal.	S^0 E.U.
Aluminum				BF_3	—	−257	61
Al	—	—	6.75	BCl_3	—	−94.6	68.6
Al^{+++}	−115.5	−126.3	−76	**Bromine**			
Al_2O_3	—	−380	12.5	$Br_2(l)$	—	—	36.7
Antimony				$Br_2(g)$	0.75	7.65	58.63
Sb	—	—	10.5	$Br_2(aq)$	0.98	−1.2	—
SbH_3	35.3	—	—	Br^-	−24.58	−28.7	19.7
Sb_2O_3	−149.0	−166	29.4	HBr	−12.54	−8.6	47.48
Sb_2O_4	−165.9	−213	30.4	BrO_3^-	5.0	−11.2	38.5
Sb_2O_5	−195.5	−230	29.9	**Cadmium**			
$SbCl_3$	−77.8	−91.4	37.0	Cd	—	—	12.3
Arsenic				Cd^{++}	−18.55	−17.6	−15.6
As	—	—	8.4	CdO	—	−65.2	13.1
AsH_3	37.7	43.6	—	$CdCl_2$	—	−93.0	31.2
As_2O_3	−137.7	−154	25.6	$CdBr_2$	—	−75.8	32.0
As_2O_5	—	−218	25.2	CdS	−33.1	−34.6	15.0
AsO_4^{---}	−153.4	−215	—	**Calcium**			
Barium				Ca	—	—	9.95
Ba	—	—	15.1	Ca^{++}	−132.7	−129.5	−11.4
Ba^{++}	−133.85	−128.4	2.3	CaO	−142.0	−151.8	9.5
BaO	—	−133	16.8	CaF_2	—	−290.2	16.4
$BaCO_3$	−271.6	−291	26.8	Calcite	−207.4	−289.5	22.2
$BaSO_4$	—	−350	31.5	**Carbon**			
Beryllium				Graphite	—	—	1.36
Be	—	—	2.28	Diamond	—	0.45	0.585
Be^{++}	−78.7	−85	—	CO	−32.81	−26.42	47.30
BeO	—	−135	3.37	CO_2	−94.26	−94.05	51.06
Bismuth				CH_4	−12.14	−17.89	44.50
Bi	—	—	13.6	C_2H_6	−7.86	−20.24	54.85
Bi_2O_3	−116.6	−137.1	36.2	C_3H_8	−5.61	−24.82	64.7
$BiCl_3$	−76.4	−90.6	45.8	n-C_4H_{10}	−3.75	−29.81	74.5
				C_2H_4	16.34	12.56	52.48
Boron				C_2H_2	50.7	54.23	45.0
B	—	—	1.7	$CH_3OH(l)$	−40.0	−57.0	30.5

* Data mainly from F. R. Bichowsky and F. D. Rossini, "The Thermochemistry of the Chemical Substances," 1936 (heats of formation); W. M. Latimer, "The Oxidation States of the Elements and their Potentials in Aqueous Solutions," 1938 (free energies and entropies); Kelley, *U. S. Bur. Mines Bull.*, **434** (1941) (entropies); see also, Chapter XIII, ref. **7**. Because of temperature differences and the variety of sources, the data are not always completely consistent; the deviations are, however, usually not greater than the experimental errors.

† Unless otherwise indicated, e.g., $SO_3(g)$, all substances are supposed to be in their stable forms at 25° C and 1 atm. It may be noted that where two of the quantities are available for any substance it is usually possible to calculate the third, provided the entropies of the elements involved are known. The heats and free energies of formation of elements in their standard states are taken as zero, by convention.

TABLE 5. (*Continued*)

Substance	ΔF^0 kcal.	ΔH^0 kcal.	S^0 E.U.	Substance	ΔF^0 kcal.	ΔH^0 kcal.	S^0 E.U.
$CH_3OH(g)$	−38.9	−48.1	56.63	**Hydrogen**			
$C_2H_5OH(l)$	−40.2	−66.4	38.4	H_2	—	—	31.21
$C_2H_5OH(g)$	−38.7	−56.3	67.6	D_2	—	—	34.62
C_6H_6	29.06	11.7	41.9	HD	—	—	34.39
$C_6H_5NH_2$	35.4	7.34	45.8	H	48.35	51.7	27.40
CCl_4	−15.6	−33.8	52.2	H^+	0.00	0.00	0.00
CS_2	17.15	15.4	36.2				
$COCl_2$	−48.96	−53.5	67.2	**Iodine**			
$HCN(g)$	27.7	30.7	48.23	$I_2(s)$	—	—	27.9
CN^-	39.14	34.9	25	$I_2(g)$	4.63	14.9	62.29
CO_3^{--}	−126.4	−160.5	−13.0	I^-	−12.33	−13.6	25.3
HCO_3^-	−140.5	−164.8	22.2	HI	0.32	5.9	49.36
Chlorine				IO_3^-	−32.25	−54.5	28.0
Cl_2	—	—	53.31	**Iron**			
HCl	−22.74	−22.06	44.66	Fe	—	—	6.5
Cl^-	−31.33	−39.9	13.50	Fe^{++}	−20.3	−20.6	−25.9
ClO_3^-	−0.25	−20.8	39.4	Fe^{+++}	−2.53	−9.3	−61
ClO_4^-	−10.7	−39.5	43.6	FeO	—	−64.3	13.4
Chromium				Fe_2O_3	—	−195	21.5
Cr	—	—	5.68	FeS	−22.9	−23	16.1
Cr^{++}	−39.4	−43	—	FeS_2	—	—	12.7
Cr^{+++}	−49.0	−65	—	**Lead**			
Cr_2O_3	—	—	19.4	Pb	—	—	15.49
CrO_4^{--}	−171.4	−207.9	10.5	Pb^{++}	−5.81	−0.2	3.9
Cobalt				PbO	−45.1	−52.5	16.6
Co	—	—	6.8	PbO_2	−52.0	−65.0	18.3
Co^{++}	−12.8	−16.5	—	Pb_3O_4	−142.2	−172.4	50.5
Co^{+++}	28.9	—	—	$PbCl_2$	−75.0	−85.7	32.6
Copper				$PbBr_2$	−62.1	−66.3	38.6
Cu	—	—	7.97	$PbCO_3$	−149.7	−168.0	48.5
Cu^+	12.04	—	—	PbS	−21.9	−22.3	21.8
Cu^{++}	15.91	−15	−26.5	$PbSO_4$	−159.5	−218.5	35.2
Cu_2O	−35.1	−42.5	24.1	**Lithium**			
CuO	−30.4	−38.5	10.4	Li	—	—	6.7
CuCl	−28.5	−34	20.8	Li^+	−70.7	−66.6	4.7
CuBr	−23.8	−27	22.8	**Magnesium**			
$CuCO_3$	−123.9	—	17.7	Mg	—	—	7.77
Cu_2S	−20.2	−19.0	28.9	Mg^{++}	−107.8	−110.2	−31.6
CuS	−11.8	−11.6	15.9	MgO	−136.4	−146.1	6.55
Fluorine				$Mg(OH)_2$	−193.3	−223	15.09
F_2	—	—	48.58	$MgCO_3$	−246.6	−268	15.7
F^-	−65.7	−78.2	−2.3	**Manganese**			
HF	−31.8	−64.5	41.53	Mn	—	—	7.6
F_2O	9.5	5.5	59	Mn^{++}	−48.6	−49.2	19.1
Gold				MnO	—	−96	14.4
Au	—	—	11.4	MnO_2	−102.9	−123	13.9
Au^{+++}	98.2	—	—	MnO_4^-	−100.6	−122.3	46.7
Au_2O_3	18.71	11	—	MnS	−46.0	−47	18.7

APPENDIX

TABLE 5. (*Continued*)

Substance	ΔF^0 kcal.	ΔH^0 kcal.	S^0 E.U.	Substance	ΔF^0 kcal.	ΔH^0 kcal.	S^0 E.U.
Mercury				KBr	−90.45	−94.1	22.6
Hg	—	—	18.5	KI	—	−78.9	24.1
Hg_2^{++}	36.85	40.2	17.7	KOH	—	−102.0	—
Hg^{++}	39.42	41.6	−6.5	KNO_3	—	−118	31.8
HgO (red)	−13.94	−21.6	16.6	K_2SO_4	—	−343	44.8
Hg_2Cl_2	−50.3	−62	47.0				
Hg_2Br_2	−42.7	−49.2	49.3	**Rubidium**			
$HgCl_2$	−42.2	−53.4	34.6	Rb	—	—	16.6
$HgBr_2$	−38.8	−40.7	38.9	Rb^+	−68.8	−61.0	28.7
HgS (red)	−8.8	−11.0	19.8	**Selenium**			
Hg_2SO_4	—	−176.5	48.0	Se	—	—	10.0
				H_2Se	15.3	18.5	—
Nickel							
Ni	—	—	7.12	**Silicon**			
Ni^{++}	−11.53	−15.2	—	Si	—	—	4.5
NiO	−53.0	−58.4	9.2	Quartz	−190.4	−203.3	10.1
				SiF_4	−351.0	−360	68.0
Nitrogen				$SiCl_4$	−134.1	−150	57.3
N_2	—	—	45.77				
N_2O	24.93	19.7	52.58	**Silver**			
NO	20.66	21.6	50.34	Ag	—	—	10.2
NO_2	12.27	8.03	57.47	Ag^+	18.44	25.2	17.54
$N_2O_4(g)$	23.4	3.1	72.7	Ag_2O	−2.59	−7.3	29.1
NOCl	16.01	12.8	63.0	AgO	2.60	−3.0	—
NH_3	−3.94	−11.03	46.03	AgCl	−26.22	−30.3	23.0
NH_4^+	−18.96	−31.5	26.4	AgBr	−22.90	−24.0	25.6
$HNO_3(l)$	—	−41.7	—	AgI	−15.8	−15.0	27.6
NO_3^-	−26.25	−49.5	35.0	$AgIO_3$	−24.1	—	37.5
NO_2^-	−8.45	−28.5	29.9	$AgNO_3$	4.96	−29.4	33.7
				Ag_2S	−9.5	−5.5	35.0
Oxygen				Ag_2SO_4	−145.9	−170.0	47.8
O_2	—	—	49.00				
O_3	39.4	34.8	57.1	**Sodium**			
$H_2O(l)$	−56.70	−68.32	16.75	Na	—	—	12.2
$H_2O(g)$	−54.64	−57.80	45.11	Na^+	−62.59	−57.5	14.0
$D_2O(l)$	−58.20	−70.41	18.08	Na_2O	—	−99	17
$D_2O(g)$	−56.06	−59.56	47.38	NaF	—	−130.0	13.1
OH^-	−37.59	−54.8	−2.49	NaCl	−91.7	−98.3	17.3
				NaBr	—	−86.7	20.1
Phosphorus				NaI	—	−69.3	22.5
P (red)	—	—	10.6	NaOH	−90.48	−102.0	—
P (yellow)	—	4.2	15.1	Na_2CO_3	−250.8	−270.0	32.5
PH_3	2.88	−2.3	50.35	$NaNO_3$	—	−111.7	27.8
PCl_3	−61.5	−76.9	52.2	Na_2SO_4	—	−330.5	35.7
PO_4^{---}	−241.0	−229	−45				
				Strontium			
Potassium				Sr	—	—	13.3
K	—	—	15.2	Sr^{++}	−133.2	−130.0	−7.3
K^+	−67.43	−60.3	24.2	SrO	—	−140.8	13.0
K_2O	—	−86	—	$SrCO_3$	−271.9	−290	23.2
KCl	−97.56	−104.3	19.76	$SrSO_4$	—	−345	28.2

TABLE 5. *(Continued)*

Substance	ΔF^0 kcal.	ΔH^0 kcal.	S^0 E.U.	Substance	ΔF^0 kcal.	ΔH^0 kcal.	S^0 E.U.
Sulfur				Tl^{+++}	49.74	28	—
Rhombic	—	—	7.62	TlCl	−44.19	−48.6	25.6
Monoclinic	0.018	0.075	7.78	**Tin**			
S^{--}	23.42	10.0	−5.5	Sn (white)	—	—	12.3
HS^{--}	2.95	−3.9	14.9	Sn^{++}	−6.28	—	−4.9
H_2S	−7.87	−5.3	49.15	Sn^{++++}	0.7	—	—
SF_6	−235.0	−262	69.6	$SnCl_4$	−110.6	−127.4	62.1
SO_2	−71.7	−70.9	59.24	SnO	—	−67.7	13.5
$SO_3(g)$	−88.5	−94.4	63.8	SnO_2	—	−138	12.5
H_2SO_4	—	−193.8	—	**Zinc**			
SO_4^{--}	−176.1	−216.3	4.4	Zn	—	—	9.95
HSO_4^-	—	−213.3	30.6	Zn^{++}	−35.18	−36.3	−25.7
SO_3^{--}	−116.4	−149	3	ZnO	—	−84	10.4
Thallium				$ZnCO_3$	−174.8	−193.3	19.7
Tl	—	—	15.4	ZnS	−43.2	−44	13.8
Tl^+	−7.76	0.8	30.5	$ZnSO_4$	—	−233.4	30.7

TABLE 6. INTEGRAL HEATS OF SOLUTION OF SALTS (1 MOLE SALT TO 200 MOLES OF WATER) AT 25° C *

Salt	ΔH	Salt	ΔH	Salt	ΔH
NH_4NO_3	7.3 kcal.	$CuSO_4 \cdot 5H_2O$	2.9 kcal.	NaOH	− 10.0 kcal.
NH_4Cl	3.9	$AgNO_3$	0.45	NaCl	1.22
$(NH_4)_2SO_4$	2.3	$NiSO_4$	− 15.1	NaBr	0.60
$Pb(NO_3)_2$	7.1	$NiSO_4 \cdot 7H_2O$	4.2	Na_2SO_4	− 0.5
$ZnCl_2$	− 15.3	$FeCl_3$	− 30.7	$Na_2SO_4 \cdot 10H_2O$	18.5
$ZnSO_4$	− 18.5	$FeCl_3 \cdot 6H_2O$	− 4.6	$NaNO_3$	5.0
$ZnSO_4 \cdot 7H_2O$	4.3	$MgCl_2$	− 35.7	KCl	4.45
$CdCl_2$	− 2.9	$MgCl_2 \cdot 6H_2O$	− 2.8	KBr	6.1
$CdSO_4$	− 10.5	$BaCl_2$	− 2.0	KNO_3	− 0.23
$CuSO_4$	− 15.9	$BaCl_2 \cdot 2H_2O$	5.0	K_2SO_4	6.2

* Data adapted from F. R. Bichowsky and F. D. Rossini, "The Thermochemistry of the Chemical Substances," 1936; the values given there refer to 18° C, but those at 25° C are not very different. The heat of solution at infinite dilution for any electrolyte may be obtained by summing the standard heats of formation of the constituent ions (from Table 5) and subtracting the heat of formation of the substance in the pure state.

INDEX

Absolute temperature, 4–5
  zero, 5, 140
    entropy, 178, 181, 183
    heat capacity, 121, 124
    unattainability, 139
Absorption coefficient, 347
Activity, 262, 269, 350–74
  in amalgams, 363–67, 369–71
  coefficient, 258, 351–54, 356
    and Debye-Hückel theory, 409–24
    determination, 381–400, 403
      boiling point, 387
      E.M.F., 392–98
      freezing point, 381–86
      isopiestic method, 387–88
      osmotic coefficient, 390–92
      solubility, 398–400, 403
    and dielectric constant, 414
    and electrical free energy, 409
    in electrolytes, 379
    and equilibrium constant, 275–76
    in gas mixtures, 263, 264
      and pressure, 265
      and temperature, 266
    and Henry's law, 353–54
    and ionic strength, 401, 410–12
    in liquid mixtures, 269
      and pressure, 270–71
      and temperature, 270, 356
    and Margules equation, 370
    mean ionic, 379
    of nonelectrolytes, 402–03
    and osmotic coefficient, 389–92
    and pressure, 270
    rational, 353, 407
    and standard states, 355, 380
    stoichiometric, 380, 412
    table, 402
    and temperature, 270, 356, 386, 387, 447
    true, 380, 412
    and valence type, 401, 412, 413
    and van Laar equation, 371
  determination, 356–74
    boiling point, 362–63
    E.M.F., 363
    freezing point, 358–61
    vapor pressure, 356, 357
  in electrolytes, 378–79
  and equilibrium constant, 274

Activity (Cont.)
  and fugacity, 262, 270, 358
  in gas mixtures, 262
    and pressure, 265
    and temperature, 266
  and Gibbs adsorption equation, 245
  in liquid mixtures, 269
    and pressure, 270
    and temperature, 270
  and Margules equation, 370
  and osmotic pressure, 372–73
  and pressure, 270
  solubility product, 398
  standard state, 350–54
    gas, 263, 263, 264, 351
    and reference state, 263, 352–54
    solute, 352–54
    solvent, 351
  and temperature, 270, 361, 362
Adiabatic combustion, 84–89
  process, 55–9
    entropy change, 147, 150
    pressure-volume relationships, 57
    temperature change, 56
    work of expansion, 58
  relationships, 162
Adsorption, 245
  Gibbs equation, 244–45
  negative, 245
Amagat's law of volumes, 31
Amalgams, activities in, 363–67, 369–71
Ammonia, heat of formation, 81
  equilibrium, 277–78
Ampere, 9
Apparent molar, heat capacity, 454–57
  heat content, 440–46, 449–52
    relative, 440–45, 451
  properties, 427–28
    and partial molar properties, 429–30
  volume, 429, 437
Atomic heat capacity, *see* Heat capacity
Avogadro number, 95, 100

Beattie-Bridgeman equation, 26
Berthelot equation, 25–26
    and entropy correction, 158
    and heat capacity correction, 169, 171
    and heat content correction, 160

Boiling point, dilute solutions, 343–45
  elevation, 344
    and activity, 362–63, 387
  and pressure, 226, 227, 231
    table, 502
  volatile solute, 344
Boltzmann constant, 100
  -Planck entropy equation, 184–85
Bond energies, 90
  and heat of reaction, 91–92
    table, 91
Bridgman, thermodynamic formulae, 212
Brønsted equation, 420

Calorie, defined, 9
  standard (15°), **7**
Calorimeter, 9
Capacity factor, energy, 6
Carbon, heat of vaporization, 90
Carnot cycle, 135, 138, 141, 148
  efficiency, 137
  on entropy-temperature diagram, 147
  theorem, 133
Cells, reversible, 330–32, 462–86
  entropy change in, 466
  general E.M.F. expression, 467–68
  heat change in, 466
  reactions, 464–66
Centigrade scale, 3
C.g.s. units, 6
Chemical change, direction, 284
  equilibrium, 273–313
  potential, 215, 217
    and activity, 262, 269, 350, 363
    of electrolytes, 379
    and fugacity, 261–62, 269
    in gas mixtures, 262–64
    of ions, 378, 407
    in liquid mixtures, 319
    and osmotic coefficient, 389
    and osmotic pressure, 372
    and partial molar free energy, 215, 427
    phases in equilibrium, 238, 244–45
    and pressure, 218–19
    significance, 238
    in surface phase, 243
    and temperature, 218–19
Chlorine, atomic heat capacity, 107
  liquid, vapor pressure, 230
  molecule, heat capacity, 98, 99, 115, 116
  vibrational partition function, 114
Clapeyron equation, 223–26, 229, 233, 235, 241
Clausius-Clapeyron equation, 227–29, 233, 292, 325, 326
  integration, 227–29

Clément and Desormes, heat capacity ratio, 59
  and Reech's theorem, 177
Closed system, 163, 204
  several phases, 216, 237, 274
Combustion, adiabatic, 84–89
  heat of, 75
  calculation, 89–90
Compressibility curves, generalized, 28
  factor, 27
    and fugacity, 256–57
    generalized treatment, 27–30
    and heat content correction, 161
  of gases, 21–31
Conservation of energy, law, 34, *see also*
  First law of thermodynamics
  and heat of reaction, 71
Consolute temperature, 337
Constants, physical, 501
Conversion factors, 11, 501
Cooling, by adiabatic expansion, 57
  by Joule-Thomson effect, 64
Corresponding states, 25
  law of, 25
Coulomb, **7**
Craft's rule, 248
Critical point, 23
  data, table, 502
  solution temperature, 337
  temperature, 22
Cubic cm. and milliliter, 6
Cycle (or cyclic process), 39
  Carnot, 135–37, 138, 141, 148
  energy change, 38–39
  entropy change, 144, 146
  heat and work, 39
  irreversible, 144, 146
    efficiency, 144
  Joule, 153

Debye, characteristic temperature, 122, 124
  heat capacity theory, 122–25
Debye-Hückel, equation, 411
  applications, 395, 398, 421–24
  for concentrated solutions, 418–20
  limiting law, for activity coefficient, 411
    and dissociation constant, 421
    and equilibrium constant, 421–24
    and solubility, 415
    tests, qualitative, 412
      quantitative, 413–16
    for freezing point, 417
    for heat capacity, 457
    for heat content, 451
    for heat of dilution, 451
    for partial molar volume, 435–37

INDEX 511

Debye-Hückel (Cont.)
  theory, 383, 407–24
    and activity coefficient, 409–10
      mean ionic, 410–11
    and dielectric constant, 414
    and electrical free energy, 409
    incomplete nature, 416, 419
    and ionic atmosphere, 408
    and osmotic coefficient, 416–17
    and standard states, 412
    and temperature, 414
Deuterium, entropy, 194, 197, 200
  heat capacity, 96, 97, 111
    rotational, 113
  rotational partition function, 109
Diamond, heat capacity, 121, 123
Diatomic molecule, heat capacity, 96, 97
  moment of inertia, table, 110
  partition function, combined, 116
    electronic, 108
    and entropy, 192
    rotational, 109–12
    vibrational, 113
  rotational energy, 110
    heat capacity, 111
    symmetry number, 109
  vibrational energy, 113, 114, 115
    frequency, 113
      table, 114
    heat capacity, 114–15
Differential, complete or exact, 16–18, 210
  heat of dilution, 83, 384, 440
    to infinite dilution, 440, 441–42
  of solution, 82, 361, 439, 449, 450
    in dilute solution, 345
    at infinite dilution, 490
    and solubility, 345
Dilute solutions, 337–46
  boiling point, 343–45
  freezing point, 340–43
  heat of solution in, 345
  Henry's law, 337–40
  homogeneous equilibria in, 279–81
  solubility and temperature, 345
Dilution, free energy of, 406
  heat of, 83, 384, 440, 441–42, 449–50
    and Debye-Hückel theory, 451
    differential, see Differential
    integral, see Integral
Disorder and entropy, 152
Dissociation constant, 421
  and Debye-Hückel theory, 421–22
  of water, 483–84
  of weak acid, 485–86
  pressure and E.M.F., 481–83
Distillation, fractional, 323, 324
  reduced pressure, 325

Duhem, see Gibbs-Duhem
Duhem-Margules equation, 318–20, 331, 332, 333, 335, 336
Dühring's rule, 231–32, 248
Dulong and Petit, heat capacity, 120
Dyne, 6

Efficiency, Carnot cycle, 137
  heat engine, 133
  irreversible cycle, 144
  reversible cycle, 137, 140
Einstein characteristic temperature, 122
  functions, 115, 193
  heat capacity theory, 121
Electrical work and heat, 9
  maximum, 45
Electrodes, oxidation-reduction, 464
  potentials, 469–80
    applications, 478–83
    and entropy, 489, 492–93
    oxidation, 471
    sign, convention, 471–72
    standard, 470–80
      table, 479
  reference, 474, 475
Electrolytes, activity coefficient, 379
  boiling point method, 387
  calculation, see Debye-Hückel
  E.M.F. method, 392–98
  freezing point method, 381–86
  and ionic strength, 401, 410–12
  isopiestic method, 387–88
  osmotic coefficient method, 390–92
  solubility method, 398–400
  table, 402
  and valence type, 401, 410–12
  chemical potential, 378
  partial molar heat capacity, 455, 457
    heat content, 444, 450
    volume, 435
Electromotive force, see E.M.F.
Electron "gas," heat capacity, 125
Electronic contribution to entropy, 191
  to heat capacity, 107
  partition function, 106
    diatomic gas, 108
    monatomic gas, 106–7
    polyatomic gas, 116
  states, 102
    monatomic gases, 105
Elements, free energy convention, 298
  heat capacity, 120–25
    at absolute zero, 121, 124
    at low temperatures, 124
  heat content convention, 74
E.M.F. and activity determination, 363–68, 392–98

E.M.F. (Cont.)
  and dissociation constant, water, 483
    weak acid, 485–86
  and electrode potentials, 472–73
  and entropy change, 466
  and free energy change, 300–02, 466, 468, 475, 478
  of galvanic (reversible) cells, 462–96
  general expression, 467–68
  and heat content change, 466
  and partial molar heat capacity, 453
    heat content, 448
  and pressure, 499
  sign, convention, 464
  standard, 302, 393, 467, 469
    and entropy change, 303, 489
    and equilibrium constant, 475–76
    and temperature, 448
Energy, 6, 36, 38, *see also* Free energy
  change and work, 38
    in cycle, 38–39
  conservation of, 34, 71
  content, 36, 38
    of ideal gas, 50–51
  dimensions, 6
  degradation and entropy change, 146
  equipartition principle, 97–98
  and heat, 6, 7, 10, 40
  intensity and capacity factors, 6
  of isolated system, 35
  kinetic, 95, 104
  and mass, 35–36
  quantum theory, 99, 106
    and partition function, 101
  rotational, 96, 97, 98, 110, 111, 117
  translational, 95, 105
  units, 6, 11
  vibrational, 96, 97, 98, 113, 114, 115, 118
    of solids, 121, 122
  and work, 6, 10, 40
  zero-point, 100
Enthalpy, 48, *see also* Heat content
Entropy, 141–52
  at absolute zero, 178, 183
  applications, 303–05, 492–93
  change, from E.M.F., 303, 466, 489
    and equilibrium, 208
    and free energy change, 302–04
    of ideal gas, 148–50, 185
    in irreversible (spontaneous) process, 144–46
    in isolated system, 146
    in phase change, 148
    in reversible process, 144
  correction, nonideal, 158
  determination, 178
  and disorder, 152

Entropy (Cont.)
  electronic contribution, 191
  of expansion, ideal gas, 148–50, 185
  extensive property, 143
  of fusion, 148, 152, 179
  gases, 180
    calculation, 188–90
    table, 198, 505–08
  of glasses, 185
  and heat capacity, 155
  of ions, standard, 488–94
    table, 492, 505–08
  of liquids, table, 179, 505–08
  of mixing, gases, 150–51
    liquids, 348
    solids, 199
  of monatomic gases, 190–92
  and nuclear spin, 194
  partial molar, 218
  and partition function, 188–96
  practical, 194
  and probability, 183–85, 189
  and pressure, 156–57
  and randomness, 152
  rotational, 193, 194–96
  of solid solutions, 185, 199
  of solids, table, 179, 505–08
  standard, 158, 179, 181, 191
    tables, 179, 198, 505–08
  statistical and thermal, 196
  statistical treatment, 183–96
  and temperature, 154–55
  -temperature diagram, 147
  thermal, 194
  and third law of thermodynamics, 178–83, 196
  of transition, 180, 182
  translational, 192
  and unavailable heat, 143
  unit, 180
  of vaporization, 148, 180
  vibrational, 193
  virtual, 194
Equations of state, 15, 18–29
  Beattie-Bridgeman, 26
  Berthelot, 25
  general, 26
  ideal gas, 15, 18
  reduced, 24–25, 174
  thermodynamic, 159–60
  van der Waals, 22
Equilibrium, chemical, 15, 273–313
  conditions, thermodynamic, 207–09, 237
    in heterogeneous systems, 216–17, 237–38
  constant, 274
    and activity, 274

Equilibrium constant (Cont.)
    and activity coefficient, 275–76
    and Debye-Hückel equation, 421–24
    and electrode potential, 476
    and E.M.F., 475, 479
    and free energy change, 274
    and fugacity, 275
    and heat of reaction, 292–95
    of isotopic exchange reactions, 313
    of metathetic reactions, 312–13
    and pressure, 286–87
    and standard free energies, 283–84, 300
    and standard state, 276
    and temperature, 286–91
        van't Hoff equation, 289
    of crystalline forms, 225
    heterogeneous chemical, 281–82, 291
    homogeneous, gaseous, 275–79
        liquid, 278–81, 290
    liquid-solid, 224, 234, 327–30
    liquid-vapor, 226–33, 234
        ideal solutions, 322–24
        nonideal solutions, 331–40
    mechanical, 15
    and partition function, 310–12
    phase, 222–47
        one-component system, 222
        polycomponent system, 237–41
    position, and pressure, 287–88
    simultaneous, 291–97
    solid-liquid, 224, 234, 327–30
    solid-vapor, 233, 234
    surface, 241–47
        conditions, thermodynamic, 243
    temperature (or thermal), 2, 3, 15
    thermodynamic, 15
    vapor-liquid, 226–33, 234
        ideal solutions, 322–24
        nonideal solutions, 331–40
    vapor-solid, 233, 234
Equipartition principle, 97
    and heat capacity, gases, 98, 111, 115, 117
        solids, 120–21
Erg, 6
    and joule, 7, 11
Euler criterion (reciprocity relationship), 210
    theorem on homogeneous functions, 215
Eutectic temperature, 241, 347
Evaporation, *see* Vaporization
Expansion, adiabatic, 55–58
    reversible, 41
    throttled, 60
    work of, 10
        reversible, 42
Extensive properties, 16
Explosion bomb, 70
    pressure and temperature, 88

First law of thermodynamics, 35, 37, 38, 68, 71, 72, 129, 202
    and thermochemistry, 68, 71, 72
Flame temperature, actual, 86
    maximum, 84–86
Fractional distillation, 323, 324
Free energy and free energy change, 202–20, 222, 242, 250–51, 273–74, 282–86, 295–310, 466–68, 475, 486–88, *see also* Activity, Chemical potential *and* Fugacity
    and activity coefficient, electrolytes, 409
    in chemical reaction, 282–83
    of dilution, 406
    and direction of chemical change, 284–85
    electrical, 408
        and Debye-Hückel theory, 409
    and electrical work, 300
    and electrode potential, 477
    and E.M.F., 300–02, 466, 468, 475, 478
    equations, addition, 297
    equilibrium conditions, 209, 222, 274
    and fugacity, 250, 258
    functions, 306–09
    in isothermal change, 205
    of mixing (solution), 375
    and net work, 203, 300
    partial molar, 215, *see also* Chemical potential
    from partition functions, 209–10, 306–10
    and phase change, 299
    and phase equilibrium, 222
    and pressure, 204
    in any process, 219
    relationships, 204
    standard, 274, 283–84
        applications, 303–05
        determination, 300
            calculation, 308
            E.M.F. method, 300–02
            entropy method, 302
            by equilibrium constant, 300
        of formation, 298
        of ions, 486–88
        tables, 299, 492, 505–08
        of reaction, 283–85
        and temperature, 295–96
    surface, 242–43
    and temperature, 204, *see also* Gibbs-Helmholtz equation
Freezing point, 327
    and composition, 328
    depression, 340, 382
        and activity, 358–61, 381–86
        and Debye-Hückel theory, 417
        and ionic strength, 417
        and osmotic coefficient, 392, 417

514 INDEX

Freezing point (Cont.)
  dilute solutions, 340–43
  and solubility, 330
Fugacity, 250–70
  and activity, 262, 264, 270, 353, 358
  and activity coefficient, 258, 263, 264
  determination, 251–58, 267
    approximate, 255
    from equation of state, 253–54
    generalized (compressibility) method, 256–58
    graphical, 256–58
    in mixtures, 267–68
  and Duhem-Margules equation, 319
  and equilibrium constant, 275
  and free energy, 250
  gas, ideal, 251
    mixtures, 261–71
    single, 250–60
  generalized curves, 259
  and Henry's law, 338–39, 353
  in ideal solution, 317–18
  liquid, mixtures, 268–71, 317–18
    single, 260
  in nonideal solution, 334
  and osmotic pressure, 373
  and phase equilibrium, 260
  and pressure, 251, 259, 264–65
  solid, 260
  and temperature, 259, 266
Fusion, entropy of, 148, 152, 179, 180
  equilibria, 224
  heat of, 76, 224, 225
    and freezing point, 340–43
    and solubility, 328–30
    and temperature, 79, 235, 359

Galvanic cells, see Cells
Gas and gases, 18–31
  compressibility, 21
    factor, 27
  constant, 19, 501
  cooling, 57, 64
  critical state, 22–23, 502
    temperatures, table, 502
  entropy, tables, 198, 505–08
  equations of state, see Equations of state
  fugacity, 250–70, see also Fugacity
  heat capacity, determination, 167
    difference ($C_p - C_v$), 54–55, 164–167
    mean, 53
    monatomic, 96
    and pressure, 51–52, 64–65, 168–70
    ratio, 59–60, 96, 171
    tables, 53, 503
    and temperature, 52–53
    and volume, 170–71

Gas and gases (Cont.)
  ideal, see Ideal gas
  Joule-Thomson effect, 60–64, 171–75
  mixtures, fugacity, 261–70, see also Fugacity
    heat changes in, 266
    ideal, 268
    partial pressures in, 20, 30
    volume changes, 31, 265
  real, 21–31
    mixtures, 30–31
  standard state, entropy, 158, 198
    free energy, 298
    heat content, 72
    throttled expansion, 60
Gay-Lussac, energy content of gas, 50
Generalized methods, for compressibility, 28
  for fugacity, 256–58
  for heat capacity correction, 170
  for heat content correction, 161
  for Joule-Thomson effect, 174–75
Gibbs, adsorption equation, 244–45
  chemical potential, 215, see also Chemical potential
  -Duhem equation, 214, 216, 318–19, 368, 370, 381, 387, 392, see also Duhem-Margules equation
  -Helmholtz equation, 205–07, 218, 451, 466, 489
  phase rule, 239
  thermodynamic potential, 203
Gram, 6

Heat capacity, 8
  apparent molar, 454–57
  atomic, solids, 120–25
    at absolute zero, 121, 125
    classical theory, 120
    Debye theory, 122
    Dulong and Petit law, 120
    Einstein theory, 121
    electron "gas" contribution, 125
    at low temperatures, 124
  of calorimeter, 9
  classical calculation, 98, 120
  constant pressure and volume, 48–49
    difference, 50, 54–55, 163–67
    maximum, 165
    ratio, 56, 96–97
      and atomicity, 60
      determination, 167
      and pressure, 60, 171
  determination, 9, 167
  dimensions and units, 8
  and entropy, 155, 178–79
  of gases, see Gas
  and internal rotation, 119

INDEX

Heat capacity (Cont.)
  and kinetic theory, 95–96
  molar, 8
  partial molar, 452–58
    and E.M.F., 453
  and partition functions, 102, 105, 111–13, 115, 117, 118–19
  and pressure, 64–65, 168–70
  quantum theory, 99–120
  ratio, 56, 60, 96–97, 167, 171
  relationships, 163–71
  rotational, 111, 117
  of solid compounds, 126
    elements, 120–25
  specific, 8
  tables, gases, 53, 503
    solids, 504
  and temperature, 52–54, 503, 504
  translational, 105
  vibrational, 114, 118
  and volume, 170–71
changes at constant pressure and volume, 47–48
of combustion, 75, 85, 89, 92
content, 48
  apparent molar, 440
    relative, 440–46, 449–52
      and Debye-Hückel theory, 450
  and cell reaction, 466
  and E.M.F., 466
  and free energy, *see* Gibbs-Helmholtz equation
  and heat of reaction, 68, 70, 73–74, *see also* Reaction
  partial molar, 218, 266, 318, 356, 361, 384, 386, 387
    in reference state, 356, 438
    relative, 438–54
    in standard state, 356
  and phase changes, 76
  and pressure, 160
    ideal gas, 52
  relative, 309
  and temperature, 54
  in throttled expansion, 61
conversion into work, 129, 130, 131, 133
  efficiency, 133, 137, 138, 140
    maximum, 138
definition, 2, 7
of dilution, *see* Dilution
and energy, 6, 7,
engine, 133
  Carnot's, 135–37
  Joule's, 153
  reversible, 133
    efficiency, 137
of formation, 72, 73–74, 91

Heat of formation (Cont.)
  of elements, 74
  and heat content, 73–74
  of ions, 494–96
  standard, 72–74, 494
  tables, 73, 492, 505–08
of fusion, *see* Fusion
of hydrogenation, 75
measurement, 9
of mixing, ideal systems, 268, 318
  nonideal, 331, 375, *see also* Solution
of reaction, 68, 70, 74, *see also* Reaction
of solution, *see* Solution
of transition, *see* Transition
unavailable and entropy, 143
of vaporization, *see* Vaporization
and work, 10
  in cycle, 39
  dependent on path, 39
  equivalence, 33
  not properties of system, 40
Helmholtz, conservation of energy, 34
  free energy, 202
  see also Gibbs-Helmholtz
Henry's law, 337–38, 345, 353
  deviations, 353–54, 365
  ionic solutions, 407, 412
  and external pressure, 339
  and gas solubility, 338–39
  and Raoult's law, 339–40, 353
Hess, law of thermochemistry, 72
Heterogeneous chemical equilibria, 281–2
  and temperature, 291
  equilibrium conditions, 216–17
  system, 14
  see also Phase equilibria
Hildebrand's rule, 232
Homogeneous equilibria, gaseous, 275–78
  and temperature, 289–90
  in solution, 278–81
    dilute, 279
  and temperature, 290
Hückel equation, 419, 420, *see also* Debye-Hückel equation
Hydrochloric acid, activity coefficients, 394
  thermal properties, 453
Hydrogen, electrode standard, 469
  entropy, 194, 197
  fugacity, 256
  heat capacity, 96, **97**, **111**
  ion, entropy, 489
    free energy, 486
    heat of formation, 494
  Joule-Thomson effect, 64
  ortho and para, 111
    heat capacity, 112
    rotational heat capacity, 111–13

Hydrogen (Cont.)
  partition function, 109, 112
  scale, electrode potentials, 469
  vibrational partition function, 114

Ice point, 3
  on absolute scale, 5
Ideal gas, 19, 159, 160
  adiabatic processes, 55–58
  constant, 19, 501
  energy content, 50–51
  entropy changes, 148–50
    of expansion, 186
  equation of state, 15, 18
  fugacity, 251, 261
  heat capacity, 96
    difference, 55, 96, 164
    and pressure, 51–62, 168
    ratio, 96
    kinetic theory, 95
  mixtures, 20, 265, 268
  solubility, 325–26
  thermometer, 4
  work of expansion, 43–44
heat of vaporization, 260, 268, 291, 318
liquid solutions, 317, 318, 320, 330
  deviations, 330–37, 427
  vapor pressures, 321
  see also Raoult's law
Indicator diagram, 135
Integral heat of dilution, 83, 440, 442–44
  of solution, 82, 439, 449, 450
  in dilute solution, 345, 450
Intensity factor, energy, 6
Intensive properties, 16
Internal pressure, 62
  rotation, energy, 119
Inversion temperature, Joule-Thomson, 63, 173
Iodine, dissociation, 311
Ionic, activity, mean, 379
  coefficient, mean, 379
    and Debye-Hückel theory, 410–11
    and osmotic coefficient, 417
  atmosphere, 408, 410
    radius, 409
  molality, mean, 379
  product, water, 483
  strength, 400, 410, 411, 412, 436
    and activity coefficient, 401, 410–12
    and equilibrium function, 421–24
    and freezing point, 417
    and osmotic coefficient, 417
Ions, activity and activity coefficient, *see* Ionic
  distance of closest approach, 418–19
  entropy, standard, 488–93

Ions, entropy (Cont.)
  applications, 492–93
  determination, 488–92
  table, 492, 505–08
free energy of formation, 486–88
  table, 492, 505–08
heat of formation, 494–96
  table, 492, 505–08
Irreversible process, 130–31
  energy degradation, 146
  entropy change, 144–46
Isentropic change, 147, *see also* Adiabatic
Isolated system, 14
  entropy change, 146
  and first law of thermodynamics, 35
Isopiestic method, activity, 387–88, 392
Isothermal change, free energy, 205
  work function, 205
  expansion, reversible, 41–43
  work of, 42–43
Isotonic, *see* Isopiestic
Isotopic exchange reaction, equilibrium constant, 313

Joule, 7
  absolute and international, 7, 11, 501
  cycle, 153
  and defined calorie, 9, 10, 501
  energy content of gas, 50–51
  mechanical equivalent of heat, 33
  -Thomson coefficient, 61–64, 171–72
    and fugacity, 259
    at low pressure, 172, 173
    sign and magnitude, 62
    for van der Waals gas, 172, 175
  effect, 60–64, 171–175
    cooling by, 64
  experiments, 51, 60
    throttled expansion, 60
  inversion temperature, 63, 173–75
    generalized treatment, 174–75

Kelvin, available energy, 203
  energy content of gas, 51
  temperature scale, 5, 139–40
  see also, Joule-Thomson
Kharasch, heat of combustion calculation, 90
Kilogram, 6
Kinetic theory, 95
  and heat capacity, 95–96
Kirchhoff, heat of reaction-temperature equation, 77–82
  application, 79
  and latent heat, 79, 234, 359
  vapor pressure equation, 229
Kopp, heat capacity of solids, 126

INDEX

Latent heat, *see* Fusion, Vaporization, etc.
Latimer, ionic entropies, 489–93
Lavoisier and Laplace, thermochemical law, 91
Law, additive pressure, 30
  additive volume, 31
  Amagat's, 31
  of conservation of energy, 34
  of corresponding states, 25
  Debye-Hückel, limiting, 411, 417, 451, 457
  Dulong and Petit, 120
  of equilibrium, 274
  Henry's, 337–38, 345, 353
  Maxwell-Boltzmann, 100
  of partial pressures, 20
  Raoult's, 317, 320–21
  thermochemical, Hess, 72
    Lavoisier and Laplace, 71
  thermodynamics, first, 35, 37, 38
    second, 129, 131, 133
    third, 178, 181–83, 194
Le Chatelier principle, 220, 287
Lewis, atomic heat capacity at constant volume, 120
  free energy, 202
  fugacity, 250
  partial molar properties, 213
  sign of electrode potential, 472
  of E.M.F., 464
Lewis-Randall, ionic strength, 400
  and activity coefficient, 401
  fugacity in gas mixtures, 268, 275
Liquid, fugacity, 260
  junction, 396
  potential, 470, 474, 475
  mixtures, activity and activity coefficients, 269–71
  fugacity, 268–71
    general equation, 334
  partially miscible, 336–37
  reference state, for activity, 351
    for heat content, 438
  solutions, activity, 269–71
    dilute, 337–46
    fugacity, 268–71, 317–20, 233–328, 331, 334
    ideal, 317, 318, 320, 330, *see also* Raoult's law
    nonideal, 330–37
      fugacity, 334
      heat change, 331
      volume change, 331, 434
    standard state for activity, 281, 351, 352
      for entropy, 180
      for heat content, 72
  -vapor equilibria, ideal solutions, 321–23
    nonideal solutions, 331–34

Liter, 6
  -atmosphere, 11
Lummer-Pringsheim, ratio of heat capacities, 59

Margules equation, 334, 336
  and activity coefficient, 370–72
  see also Duhem-Margules equation
Maximum efficiency, heat engine, 138
  work, electrical, 45
  of expansion, 44–45
  and work function, 202
Maxwell relations, 163, 211, 223
Maxwell-Boltzmann law, 100
Mayer, mechanical equivalent of heat, 33
Mechanical equivalent of heat, 33
  and heat capacities, 34
Melting point, and pressure, 224, *see also* Freezing point
Mercury, activity in amalgams, 369–70
Meter, 6
Methanol equilibrium, 304
Milliliter, 6
Mixing, entropy, gases, 150–51
  liquids, 348
  solids (solution), 199
  free energy, 375
  heat change, 318, 331, 375
  volume change, 318
Mixtures, gases, ideal, 20, 261
  real, 30–31
    activities, 262–64
    additive pressure law, 30
    additive volume law, 31
    fugacity, 261, 267–68
  liquids, ideal, 317, 318, 320, 330
  nonideal, 330–37
  partially miscible, 336–37
Molal boiling point elevation, 344
  freezing point depression, 342
Molality, 281
  and activity, 353–54
  and mole fraction, 281
  mean ionic, 379
Mole, 8
  fraction, 20
    and molality, 281
    and molarity, 279
Molecular weight, determination, 341–42, 344
Moment of inertia, diatomic molecules, 110
  and equilibrium constant, 312
  polyatomic molecules, 117
Monatomic gas, electronic partition function, 105–06
  entropy, 190–92
  heat capacities, 96

Morse equation, osmotic pressure, 376–77

Nernst heat theorem, 178
Net work, 203
    in surface, 243
Nitrogen, compressibility, 21
    entropy determination, 180
    fugacity, 252, 258
    Joule-Thomson coefficient, 63
        inversion temperature, 173
    tetroxide, dissociation, 280
Nonideal solutions, 330–37
    vapor pressures, 331–33
    see also Raoult's law
Nuclear spin, entropy, 194, 197
    partition function, 194
    quantum number, 109

Open systems, 213
    part of closed, 216
Ortho and para states, 109
    of hydrogen, 111, 112
Osmotic coefficient, 389
    and activity coefficient, 389–92
    and Debye-Hückel theory, 416–17
    and freezing point, 392, 417
    and ionic strength, 417
    and vapor pressure, 390
    pressure, 372–74
    and activity, 373–74
    and vapor pressure, 373, 376
Oxidation potentials, 471
    table, 479
Oxidation-reduction electrodes, 464
    system, 463
Oxygen, electrode potential, 477
    gas, fugacity, 255
    molecules, electronic states, 108

Partial molar, heat capacity and heat content, *see* Heat capacity *and* Heat content
    free energy, *see* Chemical potential
    properties, 213–15, 427–58
        determination, 427, 428–33
            by apparent molar properties, 429
            direct method, 428
            general methods, 432
            intercept methods, 430–31, 432
        significance, 214–15, 427
    volume, *see* Volume
    pressures, 20, 30
    law of, 20
    specific properties, 433
Partially miscible liquids, 336–37
Partition function, 101
    electronic, 106–107, 108

Partition function (Cont.)
    and energy, 101
    and entropy, 188–96
    and equilibrium constant, 310–12
    and free energy, 209, 306–10
    and heat capacity, 102
    and relative heat content, 308–09
    rotational, 103, 110, 117
    translational, 103, 104
    vibrational, 103, 114, 118
Path, thermodynamic, 37
Pauling, bond energies, 90
Perfect crystal, entropy, 178, 185, 196
Perpetual motion, first kind, 34, 39
    second kind, 132
Person, temperature and heat of reaction, 78, *see also* Kirchhoff
Phase change, entropy change, 148
    free energy change, 299
    heat change, 76, 223, *see also* Fusion, Vaporization, etc.
    equilibria, 222–47
        one-component, 222
        polycomponent, 237
    rule, 238–39
    space, 186
    surface, 241
Planck constant, 104
    quantum theory, 99
    and third law of thermodynamics, 178
    -Boltzmann entropy equation, 184–85
Polyatomic molecules, entropy and partition function, 192–96
    heat capacity, 96, 97, 116
    moments of inertia, 117
    rotational partition function, 117
    symmetry number, 117
    vibration frequencies, 118
Potential, chemical, *see* Chemical potential
    electrode, 469–83
        oxidation, 471
        table, 479
        reduction, 472
        standard, 470–83
            applications, 478–83
            and entropy, 489, 492–93
    liquid-junction, 470, 474, 475
Poynting equation, 236
Pressure, critical, 23
    table, 502
    and E.M.F., 499
    in explosions, 88–89
    internal, 62
    law of additive, 30
    partial, 20, 30
    reduced, 25
    and solubility, 339, 349

INDEX 519

Pressure (Cont.)
  and vapor pressure, 235–36
Probability and entropy, 183–85, 189
  thermodynamic, 185, 187
Process, irreversible and spontaneous, 129–30, 146
  reversible, 41–42
  spontaneous and probability, 184
  thermodynamic conditions, 207–09
Properties of system, extensive and intensive, 16

Quanta, 99
Quantum number, nuclear spin, 109
  rotational, 109
  theory, of energy, 99, 106
    of heat capacity, gases, 102–19
      solids, 121, 122

Ramsay-Young rule, 230–33
Randall, *see* Lewis-Randall
Rankine temperature scale, 5
Raoult's law, 317, 330, 338, 339, 357
  applicable to both constituents, 320–21
  deviations, 330–37, 352, 366
    and heat and volume changes, 331, 434
    and partial miscibility, 336
  and external pressure, 321
  and freezing point, 340, 342
  and Henry's law, 338–40, 353
  and solvent in dilute solution, 338, 339
Rational activity coefficient, 353, 407
Reaction, heat of, 68
  and bond energies, 90–92
  calculation, 89–92
  and conservation of energy, 71
  at constant pressure and volume, 70
  and heat content, 68, 70
    of formation, 74
  and pressure, 69, 160
  in solution, 69, 82
  standard state, 72
  and temperature, 77–82
  free energy of, 283–84
    and direction, 285
  isotherm, 282–84, 301, 468
  isotopic exchange, 313
  rate, 286
Reduced equation of state, 24–25, 27–29
  and fugacity, 258
  and Joule-Thomson inversion temperature, 174
  pressure, temperature and volume, 25, 29, 161
Reduction potential, 472
Reech's theorem, 177
Reference electrode, 474, 475

Reference (Cont.)
  state for activity, gas, 263, 264
    solute, 352–54
    for fugacity, 251, 262
    partial molar heat content in, 355–56
    for partial molar properties, 438
Refrigeration engine or refrigerator, 134, 138
  coefficient of performance, 139
Relative partial molar heat content, 361, 384, 386, 387, 438–54
  and activity coefficient, 384, 386, 387, 446
  and E.M.F., 448
  and heat of dilution, 439, 441–42, 445
  and heat of solution, 439, 449
Reversible adiabatic change, 55
  pressure and volume, 57
  temperature, 56
  work of expansion, 58
  cells, 300–02, 462, *see also* Cells
  chemical reactions, 273–82
  cycle, 135
  electrodes, 462–64, *see also* Electrodes
  expansion of gas, 41
    heat change in, 51
    maximum work, 44–45
    work of isothermal, 43–44
  heat engine, 133
    efficiency, 137, 140
  processes, 41–42
    entropy change, 144
Rotation, internal, 119
  energy of, 119
  restricted, 119
  and entropy, 197
Rotational energy, 96, 97, 98, 99, 103
  diatomic molecule, 110, 111
  entropy, 194–96
  heat capacity, 111, 112, 117
  partition function, 103
    diatomic molecule, 109–10, 194
    hydrogen, 111
    polyatomic molecule, 117
  quantum number, 109

Sackur-Tetrode equation, 190
Salt bridge, 475
Salting-out effect, 403
  term in Hückel equation, 419
Second, 6
Second law of thermodynamics, 129, 131, 133
Silver-silver chloride electrode, 474
Sodium chloride, activity coefficients, 386
  heat of dilution, 442
Solid or solids, entropy, 178–80, 182
  heat capacity, 120–25, 126

Solid, heat capacity (Cont.)
    classical theory, 120
    Debye theory, 122
    Einstein theory, 121
    Kopp's rule, 126
    at low temperatures, 124
  random orientation in, 196
  solution, separation on freezing, 327, 342–43
  standard state, for activity, 281, 351
    for entropy, 180
    for heat content, 72
Solubility, gas, ideal, 325–27
  and pressure, 339
  and temperature, 328–30
  product, 398, 415, 491
    and Debye-Hückel theory, 415–16
    by electrode potential, 480
    and free energy of solution, 491
  small particles, 246–47
  solid, 327
    and pressure, 349
    and temperature, 328–30
Solution or solutions, dilute, 279, 337–46
  boiling point, 343–44
  definition, 338
  freezing point, 340
  and Henry's law, 337–40
  and Raoult's law, 338–40
  solubility and temperature, 345
  gases in liquids, 325–27
  heat of, 82–84
    differential, 82, 345, 439, 449
    gases, 326–27
    at infinite dilution, 490
    integral, 82, 439, 449
      table, 508
    solids, 330
    and solubility, 345
  homogeneous equilibria in, 278–81
    and temperature, 290
  liquid, ideal, 317, 318, 320, 330, *see also* Raoult's law
  nonideal, 330–37
  solid, 327
    and freezing point, 342–43
Specific heat, 8, *see also* Heat capacity
Spin, nuclear, quantum number, 109
Spontaneous process, 129–30
  entropy change in, 146
  and probability, 184
  reversal of, 130–31
  thermodynamic conditions, 207–09
Standard electrode potentials, 470–80, 489, 492–93
  applications, 478–83
  determination, 474, 493

Standard electrode potentials (Cont.)
  table, 479
  E.M.F., 302, 393, 467, 469
  entropies, gases, 158, 198
    ions, 488–93
    liquids, 179, 180
    solids, 179, 180
    table, 505–08
  free energy of formation, 298
    of ions, 486–88
    tables, 299, 492, 505–08
  heat of formation, 72–74, 494
    of ions, 494–96
    tables, 73, 492, 505–08
  hydrogen electrode, 469
  states, for activity, choice of, 350–51
    gases, 262, 351
    ions, 378
    liquids, 269, 281, 317, 351
    solids, 281, 351
    solutes, 352–54
    for entropy, gases, 158, 181
      liquids, 180
      solids, 180
  and equilibrium constant, 276
  for heat of formation, 72
  and partial molar heat content, 355–56
  and reaction isotherm, 284
  and reference state for heat content, 438
  and van't Hoff equation, 289, 293
State, corresponding, 25
  equations of, 15, 18–29
    reduced, 24–25, 27–29
    generalized, 26
    thermodynamic, 159–60
  macroscopic, 187
  microscopic, 187
  standard, *see* Standard
  thermodynamic, 14
Statistical mechanics, 186
  treatment of entropy, 183–96
  weight factor, 100, 187
    for nuclear spin, 109
    for rotation, 109
Statistics, classical and quantum, 188
Steam point, 3
Stirling approximation, 188
Stoichiometric activity coefficient, 380, 412
Sublimation, equilibrium, 233–34
  heat of, 233, 234
Sulfuric acid, solutions, heat capacity, 456–57
  partial molar heat content, 442, 443–44
  and activity coefficient, 447
Surface, chemical potential, 243–44
  equilibrium condition, 243
  excess concentration, 242

INDEX

Surface (Cont.)
 forces, 241
 free energy, 242
 phase, 242
 tension, 243–46
  change with concentration, 245
  see also Adsorption
Symmetry number, diatomic molecules, 109, 312
 polyatomic molecules, 117, 196
System, 14
 closed, 163, 204, 216, 237, 274
 heterogeneous and homogeneous, 14
 isolated, 14
  and first law of thermodynamics, 35
 open, 213, 216
 properties, extensive and intensive, 16
  and thermodynamic equilibrium, 16
 simple, 15
 state of, 14

Temperature, absolute, 4–5
 characteristic (Debye), 122
  table, 124
 concept, 2
 critical, 22
  table, 502
 critical solution (consolute), 337
 difference, 2
 equilibrium, 2, 3, 15
 -entropy diagram, 147
 explosion, 88–89
 flame, 84–88
 Joule-Thomson inversion, 63, 173
 Kelvin, 139–40
 reduced, 25
 scales, 3–5, 139–40
 standard, 5
Thallium, amalgams, activity, 363–67
Thallous chloride, activity coefficient, 399–400
Thermal entropy, 194
 equilibrium, 2, 3, 15
Thermochemical equations, 69, 71
 laws, Hess, 72
  Lavoisier and Laplace, 71
Thermochemistry, 68–92
Thermodynamic or thermodynamics, change and process, 37
 conditions of equilibrium, 207–09
  of spontaneous process, 207–09
 equations of state, 159–60
 equilibrium, 15
  and properties of system, 16
  and spontaneous processes, 130
 first law, 35, 37, 38, 68, 71, 72, 129, 202
 formulae, 210–12

Thermodynamic or thermodynamics (Cont.)
 limitations, 1
 probability, 185, 187
 properties, 14–16
  complete differentials, 16–18
 reversibility, 41
 second law, 129, 131, 133
 scope, 1
 state, 14
 temperature scale, 139–40
 third law, 178, 185, 194
  tests, 181–83, 305–06
 variables, 14–16
Thermometer, 3
Thermometric scales, 3–5
 absolute, 4
 centigrade, 3
 ideal gas, 4
 Kelvin (thermodynamic), 5, 139–40
 practical, 5
 Rankine, 5
Third law of thermodynamics, 178, 194
 tests of, 181–83, 305–06
 and probability, 185
Thomson, see Kelvin and Joule-Thomson
Thornton, heat of combustion, 89
Throttled expansion, 60–61
Transition, entropy of, 180, 180
 heat of, 77, 225
 point and pressure, 225
Translational energy, 95, 99, 103
 heat capacity, 96, 105
 partition function, 103, 104–05
Trouton's rule, 232–33, 324

Univariant system, equilibrium, 223
 polycomponent, 239

Valence type and activity coefficient, 401, 412, 413
van der Waals constants, 22–25
 table, 23
 equation, 22–25
  and critical state, 22–23
  and fugacity, 254
  and heat capacity difference, 164–65
  pressure effect, 169, 171
  and internal pressure, 62
  and Joule-Thomson coefficient, 67, 172
  inversion temperature, 175
van Laar equation, 335, 336
 and activity, 371
van't Hoff, osmotic pressure equation, 377
 reaction isotherm, 282–84, 301, 468
  and direction of chemical change, 285
  and E.M.F., 468

van't Hoff (Cont.)
  temperature-equilibrium constant equation, 289
    integration, 292
Vapor-liquid equilibria, 222–24, 226–33
  in mixtures, ideal, 322–23
    and temperature, 323
    nonideal, 331–33
  pressure, and activity, 356–58
    of electrolytes, 387–88, 390–92
    of ideal liquid mixtures, 317
      curves, 321
      maximum and minimum, 333
    of nonideal liquid mixtures, 330
      curves, 331–33
    and osmotic pressure, 373, 376
    of saturated solution, 241
    of small particles, 245–46
    and temperature, 226, 227, 323
      equations, 229–33
    and total pressure, 235–37
Vaporization, entropy of, 148, 180
  equilibria, 226–33
  heat of, 76, 77, 226, 227, 228, 229, 230
    and boiling point elevation, 344
    and distillation, 324
    and gas solubility, 326–27
    ideal, 260, 268, 291, 318
    table, 502
    and temperature, 79, 235, 362
Vibrational energy, 96, 97, 98, 99, 103
  entropy, 193
  frequency, diatomic molecules, 114
    polyatomic molecules, 118
  heat capacity, gas, 114, 115, 118
    solid, 120, 121, 122
  partition function, 103, 113, 118
Volt, 7
Volume, apparent molar, 429, 437
  of electrolytes, 437
  critical, 23

Volume (Cont.)
  law of additive, 31
  partial molar, 372–73, 427, 429, 433–37
    of electrolytes, 435–36
  reduced, 25

Water, boiling point, 3
  dissociation constant, 483–84
  entropy of vapor, 197
  freezing point, 3
  -gas equilibrium and pressure, 287–88
  heat capacity of vapor, 118
  ionic product, 483–84
  molal ionization product, 485
Work, in Carnot cycle, 136
  definition, 5
  electrical, and heat, 9
  and energy, 6
    change, 38
  of expansion, 10
    and heat of reaction, 70
  function, 201
    change in isothermal change, 205
    and equilibrium condition, 208
    and maximum work, 202
    and partition function, 210
    relationships, 203
  and heat, 10
    conversion, 129, 130, 131, 133
    in cycle, 39
    dependence on path, 39
    not properties of system, 40
  maximum, in reversible process, 44–45
    and work function, 202
  sign, convention, 38

Zero, absolute, 5
  entropy of solid at, 178, 181, 183
  heat capacity of solid at, 121, 124
  unattainability of, 139
  -point energy, 100, 113